Springer Complexity

Springer Complexity is an interdisciplinary program publishing the best research and academic-level teaching on both fundamental and applied aspects of complex systems—cutting across all traditional disciplines of the natural and life sciences, engineering, economics, medicine, neuroscience, social and computer science.

Complex Systems are systems that comprise many interacting parts with the ability to generate a new quality of macroscopic collective behavior the manifestations of which are the spontaneous formation of distinctive temporal, spatial or functional structures. Models of such systems can be successfully mapped onto quite diverse "real-life" situations like the climate, the coherent emission of light from lasers, chemical reaction-diffusion systems, biological cellular networks, the dynamics of stock markets and of the internet, earthquake statistics and prediction, freeway traffic, the human brain, or the formation of opinions in social systems, to name just some of the popular applications.

Although their scope and methodologies overlap somewhat, one can distinguish the following main concepts and tools: self-organization, nonlinear dynamics, synergetics, turbulence, dynamical systems, catastrophes, instabilities, stochastic processes, chaos, graphs and networks, cellular automata, adaptive systems, genetic algorithms and computational intelligence.

The three major book publication platforms of the Springer Complexity program are the monograph series "Understanding Complex Systems" focusing on the various applications of complexity, the "Springer Series in Synergetics", which is devoted to the quantitative theoretical and methodological foundations, and the "Springer Briefs in Complexity" which are concise and topical working reports, case studies, surveys, essays and lecture notes of relevance to the field. In addition to the books in these two core series, the program also incorporates individual titles ranging from textbooks to major reference works.

Indexed by SCOPUS, INSPEC, zbMATH, SCImago.

Series Editors

Understanding Complex Systems

Hernán A. Makse · Marta Zava

The Science of Influencers and Superspreaders

Using Networks and Artificial Intelligence to Understand Fake News, Pandemics, Markets, and the Brain

Springer

Hernán A. Makse
Department of Physics
City University of New York
New York, NY, USA

Marta Zava
Department of Finance
Bocconi University
Milan, Italy

ISSN 1860-0832 ISSN 1860-0840 (electronic)
Understanding Complex Systems
ISBN 978-3-031-78057-8 ISBN 978-3-031-78058-5 (eBook)
https://doi.org/10.1007/978-3-031-78058-5

This Springer imprint is published by the registered company Springer Nature Switzerland AG
The registered company address is: Gewerbestrasse 11, 6330 Cham, Switzerland

To Laura, Loris, Ana María and Milena.

Preface

This is a book about influencers. This term is as an umbrella term to describe essential, critical, central, superspreader, and core nodes in any type of complex network. We will review network theoretical frameworks and algorithms to identify and understand the emergence of influencers not only in society but also across a spectrum of systems ranging from biology to markets. Each chapter in this book is dedicated to study a different type of influencer, in turn. We will start with the canonical influencer problem:

- **Influence Maximization Problem**: Finding the optimal set of influencers that maximizes the spreading of information in a social network.

And then apply similar concepts to describe and identify:

- Superspreaders of disease in pandemics,
- Genetic influencers in gene regulatory networks,
- Neural influencers in the brain,
- Keystone species in ecosystems, and
- Financial influencers in markets.

The influencers' theory lays out that each big data problem is unique. There is no silver-bullet solution for all big data challenges. All networks have a different structure, and therefore, influencers vary depending on the problem. In other words, a new mathematical algorithm will have to be crafted for each dataset. Some algorithms work better for social media, others for genetic networks, and others for ecological networks. In this book, we will touch upon the different applications of AI and develop tailored solutions for every significant big data system.

Organization of the Book

The book is organized as follows.

In Chap. 1, we explain the standard mathematical theory behind influencer dynamics. We explain the simplest algorithms to identify influencers for any complex network. These are based on the topological properties of the network, called centralities, which are commonly used to identify and rank essential network nodes. These include centralities involving the number of connections (degree), k-cores, PageRank, eigenvector and eigenvalue-based centralities, betweenness, closeness, and many others. We will then mathematically define the influence maximization problem, trying to capture the commonalities among the different influencer problems. We will explore the different approaches in the literature employed to solve this optimization problem, including concepts from percolation theory, network dismantling and decycling, spectral theory, spin glasses, greedy algorithms, and others. We draw upon the work published in the vast literature encompassing social, computer, mathematical, and physical sciences to survey the theoretical approaches for influencer search and identification.

In Chap. 2, social media influencers are discussed. We show how combining network theory, AI, and alternative social media data can help predict and understand opinion trends. By collecting billions of data points from social media like Twitter and Facebook, we show how a few influencers—not the traditional media—can drive the conversation and potentially change the opinions of millions of people. We will show how fake news spreads in social networks and how influencers can drive the conversation across the spectrum spanning from genuine to fake news. We show how AI is used to classify the actors and simultaneously predict opinion trends, including election outcomes with machine learning, and show how influencers can shape the outcome of elections. For instance, influencers appeared organically in social media and enormously impacted the US presidential elections in 2016 and 2020.

In Chap. 3, we will lay out how influencer marketing is done using network theories for influencer identification. We will describe how influencers emerge organically to influence people to adopt specific products and brands. Anecdotal evidence suggests that the economic status of an individual is reflected in the structure of their social ties. Here, we will show that an individual's influence on the structural integrity of the social network can be used to infer their economic wellness. This leads to targeted marketing strategies based on influencer theory, reporting a three-fold increase in response rate compared to usual targeting campaigns.

In Chap. 4, we focus on superspreaders of disease during the COVID-19 pandemic. We show how the theory of influencers fed with big data from GPS geolocalization of people obtained from mobile smartphones can be used to design digital contact tracing protocols to implement intelligent quarantines to stop the spread of COVID-19, and at the same time, minimize the disruptions to society. The spread of the coronavirus disease caused by the SARS-CoV-2 virus became a worldwide problem with devastating consequences. To slow down the spread of the disease, massive quarantines were implemented across the world, provoking enormous social and economic disruptions. Influencer theory can contribute to ameliorating this problem by providing intelligent quarantine protocols based on big data. We show that influencer theory applied to contact networks obtained from digital contact tracing can readily identify superspreader events. An optimal targeted quarantine then follows by isolating only those superspreaders (influencers) that optimally dismantle the disease's transmission chain upon isolation. This could greatly reduce the disruption to society. Thus, optimization theory applied to influencers can become a crucial tool to combat pandemics.

Chapter 5 focuses on genetic influencers in gene regulatory networks. We represent the information flow between nodes in biological networks to decipher how cellular function emerges from their interactions. In this case, influencers take the form of either essential genes for integration that hold the network together or building blocks crucial to synchronizing information. We introduce a particular form of symmetry, symmetry fibration, to identify these building blocks. Using these symmetries, we identify input trees that deconstruct the network into functional building blocks processing equivalent coexpression of genes. This has significant ramifications for the identification of disease genes and for finding molecular targets for drug development.

In Chap. 6, we focus on neural influencers in the brain. The study of the human brain, as the elemental paradigm of an efficient, robust modular system interconnected as a network, will help scientists predict the map of neural collective influencers. In this chapter, we elaborate on models of brain networks to shape the pattern of brain activations. We explore how influencer theory is applied to these networks to predict and subsequently target nodes essential for global integration in a memory network in rodents. We further show how the theory predicts the localization of essential functional areas in the human brain that are used to guide surgical interventions in the human brain with minimal impacts to cognition.

Chapter 7 studies how influencer theory can help to identify the causes for the collapse and instability of dynamical systems. These systems include ecosystems, where influencers are the keystone species, human societies, and financial systems. When keystone species located at the maximum k-core of the mutualistic ecosystem go extinct, the system reaches the tipping point of its collapse. Monitoring the k-core then proves to be a valuable method to anticipate the catastrophic collapse of many ecosystems.

Chapter 8 studies financial networks. We trace the evolution of venture capital networks, study the AI's transformative impact on mergers and acquisitions, examine institutional investors' networks in the initial public offers, and explain novel applications of natural language processing in stock market analysis. Through an interdisciplinary lens, we demonstrate how the integration of network science and AI enhances analytical capabilities and reshapes the very nature of financial markets.

How to Read This Book

The approach of this book has a strong algorithmic, computational, and physics flavor. It is based on the premise that complex human behavior cannot be inferred by studying a single actor or a small number of them. As the *leitmotiv* of complexity science enunciated by Phil Anderson in his Nobel lecture says: '*more is different*'. Human behavior is an emergent collective phenomenon arising from the interactions of their many actors, and this idea also applies to complex biological systems composed of genes, proteins, and neurons. These interactions are inevitably captured by networks, and their emergent properties appear only when we observe many of these interacting units.

Thus, understanding the emergence of collective human behavior is only possible when we observe, analyze, and predict the properties of millions or billions of individuals. This is the domain of big-data science applied to human behavior, for which a computational algorithmic approach is presented in this book. However, the foundations of the present work are well-rotted in sociology, especially in the subfield of computational sociology, studying the spread of influence and innovations. Most concepts we will develop in this book have been discussed in the sociological literature for decades and continue to be addressed.

From a social science perspective, there are excellent books that deal with the collective nature of influence. We can recommend a few, starting from the seminal book of Rogers on '*Diffusion of innovations*', to the most cited paper in sociology (by Web of Science) by Granovetter on the '*Strength of weak ties*' in social network theory, to more recent accounts in terms of modern social network data by Centola on '*How Behavior Spreads: The Science of Complex Contagions*'.

Here, we will explore emergent collective behaviors from a different viewpoint, inspired by a statistical physics approach, where emergent properties arise at what physicists call the 'thermodynamics limit'; at massively large populations. Because the topic spans a broad spectrum of research, we cannot exhaustively report every relevant work. However, we tried to organize the book to cover recent developments in several fields of broad interest.

This book is intended to inform readers at three different levels:

- First, those interested in the mathematically rigorous theories of influencers can concentrate on Chap. 1, where we explain the mathematics behind the algorithms to identify influencers.
- A second level informs data scientists interested in applying these algorithms in their own fields and day-to-day work. For those readers, we provide Chaps. 2–8 where we explain the applications of the theory of influencers to social networks (Chaps. 2 and 3), epidemiology (Chap. 4), biology (Chap. 5), neuroscience (Chap. 6), ecology (Chap. 7), and markets (Chap. 8). Appendix A provides links to freely available algorithms with ready-made codes organized by chapters to be used with Python, R, and Julia through various Github repositories.
- Third, this book is also intended to reach layman audiences in the social media, financial, marketing and politics fields, as well as audiences keen on learning

how big data, influencers, and AI can contribute to a better decision-making based on mathematically proven algorithms and advanced analytics in their field and business of interest. These readers are encouraged to glance over the more technical Chap. 1 and focus on the relevant chapters for their businesses.

New York, USA Hernán A. Makse
Milan, Italy Marta Zava

Acknowledgements

This book could not have been written without the input of countless students, postdocs, and collaborators from all sorts of fields of life. Our interest in complex networks started with conversations with then-graduate student Chaoming Song in the early 2000s about the emergence of fractal properties in networks, and it has led to the development of influencer algorithms, theories, and applications in different fields, as discussed in this book. In this journey, we have benefited from enjoyable collaborations with José Soares Andrade Jr, Francesca Arese Lucini, Paolo Boldi, Alexander Bovet, Kate Burleson-Lesser, Guido Caldarelli, Santiago Canals, Gino Del Ferraro, Martina Del Gaudio, Lazaros Gallos, Martin Golubitsky, Alireza Hashemi, Shlomo Havlin, Andrei I. Holodny, Maksim Kitsak, Ian Leifer, Qiongee Li, Omar Lizardo, Shaojun Luo, Isotta Magistrali, Flaviano Morone, David Phillips, Lucas Parra, Sen Pei, Luca Pitta, Bernardo Principi, Saulo Reis, Diego Rybski, Hernán Rozenfeld, Matteo Serafino, Mariano Sigman, Francesco Sorrentino, H. Eugene Stanley, Ian Stewart, Boleslaw K. Szymanski, Matías Travizano, Lorenzo Timpone, Giovanni Valle, Stefan Wuchty, and Zhenkun Zhou.

Disclaimer

The studies of information spreading in social media reproduced in this book typically use classifications of news outlets, social media posts, users, and influencers given in terms of different categories ranging from fake news and extremely biased news to conspiracy theories, and political orientations from left to right. These categories and classifications are a matter of opinion rather than a statement of fact. The categorizations and labels assigned to the corresponding classes are those used in the cited scientific papers and are reproduced in this book. These results have originated publicly available datasets from fact-checking organizations as credited in this book. Results are reproduced as published in scientific journals and are based on classifications of news and others from fact-checking organizations and curated lists of fake and genuine websites and outlets that have not been generated nor curated by the authors of this book. These classifications are subject to subjectivity. The views and conclusions contained in the cited articles and this book, as well as the classification of news media and individuals into fake news, extremely biased news, conspiracy theories, influencers, and categories of political orientation from the extreme left, left-leaning, left, center, right, right-leaning, to extreme right should not be interpreted as representing the opinions of the authors of this book, the publishers, nor the funders of the work. Partial funding for this work has been provided by NIH Grant R01 EB028157 and NSF Grant 2214217.

Contents

Chapter 1
Mathematical Theories of Influencers in Complex Networks

Abstract The study of influencers is dovetailed to the big data analysis performed with network theory and artificial intelligence. In a social network (e.g., Twitter, Facebook, Instagram), some individuals spread information more rapidly than any other user. In an epidemic, some individuals are superspreaders who could infect disproportionally many people. In ecosystems, keystone species are responsible for the stability and integrity of the ecological network, analogously to banks that may be too big to fail in an interconnected financial network. Essential genes and neural ensembles in the brain are crucial for integrating information in biological networks. Finding these influencers in a network is crucial to understanding the transference of information and the functionality and resilience of the system, as well as to learning about human behavior and biological function. This chapter discusses the theoretical approaches to understanding, finding, predicting, and creating influencers in general, using theoretical concepts ranging from heuristics and centralities to percolation optimization approaches, spin glasses, and beyond.

1.1 What is an Influencer Anyway?

This is a book about the theory of influencers. However, we have already argued in the preface that there is no unique way to define an influencer and that this definition depends on the context of the particular problem. A social influencer is a person who can spread information virally on social media, similar to a disease superspreader who is an influencer in the sense that they can transmit a virus to many people through the network of human contacts. A genetic influencer is a gene that is so important for the organism that its mutation could lead to the malfunction of the cell as a whole, producing disease. A neural influencer is a neuron or an ensemble of neurons essential for integrating information signals in the brain. Keystone species in ecological networks can drive the entire ecosystem to collapse upon extinction. Like banks in financial networks, they might be "too-big-to-fail".

This list can go on and on since influencers appear in many situations. Can we develop a universal theory that is valid across a dissimilar set of situations? A universal definition of influential nodes in complex networks is somewhat elusive since

H. A. Makse and M. Zava, *The Science of Influencers and Superspreaders*,
Understanding Complex Systems, https://doi.org/10.1007/978-3-031-78058-5_1

they depend on the specific dynamical process; a theory of influencers needs to be devised for each case. However, this book contends that, in the tradition of physics, there are exciting commonalities across dissimilar systems that lead to universal concepts that can be applied across different situations.

This chapter is, first, about these universal concepts that can unify vastly different systems. We will push this approach until it is clear that no further progress is possible unless the particularities of each dynamic system are considered. Starting with Chap. 2 to the end of the book, we will abandon the search for a universal theory. This approach has been compelling in understanding physical systems but may restrict our understanding of social and biological applications. We will then attempt to construct more realistic influencer theories for each case, ranging from sociological applications to biological ones.

At the most basic level, the importance of a node in a network is characterized by a series of centralities defined in terms of the topological or structural properties of each node in the network. These centralities provide scores to each node in the network. The node's importance is then obtained by ranking the nodes according to the most significant centrality score. The most straightforward influencer-finding algorithm uses this list to rank influencers from top to bottom.

We will discuss a set of centralities that capture different aspects of influencers. The most basic is the degree centrality that employs the number of connections of each node in the network to rank the nodes. This ranking then privileges the nodes called 'hubs', following the intuition that the more connections there are, the better. It has also inspired the concept of scale-free networks [1].

We will show, however, that there are plausible circumstances where the best spreaders do not correspond to the most connected people but to those nodes that have few connections, yet they are strategically located within the core of the network as identified by other centrality measures like the k-core decomposition [2]. Thus, contrary to the mainstream thought, an influencer may not necessarily be the one with the most followers in social media but could be someone with few connections who holds a series of connections that are well-connected and that eventually reconnect with the original node, creating positive feedback. In other words, the right contacts matter more than the number of connections; i.e., *"connections are not everything"* as coined by Andreas Trabesinger on the cover of the November 2010 issue of Nature Physics [2].

The low-degree influencers are called 'weak links', which is a fundamental notion in sociology termed by Granovetter as *"The Strength of Weak Ties"* [3]. According to Granovetter, strong ties (close friends) clump together, forming modules. An acquaintance (weak tie) becomes a crucial bridge (an influencer) between the two densely knit clumps of close friends. Granovetter concludes:

> Whatever is to be diffused can reach a larger number of people and traverse a greater social distance when passed through weak ties rather than strong.

Granovetter thus argues that the weak link in the network may be more crucial in transmitting information than the strong one. This idea is the *leitmotif* of our book, being valid not only for social media influencers but also for more general cases of information flow in financial and biological networks, including ecosystems and the brain. The search for these hard-to-find weak links makes the influencer problem an exciting optimization problem so hard to solve [4].

After introducing a series of basic topological rankings, we will then define the science of influencers in a more rigorous algorithmic way by introducing the Influence Maximization Problem (IMP) following the formulation of Kempe et al. [5] based on a previous formulation of the Viral Marketing Problem by Domingos and Richardson [6, 7]. We will discuss the NP-hardness of this problem and how approximated algorithmic solutions have been proposed by using greedy algorithms developed in computer science and message passing methods from the statistical physics of disordered systems [8–12]. Firstly, influencers are selected by picking vital individual spreaders one by one using a greedy approach, in which the influence of single nodes is estimated via Monte Carlo simulations [5] and compared with various centrality measures. However, influencer identification is intrinsically an NP-hard combinatorial optimization problem [5]. Therefore, a collective point of view that considers interactions among multiple spreaders is required. Studies have translated the influencer identification problem into other closely related optimization problems, such as message passing [13], belief propagation [12], optimal percolation [14, 15], network dismantling and optimal decycling [16, 17] and explosive percolation [18]. We will then discuss these approaches in turn. They have enriched our understanding by tackling the influencer identification problem with rigorous theories and further generated several sophisticated and efficient methods that apply to large-scale complex systems.

Below, we describe these different approaches to finding influencers starting with centrality rankings. This will lead us to approaches to find influencers based on the percolation theory of first and second-order phase transitions, including the optimal percolation problem leading to collective influence, network dismantling, and optimal recycling.

1.2 Topological Centralities

A first approach for influencer identification is to find a good centrality measure to rank individuals' influence. A first-level rank of influencers employs ad-hoc centrality measures based on the topological or structural properties of the underlying network. Due to their simplicity and easy algorithmic implementation, a large number of centralities were adopted to estimate nodes' influence. These metrics are independent of specific spreading processes and dynamics in the network, and can be calculated with high computational efficiency. In addition, they also shed light on the impact of network topology on spreading processes in both accelerating and confining propagations. Instead of actually simulating the dynamics of the spreading process to find the most efficient influencers or superspreaders, these centralities ignore the particular dynamics of spreading and attempt to capture the importance of each node using local or global properties of a node in the network, such as the number of connections or degree.

We are left free to choose the particular topological property that defines the influencer centrality. And there are so many to choose from. Thus, we need to validate the chosen method by comparing the results over many centralities and choose the best for each application. Even though the optimal centrality that performs best for all spreading dynamics on all underlying networks do not exist; these centrality-based approaches are still persistently used due to their simplicity and relative satisfactory performance in particular occasions.

To set the stage for the discussion, we start with common definitions that will be used across the book:

Definition 1.1 (*Network or graph*) A *network* is represented by a graph $G = (V, E)$ made of a set of nodes V and a set of links $E \subseteq V \times V$. Edges can have different weights and can be undirected or directed. In this book, we will use the terms network and graph interchangeable.

Definition 1.2 (*Adjacency matrix of the network*) The *adjacency matrix*, $\mathbf{A}$, of a network is a $N \times N$ matrix ($N = |V|$) with element $A_{ij} = 1$ if there is an edge $i \to j$, and $A_{ij} = 0$ otherwise. For a weighted network, $A_{ij} = w_{ij}$ where $w_{ij} \in \mathbb{R}$ is the weight of the edge $i \to j$. For undirected networks, $A_{ij} = A_{ji}$.

Definition 1.3 (*Degree of a node*) The *degree* k_i of node i in an undirected network is the number of its connections, $k_i = \sum_j A_{ij}$. For a directed network, the *in-degree* is the number of in-coming links $k_i^{\text{in}} = \sum_j A_{ji}$ and the *out-degree* is the number of out-going links, $k_i^{\text{out}} = \sum_j A_{ij}$. The total degree is $k_i = k_i^{\text{in}} + k_i^{\text{out}}$. In weighted undirected networks, the degree of a node is replaced by its *strength*, $s_i = \sum_j w_{ij}$ or *in-strength* and *out-strength* for weighted directed networks.

Definition 1.4 (*Walk*) A *walk* on the network is an alternating sequence of nodes and links in which every node is incident to both the links that come before and after it in the sequence.

Definition 1.5 (*Path*) A *path* on the network is a walk in which all nodes and links are distinct.

Definition 1.6 (*Connected components*) A *connected component* of an undirected graph $G(V, E)$ made of a set of V nodes (vertices) and E edges is a subgraph of G, made of a subset of V' and E' where a path exists between each pair of nodes. In directed graphs,we consider a *strongly connected components* where there is a path in both directions between all pairs of nodes, and *weakly connected components* where there exist a path in at least one direction between all pairs of nodes.

Next, we elaborate on the most used centralities. We classify the centralities in five large classes based on the feature of the graph on which they are based:

1. degree-based,
2. shortest paths-based,
3. walks-based,
4. random walk-based,
5. non-backtracking walk-based, and
6. machine learning-based approach (to be treated in Sect. 1.8).

These significant branches of influencer-finding centralities will be discussed next, together with miscellaneous centralities valid for various networks like temporal and dynamical ones, and generalizations of pair-wise graphs to hypergraphs, multiplex, and multilayers. Some of these concepts have been reviewed in [19].

1.2.1 Degree-Based Centralities: Degree and k-Core

The centrality that first comes to mind is the degree of the node in the network, such as the number of retweets in a Twitter network or the number of followers in a social network. More elaborated ones can be constructed from this simple metric that considers the network, like the k-cores. Below, we review centralities based on the degree of the node.

1.2.1.1 Degree Centrality

Intuitively, large-scale propagations are more likely to be triggered by nodes with a larger degree. Barabási and Albert [1] revealed that the existence of highly connected hubs is a common phenomenon in many real-world networks. This discovery was quantified by the measurement of heavy-tailed power-law degree distributions of many real-world networks. These networks are popularly known as 'scale-free networks,' due to their lack of a characteristic degree. The scale-free model of Barabási-Albert [1] is a popular model that generates networks with such power-law distributions. As a result, degree centrality can work as an effective and efficient metric in identifying single influencers or superspreaders by ranking the nodes in the network by their degree, from top to bottom, and using the highest degree nodes or hubs as the top influencers. As shown in [20–25], hubs not only have disproportionate importance in the propagation process but also act as the key elements that maintain network integrity. Indeed, there is a strong connection between the problem of finding the influencers or superspreaders and the problem of finding the nodes that provide integrity to the network. This problem is handled by percolation theory and will be treated in detail in later sections.

One significant algorithmic advantage of degree centrality is that it only requires local information of the node without using the network, making it applicable to many systems whose network of connections is challenging to acquire. That is, the degree of a node can be obtained by asking people, 'How many friends do you have?' rather than ‘Who are your friends?’. With almost negligible computational cost compared with other metrics, degree centrality is scalable to networks of enormous size.

Thus, a ranking based on degree can be constructed without knowing any information on how people are connected in the network. This ranking ignores the network; a purely local metric, the degree, is just enough. While this makes this rank easily applicable to any problem, it misses the core property of an influencer, i.e., an influencer connects and influences people by being strategically located in the network to spread the influence beyond its nearest neighbors. The performance of degree centrality is mainly limited by the mere local information it includes. As empirical studies indicate, most spreading phenomena proceed in a cascading fashion. As a result, the influence of a single influencer should also relate to its multi-hop neighbors, whose information is not included in the degree.

Nevertheless, there are certain circumstances where the degree centrality is found superior to some other more complex centralities [26]. Furthermore, its performance to find influencers can be enhanced by adopting an adaptive computation, a centrality called high degree adaptive, where the degree of each node is recalculated after the removal of a previously identified influencer. As a matter of fact, in later discussions, we will show that degree is the zero-order approximation to some more sophisticated metrics based on the collective influence that do take into account for the effect of the network. This can partially explain its relatively good performance in some applications.

1.2.1.2 K-Core Centrality

While the degree partially captures the relative importance of a node in the network, it misses its position relative to other nodes in the network, i.e., it ignores the network’s structure. We could then think of plausible circumstances where a high-degree node or hub is located in a peripheral network region without influence. This is shown in Fig. 1.1, where we depict a sample network with a hub (yellow node) with many connections, but almost none of its connections are connected to other nodes. Such a node will have no importance in any spreading process in the network despite being a hub. Furthermore, this hub does not contribute to the integrity of the network since its removal will cause very limited damage to the structure of the entire network.

In this simple example, we see the crux of the matter of the influencer problem:

1. “Connections aren’t everything” [2],
2. Location might be more important than connections,
3. ’Weak links’ a la Granovetter [3] might be the crucial and most difficult influencers to find,
4. Influence is a collective phenomenon [14], and
5. Influence and network integrity (percolation) is an intimately related problem.

These five observations will guide large parts of influencer theory elaborated in the next sections.

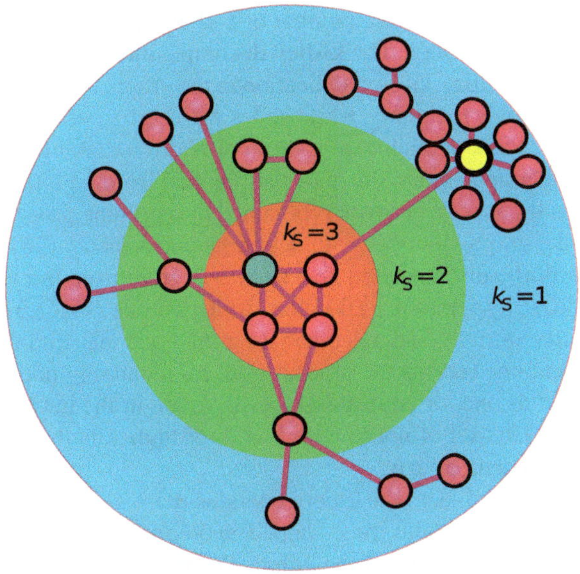

Fig. 1.1 "Connections are not everything". A hub in the periphery does not influence since it connects with poorly connected people. Instead, a relatively low-degree node in the network's core has more influence since it connects with other important nodes. The k-shell decomposition captures this structure. Nodes are assigned to k-shell layers according to their remaining degree, which is obtained by successive pruning of nodes with degrees smaller than the k_s value of the current layer. We start by removing all nodes with degree $k = 1$. After removing all the nodes with $k = 1$, some nodes may be left with one link, so we continue pruning the system iteratively until there is no node left with $k = 1$ in the network. The removed nodes and the corresponding links form a k-shell with index $k_s = 1$. Similarly, we iteratively remove the next k-shell, $k_s = 2$, and continue removing higher k-shells until all nodes are removed. As a result, each node is associated with a unique k_s index, and the network can be viewed as the union of all k-shells. The resulting classification of a node can be very different than when the degree k is used. For example, the two nodes of degree $k = 8$ (cyan and yellow nodes) in this network are in different locations: one lies in the periphery, $(k_s = 1)$ while the other is in the innermost core of the network, i.e., it has the largest k-shell index $(k_s = 3)$ or the maximum 3-core. Its influence is similar to that of its neighbors in the max 3-core, with small degree $k = 3$

A straightforward way to define the location of the node in the network is by using the k-shell decomposition, also called k-core decomposition, based on the graph's k-core. The k-core was proposed in social science by Seidman in 1983 [27]. It is defined (for undirected and unweighted graphs) as follows:

> Let $G(V, E)$ denote an undirected and unweighted graph with node (vertex) set V and edge (link) set E. A k-core of G is the maximal subgraph of G where each node in the k-core has at least degree k.

All nodes in a k-core have at least degree k, but not all nodes with degrees larger or equal to k belong to the k-core. The k-shell of a graph is the periphery of the k-core. The k-shell is composed of all the nodes that belong to the k-core but not to any (k+1)-core. The coreness, core number or k-shell index of the node, k_s, denotes the k_s-shell to which it belongs. The k-shell decomposition assigns the k-shell index k_s to each node representing its global location in the graph. Nodes with low k_s index are located at the graph's periphery, while nodes with high k_s values are located in the graph's core. Thus, the k-shell decomposition provides a ranking of the importance of the nodes in the graph, where the nodes are ranked from high to low k_s.

In practice, the k-core of a network is obtained by a pruning process; by iteratively removing from the graph all nodes with degrees smaller than k [28]. Assuming that we first remove all the disconnected nodes (zero degree) from the graph, one starts the removal process by removing all nodes of degree one (Fig. 1.3). After the first removal, nodes that initially had a degree larger than one may end up with a degree equal to one. Then, one repeats the process until no remaining node in the network has a degree equal to one, or equivalently, every node in the network has a degree at least equal to 2. This set of nodes with $k = 2$, or higher, is the 2-core. The other k-cores are obtained analogously.

The k-cores are nested; e.g., the 2-core contains the 3-core, and so on, and the 2-core is formed by the 3-core plus the 2-shell. The 0-core is the whole graph. The low k-shells are the peripheric shells of the graph. The maximum k-core of the graph is denoted by $k_{\text{core}}^{\text{max}}$ and constitutes the innermost k-core of the network, which coincides with the m value of the k-shell index k_s. It is a topological invariant of the graph, meaning it does not depend on how the nodes are labeled or the network portrayed, i.e., it is invariant under homeomorphisms.

A related concept is the k-degeneracy of the graph, defined as the maximum k for which the graph contains a non-empty k-core subgraph [29]. Thus, the degeneracy of the graph is the $k_{\text{core}}^{\text{max}}$. The maximum k-core subgraph is the inner subgraph of the network, and it is indexed by $k_{\text{core}}^{\text{max}}$ index. The degeneracy of the graph, also known as width [30] and linkage [31], it is related to the coloring number of a graph defined in [32]. The k-core number is also related to the chromatic number of the network χ (defined as the minimum number of colors to color the nodes so that no neighboring nodes have the same color), in that $k_{\text{core}}^{\text{max}}$ provides a bound for χ, i.e., $\chi \leq 1 + k_{\text{core}}^{\text{max}}$. In particular, a network is χ-colorable if it does not have a χ-core (the converse is not always true, though). Examples of k-core and k-shell decomposition can be seen in Figs. 1.2 and 1.3 in a sample network containing 3-shells, i.e., with maximum 3-core, $k_{\text{core}}^{\text{max}} = 3$.

A k-core can be composed of disconnected clusters or components. For instance, the example 3-core in Fig. 1.3c contains 3 components. These three components are disconnected in the 3-core, but they are integrated into the graph by nodes belonging to the 2-shell as shown in Fig. 1.3a. Intuitively, the node in a k-core is a node with degree k or larger that connects with other nodes of the same or higher degree, which in turn also connect with others of high degree, and so forth, until a cycle is formed that closes the loop with the original node. These feedback loops are crucial to make the high k-cores important for influencer identification and spreading processes. This implies that the high k-cores do not have dangling ends made of nodes with degrees smaller than k. The k-shell decomposition then cleans the network of those

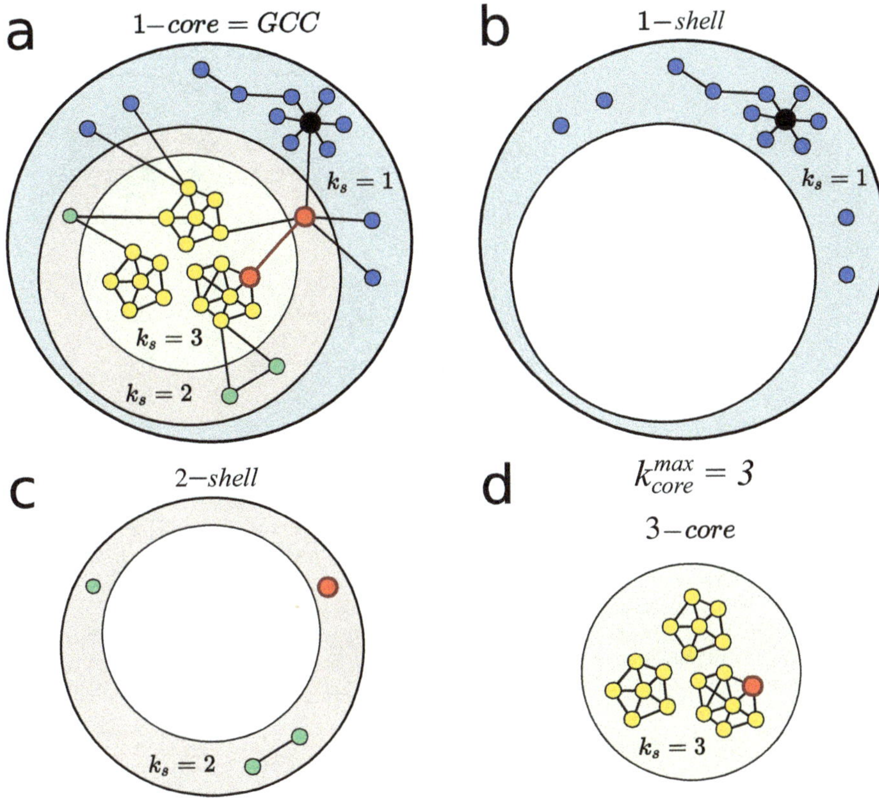

Fig. 1.2 Network structure under k-shell decomposition. **a** A sample network with 3 shells. The k-shell index k_s is not necessarily associated with other centralities. Here, the network hub in black with $k = 7$ is in the 1-shell, $k_s = 1$. The two top nodes in betweenness centrality, highlighted in red, belong to the 2-shell and the 3-shell, respectively. The 1-core is equivalent to one of the connected components. **b** The nodes with $k_s = 1$ form the 1-shell, **c** the nodes with $k_s = 2$ form the 2-shell, and **d**, the nodes with $k_s = 3$ form the 3-shell which is also the 3-core

low-degree dangling ends systematically and reveals the maximum k-core of the network, which is the most important part of spreading.

The k-core captures the idea that 'connections are not everything'. A hub can be anywhere in the k-shell decomposition of the graph according to how the hub is connected. Thus, the hubness of the nodes may not be directly related to their importance in the graph. For instance, the largest hub of the graph in Fig. 1.2a is the black node (degree = 7) that is located in the 1-shell and thus has no importance in the k-shell hierarchy, and in any spreading process in this network. On the contrary, low-degree nodes (for instance, the yellow nodes with degree = 3) are deeply located in the maximum k-core and thus are essential in the k-core hierarchy and crucial for spreading. Hubs can also be in the inner k-core, for instance, the cyan node in the sample of Fig. 1.2, which is in the maximum 3-core because it has three links in the

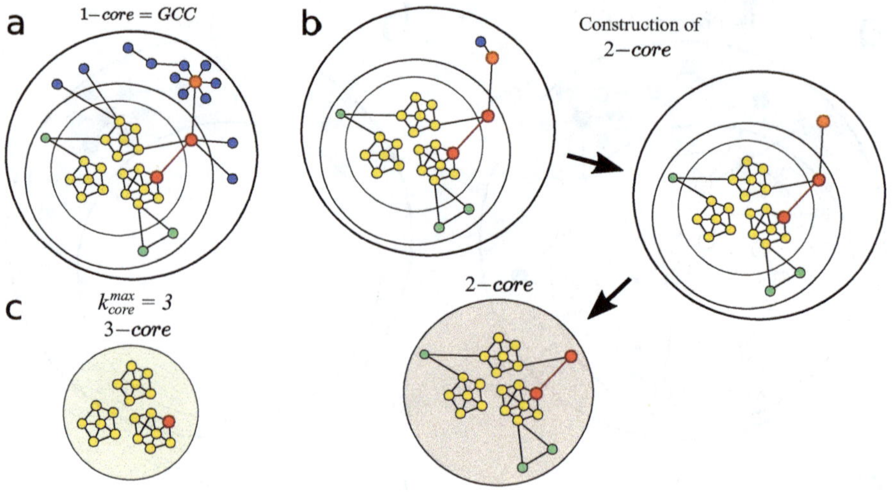

Fig. 1.3 K-cores of a network. a We start the k-shell decomposition of a network configuration where every node has at least degree $k = 1$. This set of nodes forms a 1-core and, in this case, is the largest connected component, although this is not generally valid. A network with disconnected nodes will also have a 0-core and 0-shell. **b** Then, every node with $k = 1$ is iteratively removed to obtain the 2-core. As one can see, removing these nodes changes the degree distribution. Thus, nodes are removed until all remaining nodes are left with $k \geq 2$. **c** Following the k-shell decomposition, nodes are removed until we obtain the maximum 3-core. Each k-core can be made of multiple disconnected clusters

inner k-core but is also connected to other nodes in lower k-shells that are removed by pruning. In this case, the hubness captures similar hierarchical information as the k-core, although the connections necessary for spreading are still those to the inner k-core. Compared with a degree, the k-core index considers the network's global topology. In general, the hubness and coreness are somehow related since the degree is the lower bound of the k-coreness, although coreness, being a global quantity, captures more interesting cases than the local degree.

The computational complexity of the k-shell decomposition process is $O(M)$, where M is the number of edges [33]. This means the decomposition algorithm's runtime increases linearly proportional to M. Therefore, k-shell ranking is scalable for large networks and particularly efficient when the network is sparse. Using this decomposition and taking advantage of the scalable complexity of the algorithm, visualization algorithms have been developed to display the hierarchical structure of large-scale networks using the k-core partition.

Visualization of the structure of the network using the k-shell decomposition was proposed by Batagelj et al. [34] in 1999 by ordering the nodes with the maximal k-core in the center and the lower k-shells in the periphery. One such algorithm has been developed by Alvarez-Hamelin et al. [35, 36], LaNet-vi: Large Networks Visualization tool, and it is applied to a network in Fig. 1.4 (upper panel), which was featured in [2]. The visualization utilizes Gephi [37], a popular network visualization

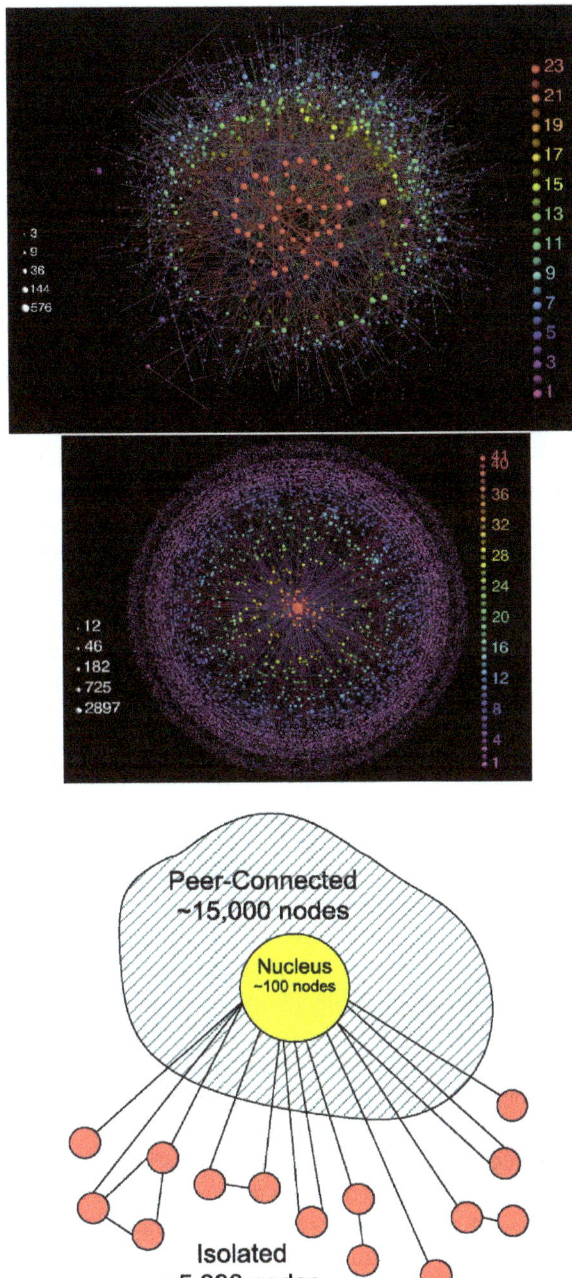

Fig. 1.4 Network k-shell decomposition. (Upper) An email network from [2] decomposed in its k-shells using the visualization algorithm of [35]. On the right of the figure is the color code of the different k-shells. The inner core at $k_{\text{core}}^{\max} = 23$ appears in red in the center of the rendering. The size of the nodes represents the degree, as given on the left of the image. (Middle) Visualization of the Internet at the autonomous system level from [40]. AS network ordered by their k-shell indices, using the software of [35]. The legend to the left denotes degree, and the legend to the right denotes k-shell index. (Lower) A schematic plot of the Medusa model decomposition of the AS-level Internet into three components. Figure reproduced from [40]. Copyright © 2007, National Academy of Sciences U.S.A

software that includes a k-core decomposition module. A k-core-base ego network visualization tool is presented in [38] to display nodes in collaboration networks with equal coreness. Using the *m*-core, defined as the maximal subgraph with edges participating in at least *m* triangles, Colomer-de-Simón et al. [39] showed the visualization of real-world networks, like the *E. coli* metabolic network as an internet kind of graph.

Carmi et al. [40] studied the internet at the level of the autonomous system subnetworks and used the k-shell decomposition and visualization algorithm of [35, 36] to uncover the structure of the Internet topology, see Fig. 1.4 (middle and lower panels). They call this structure Medusa, which is composed of a large, redundantly connected core and two components that flow data in and out from this core. One component has a fractal topology [41, 42] interconnected through peer links while the second only makes direct connections to the core.

Intuitively, superspreaders and influencers are more likely to be located in the core of the network with the largest k_s values, such that the k-core has had a major role in identifying influential spreaders in complex networks. Kitsak et al. [2] found that the k-core can be used to define the strength of influence of a node beyond the degree. They found, via numerical simulations using models of disease spreading SIR (susceptible-infected-recovered) and SIS (susceptible-infected-susceptible) [43], that spreading processes initiated by two hubs of the same degree can lead to different propagation coverage in the network, while a low degree node may propagate the spreading much further, see Fig. 1.5. This behavior was explained by the location of the source node in the k-shell decomposition of the graph. In low k-shells, the influence of a hub can be dramatically limited by its low-degree neighbors. On the contrary, even a node with moderate degree but located in the inner k-shells is capable of generating large-scale propagations with the help of its hierarchical well-connected neighbors.

Thus, the theory of k-core lays out that in a network, there are different layers of k-shells, and each node belongs to a different k-shell of the network. The k-core considers the number of connections and contacts who follow this individual and the number of contacts and connections that amplify the influence of the individual (positive feedback loop). In other words, the influencer is defined by the number of connections the individual influences and how these connections accelerate a response to the initial reaction/message. Thus, the k-core considers the node's position in the global structure of the network. The deeper we go into the network layers, the stronger the k-core becomes until it reaches the max k-core. To find the influencers, one has to uncover the layers determined by the k-core index (from the outside to the inside) to reach the maximum k-core.

The history of the k-core concept is interesting. The k-core was first introduced in social sciences by Seidman in 1983 [27], before modern approaches that use the k-core to define influence in the complex network literature. This the situation parallels other developments in modern network theory that followed ideas first developed in sociology in the late 20th century that are then re-formulated (some would say 'reinventing the wheel') under physical principles inspired by the availability of big data in the modern era of network theory in the 21st century.

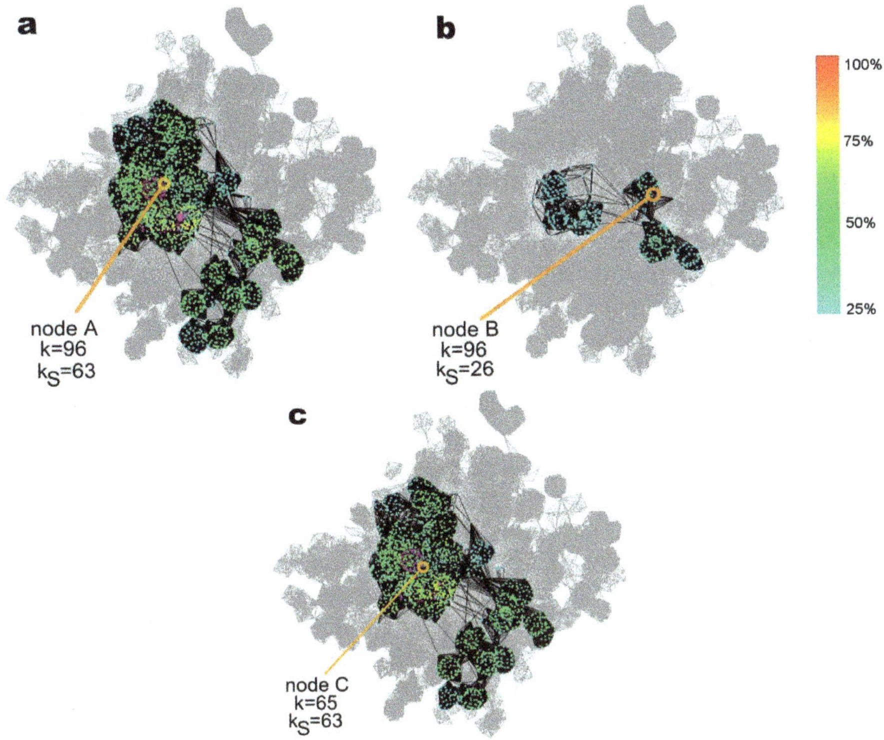

Fig. 1.5 Plausible circumstances where the hubs may not be good spreaders. a–c The extent of the efficiency of the spreading process cannot be accurately predicted based on a measure of the immediate neighborhood of the node, such as the degree k. For the contact network of inpatients from [2], the figure compares infections originating from single nodes having the same degree $k = 96$ (nodes A and B) or the same index $k_s = 63$ (nodes A and C), with infection probability $\beta = 0.035$. In the corresponding plots, the colors indicate the probability that a node will be infected when spreading starts in the corresponding origin, as long as this probability is higher than 25%. In the first case, where origin A has $k_s = 63$, spreading reaches a much wider area more frequently, in contrast to origin B ($k_s = 26$), where the infection remains largely localized in the immediate neighborhood of B. Spreading is very similar between origins A and C, which have the same k_s value, although the degree of C is much smaller than A. Figure reproduced from [2]. Copyright © 2010, Nature Publishing Group

In the original work, Seidman used the idea of cohesion as a marker of network structure to distinguish between "tightly-knitted" from "loosely-knitted" communities in networks; an idea which appeared already in the work of Bott in 1957 [44]. In turn, these concepts were inspired by the "weak links" of 1973 Granovetter's paper [3], as a way to capture influencers that are not highly connected, yet crucial for the transmission of information. The weak nodes of Granovetter's theory [3] later appeared in the concept of collective influence [14], which is developed in Sect. 1.6.1. In these theories, a node with a low degree is surrounded by hierarchical coronas of hubs at different levels, thus providing a high level of influence to the low-degree node.

Sociology studies [27] focused on the k-core from the point of view of its power to decompose or partition a graph into its shell components. Using the k_s index, one can index the network's highly interconnected partition that describes the complex system's topology. From this point of view, the k-core is of fundamental importance in the structure of any graph. Then, the k-core decomposition has been formulated as a mathematical problem by Bollobás in [45]. Bollobás showed that the k-core is a necessary (but not sufficient) condition for a subgraph to be k-colorable, and proved that non-vanishing k-cores are k-connected in random networks; a result later confirmed under less stringent conditions by Luczak [46].

In this section, we will use the k-core as a graph decomposition tool since we are mainly interested in the k-core index as a ranking for essential nodes in the graph. However, the concept of k-core and k-shell go beyond providing a graph partition. It is intimately related to the idea of percolation [47] that will be discussed in Sect. 1.4 and, in particular, to k-core percolation discussed in Sect. 1.7.1 [48, 49]. Both percolation problems have attracted much attention from mathematicians and physicists alike. The problem of k-core percolation will be addressed in Sect. 1.7.1.

A very complete account of applications of k-core, including decomposition and k-core percolation variants have been compiled by Malliaros et al. [50], see Table 1.1. The applicability of k-core decomposition to detect influential spreaders was shown in [2, 51], exemplified pictorially in Fig. 1.5. On the other hand, Borge-Holthoefer and Moreno showed that the influence of a node does not depend on its k-core location in rumor-spreading models [52]. Given the diminishing role of hubs in some dynamics [53], the k-core index has gradually been used as a reliable metric for importance, and many improvements on the original algorithm have been proposed. A drawback of the k-core index to rank influential nodes is that it is highly coarse-grained. Even for very large networks, $k_{\text{core}}^{\text{max}}$ is rarely larger than 100. As a result, each shell is composed of a large number of nodes which are left undistinguished within the same k-shell.

The superspreaders can be more accurately identified by considering more detailed topological information near the core region. In particular, Zeng and Zhang [54] distinguished the edges linked to the removed nodes (exhausted degree) and remaining nodes (residual degree) in k-core decomposition. Due to the heterogeneous distribution of exhausted degrees for nodes in the same k-shell, they suggested that the influence of these nodes also varies dramatically. Combining the residual degree and exhausted degree with a tunable parameter, a mixed degree decomposition (MDD) was proposed. MDD is a refined variant of the original k-core decomposition and achieves better performance in finding influencers on various real-world networks.

Further refinement can be reached through the k-truss decomposition proposed by Malliaros et al. [55], which is a triangle-based extension of the k-core decomposition [56]. The k-truss subgraph of a network is defined as the largest subgraph whose edges belong to at least $k - 2$ triangles. As illustrated by Malliaros et al., the k-truss subgraph makes up the most connected part of corresponding k-cores. The k-truss decomposition filters out the superspreaders in the core region. In numerical simulations of SIR models, these superspreaders affected a greater fraction of the network than others within the same coreness.

An improvement over k-trust has been proposed by Li et al. [57] to identify k-influential communities as a subgraph H^k of the network that is connected, where each node in H^k has degree k, and there is no other subgraph that satisfies these two criteria are not a subgraph of H^k and have a minimum weight lower than H^k. The improvement comes since this algorithm includes the influence of each community.

Shin et al. [58] also attacked the problem of ranking the nodes inside the largely populated k-cores. They propose the CORE-S algorithm, where they first applied the k-shell decomposition and then rank the nodes inside the k-shells with eigenvalue centralities (discussed below). They find improved results in both efficiency and accuracy by comparing their algorithm against a plain k-core decomposition, plain eigenvalue centrality, and k-trust. Along similar ideas, studies of social influence in Twitter were performed in [59] with a modified k-shell decomposition algorithm that assigns logarithmic k-shells values to users to avoid the skew distribution of the occupancy of the lower k-shells. Ref. [60] also applied a weighted k-core scheme to improve upon previous studies.

The core of a network is usually connected to nodes located in a diversity of different shells. However, some core nodes may densely connect and have limited connections to other network parts. These nodes form a core-like group substructure indistinguishable from the actual core in k-shell decomposition. As reported by Liu et al. [61], k-shell decomposition is degraded with the existence of core-like groups. Nodes in core-like groups are frequently identified as superspreaders; however, the influence of these fake superspreaders is somewhat limited by the lack of connections with other regions in the network. To filter these fake superspreaders, redundant links must be removed before the pruning process in k-shell decomposition [62]. These redundant links contribute little to global influence since they share a large proportion of overlapped neighbors. By removing redundant links in the first place, the core-like groups are fragmented before k-shell decomposition, and the effectiveness of the simulation of SIR models is confirmed.

A generalized core decomposition has been proposed by Batagelj and Zaveršnik in [63], where rather than using the degree for the pruning process, any node property function can be used leading to the p-core of the graph. The k-peak decomposition has been introduced by Govindan et al. [64] to take into account that the graph may contain regions that are sparser than others, which might be neglected in the standard k-core decomposition. The nucleus decomposition also considers the search for dense subgraph detection and how these regions are connected within the larger graph. Saríyüce et al. [65] defines the (r, s)-nucleus of a graph as the maximal subgraph where every r-clique (i.e., a complete graph of r nodes) is part of many s-cliques. The detection of dense subgraphs has been done in bipartite graphs as well. Saríyüce and Pinar [66] use graph peeling to detect dense subgraphs using the higher-order structural motif in bipartite networks [67]. A radius-bounded k-core decomposition has been proposed by Wang et al. [68] to take into account a graph projected in the space; a similar method is the city clustering algorithm of [69–71].

Different attributes of the nodes have also been considered to augment the k-core decomposition by considering similarity measures. Bonchi et al. [72] takes into account not only the degree but also the shortest paths among the nodes to define the distance-generalized core decomposition via the (k, h)-core as the maximal subgraph where each node has at least degree k and is at a shortest path smaller or equal to h within the subgraph. Similarly, Zhang et al. [73] consider other similarity measures to augment the k-core with other node attributes; e.g., a (k, r)-core satisfies both, the structural and similarity constraints.

So far, we have discussed the k-core of undirected and unweighted graphs, which are quite narrow, especially for applications. Important generalizations of k-core to other type of graphs have been proposed, as follows [50]:

Weighted undirected graphs Eidsaa and Almaas [74] introduced the s-core decomposition to generalize the k-core to weighted graphs. A link weight W_{ij} denotes the strength of the edge between i and j, and the strength of the node is $s_i = \sum_j A_{ij} W_{ij}$, where A_{ij} is the adjacency matrix. The threshold of a s_n-core is defined as $s_{n-1} = \min_i s_i$, where i belongs to nodes in the s_{n-1}-core subgraph. The s_n-core is defined by the iterative removal of all nodes with strengths $s_i \leq s_{n-1}$. The strength of node weights is recalculated after each node removal. The threshold of the s_1-core is the smallest s_i in the graph. When W_{ij} is a constant for all (i, j), the s-core decomposition is identical to the k-core decomposition, see also [75, 76].

Directed unweighted graphs In these graphs the in-degree refers to the number of incoming arcs and the out-degree to the number of outgoing arcs of each node. Giatsidis et al. [77] then introduced the D-cores with the notion of (k, ℓ)-core that defines a subgraph in which all nodes have in-degree and out-degree at least (k, ℓ), respectively. The definition of weighted s-core and (k, ℓ)-core can be combined to define the weighted directed cores.

Signed directed graphs These graphs are important since they allow two types of links, positive and negative [78–80]. We will return to this type of graph to describe biological networks with activators and inhibitors in Chaps. 5 and 6. They also describe trust networks and rumor spreading in social networks with antagonist interactions [78]. Giatsidis et al. [81] consider a directed graph of two link classes. They are called $s, t \in \{+, -\}$, but the definition applies to any class of links. Each node is characterized by four numbers specifying each class's in- and out-degree. The (k^s, ℓ^t)-core of the graph is the maximum subgraph where each node has in-degree and out-degree larger than k and ℓ, respectively, for each type of $s, t \in \{+, -\}$.

Bipartite graphs A bipartite graph can be partitioned in two disjoint sets where there are no edges between nodes in the same partition. An example is the mutualistic ecological networks that will be treated in Chap. 7. In this case, the graph can be treated as a single graph since the dynamics in the system allows for this, or with the concept of $\alpha - \beta$-core defined by Ding et al. [82] where $\alpha - \beta$ refers to the degree of each partition.

Dynamic and temporal graphs. These are graphs where nodes and links change over time, and the k-core has been adapted to these cases in [83–85]. Wu et al. [86]

considers the (k, h)-core for temporal networks where edges between nodes can appear and disappear in different intervals. Here k is the degree of the node, and h is the number of multiple temporal links between the nodes.

Multiplex Graphs These are graphs with a subset of links representing different types of interactions among the nodes. Considering that there are an ℓ-subset of interactions, then the degree is replaced by an ℓ-dimensional vector $\mathbf{k}$ defining the $\mathbf{k}$-core of the multiplex graph as the maximal set of nodes such that each node complies with the corresponding degree threshold condition in each layer of the multiplex [87–89]. The emergence of such a $\mathbf{k}$-core in random multiplex graphs with arbitrary degree distributions has been studied by Azimi-Tafreshi et al. [87] and in real networks by Osat et al. [89] showing that they emerge abruptly, yet with different properties of the single k-cores in individual network layers.

These generalizations are essential in realistic applications, such as biological networks. They are composed of directed and weighted edges of multiplex types. Indeed, the k-core found many applications across biology, physics, and sociology. Network biology studies the structure and dynamics of gene regulatory, metabolic pathways, and protein-protein interaction networks, as well as the brain and ecological networks [90]. In biological networks, the structure of protein co-occurrence networks has been studied using the k-core decomposition by Wuchty and Almaas [91]. By analyzing the protein domain co-occurrence networks of species *S. cerevisiae, C. elegans, D. melanogaster, M. musculus* and *H. sapiens* they find that the innermost cores increase with increasing degree of evolutionary development across these species. The k-cores are predominately involved in protein functions such as cell-cell contacts and signal transduction. The connected network patterns preserved through evolution may constitute nucleation kernels for the evolutionary increase in proteome complexity.

Emerson et al. [92] also studied the protein domain co-occurrence network of the human proteome, identifying the highly connected domains using the k-core decomposition and found that the inner core domains were more evolutionary conserved than the peripherical k-shells. Surprisingly, they found the inner k-cores of the network did not contain any of the significantly mutated domains that enriched cancer disease pathways. Thus, the inner k-core of the protein network has a low mutation frequency. They are highly conserved, while the outer k-shell contributes more to disease. These findings are being used for drug development. Pellegrini et al. [93] studied protein-protein interactions networks by a core and peel decomposition exploiting the k-core decomposition of the graph and identify the functional modules and protein complexes in the protein domain, outperforming other biological inspired complex prediction algorithms [94]. The weighted aspects of the k-core were studied in biological networks in [74, 95] and the bipartite structure in [96] using the k-core. Metabolic networks of plants were studied by Filho et al. [97] using a topological k-core decomposition. The maximal k-core was found to be comprised of only molecules of the primary basal metabolism and were highly conserved across species, while the metabolites acting in the secondary metabolism was found in the outer k-shells, in agreement with previous results above.

The k-core structure is closely related to the core-periphery structure [98], which is a core of highly interconnected nodes loosely connected to the nodes in the periphery [99, 100]. References [101, 102] studied the core-periphery structure of protein-protein interaction networks and found that the inner cores are functionally and structurally critical for the network.

K-core decomposition has also been applied to understand the structure of the brain. Hagmann et al. [103] analyzed diffusion MRI to build a high-resolution structural connectivity matrix of white matter tractography in the entire human brain using network methods to be discussed in Chap. 6. They perform a k-core decomposition analysis of a binary connection matrix obtained by thresholding the high-resolution fiber densities. The study of the large-scale structure of the human brain network reveals an inner core within the posterior medial and parietal cerebral cortex, as well as temporal and frontal modules. The structural core of the brain is composed of connector hubs that link all major structural modules of the brain. It contains brain regions that form the posterior components of the human default network. The study concludes that the structural core of the brain has a central role in integrating information across functionally segregated brain regions.

These results were confirmed by Lahav et al. [104] who found that the human cortical network, derived from the same diffusion spectrum imaging data as [103], is organized in a hierarchical structure reflected in the k-shell connectivity, with each hierarchy reflecting different functional roles. These k-cores are also related to rich-club structures found in the brain [105].

The role of the k-core in the structure-function relation in the brain was studied by Li et al. [106] using functional brain networks obtained from functional magnetic resonance imaging (fMRI) on subjects performing language-associated tasks. The k-core analysis reveals the most robust component of the functional language brain network [107], and also reveals fundamental differences in these networks when comparing monolingual subjects with bilingual subjects. A similar k-core analysis was performed by Arese Lucini et al. [108] to identify the influential brain areas driving the transition between the state of consciousness and subliminal states [108]. These results will be discussed in more detail in Chap. 6.

In Chap. 4, we will study the structure of the contact networks of disease spreading during the COVID-19 pandemic [109]. These studies find that the k-core is crucial to understand the structure of superspreader events in the pandemic, allowing for the use of network theory to design optimized quarantines to minimize disruptions to society. García-Algarra et al. [110] have used the k-core to rank ecosystem species. In Chap. 7, a particular type of dynamical system composed of mutualistic interactions between species, e.g., plants and pollinators, are considered on a network, and the maximum k-core will naturally appear to define the influencers (keystone species) in such a network [111]. The k-core also has applications in materials science to understand the jamming transition in granular matter [112, 113] and discontinuous shear thickening transitions in dense suspensions [114].

The literature on k-cores is vast; we have just covered some of it. The k-core will come back later in Chaps. 4, 6, 7 and 8 when we look at transmission chains of disease, the brain, ecosystems, and markets, respectively. We refer the reader interested

Table 1.1 History of theoretical developments and domain applications of k-core in mathematics, computer science, physics, biology, and sociology. Table updated and reproduced from [50]

Theoretical concepts	Year	Domain applications
Erdös-Rényi percolation [47]	1960	
von Neumann cellular automaton [307], coloring number [32]	1966	
k-degeneracy [29]	1970	
Granovetter 'strength of weak ties' [3]	1973	
Chalupa, Leath and Reich bootstrap percolation [306]	1979	
Seidman k-core definition [27]	1983	
Bollobás k-core percolation [45]	1984	
Luczak k-connected [46]	1991	
Pittel, Spencer, and Wormald k-core percolation [48]	1996	
K-core decomposition and visualization [34]	1999	
Generalized k-core [63]	2002	Cascading phenomena [208]
Main-memory computation [33]	2003	
	2005	k-core visualization [35, 36], protein networks [91]
K-core robustness [49], Bootstrap percolation [311]	2006	fingerprinting [35], jamming transition [112]
	2007	internet k-core topology [40]
Truss decomposition [56]	2008	Internet evolution [85], internet structure [36], neuroscience [103]
	2009	Protein interaction network [101]
Bootstrap percolation [254]	2010	influential spreaders [2]
Heterogeneous k-core [207], D-cores [77], disk-based computation [115], distributed computation [116], anchored k-core [117]	2011	Communities [75], influence [59], neuroscience [105]
Weighted networks [76]	2012	Visualizing triangle k-core [118], absence of influential spreaders [52]
Dynamic cores [84], weighted networks s-core [74], mixed degree decomposition [54]	2013	Network analysis [119]
Uncertain graphs [120], S-cores [81], multiplex graphs [87]	2014	Influential spreaders [51, 121]
Temporal graphs [86], density-friendly decomposition [122]	2015	Spreading [62], keyword extraction [123], influential communities [57], earthquake networks [124], cell biology [92, 102], neuroscience [125]
Incremental computation [126]	2016	Cell biology [93, 95], core-periphery [98], neuroscience [104], network decycling and dismantling [16, 17], optimal k-core percolation [127], influential spreaders [55], keyword extraction [128]
Hidden cores [129], multiplex graphs [88], parallel computation [130], density-friendly decomposition [131], k-peak decomposition [64], (k, r)-core [73], $\alpha - \beta$-core [82]	2017	Distributed clustering [132], ecology [110], influential spreaders [60]
Probabilistic cores [133], radius-bounded cores [68], bipartite peeling [66]	2018	Biology [96], metabolic networks [97], patterns and anomaly detection [58]
Distance-based cores [72]	2019	Ecology [111], network decycling and dismantling [269], jamming transition [113], consciousness [108]
	2020	Multiplex graphs [89], stock market [380], brain [106]
	2021	Brain [107]
	2022	Epidemic spreading [109], shear thickening [114]

in particular applications to these chapters for further details. Table 1.1 provides a concise history of the conceptual developments leading to the k-core and associated domain applications from computer science, mathematics, physics, sociology, biology, ecology and neuroscience reproduced and augmented from the excellent review of Malliaros et al. [50]. Efficient algorithms to compute k-core variants discussed above, including undirected, directed, weighted, signed, and bipartite graphs are mentioned in [50]. We next discuss other popular centralities.

1.2.1.3 Hirsch-Index Centrality

Hirsch index, or H-index, was initially proposed to quantify the output of a scientist [134]. Traditionally, the scientific output of an individual can be measured by the total number of papers or citations. Combined with two important features, H-index provides a better estimate of a scientist's cumulative research contributions [135]. Specifically, a scientist has index h if they have h papers with h or more citations each. Due to its applicability to ranking scientists, H-index has been introduced to complex networks to quantify the influence of a single influencer. Like its original definition, an influencer has index h if it has at least h neighbors with degrees no smaller than h. Unlike the k-shell index, the H-index condenses only local topology but includes more information than degree centrality. Simulations of the SIR model on disparate networks confirm that H-index is more reliable in superspreading identification than degree centrality and k-core index.

The metrics of degree centrality, k-shell index, and H-index are closely related. Lü et al. [136] defined the H-index operator $\mathcal{H}$ on a group of real numbers $(x_1, \cdots, x_n)$ that returns the maximum integer h such that there exist h elements in $(x_1, \cdots, x_n)$ no smaller than h. The pruning process in k-core can be viewed as a recursive calculation of the Hirsch-index h, in which a node is assigned index h if it has at least h neighbors with degree no smaller than h [134].

Considering a node i with k_i neighbors, whose degrees are $k_{j_1}, k_{j_2}, \cdots, k_{j_{k_i}}$ respectively, the H-index of i is then given by $h_i = \mathcal{H}(k_{j_1}, k_{j_2}, \cdots, k_{j_{k_i}})$. By definition, its degree centrality k_i is the zero-order H-index of node i $k_i = h_i^{(0)}$ and the sequence $h_i^{(0)}, h_i^{(1)}, h_i^{(2)}, \cdots$ converges to the k-shell index of node i, where $h_i^{(n)} = \mathcal{H}(h_{j_1}^{(n-1)}, h_{j_2}^{(n-1)}, \cdots, h_{j_{k_i}}^{(n-1)})$. In other words, degree centrality, H-index, and k-shell index are actually the initial, intermediate, and steady states of the iteration operator. In addition, new metrics can be introduced based on other orders of iteration of $\mathcal{H}$.

1.2.2 Shortest Path Centralities: Closeness and Betweenness

In contrast with the metrics that include pure topological information, some centralities try to capture specific assumptions of spreading dynamics. One assumption is that information or disease may propagate most efficiently in the network, although this may not be the case realistically. Within this assumption, one may consider the shortest paths between nodes significantly important for the spreading dynamics. As a result, several metrics have been developed using the shortest paths in networks.

1.2.2.1 Closeness Centrality

Closeness centrality [137–140] and betweenness centrality [141, 142] are dual measures of centrality developed in social sciences to quantify the influence in terms of ease of access to other nodes (closeness) and brokering power (betweenness) of individuals in social networks.

Closeness centrality quantifies the influence of a node by its average shortest distance to every other node [139, 140]. Mathematically, for a node i in the network $G(V, E)$, closeness centrality $C_C(i)$ is defined as:

$$C_C(i) = \frac{1}{\sum_{j \in V \setminus i} d(i, j)}, \tag{1.1}$$

where $d(i, j)$ stands for the shortest distance from node i to j. In networks with disconnected clusters, the distance between two nodes in different clusters are infinite. In such circumstances, closeness centrality is $C_C(i) = 0$. A proper alternative is to define the closeness centrality as the mean inverse of the harmonic distances:

$$C_C(i) = \frac{1}{n-1} \sum_{j \in V \setminus i} \frac{1}{d_{ij}}. \tag{1.2}$$

1.2.2.2 Betweenness Centrality

Another alternative is to consider the number of shortest paths instead of the mean shortest distance. In betweenness centrality, the influence of a node is quantified by the number of shortest paths passing through it [141, 142]. The betweenness centrality captures the brokering power of the node or the chance to intercept information traveling in the network. The betweenness centrality $C_B(i)$ in a connected and undirected graph is given by:

$$C_B(i) = \sum_{r \neq i \neq j \in V} \frac{\sigma_{rj}(i)}{\sigma_{rj}}, \tag{1.3}$$

where σ_{rj} is the number of different shortest paths between r and j, and $\sigma_{rj}(i)$ is the number of shortest paths among them that go through node i.

Betweenness is a centrality that captures the idea of the weak links of Granovetter [3]. Imagine a network composed of two modules where nodes are highly connected inside the modules, but only one single node connects both modules. This central node is the weak link of the networks. From the degree point of view, this central node will have a low rank since its degree is 2. However, from the point of view of betweenness centrality, this node will be ranked highly, and it is highly influential since all the paths that connect the two modules will necessarily pass through it. This example shows the power of betweenness centrality to capture the weak nodes. We will see an application of this centrality in a real-world network of human contacts transmitting COVID-19 in Chap. 4.

Betweenness and closeness centrality are the dual of each other, expressing the independence from the control of others (closeness) or the potential control over others (betweenness) [143, 144]. When considering shortest paths between h and j with different central brokers i at a fixed distance d from h ($1 \leq d < \text{dist}(h, j)$), each shortest path passes through exactly one broker and, therefore, the $\sum_{i \in V:\text{dist}(s,b)=d} \frac{\sigma_{hj}(i)}{\sigma_{hj}} = 1$.

This implies that the sum of dependencies between a sender h and a receiver j taken over possible brokers i is proportional to the distance between h and j:

$$\sum_i \frac{\sigma_{hj}(i)}{\sigma_{hj}} = \sum_{d=1}^{\text{dist}(h,j)-1} \sum_{i \in V:\text{dist}(h,i)=d} \frac{\sigma_{hj}(i)}{\sigma_{hj}} = \text{dist}(h, j) - 1. \tag{1.4}$$

Using this equation, the closeness centrality is:

$$C_C(h)^{-1} = (N - 1) + \sum_{i,j \in V} \frac{\sigma_{hj}(i)}{\sigma_{hj}}, \tag{1.5}$$

revealing its duality with betweenness centrality as a different partial sum, over i and j instead of h and j. This shows the interpretation of closeness as a measure of lack of independence on others [144]. Figure 1.6 shows an illustration of the different rankings obtained by the closeness and betweenness centrality measures using the kite graph introduced by Krackhardt [145].

Closeness and betweenness centrality can be extended to directed networks and weighted networks by considering the notion of reachability instead of distance. In weighted networks, the length of a path is regarded as the sum of the links' weights, which replaces the length of the shortest paths from an unweighted network to a weighted one [144]. Many more variants of closeness and betweenness have been developed, see Ref. [146]. The concept of betweenness centrality has also been extended by considering the number of shortest paths containing a link instead of a node in Eq. (1.3) [146]. Another extension finds the fraction of minimum spanning trees of a graph that contain a given link [147]. Versions of the closeness and betweenness centralities based on random walks instead of shortest paths have also been developed [148, 149].

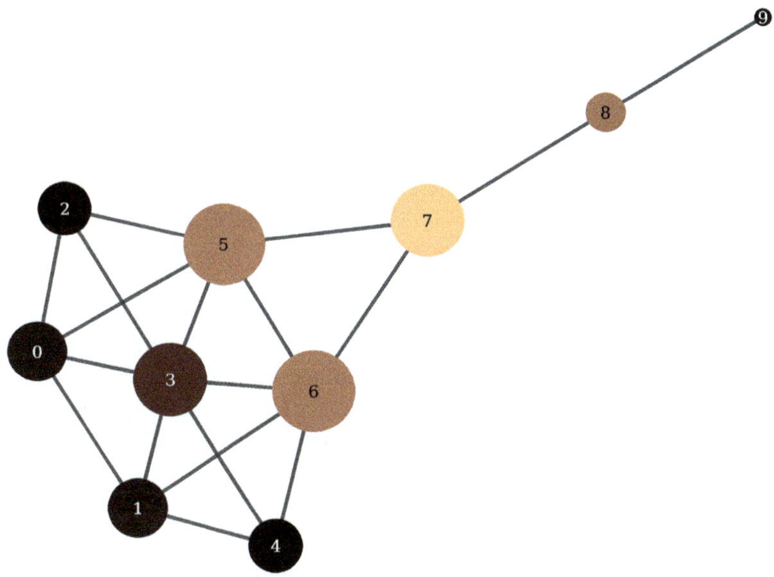

Fig. 1.6 Closeness and betweenness centrality. Differences in ranking obtained using degree, closeness, and betweenness centralities on a sample Krackhardt kite graph [145]. The closeness of a node is represented by its size, and the betweenness of a node is indicated by its color, with nodes with larger betweenness being of a lighter shade. The node with the largest degree of centrality is 3. Nodes 5 and 6 have the largest closeness centrality, and node 7 has the largest betweenness centrality. Figure reproduced from [19]. Copyright © 2021, Springer Nature

1.2.3 Walk Centralities: Eigenvector and Katz Centralities

Several widely used centrality measures, like Katz's centrality [150] and eigenvector centrality, are based on the concept of graph walks. Walks are useful to count the number of possible pathways between two different nodes; they should not be confused with random walks, which will be treated in Sect. 1.2.4 on which PageRank is based. Centrality measures based on walks try to identify central nodes from which there are the largest number of walks reaching other nodes. For an unweighted network, the ℓth power of the adjacency matrix: $\left(\mathbf{A}^{\ell}\right)_{ij}$ counts the number of walks of length ℓ existing between two nodes i and j. The number of walks of length ℓ starting from node i is then given by

$$w(\ell)_i = \sum_j \left(\mathbf{A}^{\ell}\right)_{ij}, \tag{1.6}$$

or in matrix notation

$$\mathbf{w}(\ell) = \mathbf{A}^{\ell}\mathbf{1}, \tag{1.7}$$

where **1** is the unit vector.

The number of walks of length 1 is simply the degree of the node, so the *degree centrality* of Sect. 1.2.1.1 of a node i is the first order in a walk-based centrality:

$$c_{\text{deg}}(i) = w(1)_i = \sum_j A_{ij}. \tag{1.8}$$

1.2.3.1 Eigenvector Centrality

Considering longer walks with $\ell > 1$, one can consider information about the entire network. The idea is that a node close to a node with a high centrality should also have a high centrality [150, 151]. Starting from an initial guess for the centrality of each node given by the vector $\mathbf{x}(0)$ (e.g., $\mathbf{x}(0) = \mathbf{1}$), we can propagate this initial centrality through walks of a given length. The centrality vector for walks of length ℓ is given by

$$\mathbf{x}(\ell) = \mathbf{A}^{\ell}\mathbf{x}(0). \tag{1.9}$$

Writing the initial centrality as a linear combination of the eigenvector of the adjacency matrix, the centrality converges to a vector proportional to the leading eigenvector of $\mathbf{A}$ with the largest eigenvalue, as $\ell \to \infty$. This process takes into account the entire network structure [152]. When the graph G is connected and $\mathbf{A}$ is non-negative, the Perron-Frobenius theorem ensures that the leading eigenvector is positive and that the associated largest eigenvalue is positive and simple.

Mathematically, the *eigenvector centrality* of node i is calculated iteratively from the leading eigenvector whose element i is:

$$c_{\text{eig}}(i) \propto \sum_{j=1}^{N} A_{ij} c_{\text{eig}}(j). \tag{1.10}$$

In calculations, all nodes in the network are initialized with the same value, and iterations are run until a steady state is reached. The eigenvector centrality measures the influence of an influencer with the sum of the influence of its neighbors.

This process is equivalent to the power method of computing the largest eigenvalue of the adjacency matrix, where repeated left multiplications are made to the influence vector with A. The power methods will be used again to calculate the collective influence in Sect. 1.6.1. As a result, eigenvector centrality is proportional to the right eigenvector, which corresponds to the largest eigenvalue. In the first step of iteration, eigenvector centrality recovers the same result with degree centrality, Eq. (1.8). As recursion continues, global network structure information is incorporated [151].

In most applications, the result of iterations converges quickly and the eigenvector centrality performs well. However, the eigenvector centrality tends to concentrate on a few hubs in the network, leaving other nodes almost indistinguishable with very low scores. This results from the repeated reflection of influence along the mutual connections between hubs during iterations. Martin et al. [153, 154] avoids the

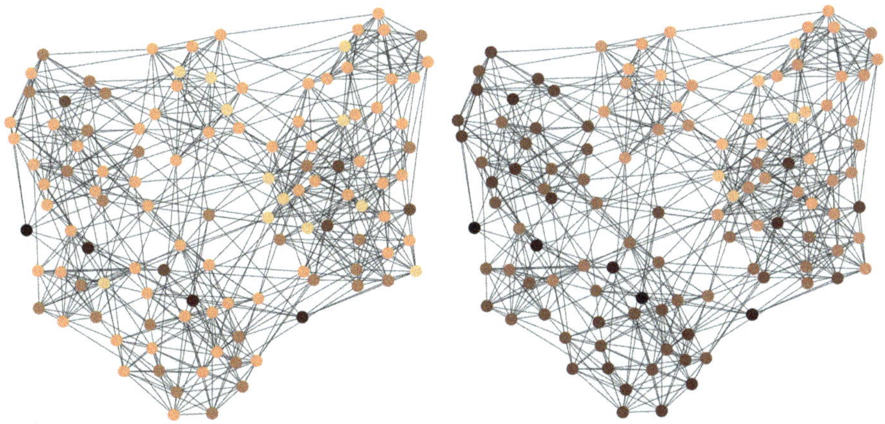

Fig. 1.7 Comparison between degree centrality and eigenvector centrality Network of American football games between Division IA colleges during the regular season Fall 2000 from [156, 157]. Degree centrality for this undirected graph is shown on the left, and eigenvector centrality is shown on the right. Lighter shades indicate a higher centrality. Several nodes have the same degree in this network, making the degree centrality unable to disentangle the importance of nodes with the same degree. Eigenvector centrality tends to be localized in regions where many high-degree nodes are close to each other, a problem that is solved with the non-backtracking matrix [153]. Figure reproduced from [19]. Copyright © 2021, Springer Nature

concentration by considering the primary eigenvector of the Hashimoto or Non-Backtracking matrix [155], which will be discussed in Sect. 1.2.5.

The eigenvector centrality can be extended to weighted networks by considering walks on the graph that are weighted by the edge weights. For directed networks, the eigenvector centrality can be computed similarly to undirected ones. However, the adjacency matrix is asymmetric for directed networks and has two sets of left- and right-eigenvectors. Depending on the application, one may prefer to use the largest left eigenvector or the largest right eigenvector if one is more interested in nodes with many incoming walks or outgoing walks, respectively. However, other types of centralities are usually preferred for directed networks, as the eigenvector centrality of nodes outside of a strongly connected component may be equal to zero. In particular, all nodes have zero eigenvector centrality in directed graphs without cycles (closed directed paths). Figure 1.7 shows a comparison of the degree centrality with the eigenvector centrality in an undirected network of American football teams from [156, 157].

1.2.3.2 Katz Centrality

In most spreading dynamics, propagation is possible along all paths, including the shortest ones. Although propagation tends to decay with the increase of the length of paths, those non-shortest pathways make up an ineligible part of spreading dynamics.

As a generalization of the closeness centrality, *Katz centrality* quantifies the influence of an influencer by the weighted number of paths originating from it [150]. Specifically, the Katz centrality of a node i is defined as:

$$C_{\text{Katz}}(i) = 1 + \sum_{\ell=1}^{\infty} \sum_{j=1}^{N} \alpha^{\ell} (A^{\ell})_{ji} , \tag{1.11}$$

where $\mathbf{A}$ is the adjacent matrix and $\alpha \in [0, 1]$ is a control parameter of the attenuation rate. Here, $\alpha = 0$ corresponds to attenuation. The unit shift does not alter the ranking of the nodes and may be seen as arising from a single walk of length zero.

This centrality based on walks tries to solve the issue of the eigenvector centrality in directed networks. The Katz centrality of node i is the weighted sum of all walks emanating from i, with the count for walks of length ℓ weighted by a factor α^{ℓ} where $0 < \alpha < 1$. In this way, the importance of longer walks is diminished compared to shorter walks. By releasing the computational burden of calculating the shortest paths between all pairs of nodes, the Katz centrality has superior computational efficiency to closeness centrality.

The factor α must be smaller than the reciprocal of the absolute value of the largest eigenvalue of $\mathbf{A}$ so that the centrality converges [150, 151, 158]. Using:

$$\mathbf{I} + \alpha\mathbf{A} + \alpha^2\mathbf{A}^2 + \alpha^3\mathbf{A}^3 + \cdots = (\mathbf{I} - \alpha\mathbf{A})^{-1} , \tag{1.12}$$

we can see that the vector $\mathbf{x}$ defining the Katz centrality of each node is also a solution of the matrix equation:

$$\mathbf{x} = (\mathbf{I} - \alpha\mathbf{A})^{-1}\,\mathbf{1} , \tag{1.13}$$

which can be rewritten as

$$\mathbf{x} = \alpha\mathbf{A}\mathbf{x} + \mathbf{1} . \tag{1.14}$$

The Katz centrality can be computed iteratively in a similar manner such as the eigenvector centrality (with Eq. (1.9)) using:

$$\mathbf{x}(t+1) = \alpha\mathbf{A}\mathbf{x}(t) + 1 . \tag{1.15}$$

The unit value is added at each iteration, guaranteeing that the Katz centrality will never be equal to zero, even in directed networks. The parameter α balances the importance of the eigenvector term compared to the constant term added at each iteration. In the limit when α approaches the reciprocal of the absolute value of the largest eigenvalue of $\mathbf{A}$, the Katz centrality approaches the eigenvector centrality on strongly connected graphs [151, 158].

Important generalizations of these walk-based centralities have been developed for hypergraphs [159, 160], multiplex networks, [161, 162], and temporal networks [163–165].

A similar metric considering arbitrary long walks was proposed in Ref. [166]. Based on the assumption that infectious walks of all lengths are independent of each other (i.e., they have no edges in common), the impact of an influencer is quantified by summarizing the probability of transmission to the end node of the walk over increasing walk lengths. With degree centrality being its first-order approximation, the impact metric performs better with increasing length of walks taken into account. In simulating infections on various real-world networks with the SIR and SIS models, it is found that the improvement can be obtained at a comparable computational cost.

In contrast with the centralities that quantify the influence of an influencer purely by its position, Klemm et al. developed the metric of dynamical influence (DI) that incorporates the interplay between dynamics and network structure [26]. For critical spreading, the influence of the initial condition of each node on the final steady state was found to be determined by the left principal eigenvector of the coupling matrix. The left principal eigenvector can be evaluated by applying higher and higher powers to the coupling matrix to a uniform vector, which essentially counts the total number of arbitrarily long walks originated from each node. Simulations on SIS and SIR models proved that DI centrality is a better predictor of spreading capabilities compared with the degree centrality, k-shell index, and betweenness centrality. Moreover, it can also be applied to other dynamical processes, including the Ising model, voter model and Kuramoto model.

From a continuous-time epidemiological perspective, Ref. [167] proposed the expected force (ExF) that also relies on possible spreading paths. Following traditional epidemiology, the spreading capacity of a seed can be quantified by the distribution of the number of infected-susceptible edges after a small number of transmission events. In calculation, all possible clusters of infected nodes after n transmission events originating from an influencer are enumerated. Then, the expected force of the influencer is calculated as the entropy of the cluster degree (infected-susceptible edges). In most applications, $n = 2$ is sufficient to get satisfactory results. In SIS and SIR dynamics, ExF performs better than accessibility, k-core, and eigenvector centrality in predicting epidemic outcomes. In addition, it can be computed independently for each node, making it applicable for extremely large or even dynamic networks.

1.2.3.3 Eigenvalue Centrality

In a large class of spreading and percolation models, the largest eigenvalue λ of the adjacent matrix plays a predominant role in phase transitions [21, 168]. As a result, the influence of an influencer can be measured by the relative change in λ upon its removal [169]. As is proposed by Restrepo et al., this dynamical importance metric takes into account the possible asymmetry of the adjacency matrix and can be easily applied to directed or weighted networks. With the left and right eigenvectors corresponding to λ into consideration, the dynamical importance incorporates global topological information. Similar ideas can also be found in the derivation of the collective influence algorithm on NB matrix [14] which will be discussed in Sect. 1.6.1 in the context of influence maximization with optimal percolation.

The *subgraph centrality* is based on the idea of counting the number of times that a node takes part in the different connected subgraphs of a network, with smaller subgraphs having higher importance [170]. The subgraphs are identified by counting closed walks of different lengths that start and end on a given node. The subgraph centrality of node i is defined as:

$$c_{\mathrm{Sub}}(i) = \sum_{\ell=0}^{\infty} \frac{(\mathbf{A}^{\ell})_{ii}}{\ell!}. \tag{1.16}$$

This centrality is more discriminating than the degree, betweenness, closeness or eigenvector centralities and gives a different ranking of nodes in real-world complex networks [170].

1.2.4 Random Walk Centralities: PageRank

If a stellar centrality outshines the rest, this is PageRank [171, 172]. PageRank is at the heart of Google's search engine. It was introduced before popular centralities like hubs, creating one of the most successful companies in history. PageRank is an influencer algorithm that ranks web pages from a network of hyperlinks. Previously, Kleinberg had developed HITS, a ranking based on hubs and authorities based on an eigenvector calculation on a co-citation matrix of the web [173].

In 1998, two computer science graduate students at Stanford, Sergey Brin and Larry Page (it is reported that the name PageRank refers to not only a webpage but also to the name of the Google co-founder) attempted to catalog all the information on the web with an altruistic mission: *"to organize the world's information and make it universally accessible and useful"* [174]. Their dream was to build a scrawler to mine every page on the web and then rank them using an influencer algorithm.

Before Google, there was Altavista, a search engine founded in 1995. Back in the day, a search done by Altavista would return quite irrelevant information ('Cortavista,' perhaps?), and the MIT bookstore was selling books with catalogs of web pages organized by topics, like the Yellow Pages of the web. In 1998, one of us traveled to Boston to visit former colleagues at BU when the epiphany happened. Talking with colleagues, we performed our first 'Google' search, and the relevant information appeared in a microsecond. At this point, we knew that a new era had been born: the democratization of digital information. We may recall other moments of similar importance in digital history: (1) around 2007, when we swapped for first time a little black box called iPhone; (2) when the cash machine in our bank swallowed a check and deposited it without asking the amount, thus heralding the commercial era of AI with the first successful commercial application; and (3) when we saw move 37 in Game 2 of the Go match between AI-powered AlphaGo and champion Lee Sedol.

In PageRank, influencers are quantified through random walks and associated eigenvalues. Like the eigenvector centrality, random walks consider the interplay between the number and the influence of neighbors. Random walks on graphs have been extensively studied in diffusive processes on networks and for the characterization of the structure of complex networks [175–178]. A random walk on a graph is defined by the trajectory of a walker that, at each time step, jumps to a neighbor of node i with equal probability. On an unweighted and undirected network, the transition probability for a walker on node i to jump to node j is equal to

$$T_{ij} = \frac{A_{ij}}{k_i}, \tag{1.17}$$

or in matrix notation

$$\mathbf{T} = \mathbf{D}^{-1}\mathbf{A}, \tag{1.18}$$

where $\mathbf{D} = \text{diag}(\mathbf{k})$ is the diagonal degree matrix. This definition can be generalized for directed networks with the degree replaced by the out-degree and to weighted networks with positive weights with the transition probability proportional to the weight of edge (i, j).

We have seen that in centralities based on walks like eigenvector and Katz centralities, the value of the centrality tends to concentrate on a few hubs in the network, leaving other nodes almost indistinguishable with very low scores. This is due to the repeated reflection of influence along the connections between a hub and its connected nodes during iterations of Eqs. (1.9) and (1.15) (see Figs. 1.7 and 1.8). To solve this problem, centrality measures based on random walks exploit the fact that if a centrality value propagates as a random walk, its value is divided across all outgoing edges of a node at each iteration. The distribution probability of finding a walker at time step n on any network node can be written as a vector $\mathbf{p}(n)$ with $\sum_i p_i = 1$. The evolution of the distribution probability is given by:

$$\mathbf{p}(n+1) = \mathbf{p}(n)\mathbf{T}. \tag{1.19}$$

On the connected undirected graph, the random walk reaches a stationary distribution satisfying:

$$\mathbf{p}^\star = \mathbf{p}^\star\mathbf{T}, \tag{1.20}$$

where the probability at each node is proportional to its degree:

$$p_i^\star = \frac{k_i}{\sum_j k_j}. \tag{1.21}$$

On undirected networks, we recover the degree centrality as the stationary state of the random walk process. On directed networks, random walks are not guaranteed to converge to a unique stationary state and a stationary distribution only exists on the strongly connected components of the graph [178].

To overcome this issue in directed networks, the *PageRank centrality* modifies the classical random walk explained above by introducing a "teleportation" probability, i.e., at each step, the walker has a probability of jumping to any other node in the network with uniform probability. PageRank was famously developed for ranking websites for the search engine Google considering this 'random surfer model' that mimics navigating through webpages by randomly clicking on hyperlinks [172].

Considering a random walker along directed hyperlinks between webpages, PageRank centrality evaluates the influence of a webpage by its probability of being visited in a search on the web by clicking hyperlinks. This probability is calculated through iteration, where the contribution that node i makes to its out-neighbor j equals to its PageRank centrality divided by the number of its out-neighbors.

The random walker can be trapped in the dangling ends of the networks leading to an artificial accumulation of Pagerank centrality. To avoid this, a probability is introduced that allows the random walker to jump to any random node in the network. Teleportation models the fact that in a web search, a user can jump to a random page with certain probability given by α. The update equation for the PageRank probability density is given by [171, 172, 178]:

$$\mathbf{p}(n+1) = \alpha \mathbf{p}(n)\mathbf{S} + \frac{1-\alpha}{N}\mathbf{1}^T\,, \tag{1.22}$$

where $\mathbf{S}$ is a transition matrix constructed from $\mathbf{A}$ such that,

$$S_{ij} = A_{ij}/k_{\text{out}}(i) \quad \text{when} \quad k_{\text{out}}(j) > 0\,. \tag{1.23}$$

For dangling nodes that have no out-going edges (i.e. $k_{\text{out}}(i) = 0$), then $S_{ij} = 1/N$. This means that the walker on such a node has a uniform probability of jumping to any other node in the network. On non-dangling nodes, at each step, walkers have a probability $\alpha < 1$ to follow an out-going edge and a probability $1 - \alpha$ to teleport to any node in the network. This modified random walk is also ergodic on networks that are not strongly connected and converges to the probability density vector, giving the *PageRank* centrality of node i as:

$$C_{\text{PR}}(i) = \alpha \sum_j C_{\text{PR}}(j) S_{ji} + \frac{1-\alpha}{N}\,, \tag{1.24}$$

which is equal to the normalized eigenvector corresponding to the largest positive eigenvalue of the so-called Google matrix [179]:

$$\mathbf{G} = \alpha \mathbf{S} + \frac{1-\alpha}{N}\mathbf{1}\mathbf{1}^T\,. \tag{1.25}$$

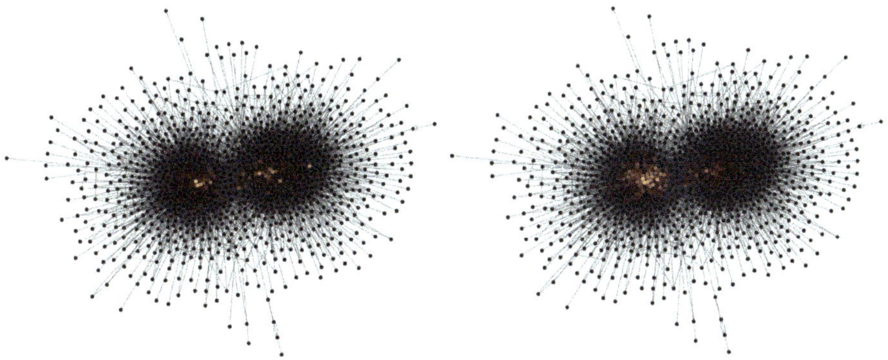

Fig. 1.8 Comparison between Katz centrality and PageRank. PageRank centrality is shown on the left, and Katz centrality on the right using a directed network of hyperlinks between weblogs on US politics from [184]. Lighter shades indicate a higher centrality. Both Katz and PageRank centrality tend to be localized in regions with many close-by hubs. Figure reproduced from [19]. Copyright © 2021, Springer Nature

In applications, the iteration of Pagerank centrality is calculated as [171, 172]:

$$p_t(i) = \frac{\alpha}{N} + (1-\alpha)\sum_j \frac{A_{ij}\, p_{t-1}(j)}{k_{\mathrm{out}}(j)}, \tag{1.26}$$

until a steady state, where $k_{\mathrm{out}}(j)$ is the number of out-neighbors of node j.

Beyond its original purpose of ranking webpages, PageRank also finds its applications in other problems, such as ranking scientists and academic papers [180], images [181] and finding important proteins in biological networks [182]. For undirected graphs, PageRank is still dominated by the degree of the nodes [183].

Figure 1.8 shows a comparison of the performance of PageRank and Katz centralities on a directed network of hyperlinks between weblogs from [184]. The similarities between Eqs. (1.9) and (1.19) as well as Eqs. (1.15) and (1.22) reveal that the degree and the PageRank centralities can be seen as similar to the eigenvector and Katz centralities, but for random walks instead of regular graph walks. In the case of random walk-based centralities, one looks for eigenvectors of transition matrices instead of adjacency matrices.

Lu et al. [185] has introduced a leader node into the network that connects all other nodes with additional bidirectional links. With the leader included, any network is strongly connected so that the random walker will not be trapped in the absence of jumping probability α. The resulting LeaderRank is a parameter-free algorithm and converges quickly. In addition, it is robust against malicious manipulations and data noise.

Another metric based on random walk is the accessibility, where the diversity of self-avoiding random walks is quantified in terms of the entropy of the probabilities of reaching destination nodes [186]. Nodes with high diversity are able to reach distinct destinations effectively through more choices of paths; thus, they can be natural candidates for superspreaders.

1.2.5 Non-backtracking Walk Centralities

We have seen that eigenvector-type centralities tend to concentrate the highly ranked influencers on a few hubs in the network while ignoring other nodes with very small scores. Most of the weight of the centrality vector localizes around one or a few hubs in the network. To address this issue of localization of the eigenvector centrality, several authors have focused on using non-backtracking walks instead of regular or random walks [153, 154, 187, 188]. Non-backtracking walks are graph walks where backtracking steps are forbidden. This means the walker cannot return to an immediately preceding node. Using this type of walk allows us to solve the localization problem as they decrease reflections between hubs and their connections during iterations of the centrality [153].

Non-backtracking walks on unweighted networks are described by the *non-backtracking matrix* , also known as the *Hashimoto matrix* $\hat{\mathcal{B}}$ [155]. For a graph $G(V, E)$ with $N = |V|$ and $M = |E|$, this matrix is a $2M \times 2M$ non-symmetric matrix where rows and columns are parametrized by the directed edges of the graph G. If the network is undirected, an equivalent directed network is considered by replacing each edge with a pair of directed edges in both directions. The element $(i \to j, \ell \to h)$ of the non-backtracking matrix $\hat{\mathcal{B}}$ is equal to one if $j = \ell$ and $i \neq h$, i.e., there is a non-backtracking path of length two from $i \to j \to h$ in the network. Otherwise, all the remaining elements are zero. More succinctly:

$$\mathcal{B}_{i\to j,\ell\to h} = \delta_{j\ell}(1 - \delta_{ih}), \tag{1.27}$$

where δ_{ij} is the Kronecker delta. Powers of $\hat{\mathcal{B}}$ enumerate non-backtracking walks similarly to powers of the adjacency matrix enumerate walks [188].

Compared with the walk allowed by the ordinary adjacent matrix, the NB matrix does not permit the immediate backtracking paths $i \to j$ and $j \to i$. As a result, the eigenvector centrality on NB matrix measures the influence of an influencer according to its neighbors in its absence. With this technique, extreme influence localization on hubs is avoided. The non-backtracking matrix is in general asymmetric with eigenvalues that are, in general, complex, however, since its entries are non-negatives, by the Perron-Frobenius theorem, its leading eigenvalue, λ, is real and non-negative, and there exists a corresponding leading eigenvector, $\mathbf{v}$, whose elements are also non-negative real numbers. If the graph G is connected and not a tree, i.e., it has at least one cycle, λ is positive [189] and satisfies the eigenvector equation:

$$\lambda\mathbf{v} = \hat{\mathcal{B}}\mathbf{v}, \tag{1.28}$$

Using these eigenvectors, Martin et al. [153] defines the *non-backtracking eigenvector centrality* of node j as:

$$c_{\mathrm{NB}}(j) = \sum_i A_{ij}\, v_{i\to j}\,. \tag{1.29}$$

Finding the leading eigenvector of $\hat{\mathcal{B}}$ can, however be computationally demanding for large graphs as the size of $\hat{\mathcal{B}}$ is usually much larger than the size of $\mathbf{A}$. However, the computation can be made much faster by computing directly c_{NB} as the first N elements of a $2N \times 2N$ matrix called the Ihara-Bass matrix as done in [190, 191]. The non-backtracking matrix have also been used to modify the Katz [187] and random walk centralities [189] and PageRank [192]. Research on non-backtracking centralities shows that they can also suffer from localization on certain networks [193, 194]. Radicchi and Castellano [154] proved that the eigenvector centrality on the NB matrix is an optimal metric for identifying single superspreaders in SIR models at criticality and further simulations also indicate that it is also reliable for noncritical spreading. We will discuss more on the non-backtracking matrix when we treat the problem of percolation in Eq. (1.42), and influence maximization through optimal percolation, leading to the collective influence algorithm in Sects. 1.5 and 1.6.1.

1.2.6 Outlook on Centralities

Before we close this section on centralities, we note that the metrics introduced here are far from being a complete set. Many metrics and methods are continuously proposed for identifying single influencers. Good review articles on centralities, network resilience, and percolation, including applications, appear in Refs. [195–197], see also [19, 198, 199].

There exists no consensus on the optimal metric that can be applied to all circumstances since there are significant differences in performance between different centralities for different spreading dynamics. As a result, one needs to relate the performance of each metric to specific dynamic processes. In subsequent chapters, we will treat this issue in more detail when we apply the centralities to particular cases.

One way to compare the centralities is by performing attacks on the network by following different centrality rankings. This is done by removing nodes and monitoring the network's largest connected component (LLC), like in a targeted percolation process to be discussed in Sect. 1.5. One calculates the fraction of nodes in the LCC as a function of the fraction of removed nodes to obtain either a critical threshold where the LLC is disintegrated or the area under the curve (AUC) of the LCC [20, 200]. The centrality with the smallest critical threshold or AUC is the best performer for network attacks.

For instance, when looking for superspreaders of disease in the COVID-19 pandemic in Chap. 4, we will see that centralities can be compared against each other by testing how they optimally break the LLC of disease transmission. When applying centralities to understand the spreading of fake news, we will be interested in finding the influencers in social media who can spread misinformation to a large number of users. This case will be treated in Chap. 2 and we will see how centralities can help us to understand these difficult problems.

A comprehensive comparison among centralities was done in [201]. In Sect. 1.8 a comparison is shown from [202] between not only centralities but also among more sophisticated methods to solve optimal percolation and dismantling problems of Sects. 1.5 and 1.6.4. The results of [201] are 'fuzzy', as there is no clear 'winner' that outperforms all centralities in all cases. We will see a progression of improving algorithms with a jump in performance with the introduction of machine learning algorithms used by Grassia et al. [202], signaling the arrival of AI to solve the influencer problem.

1.3 Influence Maximization Problem (IMP)

So far, we have defined a set of metrics and centralities to rank nodes in a network according to different topological features. Most of these centralities do not consider the specific dynamical process on the network, like disease or rumor propagation, for instance. They are purely defined in terms of the topological properties of the network. PageRank and random walks centralities are based on the dynamics of a random walk, but this is not a realistic model for information propagation in a network. A random walk simulates a 'drunkard's walk'. However, a search of a web page on the internet does not follow a random walk on the hyperlinks as in PageRank; instead, it is an optimized search to find a page in a few short steps. Even though PageRank is not based on realistic dynamics, this did not deter Google from becoming the world's leading search engine. This says a lot about the power of simple heuristics.

To develop more sophisticated measures of influence, it is important to take into account the actual dynamics of information flow in the networks, such as rumor spreading, diffusion of innovations, or epidemic processes [203–205]. These dynamics ultimately define who is an influencer and who is not. Finding optimal influencers brings up the problem of how to choose the correct dynamical model to describe the propagation of influence in the system. The process of finding influencers then relies on the type of dynamics that is used to transmit the information in the network, leading to different influencer identification problems according to different dynamics.

This opens a Pandora's box. In many applications, it may not be completely clear how to characterize the spreading of influence from one node to another node in the network, especially when dealing with social interactions. The dynamics of information transmission in a social network is a mystery, yet it can be successfully modeled differently. Some models consider that information is transmitted from person to person with a given probability, analogous to a SIR spreading model of contagion [205]. Other spreading models consider that a person adopts a behavior or idea when some friends have adopted the same idea via a threshold model. We will discuss these models in Sect. 1.3.3. Despite these modeling efforts, social influence is hard to measure quantitatively and put into a mathematical form. How can we model the process by which a person reads a tweet and decides to retweet it? Social interactions are unlike in physics, where we know precisely the interaction potential energy between positive and negative charges, and then we can build all condensed matter physics from there.

Interactions in biological systems are relatively better understood than those in sociological systems. These interactions are used to construct biological networks. They can measure the 'influence' and transmission of signals and information between genes, proteins, and metabolites that determine the function of the cell. These interactions also measure the strength of the propagation of signals from neurons to neurons by the action potentials traveling through axons and across synaptic connections and gap junctions. Likewise, genes 'spread' their influence in the gene regulatory network through regulatory interactions mediated by transcription factors. A transcription factor is a protein that can bind to the DNA binding site of a target gene to regulate its expression. These transcription factors act as activators or inhibitors, turning the genes on and off. This physical binding process between a transcription factor and the regulatory DNA domain of a target gene is well understood in terms of probabilistic mathematical models. This facilitates the definition of biological networks compared to social networks towards developing influencer theories. However, even when signal propagation and regulation are well understood, it is still an open question how these signals are integrated into the network to form the gene regulatory code in the genetic networks or the neural code in the brain. We will discuss these problems in Chaps. 5 and 6.

Ecological interactions between species that define ecosystems are understood in terms of 'benefit' being transmitted or received from species to species (predator/prey, mutualistic, etc) in an ecological network. We will discuss this case in Chap. 7. While these interactions are not physical, yet one can construct dynamic equations to predict the system behavior and the spread of influence/benefit in such a system.

These examples point to the fact that the spread of influence in complex systems cannot be described under a universal dynamics and this hampers the development of a universal theory of influencers for all complex systems. In principle, each model of propagation would give rise to different types of influencers, as the influence depends on how we model the interactions, and each case needs to be treated in particular.

Despite these issues, we will attempt to define some general features common to all influencer identification problems below. A seminal approach to the problem of influence has been developed by Kempe, Kleinberg and Tardos [5] who introduced the Influence Maximization Problem (IMP) motivated by previous work on viral marketing by Domingos and Richardson [6, 7]. In the IMP, it is assumed that we know every node's influence function. A rigorous definition of the maximization of influence can be proposed, at least in marketing and spreading information in networks. Even with this simplification, we will see that the maximization problem is hard to solve. We treat this problem next.

1.3.1 The Viral Marketing Problem

One of the first algorithmic formulations of the problem of influencer identification was done by Domingos and Richardson in 2001 [6, 7], who aimed to develop a strategy to advertise a product through viral marketing by assuming that the spread of influence travels through a network. Instead of viewing markets as a set of independent entities, they treat the market as a networked system where the potential profit contributed by a customer is mostly determined by their interactions with others. They proposed the following algorithmic problem:

Imagine that we know the network of connections in a given social network, and, most importantly, we know how to quantify the influence that each member in the network has on other members to whom they are connected. The viral marketing premise is that by targeting a selected group of few influencers (providing, for instance, free samples of a product or paying them to show a video of the product), they can trigger a cascade of influence to convince the majority of the members of the social network to acquire the product.

The *Viral Marketing Problem* refers to finding the minimal number of influencers to produce the largest extent of the cascading of influence to convince the largest number of members in the social network to acquire the product. Crucial to this problem is the assumption that the network can transmit information and influence large parts of the population. This is the same assumption that we will use in this book. We will always assume that the information, influence, or signals travel through the network of connections in a social or biological system. The theory of influencers is then designed to answer this problem.

1.3.2 Kempe-Kleinberg-Tardos Formulation of the Influence Maximization Problem

The Viral Marketing Problem posed by Domingos and Richardson was later formalized into the Influence Maximization Problem by Kempe, Kleinberg, and Tardos into a well-defined combinatorial optimization problem [5]. The formulation is built on a strong assumption: we assume that we know how the information propagates in the network. Then, the IMP consists of finding the optimal set of k influencers that would initiate the largest-scale propagation process. Formally, the problem has three inputs:

- A graph $G(V, E)$ with a set of V nodes and E edges that defines the social network where the transfer of information occurs.

- A function $\sigma(S) : S \to \mathbb{R}$ from a set of nodes $S \in N$ to the value of the influence of each node $i \in S$ in the network. This function defines the influence of each node in the network and defines the particular model chosen for the propagation of influence. According to a given model, it could be defined as the number of 'activated' users that have adopted a product in a cascading of propagation that started at the set S. More generally, it is defined as the expected number of active nodes in a diffusion process that started at a seed set S.
- A constraint or cost $k \in \mathbb{N}$ which is the number of seed influencers to initiate the diffusion process in the spreading process.

The IMP is formulated as follows: for a given number of initial spreaders $k = |S|$, we wish to find a k-node set S of maximum influence $\sigma(S)$:

$$\max_{|S| \leq k} \sigma(S) . \tag{1.30}$$

The intrinsic difficulty of this problem is rooted in the exponentially growing configuration space with k. It was proven in [5] to be among the class of the hardest optimization problems - NP-hard, and thus can only be solved approximately in polynomial time via heuristic approaches. The k-node set S is the optimal influencer set.

Even the influence function $\sigma(S)$ can be very hard to calculate efficiently, and it is an open question how to calculate this function exactly for a given model of diffusion of information. This function depends on the spreading problem that is being considered. The influence maximization problem cannot be universally defined for all dynamics and depends on the type of propagation process considered through $\sigma(S)$. Kempe et al. [5] then focused on two classes of diffusion of information models where the information propagates in the network through random processes of contagion, namely the Linear Threshold Model (LTM) and Independent Cascading Model (ICM), as explained next.

1.3.3 Models of Spreading of Influence

Since the maximization of influence depends on the type of model employed to model the diffusion of information, we next define two classes of models, the LTM and ICM, which are general enough to encompass a large type of information cascading processes, and, at the same time, it is simple enough (submodular) to allow for theoretical guarantees on their performance via greedy searches.

1.3.3.1 Linear Threshold Model (LTM)

In 1978, sociologist Mark Granovetter proposed a class of threshold models of collective behavior in society [206]. One well-studied version is the LTM. LTM is a model of propagation of opinion or information in a network where a node is activated when the number of its active neighbors exceeds a predefined threshold value. Physical models like k-core percolation, in which each node is assigned a local threshold, is a special case of LTM [207] and will be treated in subsequent chapters.

The LTM simulates a spreading of information process where an individual adopts an opinion or information under "peer pressure", that is, it becomes activated only after a given number of its neighbors are [206, 208, 209]. Linear threshold models have several different forms and differ slightly in their formulations. We define the most typical case. In a typical LTM, the diffusion of information travels on a graph G where each link has a weight: $G(V, E, \omega)$, where $\omega : V \times V \rightarrow [0, 1]$ is the weight functional. Here, we consider a null weight $\omega = 0$ iff the corresponding edge does not exist; i.e., the nodes connected by this link are not influenced by one another. Each node is assigned a predefined threshold value $\theta_i \in \mathbb{R}^+$. The spreading process in LTM is initiated by a set of seed nodes that are considered active while the rest are inactive. The dynamics then proceed step by step. In a given time step, a node is activated if the sum of weights of its active neighbors reaches at least its threshold value θ_i: That is, node i is activated if:

$$\sum_{j \in \partial i} \omega_{ij} \geq \theta_i, \tag{1.31}$$

where ∂i stands for the set of neighbors of node i in the network. After activation, the node remains active forever.

This rule implies that a node is activated if it has at least a certain number of active neighbors. Thus, LTM is a model of adoption of influence/opinion/rumor/product in a social network where the users change their opinion when they have a sufficiently large number of friends who have already adopted the opinion or product [208]. Each member has its own threshold of adoption θ_i although there are variants of this model where this can be simplified, and a common threshold is considered.

1.3.3.2 Independent Cascade Model (ICM)

The ICM is a generic model of diffusion of information by contagion in a network where an individual can be independently infected by any of their neighbors. We define the ICM introduced in [5].

In ICMs, a node can be independently infected by any of its neighbors in the network [5, 205, 210]. Spreading is initiated by a set of infectious 'seeds'. During subsequent steps, each infected node attempts to infect its susceptible neighbors independently with a certain probability until no more nodes are infected. At this point, the spreading stops. When a node is infected, it remains infected throughout the process, and each node can be infected only once during the dynamic evolution of the process.

ICMs are closely related to percolation (see Sect. 1.4): the dynamical spreading process of an ICM can be transformed to a bond percolation with a given occupation probability as shown by Newman [211]. As a result, the outcome of a dynamical ICM can be mapped to the static final state of an equivalent percolation process. This mapping bypasses the need to run dynamical models and enables the analysis of the spreading process using tools of percolation, which will be defined in subsequent chapters. Next, we discuss the attempts to use these two models to solve the IMP using greedy algorithms.

1.3.4 Greedy Algorithms for IMP

Kempe et al. [5] explores heuristic solutions to the IMP using greedy algorithms. A greedy algorithm is a heuristic that searches for the optimal local (but not global) solution at each algorithm stage. At each algorithm step, a search will make the best 'greedy' choice driven only by local information. A greedy search does not guarantee to find the global optimal solution to the problem. Yet, greedy algorithms are very popular in solving optimization problems since they provide strong empirical and theoretical performance guarantees for optimizing submodular functions.

For the general class of spreading models that include the LTM and ICM, the influence function $\sigma(S)$ satisfies the submodularity property [212, 213]. A function $\sigma(\cdot)$ is submodular if the marginal gain from adding an element to a set S is at least as high as the marginal gain from adding the same element to a superset of S. This 'diminishing return' property expresses the decrease in the gain of the output of the process after an increment of the inputs: that is, the first chocolate we try in our life tastes great, but subsequent chocolates are less and less appealing.

Intuitively, the influence of a node decreases when it becomes part of a larger group of nodes since its influence is diminished by being taken over by the other influencers in the larger group. The submodularity property is quite convenient since it allows one to prove a set of nice approximability guarantees for submodular optimization.

In 1978, Nemhauser et al. [213] proved that optimizing submodular functions with a greedy heuristic always finds a solution whose value is at least $1-(1-1/K)^K$ times the optimal value. Here, K is the size of the seed set. This bound has a limiting value of $1-1/e \approx 0.63$, which is independent of the size of network or seed set. Leveraging on this theoretical result for submodular functions, Kempe et al. [5] shows that a greedy algorithm for the LTM and ICM models is guaranteed to approximate the optimal influence set S^* within a factor of $1-1/e \approx 63\%$, i.e.,

$$\sigma(S_g) \geq (1-1/e)\,\sigma(S^*)\,, \tag{1.32}$$

where S_g is obtained from the greedy algorithm. That is, a greedy search can be used to find an approximate top influencer set whose influence on the rest of the network is at least 63% of the optimal influence achievable among all sets of size at most k. While this is a nice theoretical guarantee of the model, being 63% of the optimal solution is not ideal in most practical applications.

Thus, a greedy search can be used as a first approach to select a suboptimal set of k influential spreaders to approximate the optimal set. In this search, one needs to estimate the influence function for every node, $\sigma(S)$. This estimation represents a serious computational problem for any IMP algorithm since this influence function has to be evaluated polynomially many times on different sets S during the evolution of the algorithm. For every evaluation, this requires summing over exponentially many ($2^{|E|}$) subgraphs of $G(V, E)$. Thus, in principle, evaluating the influence function is computationally prohibited in this approach.

A way around this problem is to perform a Monte Carlo (MC) approximation of this sum by sampling the graph G. Using this MC sampling for $\sigma(S)$, a greedy hill-climbing strategy to find influencers was applied by Kempe et al. [5]. The set of influencers was then compared with the performance of different centralities discussed in Sect. 1.2, like degree centrality, distance centrality, and a random search that picks the set of influencers at random, and it is considered a null model or baseline for influencer search since no strategy can be worse than a random search.

The results of this comparison using the LTM can be observed in Fig. 1.9 reproduced from Ref. [5]. The comparison is done by choosing a set of k top influencer seed nodes following the different centralities and the greedy search, and then plotting the number of active nodes versus k obtained at the end of the diffusion of

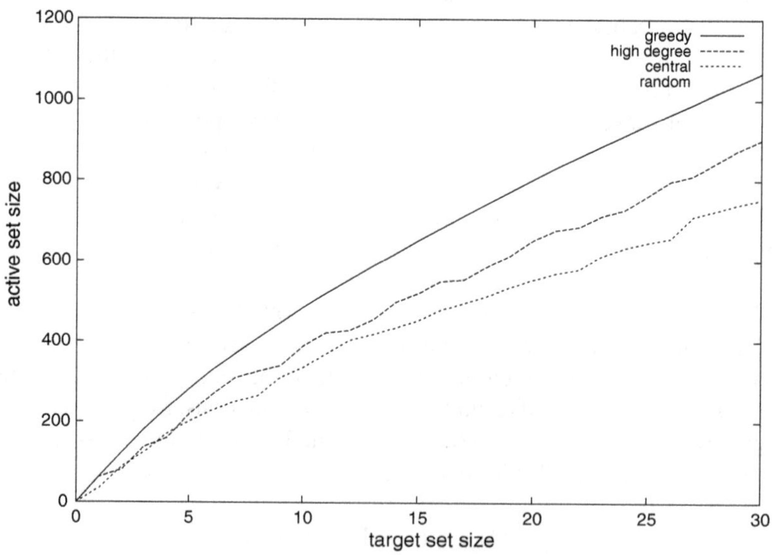

Fig. 1.9 Finding the set of k influencers for the LTM. For every size of the influencer set, k, four strategies are used to search for the optimal set. The resulting number of active sets is the influence function $\sigma(S)$ for a given set of $k = |S|$ top influencers for each algorithm. The influence function is plotted for every k to test which algorithm is better. In this case, the greedy strategy outperforms the high degree, location centrality, and a null random search model. Notice the small sets considered here. This is because greedy searches do not scale to large system sizes. Figure reproduced from [5]. Copyright © 2003, ACM

information process. The performance of the algorithm is evaluated by the largest number of active sites per given k. It is shown that the greedy algorithm outperforms the high-degree node heuristic by about 18%, and the distance centrality heuristic by over 40%; way above the value obtained by a random search. The authors conclude that this result shows that significantly better marketing results can be obtained by explicitly considering the dynamics of information in a network, rather than relying solely on structural properties of the graph.

However, the result of this basic greedy algorithm can be far from optimal in more general cases, for instance, when the cost of removing each node is not identical. In such circumstances, a naive modification of the basic greedy algorithm can be made by favoring the node with a maximum benefit-cost ratio. Yet, this intuitive generalization can perform arbitrarily worse than the optimal solution S^*. Leskovec et al. then proposed the Cost-Effective Forward (CEF) algorithm [214] that guarantees a relatively good performance. This algorithm combines the benefit-cost and unit-cost greedy algorithms and provides a $(1 - 1/e)/2$ approximation of the maximal influence. The algorithm's runtime can be partly alleviated by exploiting the sparsity of cost reductions [214]. Even though each of the two basic greedy algorithms can perform arbitrarily badly, it was proved that for a given circumstance, at least one could perform relatively well.

Even using Monte Carlo to evaluate the influence function $\sigma(S)$, these algorithms are unscalable to large-scale networks due to the heavy computational burden of MC. In practice, the number of simulations can be significantly reduced by exploiting the submodular property of the influence function. In these cases, the marginal increment of a node is monotonically decreasing with the growth of S. Then, there is no need to recompute the marginal increments for all nodes at every time step. Specifically, if the marginal increment of a node i in previous time steps is already smaller than that of another node j in the current time step, the recomputation for $\sigma(i)$ is unnecessary as it is smaller than $\sigma(j)$.

In this case, the computation would go as follows: First, the marginal influence of each node $\sigma(i)$ is initially marked valid. Before the next influencer is selected, the nodes are scanned in decreasing order of their influence. If $\sigma(i)$ for the top node i is invalid, it is recomputed and inserted into the existing order using a priority queue. If the recomputation leads to a new value that ranks at the top, it should be added into S without calculating the marginal increments for any other nodes. This algorithm is called cost-effective lazy forward (CELF) and leads to far fewer evaluations of the influence function and achieves up to a factor of 700 improvement in speed compared to CEF with equal performance. Further improvement of CELF can be achieved by using a heap data structure to record the node with the largest marginal gain among the nodes already examined in the current stage [215]. It was shown that this technique improves the efficiency of CELF by another 35–55%.

When the ICM is used to solve the IMP, the greedy algorithm can be improved by considering the mapping of ICM to bond percolation [211]. Based on this mapping, the influence function $\sigma(S)$ can be estimated with better computational efficiency without resorting to the dynamics. Chen et al. [216] generated a subgraph G' of G by selecting each edge in G independently with a transmission probability. Thus, each edge in G' belongs to a real propagation path. Then the influence function $\sigma(S)$ is the number of nodes reachable from S in G', since the influence of i can be obtained with a linear scan of the graph G' and its marginal increment to S is either 0 or $\sigma(i)$, depending on whether i is in the influence range of S or not. This procedure provides $O(N)$ speedup to the basic greedy algorithm; it can be processed with CELF to avoid unnecessary evaluations. Hu et al. [217] employed percolation theory to show that spreading processes of ICM are usually limited to a local area. Therefore, local structure can identify and quantify influential global spreaders on large-scale social networks. An efficient percolation-based greedy algorithm was then proposed.

A further improvement proposed by Chen et al. [216] uses a heuristic degree discount algorithm. This algorithm calculates a degree discounted $\sigma(i)$ by the number of its neighbors that are already included in S. For ICMs with a small propagation probability, the indirect influence between multi-hop neighbors is negligible so that we can take into account only the direct influence between immediate neighbors. This modification improves the tradeoff between performance and computational efficiency, and the performance of this algorithm nearly matches that of the basic greedy algorithm. Furthermore, it is far more efficient in combined use of the heap data structure and scalable for large-scale networks. While efficiency improves, performance does not improve beyond the crude greedy search.

Chen et al. [218] proposed another scalable variant of the greedy algorithm, which is called the maximum influence arborescence (MIA) algorithm. This variant assumes that propagations are along the maximum influence paths between each pair of nodes, defined as the path with the highest propagation probability among all possible ensembles. For a given pair of nodes, the maximum influence path is computed efficiently using the Dijkstra shortest-path algorithm [219]. The arborescence structure is the union of these paths starting or ending at a node i and this defines its local influence region, $\delta(i)$. The union of all local influence regions: $\sigma(S) = |\bigcup_{i\in S} \delta(i)|$, where $|\cdot|$ denotes the size of a set, is the global influence of a set S. All paths with probability below θ are discarded, and by adjusting θ, the size of the local influence regions can be altered so that a tradeoff between computational efficiency and performance is achieved. Since the local influence function is submodular, the basic greedy algorithm guarantees the $1 - 1/e$ bound. The linearity of local marginal influence allows for the efficient update of incremental influence during iterations. More importantly, the update is only required in a local influence region around the selected influencer.

The network's modularity, i.e., the existence of modules of highly connected nodes loosely connected with the rest, significantly impacts information diffusion [220–222]. Thus, the network's modular structure has been considered in greedy searches. The general idea is to consider the information diffusion within each community separately and disentangle it from the rest to simplify the process. Wang et al. [223] uses this idea to find influencers in mobile phone networks. In the algorithm, communities with regional information diffusion are first detected, and influential nodes are then located by selecting certain communities using a dynamic programming algorithm. The algorithm is faster than typical greedy algorithms by an order of magnitude.

1.3.4.1 Outlook on Maximization of Influence Via Greedy Algorithms

Although the basic greedy algorithm is relatively simple to implement and has a performance guarantee, it requires massive Monte Carlo simulations to estimate the influence function of each seed node-set. Thus, it is not scalable to today's large-scale social media applications. For an algorithm to have wide applicability in big data, scalable solutions should exist. This means that the runtime of the algorithm should grow with the system size N under study with a functional that is linear or close to linear, like $\sim O(N \log N)$. Furthermore, while being a mathematically rigorous result, the greedy algorithms' performance guarantee only guarantees that the greedy algorithm will find a solution within 63% of the optimal one. This is still unsatisfactory, and it leaves much room for improvement. Finally, the performance of most greedy algorithms mentioned above is guaranteed thanks to the submodular property of the influence function. However, for a general information diffusion model, and even for special cases of the LTM with fixed weights and thresholds, the influence function is not submodular, and therefore the guarantees cease to exist. Theoretical guarantees for non-submodular optimization are harder to obtain.

A variant of LTM that is not submodular appears when θ_i takes some particular values related to the degree, leading to qualitatively different classes of cascading phenomena underlying either continuous or discontinuous phase transitions. In the special case, when $\theta_i = k_i - 1$ (k_i is the degree of node i), the scale of propagation experiences a continuous phase transition [14]; this problem will be treated in Sect. 1.5. On the other hand, when intermediate values of the threshold $\theta_i = k_i - 2, k_i - 3, \ldots, 2$ are considered, then first-order discontinuous phase transitions appear in the dynamics [224]. In both cases, the influence functions are not submodular [14], and therefore, in these cases, the greedy algorithm cannot guarantee to provide a good approximation. Solutions to the IMP in LTMs without the submodular property requires an understanding of the mechanism of spreading and will be discussed in Sects. 1.5 and 1.7.4.

For these reasons, the greedy approach to IMP has not been widely adopted in real marketing applications. We require other approaches that are scalable to big data and, at the same time, have better performance. Next, we discuss approaches that can provide scalable algorithms to find influencers and have been applied to marketing and opinion spreading in social media and beyond.

But before we introduce these models of influence and their algorithmic formulations, we need to make a detour to introduce the basics concepts of percolation, k-core percolation, and the associated phase transitions in the connectivity underlying these two models. Percolation is a basic mathematical concept for which there are many excellent reviews and books [225, 226]. The reader already familiar with percolation can skip to Sect. 1.5. In the next section, we review the random percolation problem based on the non-backtracking matrix approach, since it will lead directly to optimal percolation discussed in Sect. 1.5 and the collective influence algorithm discussed in Sect. 1.6.1.

1.4 The Percolation Problem

We distinguish dynamical processes in networks into two categories: those associated with underlying continuous phase transitions and those related to discontinuous phase transitions. These two types of dynamical behaviors correspond to two fundamentally different classes of phenomena. In models with continuous phase transitions, the macroscopic properties of systems change gradually at the transition point. For instance, infectious diseases invade a population in an epidemic-free phase by infecting individuals one by one. In contrast, systems can undergo abrupt changes without warnings for models with discontinuous transitions. For instance, the extinction of one species may completely collapse an ecosystem unexpectedly and discontinuously.

In network science, the basic features of these two types of models are captured by two paradigmatic models for the emergence of network connectivity: percolation for continuous second-order phase transitions, and k-core percolation for discontinuous first-order phase transitions. Both transitions are depicted in Fig. 1.10. In subsequent sections, we will see that the influencer identification problem is related to these percolation problems. Indeed, one kind of influencers are those who provide integrity to the network and understanding the integrity and robustness of networks is a percolation problem. The consideration of percolation bridges the problem of influencer identification with percolation theories developed in statistical physics and graph theory and has inspired several elegantly designed approximative solutions to find influencers. The rapid progress in tackling the influencer identification problem is a successful example of interdisciplinary collaborative efforts.

In percolation, one is interested in the emergence or disintegration of the network's global connectivity. The feature of interest in percolation, which is also common to problems of influence and resilience on networks, is the emergence of the Giant Connected Component of the network (GCC). This concept is also essential in defining the problem of influencer identification in various dynamic processes. This is because, in any maximization of influence, one is interested in the percolation of the information through the network. This problem is necessarily related to the existence of large connected components through the network of interactions. Analogously, one is also interested in the network's resilience under the nodes' failure. And this is the quintessential dismantling/percolation problem.

The mathematical formalism used to study the emergence of connected components and their vulnerability under failure is percolation and associated k-core percolation. In this section, we first introduce percolation as the basis for defining specific influencer identification problems via optimal percolation, optimal recycling, and dismantling, and optimal immunization problem, which will be treated in subsequent sections.

Another key feature to understanding dynamical processes on the network is the sign of the interaction between the units. For instance, genes can either promote (positive interaction) or repress (negative interaction) the expression of other genes in gene regulatory networks. Neurons can excite or inhibit the firing of other neurons in neuronal networks, thereby shaping the system's global dynamics. Individuals can inhibit information propagation in signed social networks, and others are spreaders [78]. Thus, the sign of the interaction, being positive (activator/excitatory/spreaders) or negative (inhibitor/repressor), defines different types of problems.

Whether the edges in a network are considered positive or negative determines whether the concept of connected components is relevant. In percolation, all edges are taken of the same type. For dynamical systems with only positive interactions, the influencer identification problem can be mapped for some dynamical models to percolation. On the contrary, systems with positive and negative interactions do not admit straightforward definitions of the connected components unless one ignores the sign of the interactions. So, the influencer identification in these systems needs to be treated differently to consider the existence of different types of interactions.

Below, we will first consider problems with positive interactions that can be described in terms of percolation. This excludes biological systems, like the brain and genetic networks, and social signed networks [78–80], too; although in these systems, one may still apply percolation if one ignores the signs of the interactions. However, when the negative and positive interaction are taken into account, frameworks beyond percolation need to be considered. A framework to study these biological systems involves network symmetries and fibrations and will be treated in Chaps. 5 and 6.

Thus, processes involving only positive interactions are convenient to deal with because, through percolation, they can be further divided into dynamical systems with underlying continuous (second order) and discontinuous (first order) phase transitions, Fig. 1.10. In regular percolation [47] the GCC emerges continuously from zero sizes as edges/nodes are gradually included in the network. To find influencers related to continuous percolation transitions, the optimal percolation problem was proposed to search for a minimal set of nodes whose removal would optimally fragment the GCC into negligible clusters, that is, to disintegrate the network with the minimal number of removed nodes [14].

In k-core percolation [48], k-cores with non-zero size can appear abruptly through a discontinuity as nodes/edges are included in the network. For systems characterized by these discontinuous phase transitions, the problem becomes locating the influential nodes that maintain the integrity of the k-core. In ecological networks, keystone species in ecosystems were found to be located at the maximum k-core of the network [111]. We first review random percolation in Sect. 1.4.1 to set up the problem of optimal percolation in Sect. 1.5.

1.4.1 Random Percolation in Graphs

In percolation theory, the connectivity of a finite network is characterized by the fraction of nodes that belong to the largest connected component (LCC) of the network, which we denote S. The LCC becomes the giant connected component (GCC, denoted G_∞) in the thermodynamic limit of an infinitely large network. The study

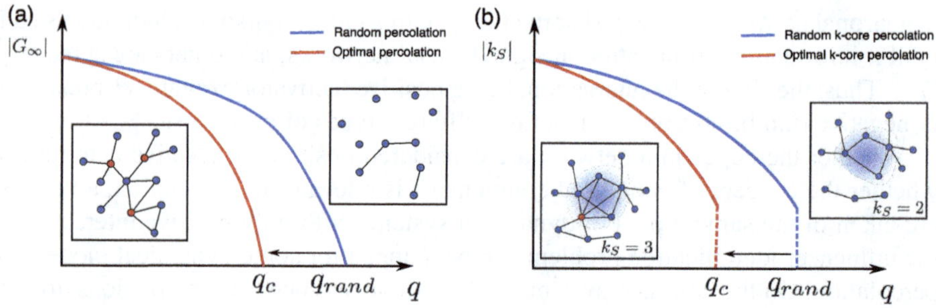

Fig. 1.10 Phase transition in percolation process. a The continuous transition of the GCC for an infinitely large network, G_∞, after the removal of q fraction of nodes. The critical values q_c^{ran} and q_c^{opt} (called q_c in this figure) are those for random and optimal percolation (Sect. 1.5), respectively. The insets show an example of a typical largest connected component in a finite network and smaller fragmented clusters. Red nodes are influencers that disintegrate the GCC after being removed. **b** The discontinuous transition of the k-core size $|k_s|$ after the removal of the q fraction of nodes. The critical values for random and optimal k-core percolation (treated in Sect. 1.7.1) are q_c^{ran} and q_c^{opt}, respectively. The insets show the removal of nodes and its effect on the k-cores. The left inset shows the 3-core in a sample network. After removing the red node, the 3-core is disintegrated, and only the 2-core is left, as shown in the right inset. Figure reproduced from [199]. Copyright © 2019, Oxford University Press

of the emergence of this component has been pioneered by the seminal work on random graph theory established by Paul Erdös and Alfréd Rényi in 1959 and 1960 [47, 227]. The random percolation process describes the emergence of G_∞ by gradually increasing the probability of connection between any pairs of nodes chosen at random.

The GCC is found to emerge continuously from zero when the network reaches an average degree $\langle k \rangle = 1$. In the inverse process, the connected component of a network can be dismantled, i.e., reduced to disconnected components whose size vanishes in the thermodynamic limit by removing a fraction $q = 1 - p$ of nodes from the network, Fig. 1.10a.

The removal process leads to a continuous phase transition at a critical value of the fraction of removed nodes. This fraction is generally denoted by q_c and called the percolation threshold, while in random percolation, we denote this threshold by q_c^{ran}. Above this threshold, only fragmented clusters remain, as shown in Fig. 1.10a. Adding nodes also results in a continuous transition at a probability generally denoted by p_c. The transition in random percolation is continuous, or second order, since the order parameter of the transition, in this case, the G_∞, decreases continuously to zero as the transition is approached below q_c^{ran}. Much work is dedicated to studying percolation in d-dimensional lattices of all sorts, like square lattices or hexagonal ones [225]. In this book, we work in infinitely dimensional percolation or mean field, which is the formation or the disintegration of a graph without embedding in any finite-dimensional space.

For a network $G(V, E)$ with $N = |V|$ nodes and $M = |E|$ edges, we can use a vector $\mathbf{n} = (n_1, \cdots, n_N)$ to represent the configuration of nodes that indicates whether a node i is removed from the network ($n_i = 0$) or not removed ($n_i = 1$). After the removal of a fraction of nodes $q = 1 - \sum_{i=1}^{N} n_i/N$, we are interested in the remaining fraction of nodes belonging to the GCC as a function of q, $G_\infty(q)$. For a finite network, the relative size of the LCC is defined as the ratio of the number of nodes in the largest connected component to the network size N. In the classical percolation theory [47], nodes are removed randomly, meaning that the nodes removed to destroy the GCC are chosen randomly from the set of nodes in the network. This random removal can be considered an attack or a failure of random nodes in the network.

The random percolation problem refers to this removal of a set of nodes chosen at random, i.e., we select these nodes independently of any of their features and each other. The random percolation transition occurs at the random percolation threshold, q_c^{ran}, of randomly removed nodes at which the GCC vanishes such that: $G_\infty(q \geq q_c^{\text{ran}}) = 0$. This transition occurs smoothly (Fig. 1.10a). For $q \geq q_c^{\text{ran}}$, the network consists of a collection of disconnected clusters of nodes whose sizes are subextensive. Alternatively, for a fraction $q < q_c^{\text{ran}}$, there are configurations of nodes that, when chosen randomly from the total number of N nodes and removed, result in a non-zero giant connected component after removal.

For random (uncorrelated) networks with a degree distribution $P(k)$, the percolation threshold q_c^{ran} has been calculated in [21, 228–230], and it is given by:

$$q_c^{\text{ran}} = 1 - \frac{\langle k \rangle}{\langle k^2 \rangle - \langle k \rangle}, \tag{1.33}$$

where $\langle k \rangle = \sum k P(k)$ and $\langle k^2 \rangle = \sum k^2 P(k)$ are the first and second moments of $P(k)$. This value predicts an extreme robustness to random attack for scale-free networks [1], i.e., $q_c^{\text{ran}} \approx 1$, for networks with a power-law degree distribution $P(k) \propto k^{-\gamma}$, with $2 \leq \gamma \leq 3$. The extreme robustness arises because to destroy the network by random attack or failure of its components, all nodes $q_c^{\text{ran}} = 1$ need to fail. However, these networks are highly vulnerable to targeted attacks [20], as will be shown below.

Networks with power-law degree distribution are ubiquitous in empirical systems [231]. The critical point q_c^{ran} upon random node removals depends on the network structure in non-random networks. More accurate estimations using the reciprocal of the largest eigenvalue of the adjacency matrix and non-backtracking matrix was developed in [232–234].

1.4.2 Second-Order Phase Transition in Random Percolation

It is interesting to elaborate on the dismantling of the GCC in percolation from the point of view of message passing techniques [8–12], and the non-backtracking matrix introduced in Sect. 1.2.5. This will allow us to introduce the optimal percolation problem in Sect. 1.5 for influencer identification. Next, we derive the critical percolation threshold from a message passing point of view.

In the limit of large networks $N \to \infty$, G_∞ represents the probability that by picking a node from the network at random, it belongs to the GCC. A given node i can be disconnected from the GCC G_∞, either because it is directly removed or because the removal of other nodes indirectly detaches it. In the former case, $n_i = 0$, while in the second one, $n_i = 1$. Thus, we see that n_i cannot tell us whether node i belongs to the GCC. We need another variable encoding the information that node i belongs or not to G_∞. This variable is the order parameter of the percolation transition, which is the probability of node i to belong to the GCC, ν_i. We define $\nu_i = 1$ if $i \in G_\infty$, and $\nu_i = 0$ otherwise. The fraction of nodes in the giant component, upon removing q nodes from the network, is then given by:

$$G_\infty(q) = \frac{1}{N}\sum_{i=1}^{N} \nu_i \,. \tag{1.34}$$

To calculate q_c^{ran}, we transform the problem into an eigenvalue problem controlling the stability of the percolation solution, which can be written explicitly in terms of $\mathbf{n}$. We first derive the relation between the vector $\boldsymbol{\nu} = (\nu_1, \ldots, \nu_N)$ and the vector $\mathbf{n} = (n_1, \ldots, n_N)$. This can be done using a message passing approach following Ref. [233].

We consider two connected nodes i and j and orient the corresponding edge from i to j. We 'virtually' remove node j (this creates a 'cavity' at j, hence the name 'cavity method') from the network and ask if node i belongs to the GCC or not by defining the auxiliary quantity $\nu_{i\to j} = 1, 0$ representing the probability of i to belong to the GCC in the absence of j. The advantage of using $\nu_{i\to j}$ instead of ν_i is that the former satisfies a close set of equations and can be solved by iteration, while the latter does not. Clearly $\nu_{i\to j} = 0$ if $n_i = 0$. So, the interesting case is when $n_i = 1$. Since j is momentarily absent from the network, the probability that i belongs to G_∞ requires that at least one among the neighbors of i different from j belongs to G_∞ when i itself is virtually removed from the network.

To calculate this probability, we employ an important approximation that allows us to solve the problem by explicitly writing down an expression for $\nu_{i\to j}$ by assuming that the network is 'locally tree-like' [233, 234]. This is an approximation of widespread use in the physics of disordered media, since it provides one of the few instances where exact results can be obtained on networks (another one is the assumption of fully connected networks). The method appears in the cavity method in the spin glass literature [8].

What is a locally tree-like network, also called a sparse network? In graph theory, a tree is a connected, undirected graph without cycles, i.e., a connected acyclic undirected graph. In a tree, there is one path connecting any two nodes. Locally tree means that these cycles do not exist locally at short path lengths in the network, yet cycles may occur for longer paths, but these cycles disappear when the network becomes infinitely large. More precisely, for a locally tree-like network, an arbitrarily large neighborhood around any node takes the form of a tree in the limit of large network size [233]. This is a pretty drastic assumption of any realistic network, not to say in social networks, which are known for being highly clustered. Indeed, the clustering coefficient was introduced to capture triangular relations that form short cycles, i.e., a friend of my friend is likely to be my friend, too, closing the loop.

Even though sparseness is an unrealistic assumption, the sparse approximation is used because it allows us to obtain exact results for the percolation threshold and many other relevant problems. This situation appears in many branches of theoretical physics, where complex physical problems can be solved only under unrealistic approximations of the mean-field type. These approximate solutions are then used as guiding principles in more realistic practical applications. One example is the spin glass problem [8], which can be solved only in mean-field theory for fully connected graphs in infinite dimensions and locally tree-like networks using the cavity method. Spin glasses, the Ising model, and percolation are the cornerstone models of statistical physics. A fully connected network and a tree are unrealistic network structures compared to natural materials in 3D. Yet, a field with significant ramifications for disordered systems and constrained optimization problems has been constructed based on such mean-field approximation. Keeping these caveats in mind, we next apply the local tree approximation to percolation to make further analytical advances.

In the Erdös-Rényi (ER) random network, the locally-tree approximation holds exactly in the limit of infinitely large graphs. This is convenient since the ER network is a generative model that is useful for checking calculations via numerical simulations. How do we construct an ER network? We built a network forming links at random. On a large scale ER random graph, each edge is present with probability $p = \langle k \rangle / N$, where $\langle k \rangle$ is the average degree and N is the number of nodes.

For locally tree-like graphs and very large system sizes N, the probabilities of finding one or more loops are suppressed by powers of $1/N$ [235–237]. Let us consider an ER random graph (in ER, the average degree is the same as the average residual degree). The total number of cycles of given length ℓ in the ER model is given on average by (the formula is valid for general sparse random graphs [238], the quantity z being in general, the average residual degree):

$$\mathcal{N}(\ell) = \left(\frac{z}{N}\right)^{\ell} \frac{1}{2\ell} N(N-1)\dots(N-\ell+1) \sim \frac{z^{\ell}}{2\ell}, \tag{1.35}$$

where the factor $1/2\ell$ comes from the fact that: (i) any one of the ℓ nodes in the loop can be taken as the starting point; (ii) the loop can be traveled in two directions. Therefore, the probability $p(\ell)$ for a node to belong to a loop of length ℓ is:

$$p(\ell) = \frac{\mathcal{N}(\ell)}{N} = \frac{z^{\ell}}{2\ell N}, \tag{1.36}$$

and thus it is of order $O(1/N)$.

The quantity $\mathcal{N}(\ell)$ enumerates the number of non-self-intersecting closed walks. Self-intersecting closed walks have a probability of order $O(1/N^2)$ since the self-intersection is encountered, on average, in a fraction $1/N^2$ of the total number of nodes. Thus, loops with many self-intersections are suppressed by a higher power of $1/N$. This argument can be generalized to random graph ensembles other than ER, provided they have a locally tree-like structure in the limit $N \to \infty$.

Thus, random networks have no short cycles within the limits of large network sizes; short cycles are absent in the thermodynamics limit. On the other hand, longer cycles are guaranteed in a finite network since random networks develop cycles of path length scaling as $\ell \sim \ln n$ (small worlds). But these cycles disappear in the limit of a large network. However, the logarithmic dependence makes this vanishing very slow, i.e., the loops become infinitely large for extremely large networks. In a random network of infinite size, we can assume that a path starting at a given node will not return to the same node, which is locally tree-like.

This approximation is used in the following message passing formula [8, 233]:

$$\nu_{i\to j} = n_i \left[1 - \prod_{k\in\partial i\setminus j} (1 - \nu_{k\to i}) \right] , \tag{1.37}$$

where $\partial i\setminus j$ is the set of nearest neighbours of i excluding j.

This equation says that the probability of being connected to the GCC is first proportional to n_i: if $n_i = 0$, then the node is not in the GCC. If $n_i = 1$, then the probability of connecting to the GCC can be factorized into the probabilities to be connected through each of the link of the node, except the one that we remove by the cavity procedure. This is the reason we evaluate the product on $\partial i\setminus j$. We can then factorize this function exactly due to the locally tree assumption since these probabilities are independent of each other, since each path emanating from node i is independent since they will never cross in the thermodynamics limit. Then, the probability to connect to the GCC is written in terms of 1 minus the probability of being not connected through each of the paths starting at the $\partial i\setminus j$ set to produce the above formula.

If we put back j in the network, we get the real probability ν_i as:

$$\nu_i = n_i \left[1 - \prod_{k\in\partial i} (1 - \nu_{k\to i}) \right] . \tag{1.38}$$

This equation is not that useful for calculating the critical value since it does not admit an iterative solution. Therefore, one works with the message passing in the cavity scheme Eq. (1.37), since this equation can be solved by iteration [8] because the solution depends on the solution itself and can be iterated until convergence.

The system defined by Eq. (1.37) always admits the solution $\{\nu_{i\to j} = 0\}$ for all $i \to j$ (regardless of the values of n_i), as can be verified by inspection. As a consequence, also $\{\nu_i = 0\}$ for all i, which in turn gives $G_\infty = 0$.

Having defined the main quantities of the percolation problem via message passing, we now turn to spectral analysis to analyze the stability of one of the phases: the disconnected one. We use stability analysis since, in this way, one can transform the problem into an eigenvalue problem, and the eigenvalues are more straightforward to calculate than the full GCC, especially the largest eigenvalue, as we show next.

Stability theory [239] tells us that the solution $G_\infty = 0$ is stable provided that the largest eigenvalue of the linear operator represented by the $2M \times 2M$ stability matrix $\hat{\mathcal{M}}$ defined on the directed links $k \to \ell, i \to j$:

$$\mathcal{M}_{k\to\ell,i\to j} = \left.\frac{\partial \nu_{i\to j}}{\partial \nu_{k\to\ell}}\right|_{\{\nu_{i\to j}=0\}}, \tag{1.39}$$

is less than one. We denote $\lambda_{\max}(\mathbf{n}; q)$ the largest eigenvalue of the stability matrix $\hat{\mathcal{M}}$. This eigenvalue depends on the vector $\mathbf{n}$ and we also add a parametric dependence on the fraction of removed nodes q. Thus, the stability of a solution set $\mathbf{n}$ of $G = 0$ is determined by the condition:

$$\lambda_{\max}(\mathbf{n}; q) < 1. \tag{1.40}$$

In general, for a fixed fraction q, there exist many possible configurations $\mathbf{n}$ that satisfy Eq. (1.40). When $q < q_c^{\text{ran}}$ each configuration $\mathbf{n}$ gives $\lambda_{\max}(\mathbf{n}; q) > 1$, since it is impossible to find a set of nodes to remove such that $G_\infty(q) = 0$, and this corresponds to the instability of the solution $\{\nu_{i\to j} = 0\}$ signaled by a value of $\lambda_{\max}(\mathbf{n}; q)$ larger than one. On the contrary, when $q > q_c^{\text{ran}}$, we have two different possibilities: there exist configurations $\mathbf{n}$ such that $\lambda_{\max}(\mathbf{n}; q) > 1$, which corresponds to node removals unable to destroy the giant component ($G_\infty(q) > 0$); on the other hand, there can be found other configurations for which $\lambda_{\max}(\mathbf{n}; q) < 1$, which corresponds to a fragmented network with $G_\infty(q) = 0$. As we approach q_c^{ran} from above, $q \to (q_c^{\text{ran}})^+$, the number of configurations $\mathbf{n}$ such that $\lambda_{\max}(\mathbf{n}; q) < 1$ (and hence $G_\infty(q) = 0$) decreases and eventually vanishes at q_c^{ran} (Fig. 1.11c).

By doing an explicit computation of Eq. (1.39) using Eq. (1.37) (valid for local tree-like random networks) we find that the matrix $\hat{\mathcal{M}}$ is given in terms of the Non-Backtracking (NB) matrix $\hat{\mathcal{B}}$ [155, 191] already introduced in Eq. (1.27) when discussing non-backtracking centralities:

$$\mathcal{M}_{k\to\ell,i\to j} = n_i \mathcal{B}_{k\to\ell,i\to j}. \tag{1.41}$$

We recall the definition:

$$\mathcal{B}_{k\to\ell,i\to j} = \begin{cases} 1 \text{ if } \ell = i \text{ and } j \neq k\,, \\ 0 \text{ otherwise.} \end{cases} \tag{1.42}$$

The stability operator $\hat{\mathcal{M}}$ is represented by a $2M \times 2M$ matrix on the $2M$ directed edges of the network. If the network is undirected, each edge is considered as two

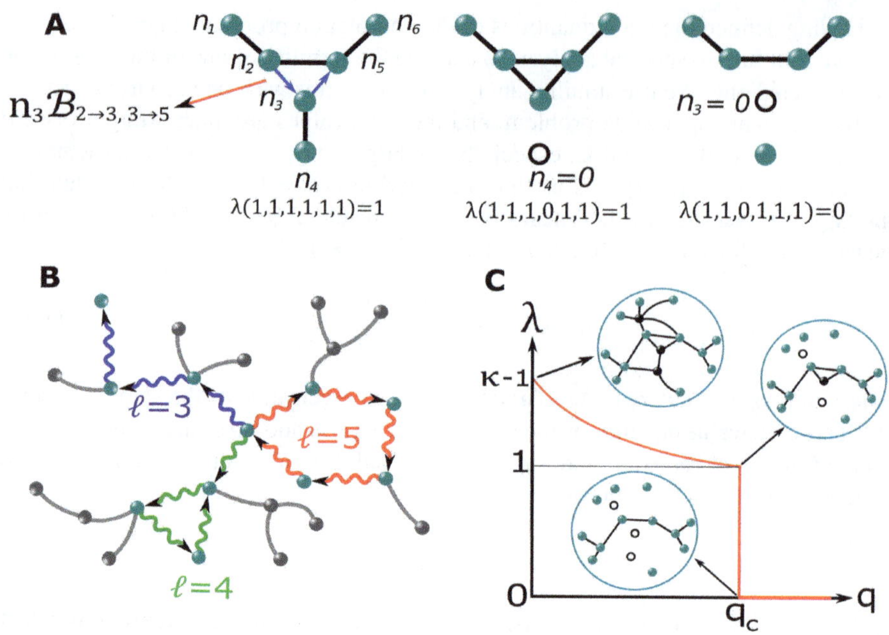

Fig. 1.11 Explanation of the non-backtracking (NB) based theory for percolation. **a** Largest eigenvalue of the NB matrix on sample networks. The network is undirected. Left panel: The action of the matrix $\hat{\mathcal{M}}$ is on the directed edges of the network. The entry $\mathcal{M}_{2\to3,3\to5} = n_3\mathcal{B}_{2\to3,3\to5} = n_3$ encodes node 3's occupancy ($n_3 = 1$) or vacancy ($n_3 = 0$). We denote the element $\mathcal{M}_{2\to3,3\to5}$ as a blue arrow in the network. The network is undirected, but the stability matrix is directed. For this network, which is explicitly written in Eq. (1.43), we find $\lambda_{\max} = 1$ by an explicit computation. Center panel: An example of maximum eigenvalue after the removal of a leaf, $n_4 = 0$, which does not decrease $\lambda_{\max}$ as compared to the network in the left. In terms of optimal percolation, this removal is non-optimal since it does not break any cycle; it just removes a leaf, which does not contribute to spreading. Right panel: A removal of $n_3 = 0$ destroys a cycle and decreases $\lambda_{\max} = 0$, and therefore it is an optimal removal since it destroys the network efficiently. Optimizing this procedure leads to optimal percolation and will be utilized to define the CI algorithm in Sect. 1.6.1 where $\lambda_{\max}$ is minimized by removing the minimum number of nodes (corresponding to the optimal influencers) that destroys all cycles in the network. **b** Examples of NB walks: walks that are not allowed to return back [190, 191]. The number of NB walks of a given length ℓ between two nodes are counted by the powers $\hat{\mathcal{B}}^{\ell}$. We show an NB open walk ($\ell = 3$), an NB closed walk with a tail ($\ell = 4$), and an NB closed walk with no tails ($\ell = 5$). **c** Schematic representation of the behavior of the largest eigenvalue of the modified non-backtracking matrix vs q, as nodes are removed from the network. This behavior is seen in random and optimal percolation. When $\lambda_{\max} \to 1$, G_∞ contains one single loop, which is then destroyed upon removing a node to become a tree. For $q > q_c$, $\lambda_{\max} = 0$ and $G_\infty = 0$. The network consists of a collection of sub-extensive trees. For $q = 0$, $\lambda_{\max} = \kappa - 1$ ($\kappa = \langle k^2\rangle/\langle k\rangle$) which is the largest eigenvalue of $\mathcal{B}$ for random networks [190] with all nodes present ($n_i = 1$). When $\lambda_{\max} = 1$, the giant component is reduced to a unicyclic graph, which can be suddenly dismantled at q_c, causing the abrupt fall of $\lambda_{\max}$ to zero. Figure adapted from [14]. Copyright © 2015, Nature Publishing Group

directed edges. For example, for the sample graph depicted in Fig. 1.11a, left panel, with $N = 6$ nodes and $M = 6$ edges, the corresponding $\hat{\mathcal{M}}$ matrix is a 12×12 matrix that reads:

$$
\begin{array}{c|cccccccccccc|}
 & 1\to 2 & 2\to 1 & 2\to 3 & 2\to 5 & 3\to 2 & 3\to 4 & 3\to 5 & 4\to 3 & 5\to 2 & 5\to 3 & 5\to 6 & 6\to 5 \\
1\to 2 & 0 & 0 & n_2 & n_2 & 0 & 0 & 0 & 0 & 0 & 0 & 0 & 0 \\
2\to 1 & 0 & 0 & 0 & 0 & 0 & 0 & 0 & 0 & 0 & 0 & 0 & 0 \\
2\to 3 & 0 & 0 & 0 & 0 & 0 & n_3 & n_3 & 0 & 0 & 0 & 0 & 0 \\
2\to 5 & 0 & 0 & 0 & 0 & 0 & 0 & 0 & 0 & 0 & n_5 & n_5 & 0 \\
3\to 2 & 0 & n_2 & 0 & n_2 & 0 & 0 & 0 & 0 & 0 & 0 & 0 & 0 \\
3\to 4 & 0 & 0 & 0 & 0 & 0 & 0 & 0 & 0 & 0 & 0 & 0 & 0 \\
3\to 5 & 0 & 0 & 0 & 0 & 0 & 0 & 0 & 0 & n_5 & 0 & n_5 & 0 \\
4\to 3 & 0 & 0 & 0 & 0 & n_3 & 0 & n_3 & 0 & 0 & 0 & 0 & 0 \\
5\to 2 & 0 & n_2 & n_2 & 0 & 0 & 0 & 0 & 0 & 0 & 0 & 0 & 0 \\
5\to 3 & 0 & 0 & 0 & 0 & n_3 & n_3 & 0 & 0 & 0 & 0 & 0 & 0 \\
5\to 6 & 0 & 0 & 0 & 0 & 0 & 0 & 0 & 0 & 0 & 0 & 0 & 0 \\
6\to 5 & 0 & 0 & 0 & 0 & 0 & 0 & 0 & 0 & n_5 & n_5 & 0 & 0
\end{array}
\, . \tag{1.43}
$$

The associated non-backtracking matrix $\hat{\mathcal{B}}$ is obtained by setting $n_i = 1$ in this matrix. The largest eigenvalue of the NB matrix $\hat{\mathcal{B}}$ is positive and simple as a consequence of the Perron-Frobenius theorem [191], and the corresponding eigenvector is such that all components are positive.

A NB walk is a random walk that is not allowed to return along the edge that it just traversed [190, 192]. We show examples of NB walks in Fig. 1.11b: an open walk of length $\ell = 3$, a NB closed walk with a tail ($\ell = 4$), and a NB closed walk with no tails ($\ell = 5$). The NB walks are the building blocks of a diagrammatic expansion to calculate $\lambda_{\max}$ [14]. Powers of the matrix $\hat{\mathcal{B}}$ count the number of non-backtracking walks of a given length in the network [191], much in the same way as powers of the adjacency matrix count the number of paths [152].

The NB matrix has received attention in the context of the detectability of communities in complex networks [190, 240]. For percolation, we have seen that the relevant eigenvalue is the largest. However, for community detection the relevant eigenvalue is the second largest since the corresponding eigenvector of the second eigenvalue can be used to label the nodes into two different communities. Since the eigenvector corresponding to the second largest eigenvalue has both positive and negative components, one can assign to one community all the nodes corresponding to the positive entries of the eigenvector, and to the other community, all the nodes labeled with the negative components. The eigenvector corresponding to the largest eigenvalue of the NB matrix does not work to separate communities since it has all positive eigenvector components so that it is insensitive to the community structure of the network.

This method can be extended to find more than two communities by iteratively applying the same scheme to each community. This community detection protocol can also be implemented using other matrices, e.g., the adjacency and the Laplacian matrices. What is crucial about this method, in comparison with other eigenvector methods for community detection is the fact that the one based on the NB operator has an optimal performance in the sense that it can detect communities down to the detectability threshold, while the other spectral methods fail before [190, 240]. The NB matrix appears when one linearizes belief propagation equations, and it was first used to detect the location of a phase transition in the 3-coloring problem by Coja-Oghlan et al. [241].

In percolation, we are not interested in the largest eigenvalue of $\hat{\mathcal{B}}$, but of $\hat{\mathcal{M}}$, which is a modified NB matrix via removal of nodes n_i. To calculate the critical threshold, we average $\lambda_{\max}(\mathbf{n}; q)$ over the ensemble of random configurations $\mathbf{n}$ satisfying $\langle n \rangle = 1 - q$. This average is calculated as:

$$\lambda_{\max}(\mathbf{n}; q) \equiv \langle \lambda_{\max}(\mathbf{n}; q) \rangle_{\mathbf{n}:\langle n \rangle = 1-q} . \tag{1.44}$$

The random percolation threshold q_c^{ran} is the solution of the equation:

$$\lambda_{\max}(\mathbf{n}; q_c^{\text{ran}}) = 1. \tag{1.45}$$

For a random ensemble of node removals in an uncorrelated network, Eq. (1.45) can be solved for the critical point, and leads to the exact solution of the critical value in random percolation, as we show in the next section. In Sect. 1.5 we will adapt this equation to solve for optimal percolation by minimizing the eigenvalue $\lambda_{\max}$ to find the optimal set of influencers.

1.4.2.1 Critical Threshold in Random Percolation

Random percolation, as opposed to optimal percolation which will be treated in the next section, corresponds to the percolation problem where one removes a set of nodes chosen at random. Random percolation chooses the set of n_i remove nodes at random from a distribution $P(\mathbf{n}; q)$ for a given q:

$$P(\mathbf{n}; q) = \prod_{i=1}^{N} (1-q)^{n_i} q^{1-n_i} . \tag{1.46}$$

This process of removal is decoupled from the non-backtracking matrix and the critical value can be obtained with relative easy [233].

When we pick the nodes at random, $P(\mathbf{n}; q)$ is factorized over the nodes, meaning that it can be written as a product over the nodes, as in Eq. (1.46). Then, the expectation value of the stability matrix $\hat{\mathcal{M}}$, Eq. (1.41), over the random set $\mathbf{n}$ becomes:

$$\mathbb{E}_{\mathbf{n}}\, \mathcal{M}_{i\to j, k\to \ell} = (1-q)\mathcal{B}_{i\to j, k\to \ell} . \tag{1.47}$$

The average over $\mathbf{n}$ of $\lambda_{\max}(\mathbf{n}; q)$ is given by:

$$\lambda_{\max}(q) = (1-q)\lambda_{\mathcal{B}}, \tag{1.48}$$

where $\lambda_{\mathcal{B}}$ is the largest eigenvalue of the NB matrix $\hat{\mathcal{B}}$ for a random (uncorrelated) network, which is [190, 240]:

$$\lambda_{\mathcal{B}} = \kappa - 1, \tag{1.49}$$

where κ is the ratio of the first two moments of the degree distribution:

$$\kappa = \frac{\langle k^2 \rangle}{\langle k \rangle}. \tag{1.50}$$

The condition for the critical point Eq. (1.45) gives rise to the prediction of the critical threshold for the percolation transition in uncorrelated random networks obtained by Karrer et al. [233] using a generating function approach. This is also the same as the famous result of Eq. (1.33) obtained by [21, 228, 230].

In terms of the critical threshold for random removal, we have:

$$q_c^{\text{ran}} = 1 - \frac{1}{\lambda_{\mathcal{B}}}. \tag{1.51}$$

In terms of the percolation threshold for the appearance of the GCC in random percolation, $p_c^{\text{ran}} = 1 - q_c^{\text{ran}}$, this equation becomes:

$$p_c^{\text{ran}} = \frac{1}{\lambda_{\mathcal{B}}}. \tag{1.52}$$

Thus, the critical percolation threshold of random percolation in a sparse network is given by the inverse of the largest eigenvalue of the non-backtracking matrix [233]. Using Eq. (1.49) for random networks we find:

$$q_c^{\text{ran}} = \frac{\kappa - 2}{\kappa - 1}, \tag{1.53}$$

as in Eq. (1.33), or equivalently,

$$p_c^{\text{ran}} = \frac{1}{\kappa - 1}. \tag{1.54}$$

Thus, when $\mathbf{n}$ is decoupled from the NB matrix in a random removal process, the leading eigenvalue of NB captures the random percolation threshold for random networks, for instance, an Erdös-Rényi graph. For a real-world graph, the largest eigenvalue can be different and not related to the first and second moment of the degree distribution. Next, we extend the random percolation problem to optimal percolation and attempt to find its critical point, which is associated with optimal influencers in the network.

1.5 Optimal Percolation

In many networks, the integrity of the system hinges on a small set of nodes, much smaller than the total size, which, if removed or deactivated in a percolation-like process, they would cause the disintegration of the whole network. We are interested in finding the minimal set of nodes in the network that, upon removal, would disintegrate the giant connected component of the system into disconnected and non-extensive components. This problem appears under many names in network science, such as the Targeted (or intentional) Attack problem [20, 22], the Network Dismantling Problem [16], the Optimal Percolation Problem [14], or the Optimal Decycling Problem [17].

In random percolation described in Sect. 1.4.1, nodes are removed without considering their importance in the network structure, i.e., they are chosen at random. In optimal percolation, nodes are chosen strategically to minimize the number of removals and maximize the damage to the GCC. For a given network, there is always a minimal set that destroys the GCC optimally. As illustrated in Fig. 1.10a, G_∞ can be dismantled well before the removal of q_c^{ran}. The minimal set corresponds to the most influential structural nodes that guarantee the global connectivity of the network [20, 22].

Finding this minimal set is a hard problem. One can always use any of the heuristics developed in Sect. 1.2 to rank the nodes and then choose the top-ranked ones that, upon removal, could dismantle the network. For example, it is known that G_∞ in scale-free networks are extremely vulnerable to targeted attacks on the hubs [20, 22, 23]. That is, if we use the high degree centrality, and then remove the nodes one by one starting from the one with the largest degree to the lowest degree, then the GCC will be disintegrated with a much smaller number of removals than q_c^{ran}. Whether this is the optimal strategy is another question.

One could use any of the centralities defined in Sect. 1.2 to define the ranking of node removals and attempt to destroy the GCC by following the different strategies, all of them ad-hoc. Since these rankings are all based on heuristics, there is no guarantee that any ranking will be better than another one, even in the same network, being a random network or a real-world network. Thus, one needs to check one by one the different rankings to find what is the best targeting strategy to disintegrate the GCC in an optimal way, i.e., with a smaller number of removals.

In any case, there will be one strategy unknown to us, which will be the optimal one, in the sense that following the removal of this particular optimal set of nodes, one can destroy the GCC by using the minimal number of node removals. This defines the optimal percolation problem.

The optimized process, deviating from the removal given by random percolation, expedites the collapse of G_∞ and reduces the critical value of q_c to an optimal value that we call q_c^{opt}, the optimal percolation threshold (Fig. 1.10a).

Considering a network $G(V, E)$, with the vector $\mathbf{n} = (n_i, \cdots, n_N)$ encoding the configuration of nodes that are removed (n_i=0) or present (n_i=1), optimal percolation aims to find not only the minimal fraction q_c^{opt} of nodes but also the corresponding optimal configuration state $\mathbf{n}^*$ such that the giant connected component G_∞ is dismantled. For $q > q_c^{\text{opt}}$, there exist a number of possible configurations (not possible to find them at random) $\mathbf{n}$ such that $G_\infty(\mathbf{n}) = \mathbf{0}$. As q decreases, fewer configurations satisfy $G_\infty(\mathbf{n}) = \mathbf{0}$, until a critical value q_c^{opt} where an optimal configuration $\mathbf{n}^*$ still exists. Below $q < q_c^{\text{opt}}$, there is no solution to $G_\infty(\mathbf{n}) = 0$.

Mathematically, the Optimal Percolation Problem attempts to find a configuration $\mathbf{n}^*$ corresponding to the minimal q_c^{opt} such that [14]:

$$q_c^{\text{opt}} = \min\{ q \in [0, 1] \mid G_\infty(q) = 0 \} . \tag{1.55}$$

Removing a configuration state $\mathbf{n}^*$ is the optimal way to dismantle the network.

For a given q, optimal percolation minimizes $G_\infty(q, \mathbf{n})$ over the configurations $\mathbf{n}$. Optimal percolation is an NP-hard combinatorial problem [5]; intractable since an explicit functional form of $G(\mathbf{n})$ versus $\mathbf{n}$ is not feasible. The solution of optimal percolation implies not only the calculation of q_c^{opt} but also the actual optimal configuration set $\mathbf{n}^*$.

The approach to finding an approximated solution of optimal percolation follows the solution of random percolation developed in Sect. 1.4.2. We transfer the problem to the study of the local stability of the solution $G_\infty = 0$, which is again given by the largest eigenvalue of the stability matrix [14, 242], which is a modified NB matrix as discussed in Sect. 1.4.2. This mapping creates a tractable problem by reducing the problem to the calculation of the largest eigenvalue of the stability matrix for which there are approximated methods that can be used to provide fast algorithms.

Most of the formulae developed for random percolation from Eqs. (1.34)–(1.40) are still valid for optimal percolation. The removal of nodes is again represented by a vector $\mathbf{n} = (n_1, \ldots, n_N)$ where a removed node $n_i = 0$ is now an optimal influencer. We still assume that we are dealing with a locally tree-like (sparse) network or a random graph, so the probability of node i for belonging to the GCC in the absence of node j is given by Eq. (1.37). The stability of the $G_\infty = 0$ solution in optimal percolation is determined by the stability matrix $\hat{\mathcal{M}}$ defined by Eq. (1.39) and given in terms of the NB matrix $\hat{\mathcal{B}}$ defined by Eq. (1.42). Again, the largest eigenvalue of the stability matrix is $\lambda_{\text{max}}(q; \mathbf{n})$ and it is a function of the configuration $\mathbf{n}$ for a given q.

The optimization in percolation enters when we try to minimize over q and $\mathbf{n}$ to obtain the optimal threshold q_c^{opt} and optimal minimal set $\mathbf{n}^*$. This is done through the condition for the stability of the null solution $G_\infty(q)$ for $q > q_c^{\text{opt}}$. This solution is stable for q values that satisfy $\lambda_{\text{max}}(q; \mathbf{n}) < 1$, as in Eq. (1.40). However, to find the optimal q_c^{opt}, we need now to minimize this condition over the configurations of removed nodes.

So far, we have been dealing with undirected networks; directed networks will be treated in Sect. 5.2.2. For undirected networks, according to the Perron-Frobenius theorem, the largest eigenvalue of $\hat{\mathcal{B}}$ is a strictly decreasing function of $\hat{\mathcal{M}}$. That is, if $\hat{\mathcal{M}} \leq \hat{\mathcal{B}}$ (meaning that the inequality holds entry by entry of the matrices) and $\hat{\mathcal{M}} \neq \hat{\mathcal{B}}$, then $\lambda(\hat{\mathcal{M}}) < \lambda(\hat{\mathcal{B}})$.

Since the stability matrix $\hat{\mathcal{M}}$ is obtained from the matrix $\hat{\mathcal{B}}$ by setting some entries to zero, $n_i = 0$, then, the optimization problem for a given q can be rephrased as finding the optimal influencer configuration $\mathbf{n}^*$ that minimizes the leading eigenvalue of the stability matrix, $\lambda_{\max}(\mathbf{n}; q)$ over all possible configurations $\mathbf{n}$ satisfying $\langle n \rangle = 1 - q$.

We call $\lambda(\mathbf{n}^*; q)$ this minimum of the largest eigenvalue for a given q:

$$\lambda(\mathbf{n}^*; q) \equiv \min_{\mathbf{n}:\langle n \rangle = 1-q} \lambda(\mathbf{n}; q) \,. \tag{1.56}$$

The optimal percolation critical threshold q_c^{opt} is then the solution of the equation:

$$\lambda(\mathbf{n}^*; q_c^{\text{opt}}) = 1. \tag{1.57}$$

Equation (1.57) is hard to solve for q_c^{opt}, since there is no explicit formula for $\lambda_{\max}(\mathbf{n}; q)$ versus $\mathbf{n}$. Furthermore, one is interested in obtaining not only the critical point but the actual set of influencers $\mathbf{n}^*$. In this case, one would need an explicit formula for $\lambda_{\max}(\mathbf{n}; q)$ as a function of $\mathbf{n}$, which is impossible to obtain. In the next section, we tackle this problem by proposing a sequence of approximations to the largest eigenvalue (and associated eigenvector) based on the Power Method [239] that converges quickly to the exact solution of the problem and allows us to determine the minimal influencers with sub-optimal yet scalable algorithms.

Finding the set of minimal influencer nodes $\mathbf{n}^*$ at q_c^{opt} is important in a number of applications. The cohesion of any network relies on this optimal set, i.e., the minimal set of influencers, which, if activated would maximize the spread of information to the whole network (e.g., via viral spreading on social media) [5, 7]; if immunized or quarantined, it would prevent the diffusion of a large scale epidemic [23, 211]; and if inactivated would prevent the integration of information into the network [20, 22].

This set is interpreted as the structural influencers maintaining the integrity of the network. For instance, in a genetic network, this set corresponds to the genes that would produce the most considerable damage to the cell when removed (mutated) from the network. This is also the optimal set that integrates the network. This will be studied more in application to neural and genetic influencers in Chaps. 5 and 6.

In terms of disease spreading in a contact network, optimal percolation can be mapped to an instance of the optimal immunization problem to detect the superspreaders of disease, see Sect. 1.5.2, or the optimal set to break the chain of transmission of the disease with minimal isolations, see Chap. 4. Regarding influencers and superspreaders of information, we show in Sect. 1.5.1 that, in the most general case, the optimal percolation set does not directly correspond to the best influencer or superspreader of information set. Whether the optimal percolation set corresponds to the superspreaders of information depends on the model used to describe the spreading process. For a particular case of the LTM with thresholds given by $\theta_i = k_i - 1$, the optimal percolation set is exactly the set of influencers for information spreading [14]. The fact that an optimal removal process is associated with the influencer problem might be counter-intuitive. We will treat this case in the next section. We will map the dynamics of opinion spreading in the LTM to the problem of message passing in percolation. This mapping formalizes the intuitive idea that influence is related to the concept of cohesion in a network.

1.5.1 Mapping of Optimal Percolation to Influence Maximization

So far, we have defined two distinct problems on networks that seem to be disconnected. First, we introduced the IMP to find influencers for viral marketing, i.e., those that transmit the information in the network optimally through a contagion process of spreading opinions. This problem is not well-defined until one chooses a model of spreading. A generic model was the LTM, which accounts for a spreading model where an actor changes their opinion by becoming activated when the number of people adopting the product in the actor's egocentric network reaches the minimum given threshold. Second, we introduced optimal percolation, which accounts for finding the minimal number of nodes in a network to destroy the giant connected component.

We show next that LTM with $\theta_i = k_i - 1$ and $w_{ij} = 1$ is mapped to a message passing problem in percolation, making an explicit connection between optimal percolation and IMP. We show that the probability of belonging to the GCC in percolation is the same as the probability of passing a message of adoption in the LTM. This opens the door to solving the IMP in LTM by mapping to optimal percolation. The optimization of influence problem in the LTM is then mapped to a structural problem where the dynamics do not play any role. This mapping between a dynamical problem and a structural one is interesting since, by this means, we avoid a critical bottleneck in the solution of the IMP, which is the calculation of the influence function $\sigma(S)$ in Eq. (1.30).

In percolation (Sect. 1.4.2), we have defined $\nu_{i\to j}$ as the probability of node i to belong to the GCC in the absence of j. To formalize the mapping from percolation to LTM information spreading process [14], we now consider $\nu_{i\to j}$ as the probability that node i is eventually not activated in LTM in absence of node j. Furthermore, in percolation, the variable $n_i = 0$ measures whether node i is removed from the network. In the LTM, the variable n_i equals zero, $n_i = 0$, if i is an initial LTM spreader, and $n_i = 1$ if not.

The mapping of LTM to percolation is valid only when we consider uniform weights $w_{ij} = 1$ for each edge (i, j) and threshold of activation $\theta_i = k_i - 1$ for each node with degree k_i. Thus, in this particular LTM, a node i becomes activated if the number of active neighbors is at least θ_i.

According to the LTM, node i is activated in the absence of node j, if the neighbouring nodes $k \in \partial i \backslash j$ are also activated. The probabilities $\nu_{i\to j}$ for spreading can be computed self-consistently through message passing equations in analogy to the percolation problem of Eq. (1.37). Indeed, the equations for computing the probability for a node not to be activated is the same as that for belonging to the GCC in percolation:

$$\nu_{i\to j} = n_i \left[1 - \prod_{k\in\partial i\setminus j} (1-\nu_{k\to i})\right] . \tag{1.58}$$

The total probability ν_i that i is not activated is obtained from Eq. (1.58) by including the contribution of $\nu_{j\to i}$ as well—analogous to Eq. (1.38):

$$\nu_i = n_i \left[1 - \prod_{k\in\partial i} (1-\nu_{k\to i})\right] , \tag{1.59}$$

and the total fraction of nodes not activated in the spreading process is analogous to Eq. (1.34):

$$G = \frac{1}{N}\sum_{i=1}^{N} \nu_i . \tag{1.60}$$

The IMP consists of maximizing the LTM spreading over the network. This is equivalent to minimizing the fraction of inactive nodes in LTM measured by G defined via $\nu_{i\to j}$ in Eq. (1.58). Optimal percolation also minimizes the same $\nu_{i\to j}$, but this quantity has another interpretation as the probability of belonging to the GCC. This reveals the dual nature of spreading in LTM (with $\theta_i = k_i - 1$) and optimal percolation. The set of nodes optimal for disintegrating the GCC in optimal percolation is the same as the set of nodes that maximizes the spreading of influence in the LTM with $\theta_i = k_i - 1$.

1.5.2 Mapping of Optimal Percolation to Optimal Immunization

When we consider the problem of disease spreading in the network, the quantity ν_i now represents the probability of node i to be infected in the disease spreading process [12, 13]. Therefore, minimizing the sum $\sum_i \nu_i$ is equivalent to reducing the size of the disease outbreak. This is performed by optimally choosing the nodes in the network to be removed/isolated/immunized, a process called optimal immunization. This process again maps to optimal percolation by considering that $n_i = 0$ if node i is an optimal immunized node and $n_i = 1$ if not.

The mapping between optimal percolation and optimal immunization is only valid when the probability of transmission of the disease is $w = 1$. When $w < 1$, the total fraction of infected individuals $n^{-1}\sum_i \nu_i$ is not the same as the giant connected component G is in percolation, and the mapping is no longer valid. The message passing equations are modified when $w \le 1$ to:

$$\nu_{i\to j} = n_i \left[1 - \prod_{k\in\partial i\setminus j} (1 - w\nu_{k\to i}) \right] . \tag{1.61}$$

The spreading of disease in the general case with $w \leq 1$ can be investigated by message passing algorithms. We will discuss this in Sect. 1.6.5 when we discuss the message passage formulation of Altarelli et al. [13] for a general formulation of the optimal immunization problem. Influence maximization in this general LTM corresponds to finding nontypical trajectories of cascading processes that deviate from the average ones. Altarelli et al. [13] analyzed the statistics of large deviations of LTM dynamics with a belief propagation algorithm, and further developed a Max-Sum algorithm to explicitly find an optimal set of seeds in terms of a predefined energy function. Other algorithms for discontinuous phase transitions in LTM will be treated in Sect. 1.7.4 using a model that interpolates between dynamics with continuous and discontinuous transitions.

In conclusion, we map percolation to a particular instance of LTM with threshold $\theta_i = k_i - 1$ and binary interactions $w_{ij} = 1$. For $\theta_i = 1$, we recover the optimal immunization problem. These two problems can be solved by studying the local stability of the solution. For intermediate values of the threshold of the form $\theta_i = k_i - 2, k_i - 3, \dots 2$, the transition at q_c is of the first order and cannot be treated by second-order percolation frameworks since a local stability criterion cannot be applied to discontinuous transitions. That is, the largest eigenvalue of the modified NB matrix is not guaranteed to provide the optimal set when the transition is of first order.

1.6 Algorithms for Finding Influencers of Spreading Dynamics with Underlying Continuous Transitions

In this section, we focus on influencer algorithms based on collective influence theory, optimal dismantling and recycling, and explosive percolation to treat problems with underlying continuous phase transitions.

1.6.1 Collective Influence (CI) Algorithm for Optimal Percolation

In this section, we discuss solutions and algorithms to the influence maximization problem based on optimal percolation. Within the message passing framework, Morone and Makse developed the Collective Influence (CI) algorithm [14, 242] that gives a good approximation to optimal percolation. Later, better algorithms based on optimal decycling and dismantling [16, 17] and explosive percolation [18] were proposed.

The optimal percolation set is obtained as an (infinite) sequence of optimized "attacks" expressed as successive approximations to the minimization of the largest eigenvalue of the modified non-backtracking matrix in Eq. (1.57). We use the Power Method [239] for this iterative scheme to calculate the leading eigenvalue. This iteration quickly converges to an approximated solution of optimal percolation. The zero-order (naive) strategy of the iteration corresponds to the lowest approximation of the theory, and it is given by the high-degree attack, Eq. (1.95). The first non-trivial collective attack strategy is equivalent to finding the ground state of a two-body spin-glass-like system; an anti-ferromagnetic Ising model with random bonds in a random external field at fixed magnetization [14]. Higher-order approximations consist of increasingly complicated many-body problems and produce better performance to dismantle the network than heuristic strategies. This solution holds under the assumption that the graph under consideration is locally tree-like.

For a given configuration $\mathbf{n}$, the leading eigenvalue $\lambda_{\max}(\mathbf{n})$ (we omit the parameter q hereafter, which is always kept fixed) is approximated the Power Method. This eigenvalue determines the growth rate of an arbitrary non-zero vector $\mathbf{w}_0$ after ℓ iterations of the matrix $\hat{\mathcal{M}}$, provided that $\mathbf{w}_0$ has nonzero projection onto the eigenvector corresponding to $\lambda_{\max}(\mathbf{n})$. Repeated iterations of a vector $\mathbf{w}_0$ leads to:

$$\mathbf{w}_\ell(\mathbf{n}) = \hat{\mathcal{M}}^\ell \mathbf{w}_0. \tag{1.62}$$

When $\ell \to \infty$, $\lambda_{\max}(\mathbf{n})$ is obtained from the Power Method as:

$$\lambda_{\max}(\mathbf{n}) = \lim_{\ell\to\infty} \left[\frac{|\mathbf{w}_\ell(\mathbf{n})|}{|\mathbf{w}_0|}\right]^{1/\ell}, \tag{1.63}$$

where

$$|\mathbf{w}_\ell(\mathbf{n})|^2 = \langle \mathbf{w}_\ell(\mathbf{n})|\mathbf{w}_\ell(\mathbf{n})\rangle = \langle \mathbf{w}_0|(\hat{\mathcal{M}}^\ell)^\dagger \hat{\mathcal{M}}^\ell|\mathbf{w}_0\rangle, \tag{1.64}$$

and $\lambda_\ell(\mathbf{n})$ is called the ℓ-dependent approximant:

$$\lambda_\ell(\mathbf{n}) = \left[\frac{|\mathbf{w}_\ell(\mathbf{n})|}{|\mathbf{w}_0|}\right]^{1/\ell}, \tag{1.65}$$

for which we derive analytical expressions. The approximation converges fast, given a large spectral gap of $\hat{\mathcal{M}}$. To do this, we enlarge the space of $\hat{\mathcal{M}}$ from $2M \times 2M$ to $N \times N \times N \times N$ by defining:

$$\mathcal{M}_{ijk\ell} = n_k A_{ij} A_{k\ell} \delta_{jk}(1 - \delta_{i\ell}), \tag{1.66}$$

where each index runs from 1 to N: $i, j, k, \ell = 1, \ldots, N$, and A_{ij} is the adjacency matrix. Practically, we represent $\hat{\mathcal{M}}$ on the nodes of the network rather than on the directed edges.

The intuition behind this scheme is that repeated iterations with $\hat{\mathcal{M}}$ increase the norm of the initial vector as the influence of nodes on non-backtracking walks of larger and larger lengths are included. The growth rate of the vector's norm is determined by the value of the largest eigenvalue, $\lambda_{\max}(\mathbf{n})$ of the modified non-backtracking matrix. As the influencer nodes are removed, the value of $\lambda_{\max}(\mathbf{n})$ decreases until the giant component is reduced to a tree (a graph without cycles) plus only one cycle when $\lambda_{\max}(\mathbf{n})$=1. The removal of the next nodes quickly destroys the giant component and $\lambda_{\max}(\mathbf{n})$ falls to zero.

The behavior of the largest eigenvalue of the stability matrix $\lambda_{\max}$ versus q in optimal percolation is the same as in random percolation case shown in Fig. 1.11c, highlighting the similarity between the two problems, although, in the optimal case, the curve is minimized towards q_c^{opt}, and in the random case it is not. That is, optimal percolation minimizes the value of q and G_∞ in Fig. 1.11c. When $q \geq q_c^{\text{opt}}$, the minimum is at $\lambda_{\max} = 0$. Then, $G_\infty = 0$ is stable (still, non-optimal configurations exist with $\lambda_{\max} > 1$ for which $G_\infty > 0$). When $q < q_c^{\text{opt}}$, the minimum of the largest eigenvalue is always $\lambda_{\max} > 1$, the solution $G_\infty = 0$ is unstable, and then $G_\infty > 0$. At the optimal percolation transition, the minimum is at $\mathbf{n}^*$ with $\lambda_{\max}(\mathbf{n}^*, q_c^{\text{opt}}) = 1$. For $q = 0$, a random network has $\lambda_{\max} = \kappa - 1$ ($\kappa = \langle k^2\rangle/\langle k\rangle$) which is the largest eigenvalue of $\mathcal{B}$ for uncorrelated networks [190] with all nodes present ($n_i = 1$). When $\lambda_{\max} = 1$, the giant component is a tree plus a loop (unicyclic graph), which is disintegrated at the transition q_c^{opt} with the abrupt fall of $\lambda_{\max}$ to zero.

The optimal influence problem of finding the optimal configuration $\mathbf{n}^*$ that minimizes the largest eigenvalue of $\hat{\mathcal{M}}$ consists of removing the nodes that are on the most non-backtracking walks in the network and keeps the giant component connected. The intuition behind the usage of non-backtracking walks in optimal percolation comes from the fact that the giant component is held together by long paths and the powers of the non-backtracking matrix allows you to find them quickly.

The computation starts with the first approximant $\ell = 1$:

$$|\mathbf{w}_1(\mathbf{n})\rangle = \hat{\mathcal{M}}|\mathbf{w}_0\rangle. \tag{1.67}$$

We use the $2M$-dimensional starting vector in the space of edges $|\mathbf{w}_0\rangle = (1, 1, \cdots, 1)^T$ which is $|w_0\rangle_{ij} = A_{ij}$ in the enlarged $N \times N$ node space with norm $|\mathbf{w}_0|^2 = \sum_i k_i = 2M$, which gives the $\ell = 1$ norm $|\mathbf{w}_1(\mathbf{n})|^2$ as:

$$|\mathbf{w}_1(\mathbf{n})|^2 = \sum_{ij} {}_{ij}\langle w_1(\mathbf{n})|w_1(\mathbf{n})\rangle_{ij} = \sum_{ij} A_{ij} z_i z_j n_i n_j, \tag{1.68}$$

where we have defined the residual degree as $z_i \equiv k_i - 1$. We obtain the first-order estimate of the eigenvalue:

$$\lambda_1(\mathbf{n}) = \left[\frac{1}{2M}\sum_{ij} A_{ij} z_i z_j n_i n_j\right]^{1/2} . \tag{1.69}$$

It is useful to give a graphical representation of the previous formula in terms of a diagrammatic expansion, which becomes indispensable for higher orders of the iteration. The interaction term in the sum in the numerator on the r.h.s of Eq. (1.69) can be represented as

$$A_{ij}(k_i - 1)(k_j - 1) n_i n_j \quad = \tag{1.70}$$

These Feynman-like diagrams are handy: a straight line that connects two sites i and j means that the variables n_i and n_j are multiplied by each other. The direction of the arrow denotes the NB walks, and the wiggly lines on the nodes represent a multiplication by the factor $z_i = k_i - 1$. The diagram is then equal to $n_i n_j z_i z_j$. The number of variables appearing in the diagram gives the order of the interaction. In this case, it is a pair-wise interaction that defines the first optimization order $\ell = 1$ as a 2-body problem. This term is interpreted as the energy of an antiferromagnetic Ising model with random bonds in a random external field at fixed magnetization, an example of a pair-wise NP-complete spin-glass. The ℓ-order term in the approximant generally describes a 2ℓ-body problem.

For $\ell \geq 2$, the problem can be mapped to a statistical mechanical system with many-body interactions, which can be recast in terms of the diagrammatic expansion, Eqs. (1.70)–(1.78).

For $\ell = 2$, $|\mathbf{w}_2(\mathbf{n})\rangle$ we have:

$$|w_2(\mathbf{n})\rangle_{ij} = \sum_{k\ell} \mathcal{M}_{ijk\ell}\, |w_1(\mathbf{n})\rangle_{k\ell} = n_j A_{ij} \sum_{\ell} A_{j\ell} n_\ell z_\ell (1 - \delta_{i\ell}), \tag{1.71}$$

and also $\langle \mathbf{w}_2(\mathbf{n})|$:

$${}_{ij}\langle w_2(\mathbf{n})| = \sum_{k\ell} {}_{k\ell}\langle w_1(\mathbf{n})|\mathcal{M}_{k\ell ij} = n_i A_{ij} \sum_{k} A_{ik} n_k z_k (1 - \delta_{kj}) . \tag{1.72}$$

The norm $|\mathbf{w}_2(\mathbf{n})|^2$ is given by:

$$|\mathbf{w}_2(\mathbf{n})|^2 = \sum_{ijk\ell} A_{ij} A_{jk} A_{k\ell} (1 - \delta_{ik})(1 - \delta_{j\ell}) z_i z_\ell n_i n_j n_k n_\ell . \tag{1.73}$$

There are two types of interactions in the sum on the r.h.s of Eq. (1.73): a 4-body and a 3-body interaction. The graphical representation of the former is:

$$A_{ij}A_{jk}A_{k\ell}z_i z_\ell n_i n_j n_k n_\ell =$$ 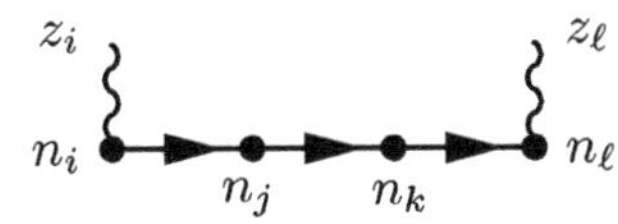

(1.74)

The diagram for the 3-body interaction is:

$$A_{ij}A_{jk}A_{ki}(z_i)^2 n_i n_j n_k =$$ 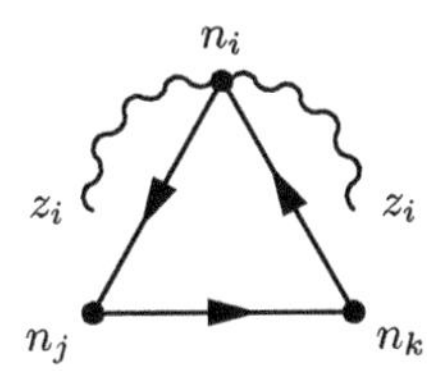

(1.75)

Then, the $\ell = 2$ approximant of the eigenvalue, $\lambda_2(\mathbf{n})$, is given by:

$$\lambda_2(\mathbf{n}) = \left[\frac{1}{2M} \sum_{ijk\ell} A_{ij}A_{jk}A_{k\ell}(1-\delta_{ik})(1-\delta_{j\ell})z_i z_\ell n_i n_j n_k n_\ell \right]^{1/4} . \tag{1.76}$$

In the next order $\ell = 3$, $|\mathbf{w}_3(\mathbf{n})|^2$ there appears a term with 6-body interactions:

$$A_{ij}A_{jk}A_{k\ell}A_{\ell m}A_{mp}z_i z_p n_i n_j n_k n_\ell n_m n_p =$$
$$=$$ 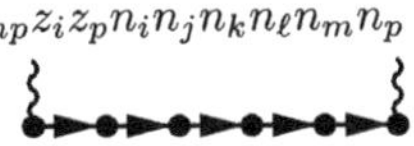 , (1.77)

terms with 5-body interactions:

terms with 4-body interactions:

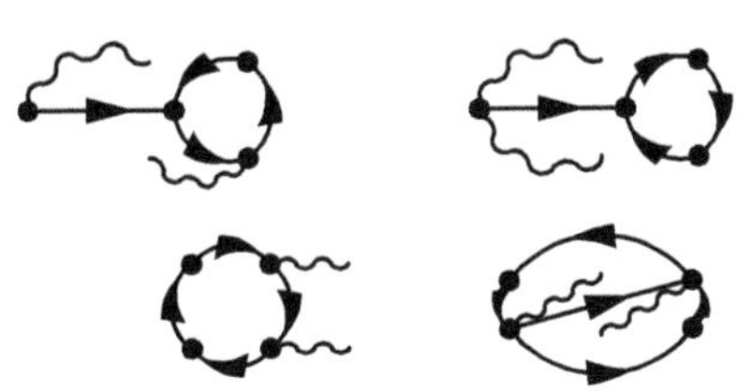

And a term with 3-body interaction:

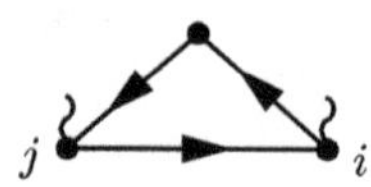

(1.78)

This diagrammatic expansion can be interpreted in terms of NB walks in the network. Nodes entering each type of interaction given by $|\mathbf{w}_1(\mathbf{n})|^2$, $|\mathbf{w}_2(\mathbf{n})|^2$, and $|\mathbf{w}_3(\mathbf{n})|^2$ are the same nodes visited by a non-backtracking walk of length $1, 3, 5$, respectively (with the possibility of traversing an edge multiple times). In general, the diagrammatic expansion of the term $|\mathbf{w}_\ell(\mathbf{n})|^2$ will then contain all the possible graphs that can be built using the nodes traversed by a NB walk of length $2\ell - 1$. Thus, we put an arrow in the diagrammatic schema of The interaction terms are used to emphasize the connection with NB walks.

The NB graphs in the expansion are built in the following way:

1. For a given ℓ we construct all NB walks of $2\ell - 1$ steps starting at one node i (with degree k_i) and ending in node j (with degree k_j). The initial and ending points are indicated by a wiggly line indicating their residual degree.
2. The initial and final nodes of the NB walk do not need to be necessarily different.
3. Loops are allowed in the NB walk. Thus, the shortest path between i and j might be smaller than the $2\ell - 1$ steps of the walk.
4. We recall that the condition for an NB random walk is that it cannot come back through the same link that it just came on, yet it can visit the same node several times.
5. As well, the NB walks are allowed to travel through the same links multiple times.
6. The The number of nodes in the NB walk is the order of the interaction.
7. The dominant graph is always a direct path (line) of length $2\ell - 1$ and 2ℓ nodes, where the shortest path between the initial and ending node is $2\ell - 1$; we will see that all the other diagrams with loops are negligible for sparse, locally tree-like random graphs.

For instance, in the case $\ell = 2$, we obtain two graphs representing NB walks of length $2\ell - 1 = 3$. The graph (1.74) represents a NB walk of length three, which traverses four distinct nodes, and thus it is a 4-body dominant interaction. The graph (1.75) also represents a NB walk of length 3, but this time, the NB walk starts and ends in the same node i, rule 2 and 3, so this term contributes with a factor z_i^2.

In the case $\ell = 3$, the leading interaction is graph (1.78) of 6-body interactions and 5 NB steps. The 3-body interaction (1.78) is a NB walk of 5 steps that starts at the node i on the right and traverses two links twice, rule 5, to end up in node j, resulting in a triangular 3-body problem.

The analytical expression for $|\mathbf{w}_1(\mathbf{n})|^2$ and $|\mathbf{w}_2(\mathbf{n})|^2$ given by Eqs. (1.68), (1.73) can be generalized to any ℓ:

$$|\mathbf{w}_\ell(\mathbf{n})|^2 = \sum_{i_1 i_2 \dots i_{2\ell}} A_{i_1 i_2} A_{i_2 i_3} \dots A_{i_{2\ell-1} i_{2\ell}} (1-\delta_{i_1 i_3})(1-\delta_{i_2 i_4}) \dots (1-\delta_{i_{2\ell-2} i_{2\ell}}) z_{i_1} z_{i_{2\ell}} n_{i_1} n_{i_2} \dots n_{i_{2\ell}} . \tag{1.79}$$

Each term of the sum in Eq. (1.79) is associated with a NB walk of length $2\ell - 1$, where edges may be crossed multiple times. A variable n_i is attached to each node the NB walk visits. The initial and ending point of the walk adds factors z_{i_1} for the starting point and $z_{i_{2\ell}}$ for the endpoint. This formula is simplified when NB walks containing loops are neglected in Eq. (1.79). In this case, the only remaining walks are the ones of length $2\ell - 1$ where each one of the 2ℓ nodes is visited only once. For example, consider a loop-free NB walk of length $2\ell - 1$ starting from node i. It visits all nodes up to a distance $2\ell - 1$, and stops at the final node, say node j. The total number of nodes visited by this NB walk is 2ℓ.

In line with the original locally tree-like assumption for the whole network, these loops can be neglected, and the local environment around any node can approximated by a tree. Under this approximation, the NB walks of length $2\ell - 1$ starting from a given node i and ending on nodes j at a distance $2\ell - 1$ away are exactly the shortest paths between i and those nodes j. Thus, under this tree-like approximation, we consider only these NB walks. These NB walks contain the products $n_{i_1} n_{i_2} \dots n_{i_{2\ell}}$ with all n_{i_α} different from each other.

Following these considerations, the leading term in $1/N$ of $|\mathbf{w}_\ell(\mathbf{n})|^2$ is:

$$|\mathbf{w}_\ell(\mathbf{n})|^2 = \sum_{i=1}^{N} (k_i - 1) \sum_{j \in \partial \mathrm{Ball}(i, 2\ell-1)} \left(\prod_{k \in \mathcal{P}_{2\ell-1}(i,j)} n_k \right) (k_j - 1) , \tag{1.80}$$

where $\mathrm{Ball}(i, \ell)$ is the set of nodes inside a ball of radius ℓ around node i, where the radius is defined taking the shortest path as the distance, $\partial\mathrm{Ball}(i, \ell)$ is the frontier (surface) of the ball, and $\mathcal{P}_\ell(i, j)$ is the set of nodes belonging to the shortest path of length ℓ connecting i and j (see Fig. 1.12a).

We see that the series expansion of the maximum eigenvalue can be written in terms of a systematic diagrammatic expansion of increasing levels of many-body interactions. Since the leading eigenvalue is related to $|\mathbf{w}_\ell(\mathbf{n})|$ via Eq. (1.63), then the minimization of $\lambda_{\max}(\mathbf{n})$ is then translated to the minimization of Eq. (1.80). We can define an energy cost function of influence, $E_\ell(\mathbf{n})$, from $|\mathbf{w}_\ell(\mathbf{n})|$ which is the quantity to minimize, as in a physical system.

Equation (1.80) contains only contributions from odd radii of the ball $(2\ell - 1)$. An analogous calculation can be done for even radii [14]. The leading contributions for the first three levels are summarized in Table 1.2. When we put together these two contributions, we find an energy cost function of the form:

$$E_\ell(\mathbf{n}) \equiv \sum_{i=1}^{N} (k_i - 1) \sum_{j \in \partial \mathrm{Ball}(i, \ell)} \left(\prod_{k \in \mathcal{P}_\ell(i,j)} n_k \right) (k_j - 1) , \tag{1.81}$$

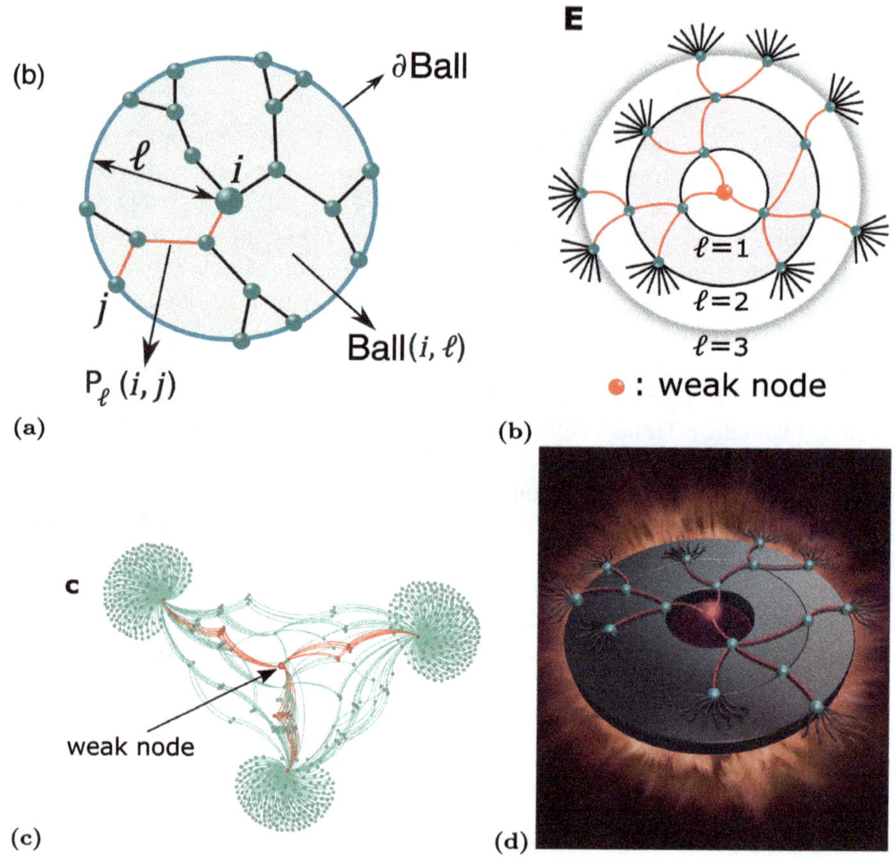

Fig. 1.12 The "strength of weak nodes" identified by collective influence a la Granovetter [3]. **a** Ball(i, ℓ) of radius ℓ around node i is shown. ∂Ball is the set of nodes on the boundary. The highlighted route is the shortest path from i to j. **b** Example of a weak node in the CI theory: a node (red) with a small number of connections surrounded by hierarchical coronas of hubs at different *ell* levels. This node is, in principle, weak due to its low connectivity. Still, it is strategically located as an important broker in the network since it connects many hubs at distances ℓ. This is the mathematical formalization of the 'strength of weak ties' of Granovetter. Most heuristic strategies miss this crucial influencer. For instance, due to their low-medium degree, they are identified very late by the high-degree ranking, i.e., after the removal of many mediocre nodes with larger degrees, or, even worse, they may be completely neglected. **c** Example of a weak node with high CI connecting hubs at $\ell = 2$ identified in the Twitter network of Fig. 1.14a by the CI algorithm. **d** Artistic rendering of a weak node with strong influence. Figure adapted from [14]. Copyright © 2015, Nature Publishing Group

Table 1.2 Leading diagrams for the first three levels of the many-body interactions contributing to the energy cost function of optimal percolation

	Even interactions	Odd interactions
Energy function	$\|\mathbf{w}_\ell(\mathbf{n})\|^2$	$\langle \mathbf{w}_\ell(\mathbf{n})\|\hat{\mathcal{M}}\|\mathbf{w}_\ell(\mathbf{n})\rangle$
Order of Interactions	$\rho = 2\ell$	$\rho = 2\ell + 1$
Leading diagram for $\ell = 0$		 one-body
Leading diagram for $\ell = 1$	 two-body	 three-body
Leading diagram for $\ell = 2$	 four-body	 five-body

where $E_\ell(\mathbf{n}) = |\mathbf{w}_{(\ell+1)/2}|^2$ for ℓ odd, and $E_\ell(\mathbf{n}) = \langle \mathbf{w}_{\ell/2}|\hat{\mathcal{M}}|\mathbf{w}_{\ell/2}\rangle$ for ℓ even.

For $\ell = 1$, $E_\ell(\mathbf{n})$ is the energy function of an Ising model with the spin variables $s_i = 2n_i - 1$ [8]. For $\ell \geq 2$, this energy corresponds to a many-body interaction problem. For example, $|\mathbf{w}_2(\mathbf{n})|^2$ leads to 4-body interactions, Eq. (1.74), and, in general, $|\mathbf{w}_\ell(\mathbf{n})|^2$ contains 2ℓ-body interactions.

Since the energy cost function Eq. (1.81) is a sum over the nodes in the network, we can use the term in the sum to define a cost function for every node in the network for a given ℓ, $E'_\ell(i)$. We rewrite:

$$E_\ell(\mathbf{n}) = \sum_{i=1}^{N} E'_\ell(i), \tag{1.82}$$

with

$$E'_\ell(i) \;=\; (k_i - 1) \sum_{j \in \partial \mathrm{Ball}(i,\ell)} \left(\prod_{k \in \mathcal{P}_\ell(i,j)} n_k \right) (k_j - 1)\,. \tag{1.83}$$

Morone and Makse [14] then use this equation to propose the Collective Influence algorithm to find optimal influencers in networks. A fast and scalable algorithm to minimize the energy $E_\ell(\mathbf{n})$ is to adaptively remove the nodes with the highest contribution to the per-node energy $E'_\ell(i)$, i.e. causing the largest decrease to the energy function. When all the nodes are present, we evaluate the energy Eq. (1.83) to $\mathbf{n} = \mathbf{1}$. For a given range of the interaction ℓ, this energy defines the Collective Influence CI of node i as [14]:

$$\mathrm{CI}_\ell(i) \;=\; (k_i - 1) \sum_{j \in \partial \mathrm{Ball}(i,\ell)} (k_j - 1). \tag{1.84}$$

The CI algorithm then finds a set of optimal influencers in the network by iteratively and adaptively removing the nodes with the largest CI, such that the largest eigenvalue of $\hat{\mathcal{M}}$ can be minimized with high efficiency. We can find and remove the one with the largest collective influence by computing this quantity for each node. After each removal, the CI score of every remaining node in the network is recalculated. Thus, the CI ranking of the nodes is produced by removing the node with the largest CI value, recomputing the CI values of its neighbors, and repeating the operation until the giant component of the network is entirely fragmented, i.e., $G_\infty \ll 1$. The optimal configuration $\mathbf{n}^*$ and q_c^{opt} are obtained from this removal process, and they are defined as a set of optimal influencers in the network according to optimal percolation.

At the beginning of the algorithm, all nodes are present, and they are assigned their $\text{CI}_\ell(i)$ values according to Eq. (1.84) for a given ℓ, and they are ranked from high CI to low CI. The node i^* with the highest $\text{CI}_\ell(i^*)$ is removed from the network, and set $n_{i^*} = 0$. Since the removal of this node changes the degree of its neighbors, we decrease their degrees by one before recomputing their $\text{CI}_\ell(i)$.

After this removal, the network consists of $N - 1$ nodes, and we can proceed as before, looking for the next node with the largest $\text{CI}_\ell(i)$. Removing the nodes one by one according to this adaptive scheme destroys the network in a nearly optimal and fast way. The algorithm can be sped up by decimating a finite fraction of nodes at each step before recomputing CI. The algorithm's performance increases for larger values of the radius ℓ of the $\text{Ball}(i, \ell)$, and for typical networks, the algorithm reaches top performance in a few layers $\ell \sim 3, 4$.

When ℓ becomes larger than the network diameter, then $\text{CI}_\ell(i) = 0$. Then, the nodes are not distinguishable by the algorithm, and thus, the method is indistinguishable from a random one. Thus, the parameter ℓ should not exceed the original network diameter. Besides, this condition is a necessary and sufficient condition for the algorithm to work since, during the process of removal, the network diameter always increases, reaching a maximum when the giant component becomes a tree just at the transition point q_c^{out}, and then suddenly drops to zero by the removal of a sublinear number of nodes.

The CI algorithm Eq. (1.84) is based on Eq. (1.81) which contains the many-body collective interactions that we refer to as "collective influence". The CI algorithm incorporates the collective effects by considering the adaptive nature of the algorithm. The adaptiveness of the CI algorithm, usually called decimation in the spin glass literature [8], is a collective way to select influential nodes since the removal of each node depends on the history of the process.

The case of zero radius $\ell = 0$ leads to $\langle \mathbf{w}_0 | \hat{\mathcal{M}} | \mathbf{w}_0 \rangle = \sum_i^N k_i(k_i - 1)n_i$ (see Eq. (1.87) below). In this case, there is no interaction in the energy cost between the nodes and the minimization of $\lambda_{\max}(\mathbf{n})$ over $\mathbf{n}$ naturally leads to the high-degree (HD) individual ranking, as discussed in Sect. 1.2.1.1 [20, 22] (see Eq. (1.89)). That is, at the zero-order CI_0, the influencers are the hubs.

Beyond this non-interacting level, the collective influence CI_ℓ for $\ell \geq 1$ has a rich topological content, and consequently can tell us more about the role played by nodes in the network than the non-interacting high-degree removal using CI_0. The augmented information comes from the sum in the r.h.s of Eq. (1.84), absent in the naive high-degree rank. This sum contains the contribution of the nodes living on the surface of the ball surrounding the central node i, and the contribution of each node to CI weighted by the factor $k_j - 1$. This means that a node placed at the center of a corona irradiating many links— the structure hierarchically emerging at different ℓ-levels as seen in Fig. 1.12b–d—can have a large collective influence, even if it has a moderate or low degree.

Thus, an advantage of CI is that it can identify seemingly 'weak nodes' with a small number of connections that are surrounded by highly connected nodes and are, therefore, on the path of many non-backtracking walks. Such 'weak-nodes' can outrank nodes with a larger degree that occupy non-influential peripheral locations in the network. Equation (1.84) can be seen as the mathematical definition of the Granovetter concept of weak node connectivity, yet with strength for communication spreading [3]. The commonly used word 'weak' in this context sounds particularly paradoxical. It is usually used as a synonymous for a low-degree node with an additional 'bridging' property, which has resisted a quantitative formulation. We provide this definition through Eq. (1.84), according to which weak nodes are, de facto, quite strong. Solving Granovetter's conundrum [3], CI quantifies the 'strength of weak ties'.

Simulation results on both synthetic networks (Fig. 1.13) and real-world social networks (Fig. 1.14) show that the CI algorithm provides a better performance in finding the minimal influencers than frequently used heuristic metrics such as high-degree (HD), high-degree adaptive (HDA), PageRank (PR), closeness (CC) and eigenvector centralities (EC) and k-core index, in this kind of networks, which include ER networks (Fig. 1.13a) as well as scale-free (SF) networks (Fig. 1.13b). Figure 1.13c shows examples of disintegration of the same SF network following HD, HDA and CI after the removal of 15% nodes. The remaining largest connected component in each case is shown in red and the disconnected components in green. CI produces a smaller giant component than the rest of the strategies.

Using a Twitter network with 469, 013 users (Fig. 1.14a), and a Mexico mobile communication network of 1.4×10^7 users from [243] (Fig. 1.14b), the CI algorithm achieves full fragmentation with a smaller set of influencers. For such massively large-scale networks, a variant of the The CI algorithm can be applied without losing performance by removing a finite fraction of nodes instead of one at each step.

1.6.1.1 CI Algorithmic Formulations

In principle, the computational complexity of the CI algorithm proposed in [14] is $O(N^2)$, with the number of nodes N. Morone et al. [242] then developed a linear-time implementation of the CI algorithm. This makes the CI algorithm highly scalable to large networks.

In practice, CI is accelerated by limiting the calculation and update of CI inside the $(\ell + 1)$-ball around the removed node. In addition, the time complexity is further reduced to $O(N \log N)$ by sorting the CI scores in a heap structure. The basic idea is that, after each node removal, we recompute CI for a $O(1)$ number of nodes and find the new largest value without thoroughly sorting the CI list. This is achieved using a max-heap data structure .

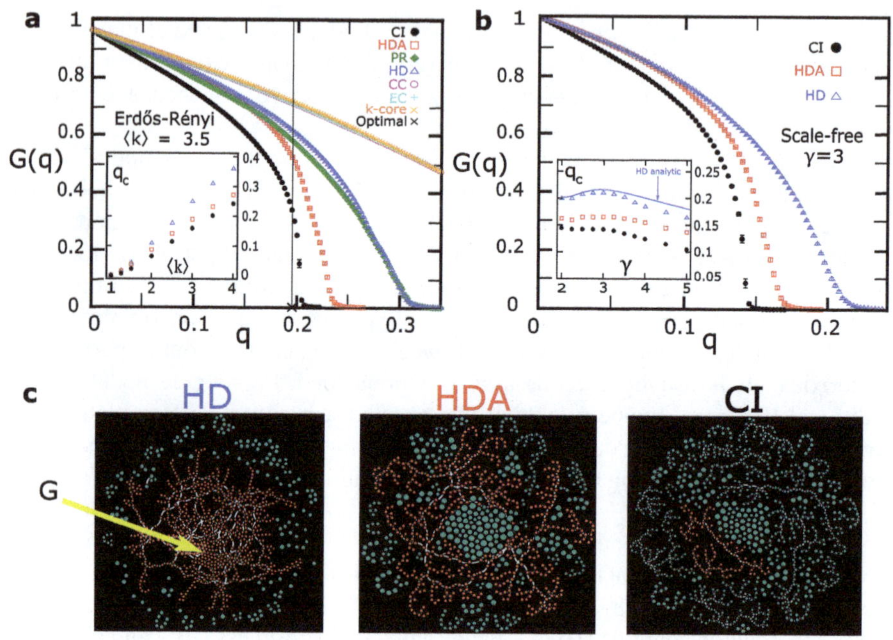

Fig. 1.13 **Performance of CI in synthetic networks.** **a** $G_\infty(q)$ in ER networks ($N = 2 \times 10^5$, $\langle k \rangle = 3.5$, error bars are s.e.m. over 20 realizations) for the true optimal solution obtained asymptotically numerically with Extremal Optimization (EO) [14] ($\times$), CI, high-degree adaptive (HDA), PageRank (PR), HD, closeness centrality (CC), eigenvector centrality (EC), and k-core. Other centralities, like betweenness, are not scalable and perform worse than HDA. CI is close to the optimal $q_c^{\text{opt}} \sim 0.193$ obtained with EO. EO can estimate the extrapolated optimal value of q_c^{opt}, but it cannot provide the optimal configuration for large systems. Inset: q_c^{opt} (obtained at the peak of the second largest cluster) for the three best methods vs $\langle k \rangle$. **b** $G_\infty(q)$ for SF network with degree exponent $\gamma = 3$, maximum degree $k_{\max} = 10^3$ and $n = 2 \times 10^5$ (error bars are s.e.m. over 20 realizations). Inset: q_c^{opt} vs γ. The continuous blue line is the HD analytical result computed in Sect. 1.6.2. **c** The removal of 15% of nodes from a SF network with $\gamma = 3$ using HD, HDA, and CI. The remaining largest connected component G is marked by red nodes and the disconnected clusters in green. CI produces a much reduced giant component. Figure adapted from [14]. Copyright © 2015, Nature Publishing Group

A heap is a binary tree encoding a hierarchical rule between the parent node at level h and its children nodes at level $h + 1$, with no hierarchy among the children. In the specific case of CI, a max-heap rule is used where each parent node of the heap stores a CI value greater or equal to those of the children, but there is no order between the left child and the right one (see Fig. 1.15a). The root node of the max-heap stores automatically the largest CI value. Heapification of a set of numbers $S = \{x_1, \ldots, x_N\}$ is a permutation Π of the elements $\{x_{\Pi(1)}, \ldots, x_{\Pi(N)}\}$ satisfying the max-heap property:

$$x_{\Pi(i)} \geq x_{\Pi(2i)} \text{ AND } x_{\Pi(i)} \geq x_{\Pi(2i+1)} \,. \tag{1.85}$$

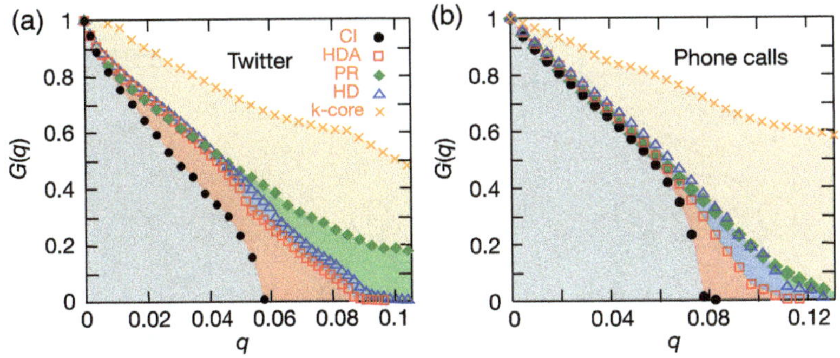

Fig. 1.14 **Performance of CI in large-scale real social networks**. Giant components $G(q)$ for **a** a Twitter network, and **b** a social network of mobile phone users in Mexico [243] computed using CI, HDA, PR, HD, and k-core strategies versus the fraction of removed nodes. In these cases, CI performs better than the rest of the heuristics. Figure adapted from [14]. Copyright © 2015, Nature Publishing Group

The function heapify(i) heapifies the CI values in the sub-tree rooted on node i to down-move node i in the heap by swapping it with the largest of its children until it satisfies the max-heap property in the final location.

The flow of the linear-time CI algorithm is exemplified in Fig. 1.15b. The code is provided in Appendix A.

- **Step 1–Computing CI**. Compute all CI values $\{\mathrm{CI}_\ell(1), \ldots, \mathrm{CI}_\ell(N)\}$ using Eq. (1.84) for a chosen ℓ. For each node i, find the nodes belonging to the frontier $\partial\mathrm{Ball}(i, \ell)$ of the ball of radius ℓ centered on i. In undirected networks, node $j \in \partial\mathrm{Ball}(i, \ell)$ is found by a breadth-first-search up to a distance ℓ from i: first visit the nearest neighbours of i at $\partial\mathrm{Ball}(i, 1)$; second, visit all the neighbours of those nodes not yet visited, arriving to $\partial\mathrm{Ball}(i, 2)$; then, continue until visiting all the nodes at $\partial\mathrm{Ball}(i, \ell)$. Use the nodes $j \in \partial\mathrm{Ball}(i, \ell)$ to evaluate $\mathrm{CI}_\ell(i)$.
- **Step 2–Building the max-heap**. Arrange $\mathrm{CI}_\ell(i)$ in a max-heap in a bottom-up fashion, from leaves to root. Practically, first, fill the heap with arbitrary values, and then we heapify all the levels starting from the lowest one. In this way, the root stores the largest CI value.
- **Step 3–Removal**. Remove the largest CI node from the network and reduce by one the degree of its neighbors. Since the largest CI is stored in the root of the max-heap, after its removal, the root has to be replaced by a new one storing the new largest CI. Replace the old root with the rightmost leaf in the heap's last level, decreasing the heap's size by one and heapifying the heap starting from the new root (see Fig. 1.15c).
- **Step 4–Updating CI values**. Only nodes inside $\mathrm{Ball}(i, \ell + 1)$ change their CI values when i is removed, while the others remain the same. Then, recompute CI for nodes affected by the removed node, only those placed at distances $1, 2, \ldots, \ell, \ell + 1$ from the removed one. The CI values in the $\ell + 1$ layer are quickly recomputed by decreasing their CI by the degree minus one. For layers $1, 2, \ldots, \ell$, use Step 1.

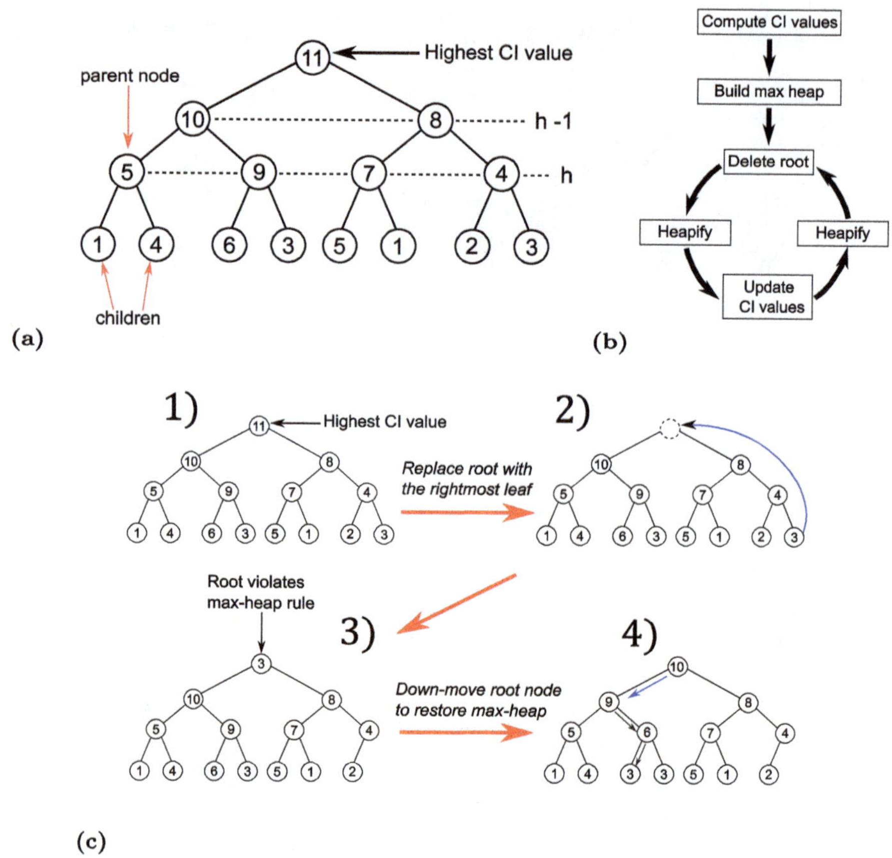

Fig. 1.15 Implementation of the linear-time CI algorithm using the max-heap data structure. **a** In the max-heap, each parent node stores a CI value larger than the ones stored by its children. No ordering prescription is imposed on the nodes belonging to the same level h of the heap. **b** Steps of CI: *(i)* Compute CI for each node, and *(ii)* allocate the CI values in the max-heap. Enter the main loop: *(iii)* Remove the highest CI node along with the root of the heap; *(iv)* heapify the heap starting from the new root (see Step 3); *(v)* update the CI values of the perturbed nodes, and heapifying the sub-trees rooted on each updated node. The loop ends when the giant component is destroyed. **c** Updating the heap: (1) the root node stores the highest CI value, node 11, which is removed. (2) The root is replaced by node 3, the rightmost leaf of the heap. (3) The new heap does not satisfy the max-heap rule, and it is (4) heapified starting from the root to restore the max-heap structure. The heapification down-moves progressively node 3 in the heap by swapping it with the largest of its children nodes until it does satisfy the max-heap property in the final location. Code provided in Appendix A. Figure reproduced from [242]. Copyright © 2016, The Authors

Since the resulting new heap may not satisfy the max-heap rule, restore the max heap structure after each change of the CI values. Under the hypothesis of locally tree-like, the new CI values of the surrounding nodes of the removed node can only be smaller than their old values (for $\ell \leq 2$ this is always true even without the tree-like hypothesis). Consequently, heapify only the sub-trees rooted on those nodes. The order of the update and heapification operations is important: each node update must be followed by the corresponding heapification before updating the next node.

- **Step 5–Algorithm halt**. In principle, halt the algorithm when the GCC has vanished. However, calculating the GCC is computationally costly requiring $O(N)$ calculations per node removal. Instead, after each removal, monitor an approximated minimum of the largest eigenvalue of the NB matrix when Nq nodes are removed from the network [14], which can be updated on runtime at nearly no additional computational cost:

$$\lambda(\ell; q) = \left(\frac{\sum_i \mathrm{CI}_\ell(i)}{N \langle k \rangle} \right)^{1/(\ell+1)}, \tag{1.86}$$

where $\langle k \rangle$ is the average degree of the network for $q = 0$. For $q = 0$ and tree-like random graphs, $\lambda(\ell; 0) = \kappa - 1$, where $\kappa = \langle k^2 \rangle / \langle k \rangle$ [190]. Removing nodes decreases $\lambda(\ell; q)$, and the network is destroyed when $\lim_{\ell \to \infty} \lambda(\ell; q = q_c^{\mathrm{opt}}) = 1$. In practice, the limit $\ell \to \infty$ is unfeasible, but for ℓ reasonably large, the relaxed condition $\lambda(\ell; q = q_c^{\mathrm{opt}}) = 1$ works well. Therefore, halt the algorithm when $\lambda(\ell; q) = 1$.

- **Step 6–Reinsertion**. CI attempts to minimize q_c^{opt}, but does not guarantees that G_∞ for $q < q_c^{\mathrm{opt}}$ is minimized. Thus, CI could give a good approximation to the minimal set $\mathbf{n}^*$, but the GCC in the $G_\infty > 0$ phase may not be a good solution. A greedy reinsertion procedure is performed to obtain the smallest possible GCC for any q. Minimizing the GCC is also useful when it is impossible to reach precisely $G_\infty = 0$ in finite systems to halt the algorithm. Reinsertion starts from set $\mathbf{n}^*$, where the network is fragmented in many clusters. Then, add back to the network a node from $\mathbf{n}^*$ that, once reinserted, joins the smallest number of clusters. Note that we do not require that the reinserted node joins the clusters of the smallest sizes, but only the minimum number of clusters, independently from their sizes. When the node is reinserted, restore the edges with its neighbors in the network (but not the ones with neighbors not yet reinserted, if any). During the reinsertion, an index $c(i)$ is used for each removed node as the number of clusters it would join together once reinserted. Nodes with the smallest $c(i)$ score are iteratively reinserted until the fraction of removed nodes decreases to q, and the giant component is minimized. The procedure is repeated until all the nodes are returned to the network. In practice, add back a finite fraction of nodes at each step; typically, adding 0.2% of nodes at each step is enough, and no improvement is found using a smaller fraction.

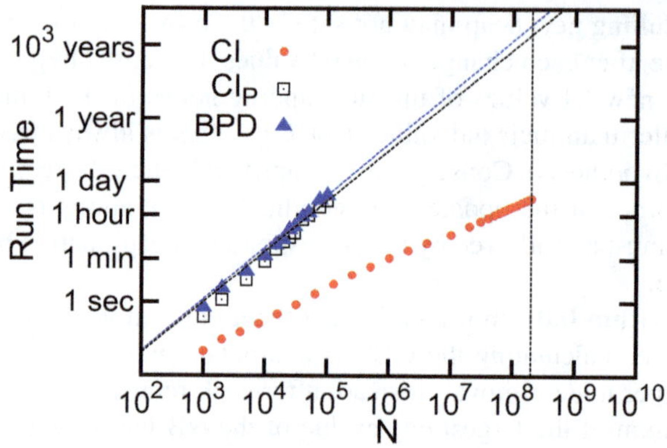

Fig. 1.16 Running time of $CI_{\ell=3}$, CI_P and BPD algorithm [17] **from Sect.** 1.6.4. ER networks $\langle k \rangle = 3$ are used, nodes are removed one by one, and the BP algorithms run the same number of iterations. CI scales linearly with system size while CI_P and BPD (using the original source-code provided in Ref. [17] at http://home.itp.ac.cn/~zhouhj/codes.html) scale quadratically with the network size N. Figure reproduced from [242]. Copyright © 2016, The Authors

The running time of this implementation of the CI algorithm is $O(N \log N)$. Steps 1 and 2 take both $O(N)$ operations, and they are performed only once. Step 3 and Step4 take each at most $O(\log N)$ operations and they are repeated $O(N)$ times. Therefore, the algorithm takes $O(N \log N + N) \sim O(N \log N)$ operations. Figure 1.16 shows this nearly linear scaling by numerical simulations on large networks up to $N = 200$ million nodes.

Given a finite radius ℓ, the CI algorithm is local. To incorporate the influence of a node at the global level, Morone et al. [242] improved the CI algorithm using a message passing approach and proposed the CI propagation algorithm (CI_P). As a variant of the CI algorithm in the limit of $\ell \to \infty$, the CI_P algorithm is able to reach the analytical optimal percolation threshold of a random cubic graphs obtained by Bau et al. [244].

Another belief-propagation variant of the CI algorithm, CI_{BP}, was also proposed in [242]. Combining the dynamics of the SIR model with message passing updating rules, CI_{BP} achieves similar performance with CI_P. However, the improvements of CI_P and CI_{BP} over vanilla CI are made at the expense of increasing computational complexity from $O(N \log N)$ to $O(N^2 \log N)$. Figure 1.16 shows the linear runtime scaling of CI with $\ell = 3$, as compared with two BP-based algorithms, the CI propagation variant CI_P and the BPD algorithm from Sect. 1.6.4 [17]. Both BP-based algorithms show a quadratic scaling, which renders the algorithms too slow to be applied to large-scale networks. For a network of 200 million nodes, while linear CI achieves the optimized set in 5 h, the BP algorithms with quadratic runtime would take $\sim 3,000$ years to accomplish the same task. This indicates the importance of using linear-runtime algorithms in large-scale influencer problems.

Kobayashi and Masuda [245] also developed an immunization algorithm for networks with community structure combining the CI algorithm and coarse-graining procedure in which communities were regarded as super nodes. From a mesoscopic scale, nodes connecting different communities are identified with a runtime that scales as $O((N^2/N_C)\log N)$ (N_C is the number of communities). The problem was also studied on multiplex networks. Osat et al. [246] showed that characteristics of multiplex networks such as edge overlap and interlayer degree-degree correlation profoundly change the properties of influencers. Neglecting the multiplex structure of a network would lead to significant inaccuracies about the prediction of its robustness.

Collective influencer identification was also applied to evolutionary game theory. Following the research of Santos et al. [247] that illustrates the importance of network structure, the significance of hub players in dominating the evolution was demonstrated [248, 249]. Based on the collective influence theory, Szolnoki and Perc [250] proposed a model in which the potency of a player to pass its strategy is quantified by CI. Considering the degree of hierarchical neighbors, greater cooperation is reached than the degree centrality. In addition, it was also found that an optimal hierarchical depth ℓ exists, which depends on the strength of the social dilemma. In contrast to the typical evolutionary game where every player adopts a strategy that leads to the highest payoff, many players may choose the most common strategy rather than the most profitable one. With these conformists, cooperation can be significantly enhanced [251]. Szolnoki and Perc [251] also found that the most favorable outcomes can emerge if the players with low CI choose to conform. On the contrary, if the influencers with high CI conform, evolutionary outcomes are even worse than selecting the conformists randomly.

1.6.2 High-Degree Influencers as Zero-Order CI Influencers

The $\ell = 0$ term in the expansion of the leading eigenvalue Eq. (1.63) corresponds to a non-interactive theory where the collective effects between influencers are absent. We will now take a closer look at this term since it reproduces the high-degree (HD) strategy of hubs targeting, and it agrees with the result of Cohen et al. [22]. This is a seminal result of importance since it demonstrated the fragility of scale-free network to targeted attacks on the hubs. Any influencer algorithm based on the dismantling of the GCC should be compared to the high-degree strategy. This strategy can be, in principle, very effective in scale-free networks.

The $\ell = 0$ problem represents the one-body interaction of the CI theory where the influencers are considered in isolation, only affected by the external field, and therefore, this strategy lacks the collective influence effects found for $\ell \geq 1$. Below, we follow the calculations in [14]. The eigenvalue is:

$$\lambda_0(\mathbf{n}) = \frac{\langle \mathbf{w}_0|\hat{\mathcal{M}}|\mathbf{w}_0\rangle}{\langle \mathbf{w}_0|\mathbf{w}_0\rangle} = \frac{\sum_i k_i(k_i-1)n_i}{\sum_i k_i} . \tag{1.87}$$

The formula for this eigenvalue is analogous to the energy of a gas of non-interactive particles in a non-negative external field, which is site-dependent: $h_i = k_i(k_i - 1)$. Then, to minimize $\lambda_0(\mathbf{n})$, it is sufficient to sort the nodes according to the external field, and the resulting strategy is the removal of the nodes with the highest fields h_i. Since the field h_i increases monotonically with the degree k_i, this strategy corresponds to high-degree strategy.

We can also calculate the critical threshold for this HD strategy, q_c^{HD}, using the stability condition $\min_{\mathbf{n}:\langle n\rangle=1-q_c} \lambda_0(\mathbf{n}) = 1$. We obtain:

$$\min_{\mathbf{n}:\langle n\rangle=1-q_c^{\mathrm{HD}}} \frac{\sum_i k_i(k_i-1)n_i}{\sum_i k_i} = 1 . \tag{1.88}$$

By setting $n_i = 0$ for the first nq_c^{HD} highest degree nodes and $n_i = 1$ for the rest to obtain:

$$\frac{1-q_c^{\mathrm{HD}}}{\langle k\rangle}\left(\langle k^2\rangle' - \langle k\rangle'\right) = 1 , \tag{1.89}$$

where the average $\langle\cdot\rangle'$ is performed over the modified degree distribution of the network with the removed nq_c^{HD} hubs, $P'(k; q_c^{\mathrm{HD}})$, which depends on q_c^{HD}:

$$q_c^{\mathrm{HD}} = \sum_{k=\zeta}^{\infty} P(k) , \tag{1.90}$$

where ζ is the lowest degree of the removed nodes compatible with q_c^{HD}. The distribution $P'(k; q_c^{\mathrm{HD}})$ is the degree distribution of the remaining nodes (i.e., the ones for which $n_i = 1$):

$$P'(k; q_c^{\mathrm{HD}}) = \frac{1}{1-q_c^{\mathrm{HD}}} P(k)\theta(\zeta - k) . \tag{1.91}$$

Equation (1.88) depends on the type of degree distribution of the network. For a scale-free network with degree distribution $P(k) \sim k^{-\gamma}$ and degree exponent γ, minimum degree m and maximum degree $k_{\max}$, the variable ζ as a function of q_c^{HD} is

$$q_c^{\mathrm{HD}} = \int_{\zeta}^{k_{\max}} P(k)\,k\,dk = \frac{\zeta^{1-\gamma} - k_{\max}^{1-\gamma}}{m^{1-\gamma} - k_{\max}^{1-\gamma}} . \tag{1.92}$$

Here, we work in the continuum limit, where the integrals replace the sums.

Equation (1.89) then provides an implicit equation for q_c^{HD}:

$$\frac{\gamma-2}{\gamma-3}\frac{m^{3-\gamma} - \zeta^{3-\gamma}}{m^{2-\gamma} - k_{\max}^{2-\gamma}} - \frac{m^{2-\gamma} - \zeta^{2-\gamma}}{m^{2-\gamma} - k_{\max}^{2-\gamma}} = 1 . \tag{1.93}$$

Equation (1.93) simplifies for an infinitely large network where $k_{\max} \to \infty$. For $\gamma \geq 2$, the result of Ref. [22] is obtained:

$$\frac{\gamma-2}{\gamma-3}\left(m - \frac{\zeta^{3-\gamma}}{m^{2-\gamma}}\right) + \frac{\zeta^{2-\gamma}}{m^{2-\gamma}} = 2\,. \tag{1.94}$$

Considering the ensemble of finite network of N nodes, where $k_{\max} = mN^{1/(\gamma-1)}$, when $n \to \infty$ then $k_{\max} \to \infty$, the relation between ζ and q_c^{HD} is $\zeta = m\,(q_c^{\text{HD}})^{1/(1-\gamma)}$. Then the HD threshold is:

$$(q_c^{\text{HD}})^{(2-\gamma)/(1-\gamma)} = 2 + \frac{\gamma-2}{\gamma-3} m\left((q_c^{\text{HD}})^{(3-\gamma)/(1-\gamma)} - 1\right)\,, \tag{1.95}$$

which is the known result for the HD attack [22]. The solution is shown in Fig. 1.17a for different values of the minimum degree $m = 1, 2, 3$.

The fragility of scale-free networks under intentional HD attack is represented by small values of q_c^{HD}. For $\gamma = 2$, Eq. (1.95) predicts a zero critical q_c^{HD} for any m, and this is interpreted as an extreme fragility of scale-free networks when the degree exponent is close to 2. This fragility arises since, for $\gamma = 2$, the network is essentially a star-graph, which is trivially destroyed by removing the central hub. This, in turn, is a consequence of the fact that the natural cut-off $k_{\max}$ diverges linearly with the system size for $\gamma = 2$, i.e., $k_{\max} \sim N$ (see Fig. 1.17a).

The fragility represented by small q_c^{HD} can be seen for other values of γ in the curve corresponding to $m = 1$, the case calculated in [22], where q_c^{HD} remains smaller than 0.06, see Fig. 1.17a. In this case, the network contains many leaves and reduces to a star at $\gamma = 2$, which is destroyed by removing the only hub, explaining the general fragility in this case. Furthermore, in this case, the network becomes a collection of dimers with $k = 1$ when $\gamma \to \infty$, which is still trivially fragile. This explains why $q_c^{\text{HD}} \to 0$ as $\gamma \to \infty$, as well. Therefore, the fragility in the case $m = 1$ has its roots in these two limiting cases. Removing the leaves ($m = 2$) results in a 2-core, which is already more robust, as seen in Fig. 1.17a. For the 3-core $m = 3$, $q_c^{\text{HD}} \approx 0.4 - 0.5$ provides a quite robust network and has the expected asymptotic limit to a non-zero q_c^{HD} of a random regular graph with $k = 3$ as $\gamma \to \infty$, $q_c^{\text{HD}} \to (k-2)/(k-1) = 0.5$. Thus, SF networks become robust in these more realistic cases, and the search for other attack strategies becomes even more important.

Nonetheless, the situation changes for finite network ensembles, where the cut-off can be very large but finite (which is the case for all networks, both real and synthetic ones). For a finite cut-off $k_{\max}$, and for $\gamma \to 2$ Eq. (1.93) gives:

$$\zeta = m + \log\left(\frac{\zeta k_{\max}}{m^2}\right)\,. \tag{1.96}$$

By expressing ζ as a function of q_c^{HD} via:

$$\zeta = \frac{m k_{\max}}{m + q_c(k_{\max} - m)}\,, \tag{1.97}$$

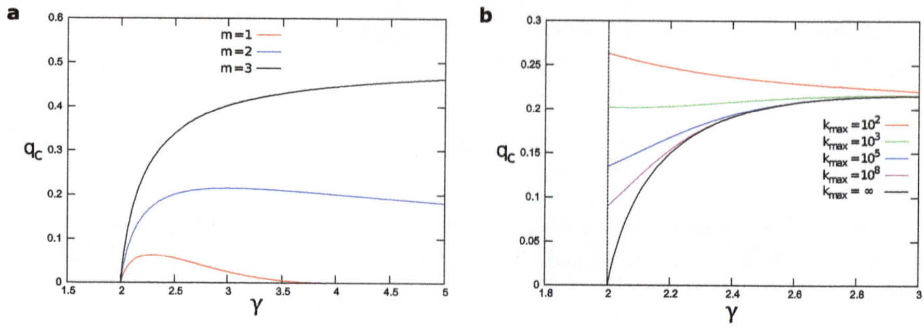

Fig. 1.17 High-degree threshold for influencers. **a** HD threshold q_c^{HD} as a function of the degree distribution exponent γ of scale-free networks in the ensemble with $k_{\text{max}} = mn^{1/(\gamma-1)}$ and $n \to \infty$ for different values of the minimum degree m: 1 (red), 2 (blue), 3 (black). **b** HD threshold q_c^{HD} as a function of the degree distribution exponent of scale-free networks with minimum degree $m = 2$ in the ensemble where k_{max} is fixed and does not scale with system size n. The curves refer to different values of the cut-off k_{max}: 10^2 (red), 10^3 (green), 10^5 (blue), 10^8 (magenta), and $k_{\text{max}} = \infty$ (black). For a typical $k_{\text{max}} = 10^3$ the network is fairly robust with $q_c^{\text{HD}} \approx 0.2$ for all γ. Therefore, searching for other, more optimal strategies to dismantle the network becomes important. Figure reproduced from [14].

then q_c^{HD} as a function of the cut-off k_{max} is given by the explicit equation:

$$\frac{mk_{\text{max}}}{m + q_c^{\text{HD}}(k_{\text{max}} - m)} = m + \log \frac{k_{\text{max}}^2}{m^2 + mq_c^{\text{HD}}(k_{\text{max}} - m)} . \tag{1.98}$$

Figure 1.17b shows the solution to Eq. (1.98) for q_c^{HD} as a function of γ. For $k_{\text{max}} \to \infty$, the asymptotic behavior is:

$$q_c^{\text{HD}} \sim \frac{m}{\log(k_{\text{max}})} \quad \text{for } k_{\text{max}} \to \infty . \tag{1.99}$$

Therefore, q_c^{HD} still vanishes when $k_{\text{max}} \to \infty$ when $\gamma = 2$, expressing the fragility of these networks but as the inverse of the logarithm of the cut-off. This is a very slow convergence, which makes questionable the claim that scale-free networks are extremely fragile under hub removal. This is true when $\gamma = 2$, while the network gets less fragile when the degree exponent gets larger than 2. As can be seen in Fig. 1.17b, even for k_{max} of the order of hundred million, q_c^{HD} is still of order 0.1 for $\gamma = 2$. For more realistic $k_{\text{max}} = 10^3$, which is typical of social networks, $q_c^{\text{HD}} \approx 0.2$ for all γ. In these situations, searching for other attack strategies beyond the hubs becomes important.

1.6.3 Limits of Applicability of Optimal Percolation to Influencer Identification

The solution of optimal percolation developed so far is valid for locally tree-like networks: in the message passing formulation Eq. (1.37), the probabilities $\nu_{k\to i}$ are assumed to be independent. This includes the thermodynamic limit of the class of random networks of Erdös-Rényi, scale-free networks [252] and the configuration model (the maximally random graphs with a given distribution [238]) which are locally tree-like and contains loops that grow as $\log N$ [253]. Nonetheless, it is generally accepted, and confirmed by many works that results obtained for tree-like graphs approximate well networks with cycles provided the density of cycles is not excessively large [8, 13, 190]. Wherever the number of cycles is abundant, the quality of the results obtained for tree-like networks deteriorates. The locally tree-like approximation has been successfully used in a plethora of problems, like spin glass models on random graphs, coloring, matching, bisection and maximum cut of graphs, and other constraint satisfaction problems [8].

We also recall that the mapping of IMP for the LTM problem to optimal percolation is valid only for a threshold $\theta_i = k_i - 1$ and $w_{ij} = 1$. Thus, CI influencers are good spreaders of information only when the dynamics follow this particular type of LTM. When any other kind of model describes the dynamics of spreading, the CI algorithm is not guaranteed to provide a good influencer solution. This is a rather limited applicability of CI. Nevertheless, in the absence of influencer solutions for other models of diffusion of information, CI can still be applied, and it will be applied to several systems in subsequent chapters and compare with other algorithms. Interesting generalizations of the LTM that cannot be approximated by CI is the case of random heterogeneous thresholds θ_i [254], and constant fixed thresholds [255].

A further interesting optimization problem is to find the optimal superspreaders under the SIR or SIS models. In general, this problem cannot be translated into optimal percolation because it does not have a transition from $G = 0$ to $G > 0$. Optimal percolation aims to find the superblockers that could be removed to make G_∞ as small as possible. Instead, superspreaders maximize the average influence if selected as seeds. Radicchi and Castellano [256] studied numerically these two cases and performed an extensive analysis over a range of real-world networks. They found that these two optimization problems are not generally equivalent, i.e., superblockers are not superspreaders. The identification of superblockers is based on the network's topology, while superspreaders are dependent on the parameters of the spreading dynamics. Although the superblockers identified by optimal percolation do not coincide with the best superspreaders in the general case, for a particular case of spreading model, the optimal superspreaders become dual of percolation minimization as discussed below.

On the other hand, Ghavasieh et al. [257] use a method grounded on statistical physics (the underlying framework has been introduced in [258]) to investigate the disintegration of empirical social, biological, and transportation systems, showing that the nodes central to information dynamics are also responsible for keeping the network integrated. The method is far from being efficient in terms of computational complexity, but it is very effective in dismantling, based on an elegant interpretation in terms of information flow.

Before concluding this section, we notice that for the network depicted in Fig. 1.11a left panel, the modified NB matrix does not depend on nodes n_1, n_4, n_6, i.e., it does not depend on the variables outside the cycle. As a consequence, its largest eigenvalue does not depend on those variables as well: $\lambda_{\max} = \lambda_{\max}(n_2, n_3, n_5)$. Therefore, removing the nodes at the end of these dangling ends does not reduce the eigenvalue $\lambda_{\max}$. This is shown in the network of Fig. 1.11a central panel, where the removal of n_4 does not change the leading eigenvalue, which is still $\lambda_{\max}(1, 1, 1) = 1$. On the contrary, by removing a node belonging to the cycle, e.g. nodes 2, 3 or 5, the network becomes a tree, and the eigenvalue drops to zero: $\lambda_{\max}(0, 1, 1) = \lambda_{\max}(1, 0, 1) = \lambda_{\max}(1, 1, 0) = 0$, as shown in Fig. 1.11a, right panel. The largest eigenvalue of the NB matrix in a tree-like network is zero. This is because a non-bactracking walker once arrived on a leaf, has no way to go.

For networks with one cycle (unicyclic graph), the largest eigenvalue is equal to one, and for networks with more than one cycle, $\lambda_{\max}$ is larger than 1. Thus, we see that a strategy to minimize the largest eigenvalue of the modified NB matrix is to destroy the cycles in the network. When the eigenvalue reaches the critical value one ($\lambda_{\max} = 1$) the network consists of a single cycle that is easily destroyed by a single removal, causing the sharp drop of the eigenvalue from one to zero. When the giant component is reduced to a tree-like topology, it can be considered as completely fragmented since any tree can be destroyed with a sub extensive number of node removals, that is, for a finite tree, just one removal is enough.

This analysis suggests that if we want to destroy a network optimally, we need to destroy the cycles. This will come in handy when we solve the optimal percolation problem to find influencers and also in immunization strategies. Moreover, it indicates that a network near the critical point of q_c^{opt} is almost tree-like, and the approximations work better and better as we approach the critical point.

The optimal disintegration of cycles is called the Optimal Decycling Problem, a problem that precedes optimal percolation and will be treated next. This problem has been studied by Mugisha and Zhou [17] and Braunstein et al. [16] who developed algorithms that outperform the CI algorithm by finding smaller influencer sets. The problem of influence maximization with continuous transitions has been studied using many approaches [16, 17, 127, 199, 234] which are reviewed next.

1.6.4 Optimal Decycling Algorithm for Network Dismantling

A way to dismantle a network is to destroy all the cycles in it, leading to the optimal decycling problem or minimum feedback vertex set (FVS) problem [16, 17]. A feedback vertex set is the set of nodes whose removal would break all the cycles in the network [259]. The optimal decycling problem is then to find the FVS with the smallest number of nodes. This defines the decycling number $\phi(G)$ of a graph

G, which is the smallest number of nodes that, when removed from G, results in a graph with no cycles [244]. Finding the decycling number is a problem that goes back at least to the work of Kirchhoff in 1847 on spanning trees [260]. For a graph G, a subset $S \subseteq G$ is said to be a decycling set of G if $G \backslash S$ is acyclic. The minimum cardinality of S is then the decycling number [244, 261].

As discussed above, the rationale behind the connection between optimal percolation and optimal decycling is that, for sparse random networks, short loops rarely exist in small connected components [235–237]. If the long loops in the giant component are cut, the network will break into small tree fragments. As indicated by Braunstein et al. [16], the optimal decycling threshold q_c^{dec} acts as an upper bound of the optimal percolation threshold q_c^{opt}. For random networks with light-tailed degree distribution (finite second moment), the minimal size of the decycling set is equal to the minimal size of the dismantling set in the limit of large system size, $N \to \infty$. The optimal decycling problem is itself an NP-hard problem, but it can be solved approximately via belief propagation algorithms.

Two approaches based on the decycling algorithm were developed in [16, 17]. Both of them apply a three-stage procedure: first, decycle the network with minimal number of nodes, then break the tree into small components and finally, reinsert some nodes into the network without increasing the size of the largest component. Compared with the CI algorithm, these two algorithms take into account the global topology of the network and achieve a better performance at the expense of algorithmic complexity.

The Belief Propagation Guided Decimation (BPD) algorithm proposed by Mugisha and Zhou [17] is based on the spin glass model of the FVS problem developed by Zhou in [262]. In order to transform the global acyclic constraint into local ones, a variable A_i, which takes the value 0, i or $j \in \partial i$, is assigned to each node. If node i is removed from the network G, $A_i = 0$. Otherwise, $A_i = i$ if it is a root of a tree or $A_i = j$ if node i has a parental node j. Given a microscopic configuration $\underline{A} = \{A_1, A_2, \cdots, A_N\}$, the fraction of removed nodes is represented by:

$$q_c^{\text{dec}}(\underline{A}) = \frac{1}{N}\sum_{i=1}^{N} \delta_{A_i}^0, \tag{1.100}$$

where δ_n^l is the Kronecker delta function ($\delta_n^l = 1$ if $n = l$ and 0 otherwise). For each edge (i, j) in the network, an edge factor $C_{ij}(A_i, A_j)$ is defined as [262]:

$$\begin{aligned} C_{ij}(A_i, A_j) = \; & \delta_{A_i}^0 \delta_{A_j}^0 + \delta_{A_i}^0 (1 - \delta_{A_j}^0 - \delta_{A_j}^i) + \delta_{A_j}^0 (1 - \delta_{A_i}^0 - \delta_{A_i}^j) \\ & + \delta_{A_i}^j (1 - \delta_{A_j}^0 - \delta_{A_j}^i) + \delta_{A_j}^i (1 - \delta_{A_i}^0 - \delta_{A_i}^j). \end{aligned} \tag{1.101}$$

The edge (i, j) is regarded as satisfied if $C_{ij}(A_i, A_j) = 1$, and unsatisfied otherwise. A configuration $\underline{A}$ is referred to as a solution of the network G given that all edges in G are satisfied. As is proven mathematically, all remaining nodes in a solution $\underline{A}$ form a legitimate subgraph consisting of several components. Each component in a legitimate subgraph is either a tree or a c-tree, which contains only one cycle. Considering all solutions of the network, a partition function of the system is defined as:

$$Z(\mu) = \sum_{\underline{A}} e^{\mu n(1-q^{dec}(\underline{A}))} \prod_{(i,j)\in G} C_{ij}(A_i, A_j), \tag{1.102}$$

where μ is the inverse of temperature. At the limit of zero temperature, the partition function is contributed exclusively by the optimal configuration $\underline{A}^*$ with the minimal fraction q_c^{dec}.

The decycling threshold is determined by the free entropy of the spin glass system:

$$q_c^{\text{dec}} = \lim_{\mu\to-\infty} \frac{1}{N\mu} \ln Z(\mu). \tag{1.103}$$

Based on the locally tree-like assumption, the marginal probability $q_i^0(t)$ for node i to be removed from the remaining network $G(t)$ can be calculated through iterations of the following Belief Propagation (BP) equations [17]:

$$q_i^0 = \left\{1 + e^{\mu}\left[1 + \sum_{k\in\partial i(t)} \frac{1 - q_{k\to i}^0}{q_{k\to i}^0 + q_{k\to i}^k}\right] \prod_{j\in\partial i(t)} \left[q_{j\to i}^0 + q_{j\to i}^j\right]\right\}^{-1}, \tag{1.104}$$

$$q_{i\to j}^0 = \frac{1}{z_{i\to j}(t)}, \tag{1.105}$$

$$q_{i\to j}^i = \frac{e^{\mu} \prod_{k\in\partial i(t)\backslash j} \left[q_{k\to i}^0 + q_{k\to i}^k\right]}{z_{i\to j}(t)}, \tag{1.106}$$

where $q_{j\to i}^0(t)$ is the probability for node j to be removed if node i is absent from $G(t)$, $\partial i(t)$ is set of neighbors of node i in $G(t)$, and $z_{i\to j}(t)$ is the normalization constant [17]

$$z_{i\to j}(t) = 1 + e^{\mu} \prod_{k\in\partial i(t)\backslash j} \left[q_{k\to i}^0 + q_{k\to i}^k\right] \times \left[1 + \sum_{\ell\in\partial i(t)\backslash j} \frac{1 - q_{\ell\to i}^0}{q_{\ell\to i}^0 + q_{\ell\to i}^\ell}\right]. \tag{1.107}$$

The BP equations are iterated each time step t for several steps and the removal probability q_i^0 is calculated for each node. The node with the highest probability q_i^0 is removed from the network even if the BP equations do not converge to a fixed point. The process stops when the network becomes acyclic. If the largest connected component remains extensive, this network can be dismantled by keeping removing nodes.

As with the CI algorithm, the BPD algorithm works well with locally-tree-like networks. For a large number of networks with abundant communities and loops,

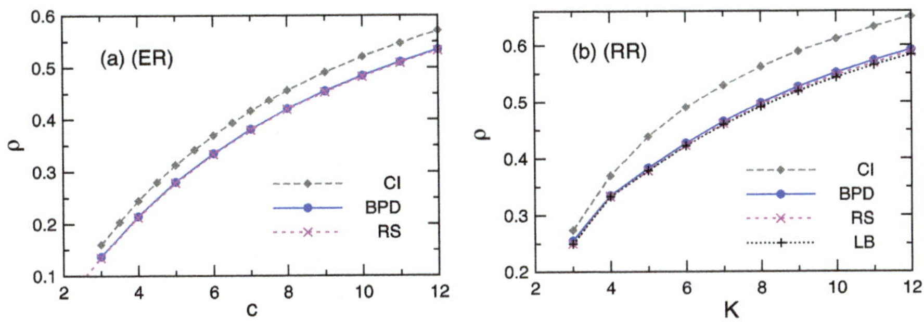

Fig. 1.18 Performance of the BPD algorithm of Mugisha and Zhou [17] **in ER and RR (random regular) networks.** **a** Fraction of removed nodes ρ to break ER random networks of mean degree c. **b** Fraction ρ versus degree K for regular random networks. Diamonds show the results obtained from the CI algorithm, crosses represent the results of the replica-symmetric (RS) mean-field theory, and the BPD results are shown in blue circles. The plus symbols in **(b)** show the mathematical lower bound (LB) on the minimum size of target nodes. The BPD algorithms produce more optimal thresholds as compared with CI, and almost the same as the theoretical results in both networks. Figure reproduced from [17]. Copyright © 2016, American Physical Society

the optimal set obtained by the FVS set is usually more than what is necessary to dismantle the network. In this case, a reinsertion process is performed without significantly increasing the size of the connected component. This process is done by a greedy algorithm, in which the nodes that cause the least increase in the largest connected component are reinserted one by one until the size of the largest component reaches a predefined threshold.

Simulations on random network ensembles and real-world networks done in [17] indicate that the BPD algorithm has better performance in finding the minimal set of influencers than the CI algorithm in the optimal percolation problem, see Fig. 1.18. For large random Erdös-Rényi (ER) networks and scale-free networks, the BPD algorithm manages to fragment the network by removing a smaller set of nodes compared with CI. In particular, the percolation threshold is close to the minimal value predicted by the replica-symmetry (RS) mean field theory [244, 262]. In the CI algorithm, the size of G_∞ decreases almost linearly with the increase of q. In contrast, G_∞ features an abrupt collapse under the BPD algorithm. This results from the intrinsic nature of the FVS problem and the efficiency of tree dismantling. With the existence of such a collapse, the BPD process can work as an efficient attack strategy, leaving no warning to the system before its total failure.

Im and Kahng [263] studied the comparative performance of CI and BPD on the dismantling efficiency in fractal networks as compared to non-fractal networks [41, 42]. Fractal networks [41, 42] are hierarchical structures where cycles have been categorized quantitatively by scaling theory [264], and the number of modules and loop-forming links can be characterized by renormalization group theory [265].

Under the sparsity assumption, the minimal set obtained for optimal percolation and dismantling coincides with the minimal set for decycling [16, 17]. While optimal percolation and dismantling are related to the existence of cycles in the network, all these results are obtained for networks that contain no local cycles, implying a kind of oxymoron. Thus, the importance of cycles in the performance of these algorithms needs to be systematically evaluated.

Im and Kahng [263] then characterize the performance of the algorithms in loopy networks with fractal scaling. They found that the BPD algorithm outperforms the CI algorithm in both types of networks, whether the network is fractal or not. The performance of the CI algorithm was comparable to that of BPD on non-fractal networks, which have high ratios of long-range shortcuts to short-range connections. However, the performance gap between the BPD and CI algorithms were larger on fractal networks.

Braunstein et al. [16] considered the optimal decycling problem from a different point of view than Mugisha and Zhou [17], and call it Network Dismantling Problem. They first noticed that a network is acyclic if and only if the network has no 2-core (or any higher core, since the k-cores are nested). Then, they mapped the decycling process to a 2-core percolation problem. Assuming a network $G(V, E)$ with $N = |V|$ nodes, a set of nodes $S \subset N$, called the decycling set, are initially removed from the network. Then the 2-core percolation can be described by the evolution of time-dependent binary variables $x_i^t(S)$ defined for every node in the network. The starting point at $t = 0$ is defined by $x_i^0(S) = 1$ for removed nodes in the decycling set, $i \in S$, and $x_i^0(S) = 0$ for $i \notin S$. Then, the evolution of $x_i^t(S)$ follows:

$$x_i^{(t+1)} = \begin{cases} 1 & \text{if } x_i^t(S) = 1 \\ \mathbb{I}\left[\sum_{j\in\partial i}\left(1 - x_j^t(S)\right) \le 1\right] & \text{if } x_i^t(S) = 0 \end{cases} \tag{1.108}$$

where the indicator function $\mathbb{I}$ is 1 if the argument is true and 0 otherwise. As x_i^t can only change from 0 to 1, the equations admit a fixed solution $x_i^*(S)$ as $t \to \infty$ given by $x_i^*(S) = 0$ iff i belongs to the 2-core of $G\backslash S$. If $x_i^*(S) = 1$ for all nodes, $G\backslash S$ contains no loops when S is the decycling set.

Braunstein et al. [16] then finds the minimal of this set by introducing the probability distribution over decycling sets S using the Boltzmann distribution from statistical physics:

$$\hat{\eta}(S) = \frac{1}{Z(\mu)} e^{\mu|S|} \prod_{i\in N} \mathbb{I}\left[x_i^*(S) = 1\right], \tag{1.109}$$

where $|S|$ is the number of nodes in S, μ is the inverse temperature, and $Z(\mu)$ is the partition function that normalizes the distribution. The minimal size of the decycling sets is obtained in the zero-temperature limit as:

$$q_c^{\text{dec}} = \lim_{\mu\to-\infty} \frac{1}{N\mu} \ln Z(\mu). \tag{1.110}$$

The partition function is difficult to compute since x_i^* depends on S in a global way. To solve this problem, Ref. [16] transformed the global constraint $\prod_{i\in V} \mathbb{I}\left[x_i^*(S) = 1\right]$ into local ones. The 2-core removal process is described by an integer $t_i(S) = \min\{t : x_i^t(S) = 1\}$ measuring the time when node i is removed from the network. For $i \in S$, we have $t_i(S) = 0$. For $i \notin S$, $t_i(S)$ depends on its neighbors as:

$$t_i(S) = \phi_i(\{t_j\}_{j\in\partial i}) = 1 + \max\nolimits_2(\{t_j(S)\}_{j\in\partial i}), \tag{1.111}$$

where the function $\max_2$ returns the second largest value on its argument. The partition function is then:

$$Z(\mu) = \sum_{\{t_i\}} e^{\mu \sum_i \psi_i(t_i)} \prod_{i\in N} \mathbb{I}[t_i < \infty]\mathbb{I}\left[t_i = \phi_i(\{t_j\}_{j\in\partial i})\right], \tag{1.112}$$

where $\psi_i(t_i) = \mathbb{I}[t_i = 0]$.

The computation of Eq. (1.112) is NP-hard. A simplification of this partition function can be performed by restricting $t_i \leq T$. All values t_i larger than T are regarded as infinity, and trees with diameters larger than $T + 1$ are considered to be part of a long cycle. For large enough T, the effect of this simplification is negligible. For locally tree-like graphs, the partition function can be computed by the cavity method [8, 266] using messages exchanged between neighboring nodes. For each $i \to j$ edge, a message $\eta_{ij}(t_i, t_j)$ as a function of activation times t_i and t_j is introduced. The messages satisfy the self-consistent Belief Propagation (BP) equations [16] for a given inverse temperature μ:

$$\eta_{ij}(t_i, t_j) \propto \sum_{\{t_k\}_{k\in\partial i\setminus j}} e^{\mu\psi_i(t_i)}\mathbb{I}\left[t_i = \phi_i(\{t_k\}_{k\in\partial i})\right] \prod_{k\in\partial i\setminus j} \eta_{ki}(t_k, t_i). \tag{1.113}$$

As the temperature approaches zero ($\mu \to \infty$), probabilities of the messages $\eta_{ij}(t_i, t_j)$ in the BP equations concentrate on the solution to Eq. (1.108) that minimizes the cost function $\sum_i \psi_i(t_i)$.

To develop an algorithm that finds the optimal decycling set, a slightly different cost function is used: $\psi_i(t_i) = \mathbb{I}[t_i = 0] + \varepsilon_i(t_i)$, where $\varepsilon_i(t_i)$ is a randomly chosen small cost. Further, the 2-core percolation process is relaxed to allow $t_i \geq 1 + \max_2(\{t_j\}_{j\in\partial i})$ in Eq. (1.108).

Defining $h_i(t_i)$ as the minimal cost to dismantle the 2-core under the condition that node i is removed at t_i, the optimal decycling set is determined by:

$$S^* = \{i \in V \mid t_i^* = 0\}, \tag{1.114}$$

where $t_i^* = \arg\min h_i(t_i)$. In calculations, $h_i(t_i)$ can be computed by the following Min-Sum algorithm, which is derived from the zero temperature limit of the BP equations. Technical details of the derivation can be found in Ref. [16]. Concretely, messages $h_i(t_i)$ are solved by iterating the following equations [16]:

$$h_i(t_i) = \psi_i(t_i) + \sum_{k\in\partial i} L_{ki}(t_i) + M_i(t_i), \tag{1.115}$$

$$h_i(0) = \psi_i(0) + \sum_{k \in \partial i} L_{ki}(T+1), \tag{1.116}$$

$$M_i(t_i) = \min\left\{0, \min_{k \in \partial i}\{R_{ki}(t_i) - L_{ki}(t_i)\}\right\}. \tag{1.117}$$

The variables L_{ij}, R_{ij}, M_{ij}, h^0_{ij} and h^1_{ij} satisfy the following equations defined on edges of the network, which are updated in each iteration:

$$L_{ki}(t_i) = \min_{t_k < t_i} h^0_{ki}(t_k), \tag{1.118}$$

$$R_{ki}(t_i) = \min\left\{h^0_{ki}(t_i), \min_{t_k > t_i} h^1_{ki}(t_k)\right\}, \tag{1.119}$$

$$M_{ij}(t_i) = \min\left\{0, \min_{k \in \partial i \backslash j}\{R_{ki}(t_i) - L_{ki}(t_i)\}\right\}, \tag{1.120}$$

$$h^0_{ij}(t_i) = \psi_i(t_i) + \sum_{k \in \partial i \backslash j} L_{ki}(t_i), \tag{1.121}$$

$$h^1_{ij}(t_i) = \psi_i(t_i) + \sum_{k \in \partial i \backslash j} L_{ki}(t_i) + M_{ij}(t_i), \tag{1.122}$$

$$h^0_{ij}(0) = \psi_i(0) + \sum_{k \in \partial i \backslash j} R_{ki}(0), \tag{1.123}$$

$$h^1_{ij}(0) = +\infty. \tag{1.124}$$

A physics-based interpretation of the above system is given in Ref. [16]. In most cases, convergence can be reached within a small number of iterations, with a computational complexity $O(MT)$ in each iteration. In case the Min-Sum equations do not converge, a reinforcement procedure is applied to damp the system as explained in [267].

While the remaining network $G \backslash S^*$ is acyclic, it may still contain some extensive tree components. These large trees are further dismantled efficiently via a greedy tree-breaking procedure with computational complexity of $O(N(\log N + T))$. In addition, for networks containing many short loops, a reverse greedy (RG) reinsertion procedure is applied similarly to the reinsertion performed in [14, 17, 242] to recover the nodes that do not increase the size of the giant component, as performed in the CI and BPD algorithms. Starting from a dismantling set, the reinsertion procedure iteratively puts back into the graph the removed nodes that result in the smallest connected component. The computational cost of this RG procedure is typically sublinear in N bounded by $k_{max} C' \log(k_{\max} C')$, where $k_{\max}$ is the maximal degree and C' is the maximum allowed component size by the RG procedure.

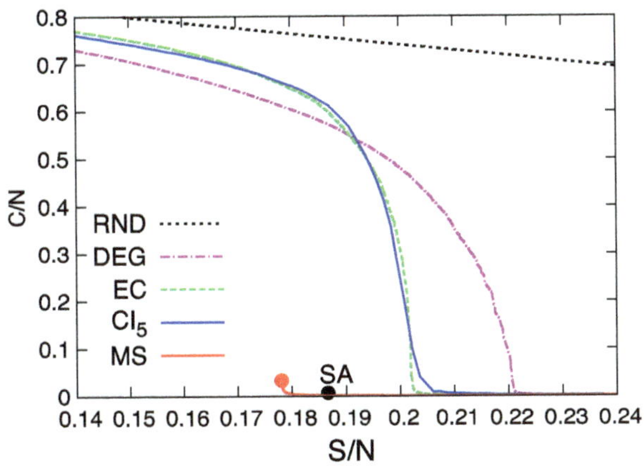

Fig. 1.19 Performance of the Min-Sum algorithm of Braunstein et al. [16] **for network dismantling via optimal decycling.** The behavior of the relative size of the largest connected component C/N as a function of the fraction of nodes S/N removed from an ER network (size $N = 78, 125$ and average degree 3.5). The performance of the Min-Sum algorithm (MS) is compared with random (RND), adaptive largest degree (DEG), adaptive eigenvector centrality (EC), adaptive CI, and simulated annealing (SA) strategies for finding the minimal decycling set. MS is shown to display the smaller q_c value of the displayed methods. Figure reproduced from [16]. Copyright © 2016, National Academy of Sciences U.S.A

Figure 1.19 shows the performance of the Min-Sum algorithms to find the minimal decycling set as compared to other approaches on an ER random network. These simulations done on both synthetic and real-world social networks demonstrate the effectiveness of this decycling algorithm. For an ER random graph of size $N = 78, 125$ and average degree 3.5, the size of G_∞ deceases to 0.032 when 17.8% of nodes are removed (red symbol in Fig. 1.19). Compared with metrics of degree centrality, eigenvector centrality, and the CI algorithm with $\ell = 5$, it was found that the three-stage algorithm is superior in dismantling the giant component. The Monte Carlo-based Simulated Annealing (SA) algorithm gives a competitive result as seen in Fig. 1.19. However, its computational complexity is much higher.

Figure 1.20 shows the comparison among the algorithms using the same Twitter network analyzed in Ref. [14] and shown in Fig. 1.14a. The Min-Sum algorithm with reverse greedy (RG) reinsertion procedure performs equally well as the simulated annealing (SA) by removing $\sim 3.4\%$ of nodes to break the giant component (smaller than 1,000 nodes). In comparison, CI needs to remove $\sim 5.2\%$ of nodes to achieve the same fragmentation performance. However, a modification of CI to reinsert the nodes using RG achieves similar performance as Min-Sum and SA with threshold $\sim 3.5\%$. It should be noted that these comparisons are made based on the calculation of the critical threshold of dismantling and not on the actual dismantling set.

A simple and faster heuristic algorithm with complexity $O(N)$, CoreHD, was developed by Zdeborová et al. [127], inspired by this decycling-based algorithm. Starting from the 2-core of a network G, CoreHD recursively removes nodes with

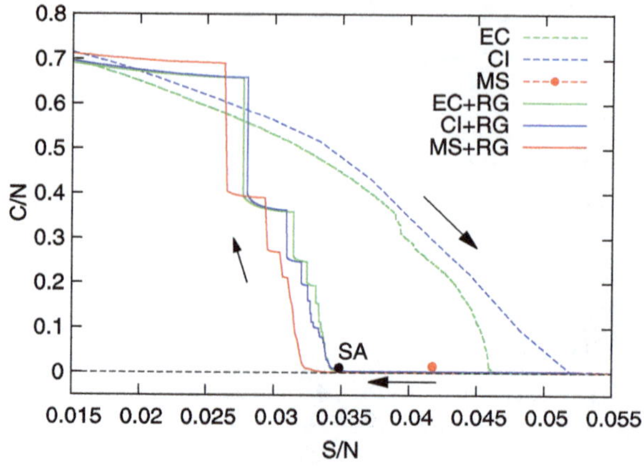

Fig. 1.20 Performance of the Min-Sum algorithm of Braunstein et al. [16] **for a Twitter network**. The same network of Fig. 1.14 from [14] is used. Plot of the fraction of nodes in the largest component after removing a fraction of nodes using different strategies: Min-Sum (MS), CI, eigenvector centrality (EC) with and without RG reinsertion. Figure reproduced from [16]. Copyright © 2016, National Academy of Sciences U.S.A

the highest degree in the 2-core until G is fully dismantled. Despite its simplicity, CoreHD has been reported to perform better than the CI algorithm. Especially for large random networks, the performance of CoreHD is close to the theoretical solution predicted by replica-symmetry and 1RSB approximation [268]. In addition, this simple algorithm is amenable to rigorous analysis, performing well even on loopy networks that are not accessible for typical message passing algorithms.

In a later work, Schmidt et al. [269] analyzed the CoreHD algorithm rigorously by mapping the node removal in CoreHD to a random process on the degree distribution of the network. They also generalized CoreHD from dismantling the 2-core to any k-core (see Sect. 1.7.5). The mapped dynamics is described by a set of coupled nonlinear ordinary differential equations to characterize the behavior of the CoreHD algorithm on random graphs. An upper bound on the size of the minimal contagious sets in random graphs were obtained, which improves the best known results [244, 268]. The CoreHD analysis also inspired an improved heuristic algorithm, called Weak-Neighbor [269], that works for dismantling the k-core and will be discussed in Sect. 1.7.5.

1.6.5 Optimal Immunization for SIR and SIS Spreading

The problem of Optimal Immunization attempts to find the optimal immunization strategy to distribute vaccines (or treatment) in a population such that an outbreak of disease can be stopped utilizing minimal costs and resources. We have seen in Sect. 1.5.2 that, in principle, this problem can be mapped to optimal percolation.

However, this mapping is only valid for a specific value of transmission probability $w = 1$ and cannot be generalized to more realistic cases. The intuition of this mapping is that by attacking the giant connected component of the transmission chain of the disease; one would be able to stop its spreading. The optimization is then directly related to finding the minimal number of people to immunize so that they are 'removed' from the network in a percolation-like process.

In principle, to optimize an immunization strategy, any of the algorithms discussed in previous sections, like the CI algorithm, BPD , or optimal decycling can be used. Furthermore, any of the topological heuristics discussed in Sect. 1.2 could be used as well to identify the minimal set of immunizators. The most straightforward targeted immunization strategy consists of choosing the immunized target by their topological features. This includes the degree, the kcore index, PageRank, eigenvector and betweenness centralities, etc, etc. More complex strategies to attack the GCC have also been attempted by using partitioning algorithms [270, 271] and on the optimization of the susceptible size [272], and will be discussed in Sect. 1.6.5.3. In fact, we will use these strategies to break the GCC of Covid-19 spreading in Chap. 4 and compare them to find the more optimal one.

Targeted immunization schemes via optimal attacks on the GCC have been studied in the past in early work by Pastor-Satorras et al. [23–25] Holme et al. [273, 274] and Cohen et al. [275]. In a seminal result that brought the attention to the relevance of broad degree distributions to spreading, Cohen et al. [21] and Pastor-Satorras and Vespignani [23] showed that the epidemic spreading in scale-free networks is highly facilitated by the hubs. Mathematically, this implies that the epidemic threshold in scale-free networks is effectively zero, meaning that these kinds of networks spread the disease to any spreading rate of the epidemic agents, even if the transmission probability of the disease is very low. Such epidemic process follows critical phenomena analogous to percolation [21].

Despite the vast use of topological heuristics and the different percolation/dismantling/decycling algorithms to find optimal immunizators, these heuristic approaches neglect the main features of the spreading dynamics of a particular disease, whether it is a SIS, SIR or any other variant models for disease spreading. The main way used to assess the effectiveness of these techniques is to monitor the size of the GCC. However, the size of the GCC may not truly reflect the extent of the epidemic. These techniques also neglect other cost functions associated with a vaccination campaign, like the true cost of each vaccine, which can widely depend on the target and the type of disease.

To overcome these limitations and to take into account a larger variety of cost functions and dynamical processes that are crucial for disease spreading, several authors have formulated the Optimal Immunization problem as a proper optimization problem [276–278]. This problem was proved to be NP-hard in the same class of the IMP [5].

In particular, Altarelli et al. [13] defined optimal immunization in the framework of a constrained optimization problem where the constraints are fixed-point equations for node/edge variables describing the stationary state of the dynamics. These constraints are incorporated into the optimization of an objective energy describing the particular optimization problem at hand. Then, a message passing approach is used to optimize the problem and design efficient algorithms. Next, we first briefly introduce the typical spreading models used in the immunization problem, and then we describe the model of Ref. [13] and compare its performance with approaches based on optimal percolation and other partitioning algorithms.

1.6.5.1 Epidemiological Models

Mathematical models of infectious diseases typically model the disease transmission by assigning different states to the individuals participating in the spreading [43, 279, 280]. The types of states considered define the spreading model. For instance, the state of each individual can be susceptible (S) to acquire the disease, infectious (I), meaning that the individual can transmit the diseases, and recovered (R), meaning that the individual has recovered after being infected. We can consider other states as well, like exposed (E), for being exposed to the diseased, but yet not infectious (I), and so forth.

In mean field formulations of disease spreading, the transmission network over which the disease is spreading is ignored, and these labels are assigned to compartments where we only keep track of the number of individuals in each compartment. Here, we are mostly interested in models defined on a transmission network so that we will assign these states to every node in the network. There are many more states that one can consider, but we will stop here for simplicity. Using these states, different modes of disease transmission give rise to different models, for instance, the susceptible-infected (SI) model, the susceptible-infected-susceptible (SIS) model , and the susceptible-infected-removed (SIR) model. The simplest is the SI model of spreading, where an individual becomes infected and infectious and remains in this state forever.

These models are widely used in the simulation [23–25, 205, 275, 281], detection [282–286] and forecast [205, 287–290] of infectious disease spreading and also information diffusion.

More specifically, we can look at a well-studied example: the SIR model as a classic model for virus spreading [205, 291]. The model describes the evolution of a spreading disease that confers immunity on its survivors, such as influenza. In a contact network, each node is an individual that can be in three possible states: susceptible (S), infected/infectious (I) and recovered/removed (R):

- Susceptible (S): this node has not been infected up to time t and can be infected at later times,
- Infectious (I): this node has been infected at time t and remains infected until it could transition to a Recovered state (R). During the infected period, the node can infect any other node in its network of transmission,
- Recovered (R): this node was infected in the past and has entered the recovered phase and cannot be infected again. That is, we model a disease that confers

immunity after recovering from the disease. This node is then removed from the spreading process.

Notice that in this model, the node becomes infected and infectious simultaneously. A more realistic model considers that a node first becomes (E) exposed to the disease and, after the latent period, becomes (I) infectious. This corresponds to the SEIR model. Figure 4.1 shows typical epidemiological profiles for SARS-CoV-1 and SARS-CoV-2, which can be modeled as SEIR processes, will be treated in more detail in Chap. 4.

At every time step, infected individuals transmit the disease to their immediate S neighbors with probability β. In the cases where transmission is successful, the node becomes Infected and Infectious simultaneously (in SIR) or later (in SEIR). Every Infectious individuals become Recovered in T_R time steps and stop transmitting the disease. The Recovered individuals can not be infected again. The fact that the individual is infectious does not mean that she has disease symptoms. As in the case of Covid-19, individuals can be infectious yet asymptomatic. This case will be treated in Chap. 4.

For a typical initial condition, all the nodes in the network are in the Susceptible state except for one node in the Infected state. After a certain number of steps, when there are no longer any Infected nodes, the disease disappears from the network. Typically, one analyzes the SIR spreading processes for different values of β and T_R.

The SIS model aims to describe spreading processes that do not confer immunity on recovered individuals. Within the framework of the SIS model, recovered individuals return to the susceptible state and can be reinfected at subsequent time steps.

We now turn to describe the solution of the immunization problem in the SIS and SIR developed by Altarelli et al. [13] using the message passing formulation. That is, how to find the minimal set of people to vaccinate or isolate/quarantine from the network when the disease spreads according to the SIR and SIS models.

1.6.5.2 Message Passing Approach to Optimal Immunization

From an optimization point of view, the message passing approach can be used in optimal immunization. Several works have shown that algorithms for combinatorial optimization based on the message passing principle outperforms, in many cases, both greedy algorithms and simulated annealing, even in a complex optimization problem involving stochastic parameters [9, 10] and dynamical rules [11, 12] message passing algorithms usually scale almost linearly with the number of edges, which makes it applicable to large real-world networks. Based on this message passing approach, Altarelli et al. [13] developed the belief propagation (BP) and max-sum (MS) algorithms for the problem of optimal immunization for the particular cases of the SIR and SIS models of spreading.

The formulation of the problem is based on the long time limit of the SIR dynamics, which is described by a set of variables $\{\nu_i\}$, $i = 1, \ldots, N$, giving the probability for each node to be infected in the stationary state after the epidemic has settled. Below we follow the formulation of the problem from Morone et al. [14]; in Altarelli et al. [13] the variable ν_i is called m_i, but we use ν_i to make contact with the notation of message passing in optimal percolation from previous sections.

Assuming that a set of nodes is initially infected with probability p (i.e., p is the probability for each node i to be initially infected), w is the transmission probability defined as the probability that a given neighbor k of node i transmits the disease to i, then the probability ν_i that node i is eventually infected in the long-time limit of SIR is:

$$\nu_i = p + (1-p)\left[1 - \prod_{k\in\partial i}(1 - w\,\nu_{k\to i})\right] . \tag{1.125}$$

The message passing equations are built using the probability $\nu_{k\to i}$ that node i is infected in the absence of its neighboring node j. Each $\nu_{i\to j}$ is associated with a directed edge $i\to j$ of the graph. Assuming a locally-tree like network, and exploiting the factorization of probabilities on the sub-branches of the tree emerging from i, $\nu_{i\to j}$ satisfies the following equation:

$$\nu_{i\to j} = p + (1-p)\left[1 - \prod_{k\in\partial i\setminus j}(1 - w\,\nu_{k\to i})\right] , \tag{1.126}$$

where $\partial i\setminus j$ denotes the neighbors of i excluding j. Although Eqs. (1.125) and (1.126) are exact only on a tree, they have also been successfully applied to study the SIR model on general random graphs [211, 292, 293].

The immunization/isolation of people in the contact network of transmission, $G(V, E)$ with $N = |V|$, is measured by a binary vector $\boldsymbol{\sigma} = (\sigma_1, \sigma_2, ..., \sigma_N)$, representing $\sigma_i = +1$ if node i is immunized, and $\sigma_i = -1$ if not. Equations (1.125) and (1.126) then become:

$$\begin{aligned}
\nu_i &= \frac{1-\sigma_i}{2}\left\{p + (1-p)\left[1 - \prod_{k\in\partial i}(1 - w\nu_{k\to i})\right]\right\} \equiv F_i(\sigma_i, \{\nu_{k\to i}\}_{k\in\partial i}) , \\
\nu_{i\to j} &= \frac{1-\sigma_i}{2}\left\{p + (1-p)\left[1 - \prod_{k\in\partial i\setminus j}(1 - w\nu_{k\to i})\right]\right\} \equiv F_{i\to j}(\sigma_i, \{\nu_{k\to i}\}_{k\in\partial i\setminus j}) .
\end{aligned} \tag{1.127}$$

The optimization problem is set in terms of the minimization of the energy cost function $E(\sigma, \nu)$:

$$E(\sigma, \nu) = \sum_{i=1}^{N} \nu_i + \mu \sum_{i=1}^{N} \frac{1+\sigma_i}{2} \equiv \sum_{i=1}^{N} e(\sigma_i, \nu_i) . \tag{1.128}$$

where μ is the chemical potential controlling the fraction of immunized nodes, as the chemical potential controls the number of particles in a real physical system. The constraint on all feasible configurations are manifested by the local update equations of ν_i.

The single site marginal $P_i(\sigma_i)$ gives the probability that node i is immunized, and it is estimated using the cavity method. Approximating the network with a tree rooted on node i, the authors of Ref. [13] derive the following equation to assess the probability distribution $P_i(\sigma_i)$:

$$P_i(\sigma_i) \simeq \int \left(\prod_{k\in\partial i} \mathrm{d}\nu_{k\to i}\mathrm{d}\nu_{i\to k} Q_{k\to i}(\nu_{k\to i}, \nu_{i\to k}) \right) \mathrm{e}^{-\beta e(\sigma_i,\nu_i)} \prod_{k\in\partial i} \delta\left[\nu_{i\to k} - F_{i\to k}\right], \tag{1.129}$$

where β is the inverse temperature, and the functions $Q_{k\to i}(\nu_{k\to i}, \nu_{i\to k})$ satisfy the following Belief-Propagation (BP) equations:

$$Q_{i\to j}(\nu_{i\to j}, \nu_{j\to i}) \simeq \sum_{\sigma_i} \int \left(\prod_{k\in\partial i\setminus j} \mathrm{d}\nu_{k\to i}\mathrm{d}\nu_{i\to k} Q_{k\to i}(\nu_{k\to i}, \nu_{i\to k}) \right) \mathrm{e}^{-\beta e(\sigma_i,\nu_i)} \prod_{k\in\partial i} \delta\left[\nu_{i\to k} - F_{i\to k}\right]. \tag{1.130}$$

These BP equations need to be iterated numerically to obtain the optimal immunization set. This is done by discretizing the function $Q_{i\to j}(\nu_{i\to j}, \nu_{j\to i})$ in a number $\mathcal{N}_{\mathrm{bin}}$ of bins, with a computational cost to update each message $Q_{i\to j}$ of order $O(\mathcal{N}^{k_i-1})$, where k_i is the degree of node i. This makes the algorithm practically unfeasible on networks having nodes with large degrees (think, e.g., to scale-free graphs). To overcome this problem, the authors of Ref. [13] use a convolution trick, which reduces the computational cost to $O((k_i - 1)\mathcal{N}_{\mathrm{bin}}^3)$.

From $Q_{i\to j}(\nu_{i\to j}, \nu_{j\to i})$, the probability distribution $\mathcal{Q}_i(\nu_i)$ that node i has been infected during the epidemics and the single spin marginal $P_i(\sigma_i)$ can be estimated, from where the average fraction of infected nodes $f(q)$ can be computed as:

$$f(q) = \frac{1}{N}\sum_{i=1}^{N} \int \mathrm{d}\nu_i\ \nu_i\ \mathcal{Q}_i(\nu_i), \tag{1.131}$$

where the average fraction of immunized nodes q is computed as:

$$q = \frac{1}{N}\sum_{i=1}^{N} \frac{1+\langle\sigma_i\rangle}{2} = \frac{1}{N}\sum_{i=1}^{N} \frac{1}{2}\left(1 + \sum_{\sigma_i} \sigma_i\ P_i(\sigma_i)\right). \tag{1.132}$$

Next, we compare the optimal immunization solution from the BP equations with the optimal percolation solution obtained by the CI algorithm. As we argued above, both problems measure different kinds of influencers, and the methods are not supposed to agree with each other. We emphasize that the proper method to deal with the optimal immunization problem is via the BP formulation of Altarelli et al. [13]. The CI algorithm gives an approximation of the immunizator set that efficiently breaks the GCC, and this set may or may not be the best to immunize the population. Nevertheless, we have two sets obtained by two independent methods, and it is interesting to test them in the same setting, i.e., how they perform to stop the spread of a disease.

Before we perform this comparison, we note another difference between the optimal immunization and percolation problems. The parameter p in the BP formalism (denoted as q in [13]) refers to the fraction of initially infected individuals. In the mapping of optimal percolation to immunization, the fraction p is assumed to be zero because, in epidemic outbreaks, the number of initiators of the epidemic is very small and typically of order $O(1)$ [294]. For instance, Sierra Leone's explosion of Ebola cases in 2014 appeared to stem from one traditional healer's funeral at which a single source infected 14 women. The SARS outbreak in 2003 started when one doctor from China infected nine other guests at a Hong Kong hotel, who then spread the virus throughout the city, eventually reaching Vietnam and Canada [295]. Another example is the patient zero-hypothesis of the AIDS epidemics [296].

On the other hand, the model of Altarelli et al. [13] is valid for $p > 0$, and, in particular, the results of Ref. [13] are illustrated for $p = 0.1$. This implies an epidemic starting with 10% of the population infected independently at the same time, which is improbable. Then, we compare both methods in two different regimes: The first case in the regime where CI can be mapped to optimal immunization (i.e., when $p \to 0$ and $w \to 1$, see Sect. 1.5.2). The second comparison is the case where $p > 0$ (we use $p = 0.1$) where the BP method has been applied in [13].

We first compare BP for $p \to 0$ and $w \to 1$. Solving numerically the BP equations requires to discretize the functions $Q_{i\to j}(\nu_{i\to j}, \nu_{j\to i})$ in a number $\mathcal{N}_{\text{bin}}$ of bins of the order $\mathcal{N}_{\text{bin}} \sim 1/p$, in order to have good numerical accuracy because the smallest possible non-zero value assumed by ν_i is $\nu_i = p$ [13]. The BP running time is of order $O(M\mathcal{N}_{bin}^3)$, M the number of edges in the graph. Since $\mathcal{N}_{\text{bin}} \sim 1/p$, the BP running time diverges as p^{-3} for $p \to 0$, and we cannot use BP directly for $p = 0$. Then, we set $p = 0.01$, as small as possible, and $w = 0.99$, as close to 1 as possible, in the BP algorithm, and also choose a close enough to the zero temperature limit, $\beta = 10$.

Figure 1.21a shows the fraction of infected nodes $f(q)$ versus the fraction of immunized nodes q obtained by CI and BP on an ER network of $N = 10^3$ nodes and average degree $\langle k\rangle = 3.5$. For these values of p and w, we cannot compute the full curve $f(q)$. The reason is that the free energy is non-convex in that region of q, the chemical potential is flat, and not all values of q in that region can be explored using the normal BP method. Physically, the fact that the thermodynamical potential becomes non-convex is the signature of a phase transition happening at a certain value of w. Then, we compare the giant component found by BP and the one obtained with CI (notice that $f(q)$ coincides with the giant component of G only in the limits $p \to 0$ and $w \to 1$, see Sect. 1.5.2).

To choose which nodes have to be removed according to BP, we run BP, and we assign to each node the value of the sign of its magnetization: $\text{sign}(\langle\sigma_i\rangle)$ (the value of the inverse temperature we chose, $\beta = 10$, is sufficiently high for the spins

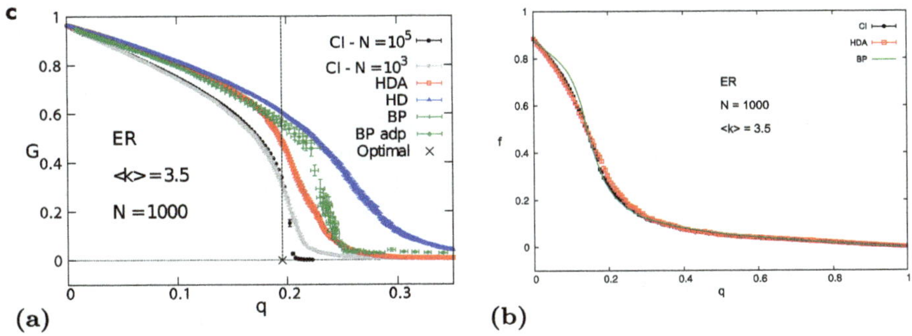

Fig. 1.21 Can we apply the optimal percolation algorithms to optimal immunization? a Comparison between the giant components obtained with CI, HDA, HD, and BP for optimal immunization [13] on an ER network of $N = 10^3$ and $\langle k \rangle = 3.5$ in the limit of the optimal immunization problem with $p \to 0$ and $w \to 1$. We also show the solution of CI from Fig. 1.13a for $N = 10^5$. We find the order of performance in CI, HDA, BP, and HD. (The average is taken over 20 realizations of the network, error bars are s.e.m.). **b** Fraction of infected nodes $f(q)$ as a function of the fraction of immunized nodes q in SIR. We use the following parameters: initial infected people $p = 0.1$, transmission probability $w = 0.5$, and an ER network of $N = 10^3$ nodes and $\langle k \rangle = 3.5$. We compare CI, HDA, and BP. All strategies give similar performance due to the large value of the initial infection p, which washes out the optimization in agreement with the results of [13]. Figure reproduced from [14]. Copyright © 2015, Nature Publishing Group

σ_i to be highly polarized). Then, node i is removed if $\text{sign}(\langle \sigma_i \rangle) = 1$, and it is not if $\text{sign}(\langle \sigma_i \rangle) = -1$. When BP does not converge, we stop the algorithm after a maximum number of iterations and use the unconverged marginals to assign the magnetizations. In this way, we can draw the full curve $G(q)$ even if BP does not converge. The results in Fig. 1.21a shows that both approaches are comparable. That is, the CI solution is a good approximation of the optimal immunization problem, and therefore, it can be used to find efficient immunizator sets. In Chap. 4 we will use this algorithm for Covid-19 spreading and comparing with other heuristics.

A second comparison between CI and BP is made for parameters where BP is well defined, and CI is not [14]; for $p \neq 0$ and $w \neq 1$. Thus, we use $p = 0.1$ and $w = 0.5$. So, this second comparison represents the opposite situation from the previous one. We compare the two methods in the following way. We use BP to compute directly the fraction of infected nodes $f(q)$ as a function of the fraction q of immunized nodes using Eqs. (1.131)–(1.132), since in this regime BP is well-defined and converges well. To compare with CI, we explicitly simulate the SIR process since we cannot directly estimate the $f(q)$. More precisely, we first identify the immunized nodes with CI, remove them from the network, and then run the SIR algorithm to obtain the final fraction of infected individuals $f(q)$.

Figure 1.21b shows the results in an ER network. As the figure shows there is little difference between BP and CI, implying that optimal percolation can be equally applied to the optimal immunization problem in this regime. This result agrees with Ref. [13], although in this regime, the large value of $p > 0$ reduces the differences between the applied strategies. Overall, the comparisons of Fig. 1.21 indicates that CI and BP provide similar solutions to immunization set.

A more elaborated version of BP was developed by Guggiola and Semerjian [268] to improve the lower bound on the immunization threshold by considering a 1-step replica symmetry breaking (1RSB), obtaining slightly larger lower bounds than those predicted by the replica symmetry (RS) approach of Altarelli et al. [13]: $q_c^{\mathrm{RS}} \leq q_c^{\mathrm{1RSB}}$, or $\theta_{\mathrm{min},0} \leq \theta_{\mathrm{min},1}$, respectively in the notation of Ref. [268].

The decycling BPD algorithm of Sect. 1.6.4, which improves the time complexity of the BP approach of [13, 268], and can be tested in SF networks, can also be applied to the optimal immunization problem. A variant of the BPD algorithm based on [262], and developed by Qin and Zhou in [297] for solving the undirected feedback vertex set problem was used and compared with CI in [14]. The algorithm of [297] was the only decycling algorithm available at the time of the comparison performed in [14]. It was shown that the CI algorithm has better performance. Later, Mugisha and Zhou developed the improved BPD algorithm of Ref. [17] discussed in Sect. 1.6.4, showing better performance than CI, as shown in Fig. 1.18.

Min [298] has used the message passing approach to calculate analytically the expected size of epidemic outbreaks originating from a single seed. It was found that while the probability of triggering an epidemic outbreak depends on the location of the seed, the final size of the outbreak is insensitive to the seed once it occurs. This approach is also applicable to weighted networks.

1.6.5.3 Equal Graph Partitioning for Optimal Immunization

In earlier work, the optimal immunization problem was treated by an equal graph partitioning (EGP) immunization strategy based on heuristic optimization of the partitioning of graphs [270, 271]. This method aims to divide the network into clusters of equal size to stop the epidemic. In EGP, the network is fragmented into small connected clusters of approximately equal size. The advantage of EGP over the targeted attack is that it saves the nodes used to isolate small clusters in a targeted attack, which is unnecessary in breaking down the network. In the EGP method, a network is first separated into two components with arbitrary size ratio by a minimal number of separators, solved using the nested dissection (ND) algorithm [299]. Then, the network can be partitioned into any desirable number of same-size clusters by applying the ND algorithm recursively. This greedy graph-partitioning strategy provides 5–50% improvement over the targeted strategy on model and real-world networks. EDG can behave well for homogeneous networks, like random regular graphs, where an equal partition could be expected to destroy the network efficiently but loses performance for heterogeneous networks like scale-free networks [14].

1.7 Algorithms to Find Influencers of Spreading Dynamics with Underlying Discontinuous Transitions

Dismantling the GCC is a drastic process since, at this point, all global communication ceases to exist in the network. However, there could be instances where the system fails from a dynamical point of view, yet the GCC of the network remains intact. In these cases, percolation, with its emphasis on the GCC, is not the right way to look at the robustness of the network. It turns out that there are other components in the networks, like the k-cores, which are able to capture these transitions. These transitions are of the first order where the order parameter presents a discontinuity at the transition, in contrast to the transitions identified in Sect. 1.6, which are of second order or continuous.

Threshold models and k-core percolation are frequently used to describe cascading processes with discontinuous phase transitions in various disciplines, for instance, failure propagation in infrastructure [300], diffusion of innovations in social networks [203], and adoption of new behaviors [301]. k-core percolation is a special case of the general class of threshold models where each node has a fixed threshold k. A difference between threshold models and ICMs is that, in threshold models, the state of a node is collectively determined by the states of all its neighbors. As a consequence, the impact of perturbing one node can propagate to a vast area of the network through long-range chains of interactions, manifested by a discontinuous phase transition in network dynamics.

In this section, we first introduce the problem of k-core percolation and its relation to dynamical models of diffusion of information with underlying discontinuous transitions. We then introduce methods developed for LTMs using belief-propagation and collective influence and then discuss algorithms designed for k-core and explosive percolation. Note that algorithms designed for LTMs are applicable to k-core percolation. Thus, we will enlarge the influencer finding problem to a richer set of problems that will then be used in subsequent chapters in applications of robustness of ecological networks and other complex systems.

1.7.1 K-Core Percolation and Linear Threshold Models of Diffusion of Information

We have already introduced the process of k-core decomposition in Sect. 1.2.1.2. Let's remind ourselves of some of its properties that will be useful in introducing the associated problem of k-core and then percolation. The k-core decomposition classifies the nodes in a network into layers with increasingly dense connections. The k-core is defined as the largest subgraph whose nodes have at least k links [27]. The 1-core of a network corresponds to the connected components of the network, and the 2-core incorporates loops. We assign each node a unique k-shell index k_s that indicates the highest k-core it belongs to. This index, called the coreness, is obtained through k-shell decomposition by pruning. Nodes with low k_s values are located at the network's periphery, while the core consists of nodes with high k_s values.

K-core percolation refers to the emergence of the different k-cores components in a graph as we add links to the graph in a random fashion [45, 48, 49]. This process leads to interesting physics: the emergence of the k-cores (with $k \geq 3$) is through abrupt, discontinuous phase transitions, similar to phase transitions in liquids and solids. The k-core structure provides higher-order information on network connectivity beyond the giant connected component, which is captured by the 1-core. Discontinuities of order parameters at the transition are the hallmarks of first-order phase transitions in physics, which are interesting since they can capture the dynamics of catastrophic and unpredictable events observed in many dynamical systems [300, 302–304]. One such a case will be treated in Chap. 7 where we will show that the k-cores determine the stability and vulnerability of ecological networks and their monitoring can help us to predict the tipping point of ecosystem collapse [111].

K-core percolation also considers the inverse problem of the removal of nodes with the subsequent breakdown of the giant k-cores, as one removes links or nodes from the graph randomly. Again, this process fundamentally differs from random percolation since the k-core size undergoes a discontinuous, first-order phase transition as nodes are removed from the network. For example, in Fig. 1.10b, the left inset illustrates the 3-core of the network. Upon removing the red node, the 3-core is destroyed, with only the 2-core left, as shown in the right inset. In this example, the 3-core disappears abruptly from a non-zero size. Such discontinuous transition arises from the threshold rule of k-core percolation, and lies at the heart of explanations of catastrophic cascading failures in real-world systems [111, 300, 305].

The k-core percolation transition in random networks has been investigated analytically by Pittel, Spencer, and Wormald [48], who showed that the emergence of the 3-cores and higher k-cores in a random network occurs abruptly through first-order transitions happen when the average degree of the graph exceeds a particular critical value. For the case of a random uncorrelated graph [227], the critical values were calculated exactly in [48].

Figure 1.22 shows the exact predictions of [48]. The 1-core contains all the connected components. The giant 2-core is related to the giant component studied in percolation and appears continuously at an average degree $c_2 = 1$, a result that is related to the emergence of the GCC as calculated in the classical result Erdös and Rényi for random percolation [47]. Beyond these two initial cores, Pittel et al. showed that for $k \geq 3$, the giant k-cores appear suddenly through first-order transitions at sharply defined values of the average degree denoted c_k, where the size of the corresponding k-core jumps from zero to a finite value, usually large compared to the total size of the network. These giant k-cores appear suddenly and dominate the entire structure of the network at the expense of the lower k-cores.

For instance, the 3-core appears suddenly at $c_3 \approx 3.35$, jumping from zero to an occupancy (number of nodes in the k-core divided by the total number of nodes) of $p_3 \approx 0.27$. Subsequently, the 4-core appears at $c_4 \approx 5.14$ with occupancy $p_4 \approx 0.43$, while the 5-core appears at $c_5 \approx 6.81$ with occupancy $p_5 \approx 0.55$, and so on. The

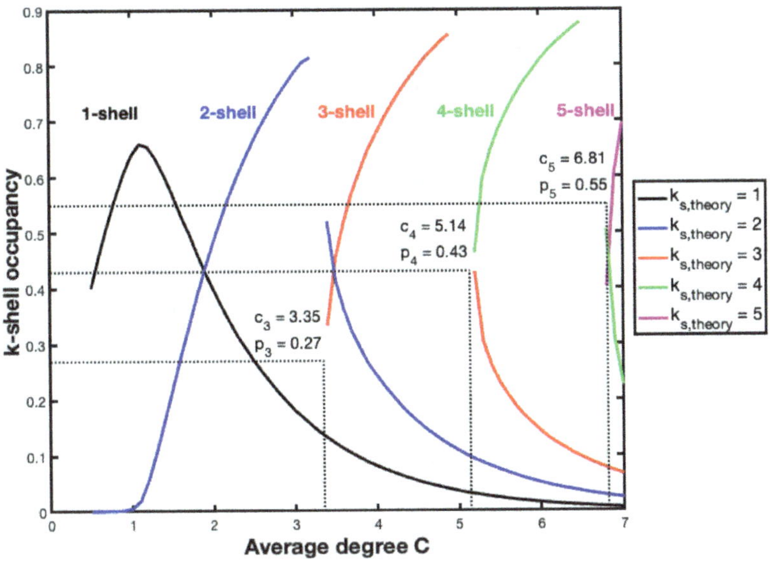

Fig. 1.22 Emergence of the giant k-cores in a random graph. The dotted lines show the exact calculations of the thresholds and occupancies of the k-cores at each emerging transition obtained in the classical result of Pittel et al. [48]. The solid lines show numerical simulations in a finite ER network of 2, 000 nodes.The 2-core emerges continuously at average degree $c_2 = 1$ as given by the classical calculations of Erdös-Rényi for random percolation [47]. All the subsequent k-cores appear in succession and abruptly at the average degrees and occupancies predicted by the theory. Notice that we measure the occupancy of each k-shell, not the k-core. However, when a new k-shell for instance, when the 3-shell appears, it is automatically the maximum core of the network. Thus, this emergent 3-shell is also the 3-core. That is the reason we refer to this process as the k-core emergence, and not k-shell emergence

sudden emergence of k-cores is quite an amusing phenomenon that has attracted much attention across physics and mathematics.

To understand the behavior of the k-core occupancy near these critical points, Ref. [113] used a set of Erdös-Rényi networks with $N = 2,000$ nodes, and average degree c ranging from $c = 0.5$ to $c = 7.0$ in steps of 0.1 to investigate numerically the emergence of the k-cores in the context of the jamming transition of packings [112, 114]. They performed a k-shell decomposition of each network, finding both the occupancy of every shell in the network, and the value of $k_{\text{core}}^{\text{max}}$. For each value of c, 1,000 ER networks are generated, and the values for $k_{\text{core}}^{\text{max}}$ and shell occupancy are averaged over these. The results are shown in Fig. 1.22 as solid curves in black, blue, red, green, and cyan, corresponding, respectively, to the 1 through 5-cores. At theoretically-predicted c_k and fractional k-shell occupancies p_k [48], denoted in Fig. 1.22 by the dotted lines, the networks undergo the predicted transitions wherein a $k + 1$-core (i.e., a new value of $k_{\text{core}}^{\text{max}}$) emerges. For $k \geq 3$-cores the occupancy of the former innermost k-core falls sharply and discontinuously to a minimum while the occupancy of the new core sharply and discontinuously increases.

In physics, the k-core was introduced as a bootstrap percolation by Chalupa, Leath and Reich [306] in 1979 to describe ordering in magnetic materials. Bootstrap percolation describes a discrete time-spreading process on a graph. Initially, nodes in the graph can take two values, active or inactive. At the beginning of the dynamics, an inactive node becomes active if it has at least k active neighbors, where k is a fixed parameter of the model, while active nodes remain in this state forever. The activation rule is iteratively applied until the system reaches a final state with no further changes. Bootstrap is a representation of a cellular automaton [307] , which may be thought of as the monotone version of the Glauber dynamics of the Ising model of ferromagnetism [308]. Although the original paper of bootstrap percolation [306] did not mention its relation with k-core, it was later shown by Gottschau [309] that the cluster that survives the spreading process with threshold k in bootstrap percolation is the k-core of the graph.

After these initial developments, the k-core has been applied broadly to a variety of systems. An account of these applications has been compiled by Malliaros et al. [50] and can be seen in Table 1.1. Its relevance to critical phenomena in physical processes and networks [310] has been extensively studied by Dorogovtsev, Goltsev and Mendes [49, 311] and Baxter et al. [207, 254] and Cellai et al. [312, 313], where both processes, k-core, and bootstrap percolation, are studied. It is shown by Dorogovtsev et al. [49, 310] that the k-core provides information about the robustness of random networks

The k-core and bootstrap also describe epidemic and information diffusion processes modeled by threshold models of rumor-spreading and cascading behaviors [208, 209, 255, 314, 315]. The LTM is related to the zero-temperature limit of the random-field Ising model [316, 317] and bootstrap percolation [11, 49, 306, 318]. A variant has been introduced by Watts [208] in which the activation condition is given by a minimal fraction of active neighbors instead of a minimal fixed number of neighbors k. Watts [208] modeled the global cascade on random networks using a linear threshold model and derived the critical condition for the discontinuous transition. Epidemic spreading models are also studied where the initial set of active nodes are infected and the disease spreads by infecting at each time step nodes with at least k already-infected neighbors.

Several works have explored the phase diagram and mechanism of transitions in k-core percolation and threshold models [49, 111, 208, 209, 254, 311]. Goltsev et al. [311] consider k-core percolation as a mixed transition with a discontinuous emergence of k-core as well as a continuous emergence of GCC. The authors demonstrated the crucial role of the 'corona,' a subset of nodes in the k-core that have exactly k neighbors: a random removal of even one node from the corona will trigger the collapse of a vast region of the k-core around the removed node. This corona has been shown to be important in the emergence of extreme opinions in society by Ramos et al. [319]. Baxter et al. [254] further gave the condition for the appearance of both the continuous transition of GCC and the discontinuous transition of k-core in networks with arbitrary degree distribution and threshold. Mathematically, the discontinuous phase transition is explained by the bifurcation in the dynamical system, where the stable solution to the equations describing the self-consistent relations in model statistics jumps abruptly from non-zero to zero value. Such bifurcation-induced transition is also responsible for the global cascade and vulnerability in interdependent networks and network of networks [300, 320, 321].

Having defined the concepts of k-core and bootstrap percolation and using the LTM and ICM spreading models from Sect. 1.3.3, we will next explain algorithms to find influencers based on the greedy approach, belief propagation, collective influence, and explosive percolation.

1.7.2 Greedy Algorithms in Discontinuous Transition in LTM

For threshold models, the influencer identification problem is to search for a given number of seeds that can lead to the maximal number of activated nodes. The greedy algorithm discussed in Sect. 1.3.4 applies to more general forms of the LTM. When weights and thresholds are drawn from a uniform probability distribution, we recall that the influence function is submodular [5]. This allows greedy algorithms to approximate the optimization with optimal guarantees.

Chen et al. [218] gave an efficient approximation of the influence of an individual node in a local subgraph G'. Using the randomized algorithm of Cohen [322], the influence of a set S is quantified by the number of nodes reachable from S in the subgraph G'. Although computing the exact influence in a network is #P-hard, an approximation based on a directed acyclic graph (DAG) can be finished within a linear time. A local DAG (LDAG) is considered instead of DAG to accelerate the calculation further. The construction of LDAG is done by including a majority of the influence while discarding the nodes with a small influence. Similar to Ref. [218], a threshold controls the size of LDAG, so the tradeoff between efficiency and accuracy can be tuned. Using the LDAG, the incremental influence of each node can be quantified with excellent efficiency. As a result, the LDAG algorithm is scalable to networks with millions of nodes and is among the best greedy algorithms in performance.

The LDAG algorithm assumes that the influence of a node is mainly within its LDAG. However, the paths that are ignored may still have a high influence, compromising the algorithm's performance. The algorithm also depends on the proper choice of the LDAG, which is an NP-hard problem itself. To avoid these problems, Goyal et al. [215] developed the SIMPATH algorithm in which the influence of a node is quantified by the number of simple paths starting from it. This problem is also #P-hard. However, it can be well approximated by enumerating paths within a small neighborhood. The influence of a set S is then the sum of the influence of each node in a locally induced subgraph. Similar to the arborescence structures constructed in Ref. [218], a tuning parameter is introduced to control the size of this neighborhood, which provides a trade-off between the accuracy and computational efficiency. A vertex cover optimization was introduced to reduce the number of estimation calls in SIMPATH so that only the influence of nodes in the vertex cover set is computed. The rest of the nodes' influence can be derived from their neighbors. The SIMPATH algorithm is more efficient, consumes less memory, and produces seed sets that have a larger influence than the basic greedy algorithm.

Karampourniotis et al. [323] proposed two new metrics to rank nodes in LTMs. The first one, termed Balanced Index prefers nodes with high resistance to adoption and those with large out-degree by tuning the weights between three features–degree, susceptibility to new activation, and the impact of activation on its neighbors. The other one, termed Group Performance Index (GPI) evaluates the performance of each node as a seed in a randomly chosen seed set. The score is inversely proportional to the number of seeds leading to a desired spread. The random sampling approach avoids the prohibitive computational burden of look-ahead algorithms and reduces the complexity to $O(N^2)$ in sparse networks. If a limited sphere of influence is used in the computation of spreading, the complexity can be further reduced to $O(N)$. In LTMs, the GPI algorithm outperforms other heuristic greedy strategies for different seed sizes, threshold distribution, and network assortativity.

The performance of most greedy algorithms mentioned above is guaranteed thanks to the submodular property of the influence function. However, for a general LTM with fixed weights and thresholds, the influence function is not always submodular [5]. We have already discussed two classes of LTM that are not submodular and lead to continuous or discontinuous phase transitions in previous sections. Solutions to the influencer problem in LTMs without submodular property require going beyond greedy searches through a better understanding of the physical mechanism of the underlying spreading process, as discussed next.

1.7.3 Belief-Propagation Algorithms for LTM

To solve the influence maximization problem on a general LTM, Altarelli et al. [11, 12] used the Belief-propagation approach that we discussed already in Sects. 1.6.4 and 1.6.5.2, by regarding it as a nontypical trajectory that deviates from the average behavior of dynamics initiated by randomly chosen seeds. To explore the dynamical properties of these nontypical trajectories on general LTMs, Ref. [11] uses a BP algorithm that could estimate statistical properties of nontypical trajectories and found the initial conditions that lead to cascading with desired properties [12]. In contrast to ICMs, the trajectory of a given LTM is determined solely by its initial condition. Due to the irreversibility of LTM dynamics, the spreading process can be parameterized by a configuration $\mathbf{t} = (t_1, t_2, ...t_N)$, where $t_i \in \mathbf{T} = \{0, 1, ...T, \infty\}$ is the activation time of node i. The dynamical rules in LTM can be represented by the constraint on the activation time of a node and its neighbors [12]: $t_i = \phi_i(\{t_j\})$, where

$$t_i = \phi_i(\{t_j\}) = \min\left\{t \in \mathbf{T} : \sum_{j \in \partial i} \omega_{ji} \mathbb{I}[t_j < t] \geq \theta_i \right\}. \tag{1.133}$$

A statistical physics approach is followed by considering the Boltzmann distribution:

$$P(\mathbf{t}) = \frac{1}{Z} e^{-\beta \varepsilon(\mathbf{t})} \prod_{i \in V} \psi_i(t_i, \{t_j\}), \tag{1.134}$$

where

$$\psi_i(t_i, \{t_j\}) = \mathbb{I}[t_i = 0] + \mathbb{I}[t_i = \phi_i(\{t_j\})], \tag{1.135}$$

and the partition function is:

$$Z = \sum_{\mathbf{t}} e^{-\beta \varepsilon(\mathbf{t})} \prod_{i \in V} \psi_i(t_i, \{t_j\}). \tag{1.136}$$

The energy cost function is:

$$\varepsilon(\mathbf{t}) = \sum_i \varepsilon_i(t_i) = \sum_i \mathbb{I}[t_i = 0] - \epsilon \mathbb{I}[t_i < \infty]. \tag{1.137}$$

For $\varepsilon = 0$, the distribution degenerates to the spreading dynamics initiated by a random set of seeds.

The factor graph that describes the constraints of a configuration contains many short loops. A dual-factor graph is constructed to avoid these loops with a variable node (t_i, t_j) introduced to each edge (i, j). The obtained dual factor graph is locally tree-like if the original network is, and allows for the application of the cavity method. The cavity marginals $H_{ij}(t_i, t_j)$ are defined as the probability that nodes i and j get activated at t_i and t_j respectively. Using replica symmetric, the recursive relation of the cavity marginal $H_{ij}(t_i, t_j)$ on the dual factor graph defines the following belief propagation (BP) equations [12]:

$$H_{ij}(t_i, t_j) \propto e^{-\beta \varepsilon_i(t_i)} \sum_{\{t_k\}} \psi_i(t_i, \{t_k\}) \prod_k H_{ki}(t_k, t_i). \tag{1.138}$$

The marginal probability that node j is activated at time t_j is:

$$P_j(t_j) \propto \sum_{\{t_i\}_{i \in \partial j}} e^{-\beta \varepsilon_j(t_j)} \psi_j(t_j, \{t_i\}) \prod_{i \in \partial j} H_{ij}(t_i, t_j). \tag{1.139}$$

The solution of these BP equations is obtained through numerical iteration and reveals the statistics of non-typical trajectories, such as the entropy and distribution of activation time. More importantly than calculating the critical thresholds is to obtain the actual influencer set. To obtain the optimal set of seeds, Max-Sum equations can be derived by setting the inverse temperature $\beta \to \infty$ in the energy function following the methods in [12], leading to the Max-Sum (MS) algorithm. Using random regular graphs and real-world networks, they find that the Max-Sum algorithm outperforms linear programming algorithm, greedy algorithm, and simulated annealing algorithms.

Guggiola and Semerjian [268] extended this to consider replica symmetry breaking and studied the minimal contagious set problem (i.e., the minimal seed set that can activate the entire graph) for LTM dynamics with and without a constraint on the maximal activation time T. In this theoretically impressive work, authors aim to find the theoretical limit of the minimal contagious set in random regular graphs using the cavity method and replica symmetry breaking. A survey propagation like algorithm based on the work of [324] is then applied on single instances of random regular graphs to find the exact seed set. It was found that the survey propagation algorithm achieves near-optimal performance in the limit of small activation times. We refer to Ref. [268] for more details on the survey propagation algorithm.

1.7.4 Collective Influence Algorithm for LTM

While the collective influence theory of Sect. 1.6.1 has been developed for percolation problems with underlying continuous transitions, some ideas from this theory can be used to deal also with the influence maximization problem in general LTMs with discontinuous transitions [224]. The discontinuity at the transition in LTMs does not allow to apply linear stability theory as in percolation. Thus, the stability analysis of the NB matrix via its largest eigenvalue developed in Sect. 1.6.1 cannot be applied to the general LTM. However, certain aspects of the collective influence theory can still be used in this discontinuous case. In particular, the message passing formalism is still valid and leads to an approximate algorithm to solve the problem of finding the minimal seed in LTM.

For a network $G(V, E)$ with $N = |V|$ nodes and $M = |E|$ links, the vector $\mathbf{n} = (n_1, n_2, \cdots, n_N)$ measures whether a node i is chosen as a seed ($n_i = 1$) or not ($n_i = 0$). We recall the LTM spreading rule. Starting from a q fraction of active seeds, it evolves according to the threshold rule: a node i becomes active if it has at least θ_i active neighbors, where the threshold θ_i is an integer ranging from 1 to the degree of node i. As usual, the influence maximization problem in LTM aims to find the optimal set of seeds so that the size of the active population is maximized.

The following formalism has been developed by Pei et al. [224]. Analogous to the message passing formalism in optimal percolation for each link $i \to j$, we introduce a cavity variable $\nu_{i\to j}$ that measures the probability of i being active when node j is removed from the network. If $n_i = 1$, then $\nu_{i\to j} = 1$. Otherwise, $\nu_{i\to j} = 1$ only when there are at least θ_i active neighbors excluding j.

Since there exist many possible choices of these θ_i neighbors, we define $P^{\theta_i}_{\partial i\backslash j}$ as the set of all combinations of θ_i nodes selected from $\partial i\backslash j$ (the set of nearest neighbors of i excluding j). If k_i is the degree of i, there are $\binom{k_i-1}{\theta_i}$ combinations, so the set $P^{\theta_i}_{\partial i\backslash j}$ contains $\binom{k_i-1}{\theta_i}$ elements, denoted by P_h, $h = 1, \cdots, \binom{k_i-1}{\theta_i}$. Each element P_h has the form $P_h = \{p_{h_1}, \cdots, p_{h_{\theta_i}}\}$ where $\{p_{h_1}, \cdots, p_{h_{\theta_i}}\}$ are the θ_i nodes

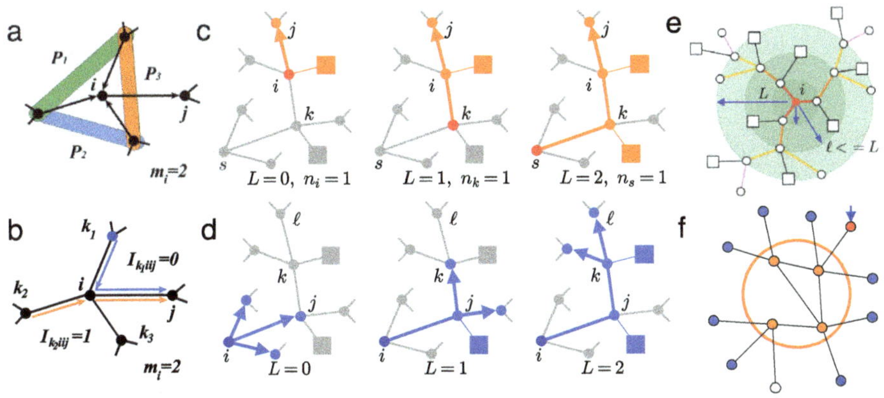

Fig. 1.23 Subcritical paths and collective influence of spreaders in the LTM. a Three combinations of neighbors P_1, P_2 and P_3 corresponding to $\nu_{i\to j}$ in the message passing equation. Node i has a threshold $\theta_i = 2$. The full activation of at least one combination will lead to $\nu_{i\to j} = 1$. **b** For link $i \to j$ with an active neighbor k_1 and inactive ones k_2 and k_3, $I_{k_1\to i, i\to j} = 0$ since i has $0\,(< \theta_i - 1)$ active neighbors, excluding k_1 and j, while $I_{k_2\to i, i\to j} = 1$, since i has $1\ (= \theta_i - 1)$ active neighbors k_1, excluding k_2 and j. **c** Subcritical paths ending with link $i \to j$ for $L = 0,\ 1,\ 2$. Red dots are seeds, and squares represent $\theta_i - 1$ active neighbors attached to subcritical nodes. Thick links highlight subcritical paths. **d** The contribution of seed i to $\|\nu_{\to}\|$ exerted through subcritical paths of length $L = 0,\ 1,\ 2$. **(e)** Calculation method of the collective influence for LTM: CI-TM$_L(i)$. Subcritical paths starting from i with length $\ell \le L$ are displayed in different colors. **(f)** An example of a subcritical cluster. Assuming a uniform threshold $\theta_i = 3$, nodes inside the circle are subcritical since they all have 2 active neighbors, represented by blue nodes. Activation of the red node will trigger a cascade covering all subcritical nodes. Figure reproduced from [224].

in the h-th combination. Figure 1.23a illustrates all three combinations P_1, P_2 and P_3 corresponding to $\nu_{i\to j}$ for node i with a threshold $\theta_i = 2$. If at least one combination is fully activated, we have $\nu_{i\to j} = 1$.

As we can see, the situation is more complicated than in the case of message passing in optimal percolation. The combinatorial nature of $P^{\theta_i}_{\partial i\setminus j}$ is responsible for the first-order nature of the transition in the general LTM. Thus, dealing with models exhibiting discontinuous phase transitions presents a set of challenges that are radically different from those that are instead faced in continuous transitions, like optimal percolation, which can be solved by linearization and stability analysis around the critical point due to its continuous nature. Indeed, we may argue that, in statistical physics, the theory of second-order phase transition is more developed than that of first-order transitions and there are more analytical tools available to deal with continuous transitions rather than discontinuous ones. The present mathematical framework, interpolating between both situations, offers the possibility to explore the complexity of the situation in a single framework.

As usual, under the assumption of a locally tree-like network, $\nu_{i\to j}$ satisfies the following self-consistent message passing equation:

$$\nu_{i\to j} = n_i + (1-n_i)\left[1 - \prod_{P_h \in P^{\theta_i}_{\partial i \setminus j}} \left(1 - \prod_{p \in P_h} \nu_{p\to i}\right)\right]. \tag{1.140}$$

The final state of i is determined by:

$$\nu_i = n_i + (1-n_i)\left[1 - \prod_{P_h \in P^{\theta_i}_{\partial i}} \left(1 - \prod_{p \in P_h} \nu_{p\to i}\right)\right]. \tag{1.141}$$

Equations (1.140)–(1.141) describe the general case of LTM, defined by a generic θ_i. We have seen in Sect. 1.5.1 that for the special choice of threshold $\theta_i = k_i - 1$ the problem is mapped to percolation with a second-order transition. This is because there is only one combination in $P^{\theta_i}_{\partial i \setminus j}$, and the transition becomes continuous. In this case, Eq. (1.140) recovers the message passing equation of optimal percolation Eq. (1.58). The other extreme where we set $\theta_i = 1$ is equivalent to the problem of optimal immunization as shown in Sect. 1.5.2. For any other choice of θ_i the multiplicity of activated neighbors leads to a first-order transition and cascading, as discussed here. This model represents an interesting case where we can observe a transition from smooth transitions to abrupt ones by tuning the threshold.

Different from the case of optimal percolation in Ref. [14], the zero solution $\{\nu_{i\to j} = 0\}$ is not a fixed point of the message passing equation Eq. (1.140). As a consequence, the stability analysis around zero solution in Ref. [14] is no longer valid for LTMs, and the problem is not related anymore to the NB matrix. However, the solution of Eq. (1.140) can be approximated through iteration of the linearized system.

By linearizing Eq. (1.140), it was found that the subsequent activation of nodes in each iteration only depends on the number of subcritical nodes, defined as the nodes with $\theta_i - 1$ active neighbors (i.e., nodes whose activation can be triggered by one more active neighbor) in the work of Goltsev et al. [311] and Dorogovtsev et al. [49]. Moreover, subcritical nodes can form long subcritical paths that generate a long-range cascade of activation, which is core to the discontinuous transition in LTM dynamics. The influence of spreaders is then fixed by the subcritical paths along which cascades propagate: the number of subcritical paths attached to each spreader determines its contribution to global cascades.

Following this idea, the Collective Influence in Threshold Model (CI-TM) was proposed by Pei et al. [224] that recursively selects nodes with the largest CI-TM score. The CI-TM score enumerates the number of subcritical paths starting from each node, and uses that to quantify nodes' spreading capability.

Equation (1.140) is a nonlinear function of $\nu_{\to} = (\cdots, \nu_{i\to j}, \cdots)^T$:

$$\nu_{\to} = \mathbf{n}_{\to} + \mathbf{G}(\nu_{\to}). \tag{1.142}$$

Here, $\mathbf{n}_{\rightarrow} = (\cdots, n_{i\rightarrow j}, \cdots)^T$ in which $n_{i\rightarrow j} = n_i$ for link $i \rightarrow j$, and $\mathbf{G} = (\cdots, G_{i\rightarrow j}, \cdots)^T$ where $G_{i\rightarrow j}$ is the nonlinear function of vector $\nu_{\rightarrow}$ for link $i \rightarrow j$.

Given the initial configuration of seeds $\mathbf{n}$, the final state of $\nu_{\rightarrow}$ is fully determined by the self-consistent Eq. (1.142). However, this equation cannot be solved directly (like in the case of optimal percolation) due to the exponentially growing number of combinations in $P^{\theta_i}_{\partial i\setminus j}$. Assuming a small number of seeds, the solution can be estimated approximately by iteration. Equation (1.142) is treated as:

$$\nu^{t+1}_{\rightarrow} = \mathbf{n}_{\rightarrow} + \mathbf{G}(\nu^{t}_{\rightarrow}) \tag{1.143}$$

with initial condition $\nu^0_{\rightarrow} = \mathbf{n}_{\rightarrow}$.

The calculation can be simplified by linearizing the nonlinear function $G_{i\rightarrow j}$ and considering $G'_{i\rightarrow j}(\nu_{\rightarrow}) = (\cdots, \frac{\partial G_{i\rightarrow j}}{\partial \nu_{k\rightarrow \ell}}, \cdots)$. From Eq. (1.140), we have $\frac{\partial G_{i\rightarrow j}}{\partial \nu_{k\rightarrow \ell}} = 0$ for $\ell \neq i$, while for $\ell = i$ and $k \neq j$, we have:

$$\begin{aligned} \frac{\partial G_{i\rightarrow j}}{\partial \nu_{k\rightarrow i}} = (1 - n_i) \prod_{P_h \in P^{\theta_i}_{\partial i\setminus j}, k\notin P_h} \left(1 - \prod_{p\in P_h} \nu_{p\rightarrow i}\right) \\ \sum_{P_h \in P^{\theta_i}_{\partial i\setminus j}, k\in P_h} \left[\left(\prod_{p\in P_h\setminus k} \nu_{p\rightarrow i}\right) \prod_{P'_h \neq P_h, k\in P'_h} \left(1 - \prod_{p\in P'_h} \nu_{p\rightarrow i}\right)\right]. \end{aligned} \tag{1.144}$$

Despite the complexity of this equation, it is determined only by the quantity:

$$a_{k\rightarrow i, i\rightarrow j} = \sum_{p\in \partial i\setminus(k,j)} \nu_{p\rightarrow i}, \tag{1.145}$$

representing the number of i's active neighbors excluding k and j when i is absent from the network, as well as the matrix $I_{k\rightarrow \ell, i\rightarrow j}$ for links $k \rightarrow \ell$ and $i \rightarrow j$:

$$I_{k\rightarrow \ell, i\rightarrow j} = \begin{cases} 1 & \text{if } \ell = i, k \neq j, a_{k\rightarrow i, i\rightarrow j} = \theta_i - 1, \\ 0 & \text{otherwise} \end{cases} \tag{1.146}$$

The definition of $I_{k\rightarrow \ell, i\rightarrow j}$ generalizes the Hashimoto non-backtracking matrix $\hat{\mathcal{B}}$, Eq. (1.42), that controls the stability of the null solution in optimal percolation. In the case of $\theta_i = k_i - 1$, our quantity $I_{k\rightarrow \ell, i\rightarrow j}$ is transformed to the corresponding element of the NB matrix $\mathcal{B}_{k\rightarrow \ell, i\rightarrow j}$ in optimal percolation [224]. Thus, we see how the framework to treat an optimization problem characterized by a second-order phase transition can be somehow extended to a process with a first-order transition.

Furthermore, $I_{k\rightarrow \ell, i\rightarrow j}$ is closely related to subcritical nodes [49, 311]. Since a node i is subcritical if it has $\theta_i - 1$ active neighbors, one more activation of its neighbors will cause i to be activated. From Eq. (1.146) we know that $I_{k\rightarrow \ell, i\rightarrow j} = 1$ only if the links $k \rightarrow \ell$ and $i \rightarrow j$ are connected, non-backtracking, and additionally, node

i is subcritical in the absence of node k and j. In Fig. 1.23b, node i has an active neighbor k_1 and two inactive ones k_2 and k_3. By definition, for a threshold $\theta_i = 2$, we conclude $I_{k_1 \to i, i \to j} = 0$ since i has no active neighbor excluding k_1 and j, while $I_{k_2 \to i, i \to j} = 1$ because i has 1 $(= \theta_i - 1)$ active neighbor excluding k_2 and j. Furthermore, a subcritical path of length L is defined as the path $i_1 \to i_2 \to \cdots \to i_L \to i \to j$ such that $n_{i_1} = 1, n_{i_2} = 0, \cdots, n_i = 0, I^0_{i_1 \to i_2, i_2 \to i_3} = 1, \cdots, I^{L-1}_{i_L \to i, i \to j} = 1$, and any two consecutive links are non-backtracking. The superscript t in $I^t_{k \to \ell, i \to j}$ indicates to compute $I_{k \to \ell, i \to j}$ using the linearized system at t-th iteration in Eq. (1.143) [224]. If $i_1 = i$, the path's length $L = 0$. The subcritical paths of length $L = 0$, 1 and 2 are illustrated in Fig. 1.23c.

To quantity $\|\nu_{\to}\| = \sum_{ij} \nu_{i \to j}$ quantifies the active population in LTM. Starting from $\|\nu_{\to}\| = 0$ when no seed is selected, $\|\nu_{\to}\|$ increases as more seeds are activated. Therefore, the collective influence of a given number of seeds can be optimized by maximizing $\|\nu_{\to}\|$. The concept of a subcritical path allows the introduction of a linearly scalable algorithm for massively large-scale networks.

For each element in $\nu_{\to}$, the contribution of a seed i to $\|\nu_{\to}\|$ is composed of all its collective contributions to every potential element exerted through the subcritical paths attached to i. The seed's contribution to $\|\nu_{\to}\|$ defines its Collective Influence in Threshold Model (CI-TM) to find the best influencers. The definition of node i's influence CI-TM in an area of length L is then:

$$\begin{aligned} \text{CI-TM}_L(i) &= \text{number of subcritical paths starting} \\ &\quad \text{from } i \text{ with length } 0 \le \ell \le L. \end{aligned} \tag{1.147}$$

Computing CI-TM$_L$ is equivalent to iteratively visiting subcritical neighbors of each node layer by layer within L radius. As in CI, the zero-order CI-TM is the degree of optimal percolation. For the trivial case of subcritical paths with length $L = 0$, we define CI-TM$_0(i) = k_i$. The first panel of Fig. 1.23d illustrates CI-TM$_0(i) =$ 3 for node i. For $L \ge 1$, subcritical paths define CI-TM. For instance, for $L = 1$ (Fig. 1.23d), we have:

$$\text{CI-TM}_1(i) = k_i + \sum_{j \in \partial i} (1 - n_j) \sum_{k \in \partial j \setminus i} I^0_{i \to j, j \to k} \,. \tag{1.148}$$

Figure 1.23e illustrates the calculation of node i's CI-TM for $L = 2$, in which subcritical paths with length $\ell \le L$ are distinguished by colors. To optimize the influencer set in quasi-linear runtime, $O(N \log N)$, an adaptive greedy approach is used [224] by selecting the nodes with maximal CI-TM, and making use of the data structure of the heap that takes $O(\log N)$ time [242].

In the numerical simulations seen in Fig. 1.24, the CI-TM algorithm outperforms the greedy algorithm and several heuristic centralities and achieves comparable performance to the Max-Sum (MS) algorithm of Altarelli et al. [12] in synthetic random networks. Figure 1.24a shows the active giant component $Q(q)$ versus the fraction of seed q on ER networks ($t = 0.5$, $N = 2 \times 10^5$, $\langle k \rangle = 6$). Similar to bootstrap perco-

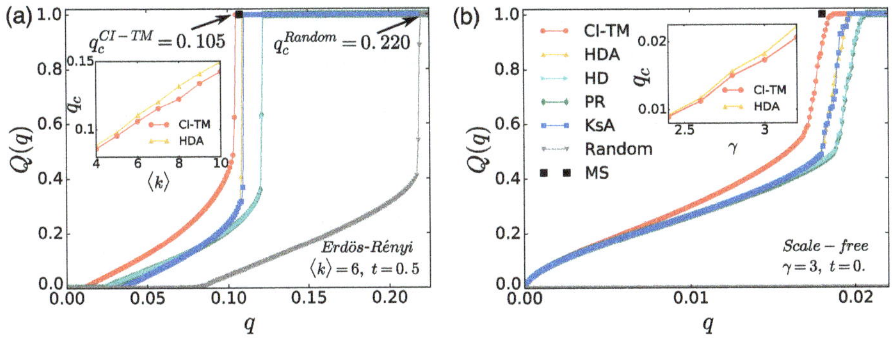

Fig. 1.24 Performance of the CI-TM algorithm in ER and SF networks. a Size of active giant component $Q(q)$ as a function of the fraction of seed q for ER networks ($N = 2 \times 10^5$, $\langle k \rangle = 6$). The CI-TM algorithm is compared with high-degree adaptive (HDA), high-degree (HD), PageRank (PR), k-core adaptive (KsA), Random, and the Max-Sum (MS) algorithm [12]. Inset shows the critical values q_c identified by HDA and CI-TM for different mean degrees. **b** Results for SF networks ($N = 2 \times 10^5$, $\gamma = 3$). Inset presents the critical values q_c for different power-law exponents γ. Figure adapted from [224]. Copyright © 2017, The Authors

lation on homogeneous networks [254], $Q(q)$ first undergoes a continuous transition from $Q(q) = 0$ to nonzero, and then a first-order transition at a higher value of q_c. Compared with heuristics, CI-TM algorithm achieves a larger active population for a given number of seeds with an earlier continuous transition, and it also activates the total population with a smaller seed set. A random selection represents the average behavior of cascade initiated by randomly chosen seeds, with a critical value $q_c^{\text{ran}} = 0.220$. Although the original k-core ranking has an unsatisfactory performance for multi-source spreading, an adaptive version KsA gives a better result similar to HDA. Similar results are obtained in SF networks, Fig. 1.24b.

1.7.5 Influencers for Optimal k-Core Percolation

The influencers for k-core percolation are those that, upon removal from the network, would disintegrate a given k-core in an optimal way [269]. Similar to optimal percolation, the configuration of node removal $\mathbf{n}$ can be optimized to induce an early transition. As exemplified in Fig. 1.10b, for random k-core percolation, at the critical point q_c^{ran}, we have $\lim_{q \to (q_c^{\text{ran}})^-} |k_s(q)| > 0$, $\lim_{q \to (q_c^{\text{ran}})^+} |k_s(q)| = 0$ and $|k_s(q)| = 0$ for $q > q_c^{\text{ran}}$, where $|k_s|$ is the k-core size. The optimal k-core percolation problem is to find the optimal configuration $\mathbf{n}^*$ for which $q_c^{\text{opt}} = \min\{q \in [0, 1] \mid \lim_{q' \to q^+} k_S(q') = 0\}$.

Since k-core percolation is a special case of LTMs, influence maximization algorithms developed for general LTMs can be extended to work for optimal k-core percolation. Consider a special case of the LTM described in Sect. 1.3.3.1 with all weights $w_{ij} = 1$, and $\theta_i \in \mathbb{N}$. In this LTM, a node is activated if it has at least θ_i of its neighbors activated, remaining activated forever. After the dynamics have reached the final steady state, the inactive nodes have less than θ_i active neighbors. Consider that k_i^{ac} and k_i^{inac} are the number of activated and inactivated neighbors of i, respectively at the end of the process ($k_i^{\mathrm{inac}} + k_i^{\mathrm{act}} = k_i$, where k_i is the degree). Then, an inactive node at the end of the process has $k_i^{\mathrm{inac}} > k_i - \theta_i$, or at least $k_i - \theta_i + 1$ inactive neighbors. Since the dynamics are irreversible, activating a node with θ_i or more active neighbors is equivalent to removing an inactive node with fewer than $k_i - \theta_i$ nodes from the network [269]. Therefore, dismantling the k-core is equivalent to the network activation following an LTM with a threshold given by $\theta_i = k_i - k + 1$. That is, the set of nodes that dismantles a given k-core can be used as the seed nodes that will activate the whole graph in LTM dynamics with a threshold chosen as $\theta_i = k_i - k + 1$ [269]. Therefore, finding the minimal set of nodes to dismantle the k-core is analogous to finding the minimal set to maximize the spreading in the analogous LTM model. The problem of finding the smallest set to remove the k-core is also called the Minimal Contagious Set problem [268, 325–329], which is a decision problem proven to be NP-hard in a general graph [326].

Several theoretical works in statistical physics and combinatorial optimization have explored the lower and upper bounds on the size of the minimal set to destroy the k-core. These bounds are useful to assess how far a particular algorithm is from the theoretical limit for the size of minimal contagious. As discussed in the Sect. 1.6.4, destroying the 2-core of the network is equivalent to decycling, a fact that was utilized in [127] to develop the CoreHD algorithm to decycle (and dismantle) the network.

Bau et al. [244] studied the decycling numbers of random regular graphs. For a random cubic graph G of order n, where all nodes have degree 3; it was proven that the decycling number (the smallest number of nodes that can be removed from G so that the resulting graph contains no cycles) is

$$\phi(G) = \lceil n/4 + 1/2 \rceil. \tag{1.149}$$

For a random d-regular graph G of order n and $d \geq 4$, Bau et al. [244] proved that $\phi(G)/n$ is bounded below and above asymptotically almost surely by certain constants that depend solely on d. In particular, the lower and upper bounds can be calculated by solving an algebraic equation and a set of differential equations, respectively. Janson and Thomason [330] found that, for sparse random graphs or random regular networks of N nodes and $N \to \infty$, the number of nodes that must be removed so that no component with more than k nodes exists is essentially the same for all values of k if $k \to \infty$ and $k = O(N)$. Reichman [327] showed that the size of a contagious set is bounded from above by $\sum_{v \in V} \min\left\{1, \frac{k}{d(v)+1}\right\}$ in the destruction of k-core ($d(v)$ is the degree of node v).

Later, Guggiola and Semerjian [268] used the cavity method with replica symmetry breaking to obtain several conjectures on the size of minimal contagious sets for k-core percolation in random regular graphs. In particular, Ref. [268] conjectured that the minimal contagious set size is 1/6 for 5-regular random graphs with a activation threshold of 3, and 1/4 for 6-regular with activation threshold 4. In addition, they also proposed the conjecture for $(k+1)$-regular networks with the threshold k that the minimal contagious set size is $1 - 2\ln k/k - 2/k + O(1/(k \ln k))$. According to this conjecture, the minimal contagious set size for 3-regular (cubic) random graphs with an activation threshold of 2 is 1/4, which is in agreement with the decycling number of cubic random graphs $\phi(G)/n \to 1/4$ obtained in Ref. [244]. Sun et al. [331] also proposed a lower bound of the network dismantling problem by analyzing specific 2-core subnetworks of many real-world networks that have a heterogeneous degree distribution. Coja-Oghlan et al. [329] explored the minimal contagious set problem on graphs with expansion properties.

Schmidt et al. [269] studied the minimal contagious sets for k-core percolation in random networks. In this work, the authors proposed a generalized CoreHD algorithm called Weak-Neighbor, in which nodes with the highest degree and low average degree of the neighbors in the k-core are recursively removed until the k-core completely collapses. To achieve this removal, they tested two schemes yielding similar performance:

- The nodes are removed in order according to $\arg\max_i k_i - s_i$, where s_i is the average degree of the neighbors of i.
- Separate the two steps by first selecting the set $\mathcal{M} \leftarrow \{i \in \mathcal{V}_C \mid i = \arg\max_i [k_i]\}$ and then update it according to $\mathcal{M} \leftarrow \{i \in \mathcal{M} \mid i = \arg\min_i \left[\sum_{j\in\partial i} k_j\right]\}$. Here $\mathcal{V}_C$ is the set of nodes in the k-core.

To analyze the properties of this algorithm, the generalized CoreHD-guided k-core removal was translated to a random process on the degree distribution of the graph using the Wormald approach [332, 333], which is a simple and rigorous numerical method for tracking the macroscopic dynamics of the peeling algorithm when the maximum degree is bounded and the degree distribution remains positive.

To describe the complexity of the algorithm on a random graph, the running time of the algorithm process was characterized by a set of nonlinear ordinary differential equations. The algorithm has $O(N)$ complexity for networks with bounded degrees, where N is the network size. By analyzing the stopping time, new upper bounds on the minimal contagious set are obtained, which improve the best previously known bounds from Ref. [244, 268]. This approach applies not only to random regular graphs but also to random networks generated from the configuration model with a given degree distribution.

Figure 1.25 compares the performance of the optimal k-core algorithms: CoreHD and Weak-Neighbor with the CI-TM algorithm of Eq. (1.147) with $L = 10$, CI-TM_{10} (beyond $L = 10$ resulted in negligible improvements) on Erdös-Rényi graphs. While the CI-TM outperforms a set of basic heuristics (Fig. 1.24), Weak-Neighbor algorithm improves over the generalized CoreHD algorithm and CI-TM algorithm in a range of k-core percolation processes in random regular graphs, although all algorithms perform similarly well, with the improvement of Weak-Neighbor, which is increasing for large k-cores. The authors of Ref. [269] concluded that CI-TM may not formulate an optimal use of the L neighboring to solve the problem: the

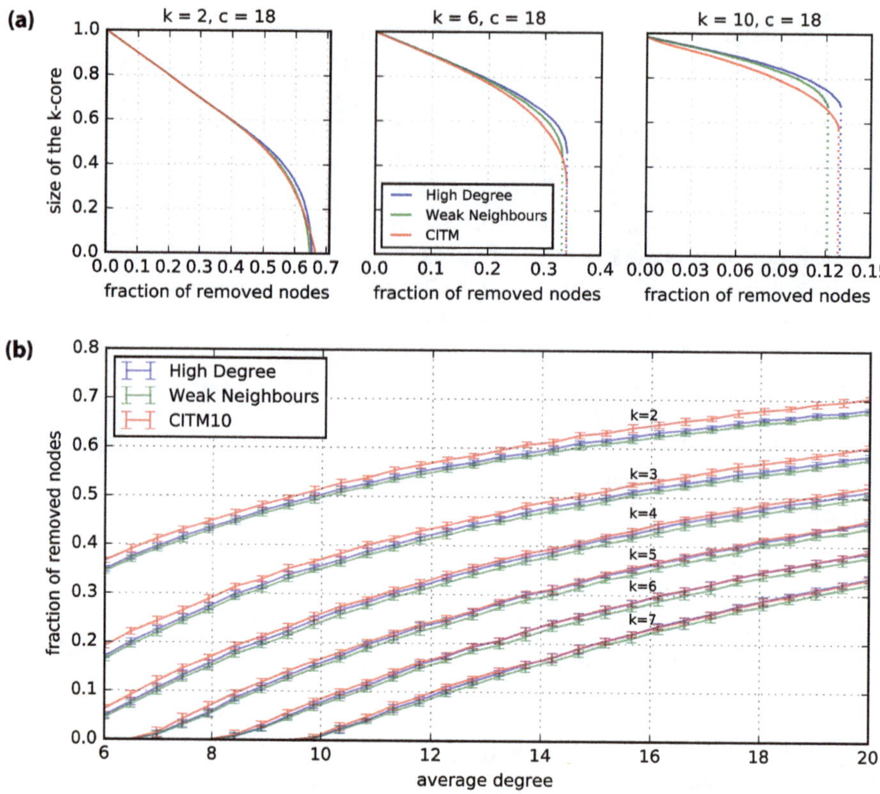

Fig. 1.25 Performance of the optimal k-core percolation algorithms: CoreHD, Weak-Neighbor, and CI-TM on ER graphs. a Typical k-core attack transitions of the three algorithms for different k-cores from 2 to 10 and fix average degree. CI-TM$_{10}$ is run with $L = 10$ set to the size of the graph, beyond 10 results in negligible improvements. Weak-Neighbor performs better than the other algorithm, although tiny margins separate the three algorithms. For larger k-cores CI-TM and CoreHD become closer in performance. However, the gap with Weak-Neighbor becomes larger. **b** Comparison of the three algorithms for different k-cores and varying average degrees for the critical threshold. Figure reproduced from [269]. Copyright © 2019, American Physical Society

search for an optimized strategy that uses only local information up to a distance remains an open problem. Further results on random regular graphs are summarized in Table 1.3 for the three algorithms and different k-cores, as well as a comparison with the conjectured optimal non-constructive cavity method results of Guggiola and Semerjian [268], showing that the algorithmic results deviate at large k-cores from the conjectured optimal solutions.

Table 1.3 Performance results of the critical threshold to dismantle the k-cores on random regular graphs with degree as indicated. Results are shown for CI-TM$_{10}$, CoreHD, Weak-Neighbor (with arg max$_i$ $k_i - s_i$), and the conjectured optimal non-constructive results obtained by the cavity method in [268]. Table reproduced from [269]

degree	k-core	CI-TM$_{10}$ [224]	CoreHD [127]	Weak-Neighbor [269]	Cavity [268]
3	2	.254	.2500	.2500	.25000
4	2	.365	.3462	.3376	.33333
	3	.077	.0963	.0744	.04633
5	2	.437	.4111	.3965	.37847
	3	.205	.2084	.1876	.16667
	4	.032	.0477	.0277	.01326
6	2	.493	.4606	.4438	.42270
	3	.296	.2811	.2644	.25000
	4	.121	.1401	.1081	.07625
	5	.019	.0281	.0134	.00582
7	2	.540	.5006	.4831	.46023
	3	.362	.3376	.3206	.30009
	4	.207	.2115	.1813	.15006
	5	.081	.1010	.0686	.04289
	6	.013	.0185	.0077	.00317

1.7.6 Explosive Percolation-Based Immunization

An approach for optimal immunization called Explosive Immunization (EI) was developed by Clusella et al. [18] based on explosive percolation (EP), which features an explosive emergence of the giant component at the critical point [334–337]. To obtain an explosive percolation transition, Achlioptas et al. [334] modified the ordinary percolation process of edge addition, wherein, at each step, two candidate edges are chosen randomly, but only one of them is occupied. Given the weight of a node is measured by the size of the connected component it belongs to the edge with the minimal sum or product of nodes' weights are selected, called the min-cluster and min-product rule, respectively. These rules favor the connection between small components, thereby suppressing the generation of an extensive GCC in ordinary percolation. The mechanism of explosive percolation was further illustrated by Friedman et al. [338], who stressed the importance of a "powder keg" in the explosive phase transition. Consisting of a large number of sub-extensive components, the existence of a powder keg leads directly to explosive percolation, even if each individual component in it is negligible.

The Explosive Immunization algorithm [18] starts from a configuration where all nodes are removed from the network ($q = 1$). Then less "dangerous" nodes are progressively unvaccinated. The procedure is performed in two schemes for $q > q_c^{\mathrm{EI}}$

and $q < q_c^{\mathrm{EI}}$, each of which uses a score to rank nodes in terms of their suitability to be unvaccinated. Similar to the construction of EP, each time step, m candidates (typically $m \approx 10^3$) are randomly selected. For $q > q_c^{\mathrm{EI}}$, the node with the lowest blocking ability (the weakest blocker) is put back into the network. The blocking ability is quantified by a score $\sigma_i^{(1)}$:

$$\sigma_i^{(1)} = k_i^{(\mathrm{eff})} + \sum_{\mathcal{C} \subset \mathcal{N}_i} (\sqrt{|\mathcal{C}|} - 1), \tag{1.150}$$

where $\mathcal{N}_i$ is the set of all components connected to node i and $|\mathcal{C}|$ is the size of a component $\mathcal{C}$. $k_i^{(\mathrm{eff})}$ measures the local connectivity of node i and can be determined by a set of closed equations [18]:

$$k_i^{(\mathrm{eff})} = k_i - L_i - M_i(\{k_j^{(\mathrm{eff})}\}_{j \in \partial i}), \tag{1.151}$$

where k_i is the degree of node i, L_i is the number of leaves in the neighborhood of node i and M_i returns the number of strong hubs. The strong hubs are defined recursively as nodes with $k_i^{(\mathrm{eff})}$ larger than a threshold value (set as 6 in applications). The terms L_i and M_i are subtracted from k_i since leaves have no contribution to connectivity and hubs are more likely to be removed in explosive immunization.

In the first part of the EI algorithm, the node with the lowest $\sigma_i^{(1)}$ score among m candidates is unvaccinated in each iteration. This process eventually reaches a critical fraction of immunized nodes q_c^{EI} where the G_∞ size exceeds a small threshold value. In the region of $q < q_c^{\mathrm{EI}}$, however, the same procedure will lead to an abrupt jump of the G_∞ size when two large components are joined together. As a consequence, in the second part at $q < q_c^{\mathrm{EI}}$, another score $\sigma_i^{(2)}$ is used to suppress such explosive growth of the giant component. The definition of $\sigma_i^{(2)}$ reads [18]

$$\sigma_i^{(2)} = \begin{cases} \infty & \text{if } G_\infty \subsetneq \mathcal{N}_i, \\ |\mathcal{N}_i| & \text{else, if } \arg\min_i |\mathcal{N}_i| \text{ is unique,} \\ |\mathcal{N}_i| + \epsilon |\mathcal{C}_2| & \text{else,} \end{cases} \tag{1.152}$$

where $|\mathcal{N}_i|$ is the number of components connected to i, $\mathcal{C}_2$ is the second largest component in $\mathcal{N}_i$, and ϵ is a small positive number. According to the score $\sigma_i^{(2)}$, the selection is made only among the neighborhood of G_∞. The candidate with the smallest number of neighboring components is favored; if it is not unique, the one with the smallest $|\mathcal{C}_2|$ is selected. This process proceeds recursively until the fraction of vaccinated nodes q reaches the expected value.

Using the Newman-Ziff percolation algorithm in identifying susceptible components [339], the explosive immunization algorithm is computationally efficient, which scales as $O(N \log N)$. In addition, it can be accelerated further by considering a small number of candidates. Simulations on both synthetic and real-world networks indicate that the explosive immunization algorithm outperforms the CI algorithm. It achieves the smallest percolation threshold q_c^{EI} except for the belief propagation algorithms in Ref. [16, 17]. Specially, considering an alternative order parameter that quantifies the average size of all clusters:

$$\overline{|\mathcal{G}|} = \frac{1}{N} \sum_i |\mathcal{G}_i(q)|^2, \tag{1.153}$$

where $|\mathcal{G}_i(q)|$ denotes the sizes of all connected component after vaccinating q fraction of nodes, the explosive immunization algorithm is even superior to the BP algorithms proposed in Ref. [16, 17].

1.8 Machine Learning to the Rescue

Last but not least, a promising line of research that could render the above approaches obsolete has been developed by Grassia, De Domenico, and Mangioni [202].

We have seen how topological centralities attempt to capture different features of the network, being the degree, walks, or k-core location, that could better identify an influencer. These features are selected by a scientist (human), and the chosen centrality then provides the sought ranking. Grassia et al. [202] posed the question: why don't we let a machine find the best feature that could uncover the optimal influencer set to produce the best-ranking without the strong assumptions and biases made by humans when they favor a given type of centrality feature? Think of why we are always biased to choose features that would favor those with a large number of connections or the most popular ones.

Machine learning then could have the answer to the influencer problem. Grassia et al. [202] have trained a machine to solve the problem of dismantling a network with minimal sets of nodes. The authors are interested in finding more efficient algorithms than human-based heuristic centralities and models. Figure 1.26 displays the machine learning pipeline.

Previously, Ma et al. [340] have also developed influence maximization techniques that go beyond the structural properties of the networks. The model exploits machine learning for feature extraction and selecting influential nodes in large graphs. They design generalized machine learning schemes that, in addition to topological features, also employ node-based and flow-based attributes facilitated by activity patterns in empirical data sets in online social networks. Gao et al. [341] also developed a social influencer predictor using heterogeneous Graph Neural Networks (GNN).

The general idea is to train a machine using a training set of small networks that can be dismantled optimally by exhaustive enumeration, and then compute the node's feature that allows for optimal dismantling. The model is then fed to a larger target network to be dismantled by assigning each node a probability that it belongs to the (sub)-optimal dismantling set. Using this probability, nodes are then ranked as influencers, and then they are removed from the network to test their dismantling power. The set that destroys the network with the minimal number of removals is the

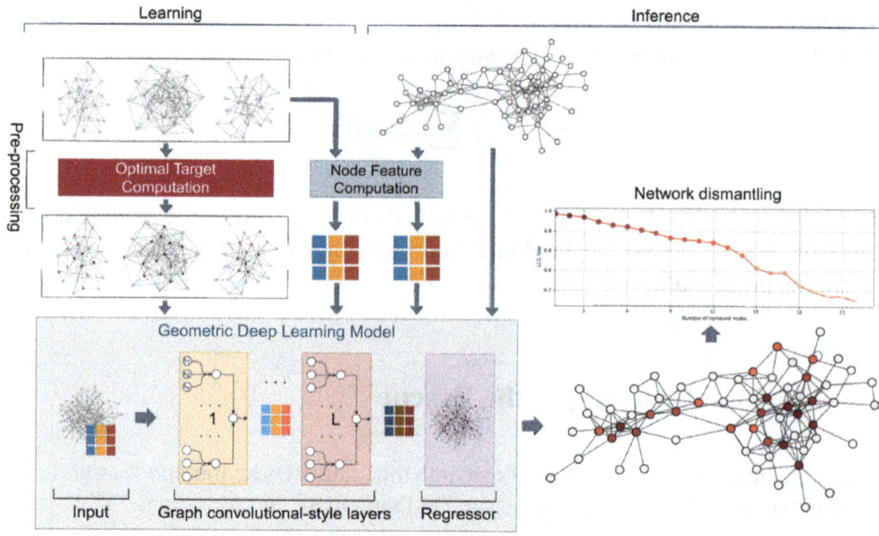

Fig. 1.26 Machine Learning for Influencers. Grassia et al. [202] proposed a machine learning pipeline to learn first the features of the influencers from small networks that can be dismantled by brute force and then predict influencers that could optimally dismantle a network in large graphs. This method could render the centrality = based influencer finder problem obsolete, as now a machine can be shown to find much better features of influencers than human-chosen centralities based on the topological features of the graph. Figure reproduced from [202]. Copyright © 2021, The Authors

sub-optimal one (sub-optimal here refers to the fact that this is an NP-hard problem and the optimal solution is not achievable by any algorithm in non-exponential time). The particular architecture used in this study consists of graph convolutional-style layers (Graph Attention Network (GAT) layers) coupled with linear layers followed by a regressor (a multilayer perceptron), as shown in Fig. 1.26.

In computational experiments, empirical networks are dismantled spanning biological, social, infrastructure, communication, trophic, and technological systems, and compare the goodness of the machine learning strategy, named GDM (Graph Dismantling with Machine Learning) against state-of-the-art dismantling methods described in Sect. 1.6.4: Generalized Network Dismantling (GND, a partitioning algorithm) [342], CoreHD [127] Min-Sum (MS) [16]; Explosive Immunization (EI) [18] described in Sect. 1.7.6; and Collective Influence (CI) algorithm [14] described in Sect. 1.6.1.

Figure 1.27 shows the results of the dismantling process in a sample network from [343], showing that machine learning GDM outperforms the methods mentioned above. The dismantling of the largest connected component (LLC in Fig. 1.27a) shows an interesting progression in performance given by the time evolution of previous algorithms: The CI algorithm, one of the first algorithms proposed in 2015 [14] is the worst performer, necessitating more than 27 removals to break the LLC. Next in line are the CoreHD and MS algorithms, proposed in 2016, and producing

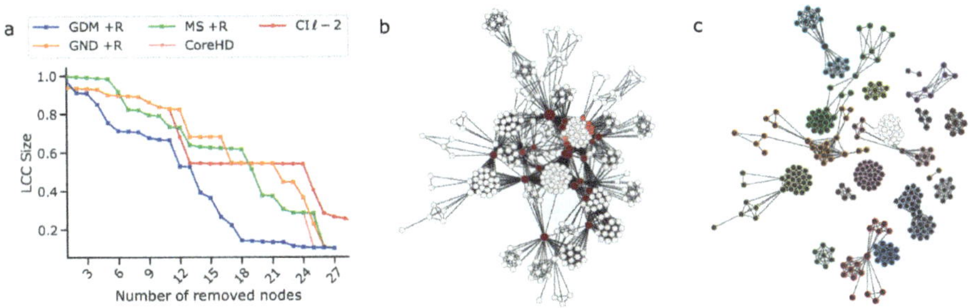

Fig. 1.27 Dismantling a corruption network with machine learning. A corruption network built from scandals in Brazil is used to test the performance of machine learning against other algorithms in the dismantling problem. **a** GDM+R is the proposed machine learning algorithm with reinsertion of the nodes from Grassia et al. [202], which disintegrates the network faster than other methods. The CI algorithm [14] is the worst performer. **b** and **c** show the structure before and after dismantling with GDM, respectively. The color of the nodes represents (from dark red to white) the attack order, while their size represents their betweenness value. In the attacked network, darker nodes do not belong to the largest connected component, and their contour color represents the component they belong to. Figure reproduced from [202]. Copyright © 2021, The Authors

26 removals to break the LLC. The next line is GND, which was proposed in 2019 and performs better than its ancestors by dismantling the LLC with 25 removals. The quantum leap is obtained with GDM, proposed in 2021, and reducing the LLC with a smaller set of 18 nodes. We wonder, which one will be next in this sequence of better and better algorithms? Is the actual optimal set already on the horizon? The search for the optimal set is still open, yet the algorithms are improving as time passes.

Performing a thorough analysis over a large number of empirical systems confirms the results of Fig. 1.27. Figure 1.28 shows that machine learning outperform other methods under study, and it is able to identify higher-order topological patterns that are transparent to other methods and centralities, and are not obvious to a human observer.

Artime and De Domenico [344] studied percolation on networks by adding metadata about nodes. The idea is that in many applications one wants to dismantle the network following more sophisticated protocols, for instance, by combining topological properties and non-topological node information when available. A theoretical framework– named feature-enriched percolation– was developed to account also for dynamical features of nodes. This allows to study critical behavior of networks beyond purely structural indicators. Applications using the Bayesian machine scientist developed by Guimerà et al. [345] help to find an analytical form for a complicated relation, that later is used as feature in the theory. This is an example of how explainable machine learning can be used to assist researchers.

This research exemplifies the power of AI applied to the influencer problem. This machine-assisted analysis is more efficient than many centrality models. We expect that in the future, machine learning will turn not only into a revolutionary approach for finding influencers, but also will shed light on applications for related problems

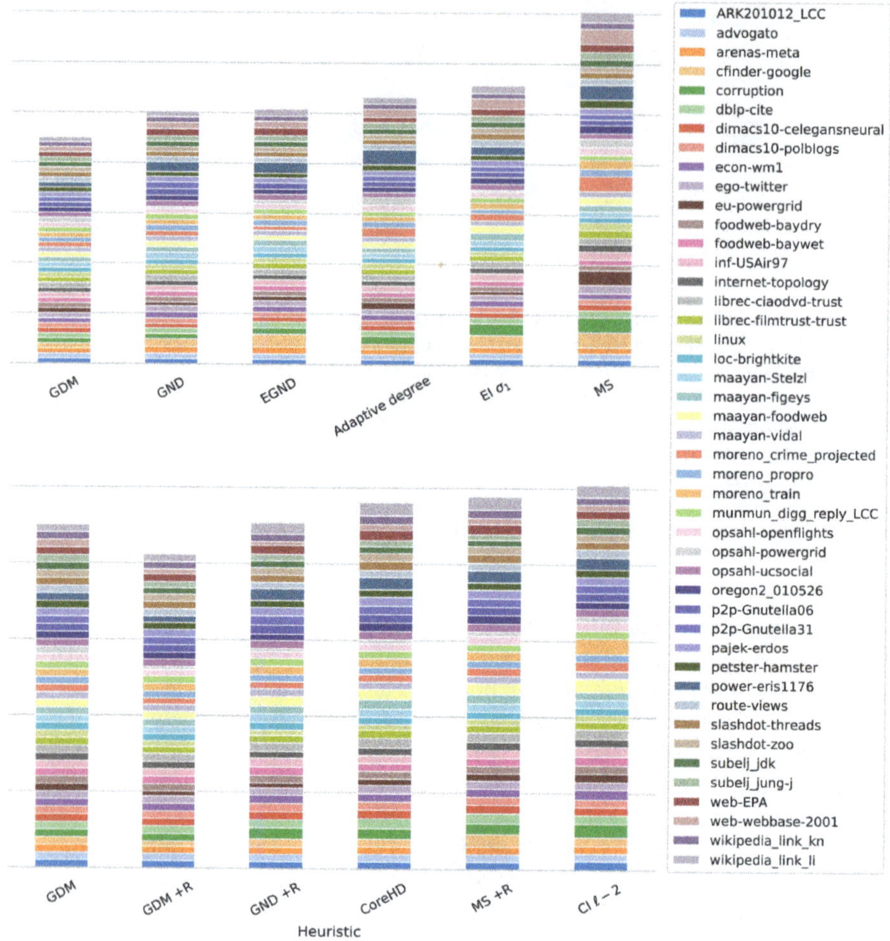

Fig. 1.28 **Performance of the machine learning GDM approach against other methods**. The figure shows the per-method cumulative area under the curve (AUC) of the LLC curve over a large number of real-world network dismantling shown on the right. The lower the column, the better the performance. The GDM+R with reinsertion is the best performer, confirming the results for the particular network in Fig. 1.27. Figure reproduced from [202]. Copyright © 2021, The Authors

like quantitative methods to assess systemic risks and early detection of system's instabilities and tipping points.

1.9 Outlook for Theory of Influencers

We have seen how the identification of influencers can be attacked from many different points of view. For instance, the spreading and disintegration problem can be mapped to optimal percolation and this problem can be solved with CI by optimizing the spectral radius of the NB matrix; the BPD and network dismantling algorithms to optimally remove cycles in the network; the EI algorithm to gradually identify less vital nodes so that an explosive collapse of the network would occur if the remaining critical nodes are attacked, the EGP and GND algorithms to recursively partition the network into equal-size components with different techniques, or the large deviation theory of percolation considering the rare events deviating from random percolation. These algorithms can tackle the problem of influencers, yet with different performances and scalability according to different applications.

A review of algorithms for optimal percolation and dismantling has been published by Artime et al. [346]. The studied algorithms are summarized in Table 1.4 together with their computational complexity. This study includes a comparison among the most popular algorithms and a repository containing scripts and data of the studied algorithms available at https://networkdismantling.github.io and https://github.com/NetworkDismantling/review. We reproduce the results of this comparison in Fig. 1.29 on a particular real-world network. Artime et al. also provides an integrated code to compare the performance of the algorithms systematically.

In terms of implementation, CI is the easiest to implement since it consists of a single explicit formula, Eq. (1.84), which is easily implemented algorithmically. It is also intuitive to interpret since it has a simpler interpretation in terms of weak nodes. On the other hand, algorithms based on BP, network decycling and dismantling and

Table 1.4 Summary of the main algorithms reviewed by Artime et al. [346] and shown in Fig. 1.29 for targeted attacks including the computational complexity for sparse networks in terms of the number of nodes $|V|$ and edges $|E|$ of the network. T is a fixed parameter of the Min-Sum algorithm, $\epsilon > 0$, e is the ensemble size, and h is the number of attention heads employed. Table reproduced from [346]

Algorithm	Type	Computational complexity	Refs.
Collective Influence (CI)	Optimal percolation	$O(\|V\|log\|V\|)$	[242]
Belief Propagation-guided Decimation (BPD)	Message passing-based decycling	$O(\|E\|T)$	[17]
Min-Sum	Message passing-based decycling	$O(\|E\|T) + O(\|V\|(\log(\|V\|) + T))$	[16]
Generalized Network Dismantling (GND)	Spectral partitioning	$O(\|V\| \log^{(2+\epsilon)}(\|V\|))$	[342]
Ensemble GND (EGND)	Spectral partitioning	$O(e) \cdot O(GND)$	[347]
CoreHD	Degree-based decycling	$O(\|V\|)$ (on sparse networks)	[127]
Explosive Immunization (EI)	Explosive percolation	$O(\|V\| \log(\|V\|))$	[18]
GDM	Machine learning	$O(h(\|V\| + \|E\|))$	[202]
CoreGDM	Machine learning	$O(h(\|V\| + \|E\|))$	[348]
FINDER	Machine learning	$O(\|E\| + \|V\|(1 + \log(\|V\|))$	[349]

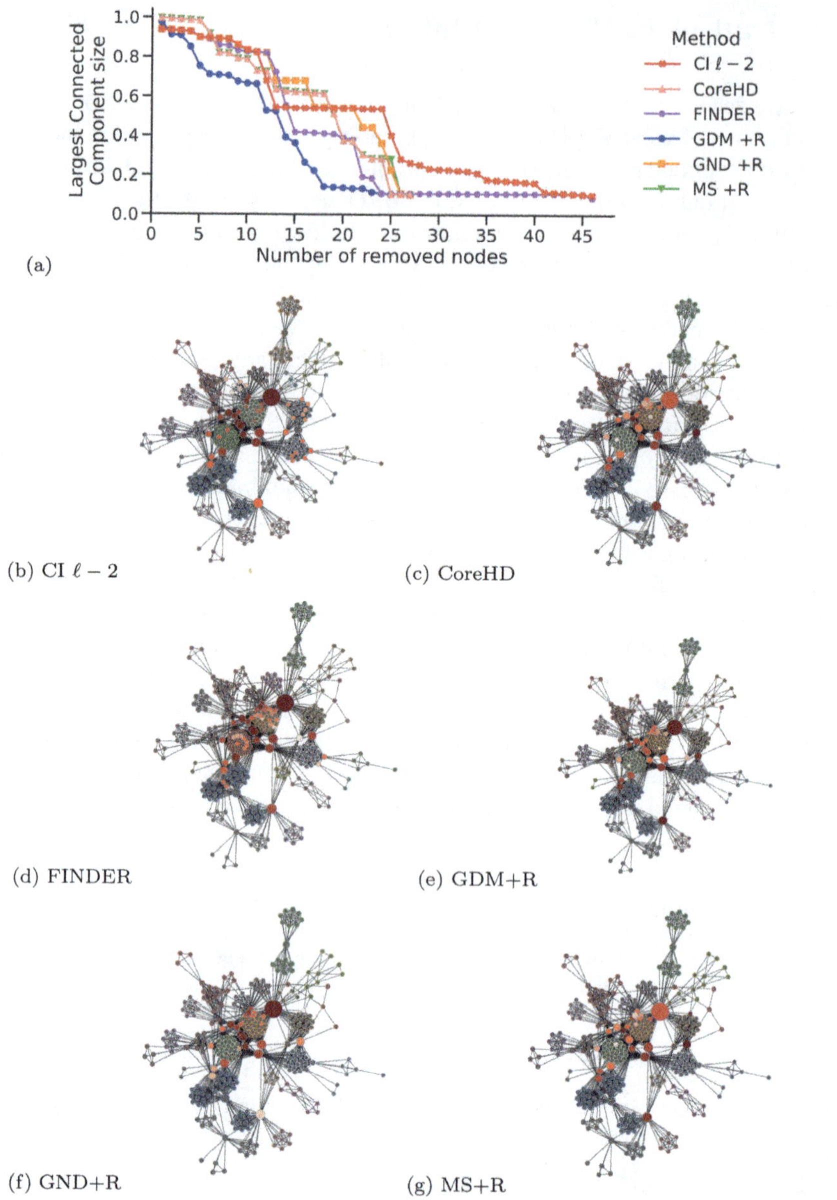

(a)

(b) CI $\ell - 2$ (c) CoreHD

(d) FINDER (e) GDM+R

(f) GND+R (g) MS+R

Fig. 1.29 A comparison of state-of-the-art algorithms for dismantling networks performed by Artime et al. [346]. **a** The cutting-edge algorithms explained in Table 1.4 compared in terms of their ability to dismantle the system with the smallest number of removal as in optimal percolation on a particular network, the Brazilian corruption network from [350]. Plots from **b–g** represent different algorithms as indicated, see Table 1.4. R means with reinsertion. Plotted is the size of the largest connected component (LCC). The removed nodes are represented in orders by their colors from dark red to white in the attack order (nodes in gray are not removed). Their size represents their betweenness centrality. The contour color of the remaining nodes represents the cluster to which they belong after the attack. Figure reproduced from [346]

large deviations of percolation need to iterate BP and Min-Sum equations to find solutions; these solutions are implicit, being more complex from the point of view of algorithmic implementation and interpretation.

Figures 1.18, 1.20, 1.21, and 1.25 shows a comparison of these algorithms. Most algorithms are close to the putative optimal configuration. The improvements found are minor and, thus, may not be of relevance in real-world applications. Some improvements are obtained at the expense of complex algorithmic developments.

Comparisons among algorithms are made using the value of the critical point q_c^{opt}. However, q_c^{opt} is not necessary for real applications since one is interested in the actual configuration of influencers. Minor improvements in the critical value that do not translate to an improvement in the configuration of influencers have little practical applications.

A commonality of these approaches is that an essential global optimization problem is transformed into another problem that can be solved locally. For instance, CI defines a centrality based on local structure; the BP equations in the BPD and network dismantle algorithms incorporate local constraints compatible with the global constraints; the score calculation in the EI algorithm depends on local connectivity and the EGP and GND algorithms are designed to recursively partition smaller local clusters. We summarize the features of these algorithm in Table 1.5.

The features of the algorithms developed for the IMP are summarized in Table 1.5 too. First, greedy approaches are proposed to estimate the influence of each node using either Monte Carlo simulation or local structural information. Following this idea, improvements are designed in two directions: avoiding unnecessary simulations or developing better local proxies for influence. While the performance of greedy algorithms is guaranteed for dynamics with submodular property, its guarantee is far from the optimal and there exist many problems that are not submodular and for which there are no optimal guarantees. The advantage of message passing algorithms is that they do not require submodularity. This approach calculates the spreading outcomes by solving a set of BP equations, thus considering the problem from a global viewpoint.

Despite significant advances in influencer identification, many ongoing problems and directions must be addressed in future works. First, there is a gap between the state-of-the-art algorithms' results and the theoretical limit in structures such as random regular networks. This provides a theoretical room for improvement in algorithm design. Second, the topological structure of real-world complex networks can be much more complicated than the case considered in most algorithms, which are based on the assumption of sparseness. The comparative analysis of Wandelt et al. [201] found that the algorithms perform well only on specific network types. Further, connections may be time-varying in temporal networks [351] or possess complicated interlayer interactions in multiplex networks [352, 353]. Third, in many systems, links are often of different types with distinct functions. These systems cannot be described by the simple network structure discussed here, and they do not even admit a formal definition of influencers. In future works, these open problems will be explored in more detail.

Table 1.5 List of the main influencer algorithms discussed in this chapter, beyond the centralities discussed in Sect. 1.2

Year	Name	Description and complexity
2003	Influence Maximization Problem (IMP) [5]	MC and Greedy approach, submodular function, performance guaranteed within a factor $(1 - 1/e)$
2007	Cost-Effective Forward algorithm (CEF) [214]	Consider non-unit cost, performance guaranteed within a factor $(1 - 1/e)/2$
	Cost-Effective lazy forward (CELF) [214]	Fewer Monte Carlo simulations, higher efficiency, heap structure
2008	Equal graph partitioning (EGP) [270]	Recursively partition networks into clusters of similar size, avoid breaking small clusters
2009	Degree discount algorithm [216]	Direct influence between immediate neighbors
2010	Maximum influence arborescence (MIA) [218]	Maximum influence path, Dijkstra shortest-path algorithm, arborescence structure, tradeoff between computational efficiency and performance
	Community-based algorithm [223]	Community detection, dynamic programming algorithm
2011	SIMPATH algorithm [215]	Enumerating simple paths, look ahead optimization, trade-off between accuracy and efficiency
2013	Belief-propagation algorithm [12]	Large deviations of LTM dynamics, consider cost and revenue, Max-Sum equations, solved by iteration until convergence
2014	message passing approach [13]	Belief propagation, Max-Sum algorithm, SIR and SIS model, solved through iteration
2015	Collective Influence (CI) [14]	Stability of the NB matrix, greedy approach, easy to interpret and implement, adapted to real-world problems. $O(N \log N)$
	Survey propagation [268]	Near-optimal performance, solved by iteration until convergence, applied to random regular graphs
2016	Network dismantle [16]	Optimal decycling, belief-propagation approach, Min-Sum algorithm, solved by iteration until convergence, compute the optimal node set simultaneously. $O(MT)$ per iteration
	Belief-propagation guided decimation (BPD) [17]	Minimum feedback vertex set, spin glass model, belief-propagation approach, no convergence needed in BP iteration, select nodes iteratively. $O(N \log N)$
	Explosive Immunization (EI) [18]	Explosive percolation, iteratively select less important nodes from candidates, based on the score defined for each node. $O(N \log N)$
	CoreHD [127]	Remove high-degree nodes in 2-core, perform well on loopy networks. $O(N)$ complexity
2017	Collective Influence in Threshold Model (CI-TM) [224]	Greedy approach, based on linearized message passing equations, uses subcritical clusters to estimate influence. $O(N \log N)$
2018	message passing approach [298]	Expected size of epidemic outbreaks, insensitive to origin, applicable to weighted networks
2019	Generalized network dismantle (GND) [342]	Consider the costs of removing nodes, node-weighted partition, recursive equal-size partition, solved by spectral properties of a Laplacian and weighted vertex cover. $O(N \log^{2+\epsilon}(N))$
	Weak-Neighbor [269]	Improvement over the generalized CoreHD algorithm and CI-TM algorithm. $O(N)$
2021	Graph Dismantling with Machine Learning (GDM) [202]	Machine learning to find influencers, GAT coupled with a multilayer perceptron

Overall, the central message of this chapter is that there is no single way to choose a centrality. In principle, the influencers depend on the kind of graph, whether they are directed or undirected, unweighted or weighted, and the dynamics. We could consider whether one is looking for nodes with the most brokering power, nodes with the most influence, or the minimal set of nodes that can spread information to the whole network. The differences in ranking obtained from different centrality measures should be compared to better understand the significance of each ranking in a given context.

Most importantly, the centralities depend on the type of interactions that describe the system. In this chapter, we have covered mainly interactions that can be abstracted into a graph. A graph captures only pair-wise interactions and does not capture many-body higher-order interactions where more than two nodes act at the same time. This is an important generalization of the influencer problem, which we have not addressed in this chapter, and requires more development in the future. Indeed, the pair-wise assumption engraved in the graph, which says that the nodes interact in pairs and, therefore, can be represented with an adjacency matrix, is too contrived to describe all processes in nature. It would be more natural to assume that nodes in a system can interact in groups rather than in pairs. This leads to the consideration of an hypergraph, which generalized the notion of a graph in which an edge can join any number of nodes, in contrast to a pair-wise graph where an edge joins only two nodes. Thus, a graph is a particular realization of a hypergraph, the latter being more general and allowing many more interesting problems to be taken into account. Multiplex networks are also important generalizations of graphs where nodes are allowed to have a set of different kinds of interactions. Temporal and dynamical networks, like those we will consider for the transmission chain of COVID-19 in Chap. 4, are important in applications as well.

Research is currently exploding in all these directions to understand the dynamics and structural properties of all these variants of networked systems beyond graphs. The development of centralities for these extended structures is an exciting field of current research. A graph is too constrained to capture the dynamics of biological networks, where proteins, for instance, act in complexes. There are many ongoing interesting works along these directions on hypergraphs [159, 160, 354–356], on simplicial complexes, which consider higher-order simplices rather than nodes; an excellent introduction to this topic is the book of Bianconi [357, 358].

Centralities in temporal networks or time-evolving networks are being developed in [163–165, 359, 360]. Multiplex network centralities are being studied in [87–89], and multilayer networks in [161, 162, 352, 353, 361–363]. De Domenico et al. [364] introduced a framework based on random walks to study the navigability of interdependent/multilayer networks in case of random failures to study functional dismantling. They attack a multilayer network, according to specific protocols (even using actual disruptions in the London tube) and provide an analytical method to quantify the resilience to failures in terms of navigability.

Unfortunately, we have just scrapped the surface on all these significant developments; we expect that in the coming years influencer theory will be further developed in all these different exciting directions.

Spatial networks are also important generalizations of a graph, where the nodes in the graph are associated to a geometric space [365], like neurons in the brain, internet routers, and power grids. According to de Arruda et al. [366], in non-spatial networks, k-shell index and degree centrality are more relevant to epidemic spreading, while average neighborhood degree, closeness, and accessibility suit better for rumor dynamics [281]. The difference between non-spatial and spatial networks illustrates that the interplay between topology and dynamics should be taken into account in influencer identification, an idea that we will develop in subsequent chapters. Hyperbolic geometric networks [367], as generalizations of random geometric networks, are also important spatial structures that require influencer-finding developments.

This vast research landscape is further extended to studies of the properties of centrality measures based on non-backtracking walks [194] and centralities in signed networks [368]. Research is also active in the development of centralities for specific applications based on network dynamics including centralities for social networks based on game-theoretical concepts [369–371], centralities for detecting vulnerabilities in power grids based on networks of oscillators [372–374], and centralities for identifying key genes in gene regulatory networks based on the propagation of biological signals [375–378] to be discussed in Chap. 5.

Despite their simplifications and contrived natures, the basic topological centralities defined on graphs are still in use due to the basic universal principles behind them and their high computational efficiency. In the following chapters, we will move on from topological centralities and bring the dynamics into consideration. This will open the influencers problem to be applied to concrete applications, as we will see in subsequent chapter. Yet, the main approximation considered is that the system interacts through a graph.

And now, what do we do with all this machinery? Imagine that someone comes to you with the following problem: find the influencers! The first step is to build the network. Then, we can pick a set of suitable influencer finding algorithms according to the characteristics of the network and dynamics. We can not tell in advance which algorithm will be better suited. It all depends on the type of network and the type of interactions that define the links in the network.

In next chapters we will study some urgent applications of influencer science in social and biological settings. These study-cases will give the reader an idea of at least the class of algorithms that can be applied to each case. For instance, when trying to break the chain of transmission of COVID-19 in Chap. 4, the most suitable algorithms will belong to the class of percolation or dismantling algorithms, since a strategy to stop the spreading of the disease is to break the giant connected component. In this case, we will investigate CI, kcore, betweenness centrality, high degree, PageRank, or the decycling and dismantling more sophisticated ones. A point that it should be emphasized again and again, is that, in real life situations, one is dealing with billions of interactions and millions, perhaps also billions of nodes in a network, for instance, studying Facebook or the entire human population. Tackling these massively large problems requires the use of near-linear scalable algorithms.

In principle, this means that the runtime of the algorithm should scale polynomially in the number of nodes, i.e., $O(N^\alpha)$, where α is some exponent. This, in principle, rules out the use of the set of algorithms that are really intractable, scaling exponentially or even factorially with N. However, in real life problems, being polynomial is not enough. Experience tells us that in real applications any algorithm that scales quadratically and above ($\alpha \geq 2$) will have no use in tackling real-world problems. Thus, algorithms that scale at most as $O[N \log(N)]$ are mostly desirable. The logarithmic dependence comes about in many problems, for instance, ordering a set is logarithmic and replaces a quadratic solution with one that is quite tractable. Even when the algorithm may scale nearly linearly, they could have pre-factors that depend on the parameters of the model that may render the algorithms slow in real systems, like in the case studied in Sect. 1.6.5.2. Anything slower than a strictly nearly linear complexity will most probably have no practical applications in real-life.

In Chap. 5, we will see that taking into account the dynamics opens up more exciting problems than percolation. Yet, these problems are not amenable to simple solutions like CI, and sophisticated numerical schemes need to be used to solve them. These schemes may not scale well, thus one is always attracted again to the scalable solutions using simple structural heuristics, like CI. An excellent example where the dynamical solution introduces novel physics into the problem of finding influencers will be explored in the mutualistic ecosystems studied in Chap. 7. Due to the simplicity of the interactions (all positive), the dynamical system can be solved precisely, and the influencers emerge from the structure of the k-core quite naturally.

Since different algorithms give different set of influencers, it is advisable to study the overlap between the influencer sets provided by these algorithms. The ultimate validation should be done against experimental data. For instance, influencers in social media identified by structural heuristics should be verified by their reach to their audience. We will discuss some KPIs in Chap. 3. These audiences are global, and validations should not be confined to the immediate echo-chamber of the influencers. That is, influencers should be tested as a collective phenomenon, since we are looking for a minimal set that could reach the entire population, so there could be a lot of overlaps in the reached audiences among them. A global maximization of influence tries to avoid these overlaps between influencer reaches as much as possible. So, experimental validations of the influencer status should be performed with care, see Teng et al. [379].

A more straightforward validation is percolation. One can remove the influencers one by one from the network (this can be done adaptably, that is recomputing the list of influencers every time that one of them is removed) and monitor the size of the GCC. Chap. 4, Sect. 4.4 provides an example. The algorithm with the list that will destroy the GCC with the minimal removals could be thought as the most optimal for the given application. Again, this is just an approximate validation procedure since the mapping to optimal percolation occurs only under certain circumstances as discussed in Sect. 1.6.3.

A second layer of validation is the test of influencers in the real world and this validation is highly dependent on the problem at hand. For instance, a real-world validation has been performed by utilizing a list of influencers in a real marketing campaign, and it was shown that the response rate in this marketing campaign was increased by a factor of 3x when the influencer algorithms were used as compared to picking influencers at random [243]. We describe this study in Sect. 3.3.1. In general, the reader is left to use her/his own creativity to treat and validate each problem at hand. Good luck!

References

1. Barabási, A.-L., Albert, R.: Emergence of scaling in random networks. Science **286**, 509–512 (1999)
2. Kitsak, M., Gallos, L.K., Havlin, S., Liljeros, F., Muchnik, L., Stanley, H.E., Makse, H.A.: Identifying influential spreaders in complex networks. Nat. Phys. **6**, 888–893 (2010)
3. Granovetter, M.S.: The strength of weak ties. Am. J. Soc. **78**, 1360–1380 (1973)
4. MIT Technology Review: Emerging Technology from the arXiv. In: The Emerging Science of Superspreaders (And How to Tell If You're One Of Them) (13 May 2014). https://www.technologyreview.com/2014/05/13/13350/the-emerging-science-of-superspreaders-and-how-to-tell-if-youre-one-of-them/. Accessed Apr 2022
5. Kempe, D., Kleinberg, J., Tardos, E.: Maximizing the spread of influence through a social network. In: Proceedings of 9th ACM SIGKDD International Conference on Knowledge Discovery and Data Mining, pp. 137–143 (2003)
6. Domingos, P., Richardson, M.: Mining the Network Value of Customers. In: Seventh International Conference on Knowledge Discovery and Data Mining (2001)
7. Domingos, P., Richardson, M.: Mining knowledge-sharing sites for viral marketing. In: Proceedings of 8th ACM SIGKDD International Conference on Knowledge Discovery and Data Mining, pp. 61–70 (2002)
8. Mézard, M., Montanari, A.: Information, Physics, and Computation. Oxford University Press, USA (2009)
9. Altarelli, F., Braunstein, A., Ramezanpour, A., Zecchina, R.: Stochastic optimization by message passing. J. Stat. Mech. P11009 (2011)
10. Altarelli, F., Braunstein, A., Ramezanpour, A., Zecchina, R.: Stochastic matching problem. Phys. Rev. Lett. **106**, 190601 (2011)
11. Altarelli, F., Braunstein, A., Dall'Asta, L., Zecchina, R.: Large deviations of cascade processes on graphs. Phys. Rev. E **87**, 062115 (2013)
12. Altarelli, F., Braunstein, A., Dall'Asta, L., Zecchina, R.: Optimizing spread dynamics on graphs by message passing. J. Stat. Mech. P09011 (2013)
13. Altarelli, F., Braunstein, A., Dall'Asta, L., Wakeling, J.R., Zecchina, R.: Containing epidemic outbreaks by message passing techniques. Phys. Rev. X **4**, 021024 (2014)
14. Morone, F., Makse, H.A.: Influence maximization in complex networks through optimal percolation. Nature **524**, 65–68 (2015)
15. Kovács, I.A., Barabási, A.-L.: News & views: destruction perfected. Nature **524**, 38–39 (2015)
16. Braunstein, A., Dall'Asta, L., Semerjian, G., Zdeborová, L.: Network dismantling. Proc. Natl. Acad. Sci. U.S.A. **113**, 12368–12373 (2016)
17. Mugisha, S., Zhou, H.-J.: Identifying optimal targets of network attack by belief propagation. Phys. Rev. E **94**, 012305 (2016)
18. Clusella, P., Grassberger, P., Pérez-Reche, F.J., Politi, A.: Immunization and targeted destruction of networks using explosive percolation. Phys. Rev. Lett. **117**, 208301 (2016)
19. Bovet, A., Makse, H.A.: Centralities in complex networks. In: Meyers, R.A. (eds.) Encyclopedia of Complexity and Systems Science. Springer, Berlin, Heidelberg (2021)
20. Albert, R., Jeong, H., Barabási, A.-L.: Error and attack tolerance in complex networks. Nature **406**, 378–382 (2000)

21. Cohen, R., Erez, K., ben-Avraham, D., Havlin, S.: Resilience of the Internet to random breakdown. Phys. Rev. Lett. **85**, 4626–4629 (2000)
22. Cohen, R., Erez, K., ben-Avraham, D., Havlin, S.: Breakdown of the Internet under intentional attack. Phys. Rev. Lett. **86**, 3682–3685 (2001)
23. Pastor-Satorras, R., Vespignani, A.: Epidemic spreading in scale-free networks. Phys. Rev. Lett. **86**, 3200–3203 (2001)
24. Pastor-Satorras, R., Vespignani, A.: Epidemic dynamics and endemic states in complex networks. Phys. Rev. E **63**, 066117 (2001)
25. Pastor-Satorras, R., Vespignani, A.: Immunization of complex networks. Phys. Rev. E **65**(3), 036104 (2002)
26. Klemm, K., Serrano, M.A., Eguíluz, V.M., San Miguel, M.: A measure of individual role in collective dynamics. Sci. Rep. **2**, 292 (2012)
27. Seidman, S.B.: Network structure and minimum degree. Soc. Netw. **5**, 269–287 (1983)
28. Batagelj, V., Zaveršnik, M.: Fast algorithms for determining (generalized) core groups in social networks. Adv. Data Anal. Classif. **5**, 129–145 (2011)
29. Lick, D.R., White, A.T.: k-degenerate graphs. Canad. J. Math. **22**, 1082–1096 (1970)
30. Freuder, E.C.: A sufficient condition for backtrack-free search. J. ACM **29**, 24–32 (1982)
31. Kirousis, L.M., Thilikos, D.M.: The linkage of a graph. SIAM J. Comput. **25**, 626–647 (1996)
32. Erdös, P., Hajnal, A.: On chromatic number of graphs and set-systems. Acta Mathematica Academiae Scientiarum Hungarica **17**, 61–99 (1966)
33. Batagelj, V., Zaversnik, M.: An $O(m)$ algorithm for cores decomposition of networks (2003). arXiv:cs.DS/0310049
34. Batagelj, V., Mrvar, A., Zaveršnik, M.: Partitioning approach to visualization of large graphs. In: International Symposium on Graph Drawing, pp. 90–97. Springer (1999)
35. Alvarez-Hamelin, J.I., Dall'Asta, L., Barrat, A., Vespignani, A.: Large scale networks fingerprinting and visualization using the k-core decomposition. In: NIPS'06, Advances in Neural Information Processing Systems, pp. 41–50 (2006). https://lanet-vi.fi.uba.ar
36. Alvarez-Hamelin, J.I., Dall'Asta, L., Barrat, A., Vespignani, A.: K-core decomposition of the internet graphs: hierarchies, self-similarity and measurement biases. Netw. Heterog. Media **3**, 371 (2008)
37. Bastian, M., Heymann, S., Jacomy, M.: Gephi: an open source software for exploring and manipulating networks. In: International AAAI Conference on Weblogs and Social Media, vol. 8, pp. 361–362 (2009)
38. Giatsidis, C., Berberich, K., Thilikos, D.M., Vazirgiannis, M.: Visual exploration of collaboration networks based on graph degeneracy. In: Proceedings of the 18th ACM SIGKDD International Conference on Knowledge Discovery and Data Mining, pp. 1512–1515. ACM (2012)
39. Colomer-de-Simón, P., Serrano, M.A., Beiró, M.G., Alvarez-Hamelin, J.I., Boguñá, M.: Deciphering the global organization of clustering in real complex networks. Sci. Rep. **3**, 2517 (2013)
40. Carmi, S., Havlin, S., Kirkpatrick, S., Shavitt, Y., Shir, E.: A model of Internet topology using k-shell decomposition. Proc. Natl. Acad. Sci. U.S.A. **104**, 11150 (2007)
41. Song, C., Havlin, S., Makse, H.A.: Self-similarity of complex networks. Nature **433**, 392–395 (2005)
42. Song, C., Havlin, S., Makse, H.A.: Origins of fractality in the growth of complex networks. Nat. Phys. **2**, 275–281 (2006)
43. Anderson, R.M., May, R.M.: Infectious Diseases of Humans: Dynamics and Control. Oxford University Press, Oxford (1992)
44. Bott, E.: Family and Social Network. Tavistock, London (1957)
45. Bollobás, B.: Graph Theory and Combinatorics. In: Bollobás, B. (ed.) Proceedings of Cambridge Combinatorial Conference in honour of Paul Erdös, pp. 35–57. Academic Press, NY (1984)
46. Luczak, T.: Size and connectivity of the k-core of a random graph. Discret. Math. **91**, 61–68 (1991)

47. Erdös, P., Rényi, A.: On the evolution of random graphs. Publ. Math. Inst. Hung. Acad. Sci. **5**, 17–61 (1960)
48. Pittel, B., Spencer, J., Wormald, N.: Sudden emergence of a giant k-core in a random graph. J. Comb. Theory, Ser. B **67**, 111 (1996)
49. Dorogovtsev, S.N., Goltsev, A.V., Mendes, J.F.F.: K-core organization of complex networks. Phys. Rev. Lett. **96**, 040601 (2006)
50. Malliaros, F., Giatsidis, C., Papadopoulos, A., Vazirgiannis, M.: The core decomposition of networks: theory, algorithms and applications. VLDB J. **29**, 61–92 (2020)
51. Pei, S., Muchnik, L., Andrade, J.S., Jr., Zheng, Z., Makse, H.A.: Searching for superspreaders of information in real-world social media. Sci. Rep. **4**, 5547 (2014)
52. Borge-Holthoefer, J., Moreno, Y.: Absence of influential spreaders in rumor dynamics. Phys. Rev. E **85**, 026116 (2012)
53. Quax, R., Apolloni, A., Sloot, P.M.A.: The diminishing role of hubs in dynamical processes on complex networks. J. R. Soc. Interface **10**, 20130568 (2013)
54. Zeng, A., Zhang, C.-J.: Ranking spreaders by decomposing complex networks. Phys. Lett. A **377**, 1031–1035 (2013)
55. Malliaros, F.D., Rossi, M.-E.G., Vazirgiannis, M.: Locating influential nodes in complex networks. Sci. Rep. **6**, 19307 (2016)
56. Cohen, J.: Trusses: cohesive subgraphs for social network analysis. National Security Agency Technical Report (2008)
57. Li, R.-H., Qin, L., Yu, J.X., Mao, R.: Influential community search in large networks. In: Proceedings of VLDB Endowment, vol. 8, pp. 509–520 (2015)
58. Shin, K., Eliassi-Rad, T., Faloutsos, C.: Patterns and anomalies in k-cores of real-world graphs with applications. Knowl. Inf. Syst. **54**, 677–710 (2018)
59. Brown, P., Feng, J.: Measuring user influence on Twitter using modified k-shell decomposition. In: The Social Mobile Web, volume WS-11-02 of AAAI Workshops. AAAI (2011)
60. Al-garadi, M.A., Varathan, K.D., Ravana, S.D.: Identification of influential spreaders in online social networks using interaction weighted k-core decomposition method. Phys. A **468**, 278–288 (2017)
61. Liu, Y., Tang, M., Zhou, T., Do, Y.: Core-like groups result in invalidation of identifying super-spreader by k-shell decomposition. Sci. Rep. **5**, 9602 (2015)
62. Liu, Y., Tang, M., Zhou, T., Do, Y.: Improving the accuracy of the k-shell method by removing redundant links-from a perspective of spreading dynamics. Sci. Rep. **5**, 13172 (2015)
63. Batagelj, V., Zaveršnik, M.: Generalized cores (2002). arXiv:cs.DS/0202039
64. Govindan, P., Wang, C., Xu, C., Duan, H., Soundarajan, S.: The k-peak decomposition: mapping the global structure of graphs. In: Proceedings of the 26th International Conference on World Wide Web, WWW'17, pp. 1441–1450 (2017)
65. Saríyüce, A.E., Seshadhri, C., Pinar, A., Catalyürek, U.V.: Finding the hierarchy of dense subgraphs using nucleus decompositions. In: Proceedings of the 24th International Conference on World Wide Web, WWW'15, pp. 927–937 (2015)
66. Saríyüce, A.E., Pinar, A.: Peeling bipartite networks for dense subgraph discovery. In: Proceedings of the Eleventh ACM International Conference on Web Search and Data Mining, WSDM, pp. 504–512 (2018)
67. Benson, A.R., Gleich, D.F., Leskovec, J.: Higher-order organization of complex networks. Science **353**, 163–166 (2016)
68. Wang, K., Cao, X., Lin, X., Zhang, W., Qin, L.: Efficient computing of radius-bounded k-cores. In: 2018 IEEE 34th International Conference on Data Engineering (ICDE), pp. 233–244 (2018)
69. Rozenfeld, H.D., Rybski, D., Gabaix, X., Makse, H.A.: The area and population of cities: new insights from a different perspective on cities. Am. Econ. Rev. **101**, 560–580 (2011)
70. Rozenfeld, H.D., Rybski, D., Andrade, J.S., Jr., Batty, M., Stanley, H.E., Makse, H.A.: Laws of population growth. Proc. Nat. Acad. Sci. U.S.A. **105**, 18702–18707 (2008)
71. Makse, H.A., de Andrade, J.S., Batty, M., Havlin, S., Stanley, H.E.: Modeling urban growth patterns with correlated percolation. Phys. Rev. E **58**, 7054–7062 (1998)

72. Bonchi, F., Khan, A., Severini, L.: Distance-generalized core decomposition. In: Proceedings of the 2019 ACM SIGMOD International Conference on Management of Data (2019)
73. Zhang, F., Zhang, Y., Qin, L., Zhang, W., Lin, X.: When engagement meets similarity: Efficient (k, r)-core computation on social networks. In: Proceedings of the VLDB Endowment, vol. 10, pp. 998–1009 (2017)
74. Eidsaa, M., Almaas, E.: S-core network decomposition: a generalization of k-core analysis to weighted networks. Phys. Rev. E **88**, 062819 (2013)
75. Giatsidis, C., Thilikos, D.M., Vazirgiannis, M.: Evaluating cooperation in communities with the k-core structure. In: ASONAM'11: Proceedings of the International Conference on Advances in Social Networks Analysis and Mining, pp. 87–93 (2011)
76. Garas, A., Schweitzer, F., Havlin, S.: A k-shell decomposition method for weighted networks. New J. Phys. **14**, 083030 (2012)
77. Giatsidis, C., Thilikos, D.M., Vazirgiannis, M.: D-cores: measuring collaboration of directed graphs based on degeneracy. In: ICDM'11: Proceedings of the 11th IEEE International Conference on Data Mining, pp. 201–210 (2011)
78. Leskovec, J., Huttenlocher, D., Kleinberg, J.: Signed networks in social media. In: Proceedings of the SIGCHI Conference on Human Factors in Computing Systems (ACM, 2010), pp. 1361–1370 (2010)
79. Chiang, K.-Y., Hsieh, C.-J., Natarajan, N., Dhillon, I.S., Tewari, A.: Prediction and clustering in signed networks: a local to global perspective. J. Mach. Learn. Res. **15**, 1177–1213 (2014)
80. Tang, J., Chang, Y., Aggarwal, C., Liu, H.: A survey of signed network mining in social media. ACM Comput. Surv. (CSUR) **49**, 42 (2016)
81. Giatsidis, C., Cautis, B., Maniu, S., Thilikos, D.M., Vazirgiannis, M.: Quantifying trust dynamics in signed graphs, the s-cores approach. In: Proceedings of the 2014 SIAM International Conference on Data Mining, Philadelphia, Pennsylvania, USA, Apr 24–26, 2014, pp. 668–676 (2014)
82. Ding, D., Li, H., Huang, Z., Mamoulis, N.: Efficient fault-tolerant group recommendation using alpha-beta-core. In: Proceedings of the 2017 ACM on Conference on Information and Knowledge Management, CIKM'17, pp. 2047–2050. New York, NY, USA. ACM (2017)
83. Miorandi, D., De Pellegrini, F.: K-shell decomposition for dynamic complex networks. In: Proceedings of the 8th International Symposium on Modeling and Optimization in Mobile, Ad Hoc and Wireless Networks (WiOpt, IEEE), pp. 488–496 (2010)
84. Saríyüce, A.E., Gedik, B., Jacques-Silva, G., Wu, K.-L., Catalyürek, U.V.: Streaming algorithms for k-core decomposition. In: Proceedings of the VLDB Endowment, vol. 6, pp. 433–444 (2013)
85. Zhang, G.-Q., Zhang, G.-Q., Yang, Q.-F., Cheng, S.-Q., Zhou, T.: Evolution of the internet and its cores. New J. Phys. **10**, 123027 (2008)
86. Wu, H., Cheng, J., Lu, Y., Ke, Y., Huang, Y., Yan, D., Wu, H.: Core decomposition in large temporal graphs. In: BigData, pp. 649–658. IEEE (2015)
87. Azimi-Tafreshi, N., Gómez-Gardeñes, J., Dorogovtsev, S.N.: k-core percolation on multiplex networks. Phys. Rev. E **90**, 032816 (2014)
88. Galimberti, E., Bonchi, F., Gullo, F.: Core decomposition and densest subgraph in multilayer networks. In: Proceedings of the 2017 ACM on Conference on Information and Knowledge Management, CIKM'17, pp. 1807–1816. New York, NY, USA, ACM (2017)
89. Osat, S., Radicchi, F., Papadopoulos, F.: k-core structure of real multiplex networks. Phys. Rev. Res. **2**, 023176 (2020)
90. Buchanan, M., Caldarelli, G., De Los Rios, P., Rao, F., Vendruscolo, M. (eds.): Networks in Cell Biology. Cambridge University Press, Cambridge (2010)
91. Wuchty, S., Almaas, E.: Evolutionary cores of domain co-occurrence networks. BMC Evol. Biol. **5**, 24 (2005)
92. Emerson, A.I., Andrews, S., Ahmed, I., Azis, T.K., Malek, J.A.: K-core decomposition of a protein domain co-occurrence network reveals lower cancer mutation rates for interior cores. J. Clin. Bioinforma. **5**, 1 (2015)

93. Pellegrini, M., Baglioni, M., Geraci, F.: Protein complex prediction for large protein protein interaction networks with the core & peel method. BMC Bioinform. **17**, 372 (2016)
94. Srihari, S., Leong, H.W.: A survey of computational methods for protein complex prediction from protein interaction networks. J. Bioinform. Comput. Biol. **11**, 1230002 (2013)
95. Eidsaa, M.: Core Decomposition Analysis of Weighted Biological Networks. Ph.D thesis, NTNU (2016)
96. García-Algarra, J., Pastor, J., Mouronte, M.L., Galeano, J.: A structural approach to disentangle the visualization of bipartite biological networks. Complexity **2018**, 6204947 (2018)
97. Filho, H.A., Machicao, J., Bruno, O.M.: A hierarchical model of metabolic machinery based on the kcore decomposition of plant metabolic networks. PLoS ONE **13**, 1–15 (2018)
98. Verma, T., Russmann, F., Araujo, N., Nagler, J., Herrmann, H.J.: Emergence of core-peripheries in networks. Nat. Commun. **7**, 10441 (2016)
99. Csermely, P., London, A., Wu, L.-Y., Uzzi, B.: Structure and dynamics of core/periphery networks. J. Comp. Net. **1**, 93–123 (2013)
100. Borgatti, S.P., Everett, M.G.: Models of core/periphery structures. Soc. Net. **21**, 375–395 (2000)
101. Luo, F., Li, B., Wan, X.-F., Scheuermann, R.H.: Core and periphery structures in protein interaction networks. BMC Bioinform. **10**, s8 (2009)
102. Isaac, A.E., Sinha, S.: Analysis of core-periphery organization in protein contact networks reveals groups of structurally and functionally critical residues. J. Biosci. **40**, 683–699 (2015)
103. Hagmann, P., Cammoun, L., Gigandet, X., Meuli, R., Honey, C.J., Wedeen, V.J., Sporns, O.: Mapping the structural core of human cerebral cortex. PLoS Biol. **6**, e159 (2008)
104. Lahav, N., Ksherim, B., Ben-Simon, E., Maron-Katz, A., Cohen, R., Havlin, S.: K-shell decomposition reveals hierarchical cortical organization of the human brain. New J. Phys. **18**, 083013 (2016)
105. van den Heuvel, M.P., Sporns, O.: Rich-club organization of the human connectome. J. Neurosci. **31**, 15775–15786 (2011)
106. Li, Q., Del Ferraro, G., Pasquini, L., Peck, K.K., Makse, H.A., Holodny, A.I.: Core language brain network for fMRI language task used in clinical applications. Netw. Neurosci. **4**, 134–154 (2020)
107. Li, Q., Pasquini, L., Del Ferraro, G., Gene, M., Peck, K.K., Makse, H.A., Holodny, A.I.: Monolingual and bilingual language networks in healthy subjects using functional MRI and graph theory. Sci. Rep. **11**, 10568 (2021)
108. Arese Lucini, F., Del Ferraro, G., Sigman, M., Makse, H.A.: How the brain transitions from conscious to subliminal perception. Neuroscience **411**, 280–290 (2019)
109. Serafino, M., Monteiro, H.S., Luo, S., Reis, S.D.S., Igual, C., Lima Neto, A.S., Travizano, M., Jr. Andrade, J.S., Makse, H.A.: Digital contact tracing and network theory to stop the spread of COVID-19 using big-data on human mobility geolocalization. PLoS Comput. Biol. **18**, e1009865 (2022)
110. García-Algarra, J., Pastor, J.M., Iriondo, J.M., Galeano, J.: Ranking of critical species to preserve the functionality of mutualistic networks using the k-core decomposition. PeerJ **5**, e3321 (2017)
111. Morone, F., Del Ferraro, G., Makse, H.A.: The k-core as a predictor of structural collapse in mutualistic ecosystems. Nat. Phys. **15**, 95–102 (2019)
112. Schwarz, J., Liu, A.J., Chayes, L.: The onset of jamming as the sudden emergence of an infinite k-core cluster. Europhys. Lett. **73**, 560 (2006)
113. Morone, F., Burleson-Lesser, K., Vinutha, H.A., Sastry, S., Makse, H.A.: The jamming transition is a k-core percolation transition. Phys. A **516**, 172–177 (2019)
114. Sedes, O., Makse, H.A., Chakraborty, B., Morris, J.F.: K-core analysis of shear-thickening suspensions. J. Fluid Mech. **7**, 024304 (2022)
115. Cheng, J., Ke, Y., Chu, S., Ozsu, M.T.: Efficient core decomposition in massive networks. In: ICDE, pp. 51–62 (2011)
116. Montresor, A., De Pellegrini, F., Miorandi, D.: Distributed k-core decomposition. In: PODC, pp. 207–208 (2011)

117. Bhawalkar, K., Kleinberg, J., Lewi, K., Roughgarden, T., Sharma, A.: Preventing unraveling in social networks: the anchored k-core problem. In: ICALP'11: Proceedings of the 39th International Colloquium Conference on Automata, Languages, and Programming, pp. 440–451 (2011)
118. Zhang, Y., Parthasarathy, S.: Extracting analyzing and visualizing triangle k-core motifs within networks. In: ICDE'12: Proceedings of the 2012 IEEE 28th International Conference on Data Engineering, pp. 1049–1060 (2012)
119. Adiga, A., Vullikanti, A.K.S.: How robust is the core of a network? In: Blockeel, H., Kersting, K., Nijssen, S., Železný, F. (eds.): Machine Learning and Knowledge Discovery in Databases, pp. 541–556. Springer, Berlin, Heidelberg (2013)
120. Bonchi, F., Gullo, F., Kaltenbrunner, A., Volkovich, Y.: Core decomposition of uncertain graphs. In: KDD, pp. 1316–1325 (2014)
121. Lin, J.-H., Guo, Q., Dong, W.-Z., Tang, L.-Y., Liu, J.-G.: Identifying the node spreading influence with largest k-core values. Phys. Lett. A **378**, 3279–3284 (2014)
122. Tatti, N., Gionis, A.: Density-friendly graph decomposition. In: WWW, pp. 1089–1099 (2015)
123. Rousseau, F., Vazirgiannis, M.: Main core retention on graph-of-words for single-document keyword extraction. In: ECIR'15: Proceedings of the 37th European Conference on Information Retrieval, pp. 382–393 (2015)
124. He, X., Zhao, H., Cai, W., Li, G.-G., Pei, F.-D.: Analyzing the structure of earthquake network by k-core decomposition. Phys. A **421**, 34–43 (2015)
125. Wood, C.I., Hicks, I.V.: The minimal k-core problem for modeling k-assemblies. J. Math. Neurosci. **5**, 14 (2015)
126. Sarı́yüce, A.E., Gedik, B., Jacques-Silva, G., Wu, K.-L., Catalyürek, Ü.V.: Incremental k-core decomposition: algorithms and evaluation. VLDB J. **25**, 425–447 (2016)
127. Zdeborová, L., Zhang, P., Zhou, H.-J.: Fast and simple decycling and dismantling of networks. Sci. Rep. **6**, 37954 (2016)
128. Tixier, A., Malliaros, F.D., Vazirgiannis, M.: A graph degeneracy-based approach to keyword extraction. In: Proceedings of the 2016 Conference on Empirical Methods in Natural Language Processing. Association for Computational Linguistics, pp. 1860–1870 (2016)
129. Strouthopoulos, P., Papadopoulos, A.N.: Core discovery in hidden graphs. In: Data and Knowledge Engineering, CoRR (2017). arXiv:1712.02827
130. Kabir, H., Madduri, K.: Parallel k-core decomposition on multicore platforms. In: IPDPS Workshops, pp. 1482–1491. IEEE Computer Society (2017)
131. Danisch, M., Chan, T.-H.H., Sozio, M.: Large scale density-friendly graph decomposition via convex programming. In: Proceedings of the 26th International Conference on World Wide Web, WWW'17, pp. 233–242 (2017)
132. Cheng, S.-T., Chen, Y.-C., Tsai, M.-S.: Using k-core decomposition to find cluster centers for k-means algorithm in graphx on spark. In: Proceedings of the 8-th International Conference on Cloud Computing, GRIDs, and Virtualization, pp. 93–98 (2017)
133. Peng, Y., Zhang, Y., Zhang, W., Lin, X., Qin, L.: Efficient probabilistic k-core computation on uncertain graphs. In: 2018 IEEE 34th International Conference on Data Engineering (ICDE), pp. 1192–1203 (2018)
134. Hirsch, J.E.: An index to quantify an individual's scientific research output. Proc. Natl. Acad. Sci. U.S.A. **102**, 16569–16572 (2005)
135. Hirsch, J.E.: Does the h-index have predictive power? Proc. Natl. Acad. Sci. U.S.A. **104**, 19193–19198 (2007)
136. Lü, L., Zhou, T., Zhang, Q.-M., Stanley, H.E.: The H-index of a network node and its relation to degree and coreness. Nat. Commun. **7**, 10168 (2016)
137. Bavelas, A.: Communication patterns in task-oriented groups. J. Acoust. Soc. Am. **22**, 725–730 (1950)
138. Beauchamp, M.A.: An improved index of centrality. Behav. Sci. **10**, 161–163 (1965)
139. Sabidussi, G.: The centrality index of a graph. Psychometrika **31**, 581–603 (1966)
140. Dangalchev, C.: Residual closeness in networks. Phys. A **365**, 556–564 (2006)

141. Freeman, L.C.: Centrality in social networks: conceptual clarification. Soc. Netw. **1**, 215–239 (1978)
142. Friedkin, N.E.: Theoretical foundations for centrality measures. Am. J. Sociol. **96**, 1478–1504 (1991)
143. Freeman, L.C.: The gatekeeper, pair-dependency and structural centrality. Qual. Quant. **14**, 585–592 (1980)
144. Brandes, U., Borgatti, S.P., Freeman, L.C.: Maintaining the duality of closeness and betweenness centrality. Soc. Netw. **44**, 153–159 (2016)
145. Krackhardt, D.: Assessing the political landscape: Structure, cognition, and power in organizations. Adm. Sci. Q. **35**, 342–369 (1990)
146. Brandes, U.: On variants of shortest-path betweenness centrality and their generic computation. Soc. Netw. **30**, 136–145 (2008)
147. Teixeira, A.S., Monteiro, P.T., Carriço, J.A., Ramirez, M., Francisco, A.P.: Not seeing the forest for the trees: size of the minimum spanning trees (MSTs) forest and branch significance in MST-based phylogenetic analysis. PLoS ONE **10**, e0119315 (2015)
148. White, S., Smyth, P.: Algorithms for estimating relative importance in networks. In: Proceedings of the ninth ACM SIGKDD International Conference on Knowledge Discovery and Data Mining, pp. 266–275 (2003)
149. Newman, M.E.J.: A measure of betweenness centrality based on random walks. Soc. Netw. **27**, 39–54 (2005)
150. Katz, L.: A new status index derived from sociometric analysis. Psychometrika **18**, 39–43 (1953)
151. Bonacich, P.: Factoring and weighting approaches to status scores and clique identification. J. Math. Sociol. **2**, 113–120 (1972)
152. Newman, M.: Networks: An introduction. Oxford University Press, New York (2010)
153. Martin, T., Zhang, X., Newman, M.E.J.: Localization and centrality in networks. Phys. Rev. E **90**, 052808 (2014)
154. Radicchi, F., Castellano, C.: Leveraging percolation theory to single out influential spreaders in networks. Phys. Rev. E **93**, 062314 (2016)
155. Hashimoto, K.: Zeta functions of finite graphs and representations of p-adic groups. Adv. Stud. Pure Math. **15**, 211–280 (1989)
156. Girvan, M., Newman, M.E.J.: Community structure in social and biological networks. Proc. Natl. Acad. Sci. U.S.A. **99**, 7821–7826 (2002)
157. Evans, T.S.: Clique graphs and overlapping communities. J. Stat. Mech. P12037 (2010)
158. Bonacich, P.: Power and centrality: a family of measures. Am. J. Sociol. **92**, 1170–1182 (1987)
159. Bonacich, P., Holdren, A.C., Johnston, M.: Michael Hyper-edges and multidimensional centrality. Soc. Netw. **26**, 189–203 (2004)
160. Benson, A.R.: Three hypergraph eigenvector centralities. SIAM J. Math. Data Sci. **1**, 293–312 (2019)
161. Solá, L., Romance, M., Criado, R., Flores, J., García del Amo, A., Boccaletti, S.: Eigenvector centrality of nodes in multiplex networks. Chaos **23**, 033131 (2013)
162. De Domenico, M., Solé-Ribalta, A., Omodei, E., Gómez, S., Arenas, A.: Ranking in interconnected multilayer networks reveals versatile nodes. Nat. Commun. **6**, 1–6 (2015)
163. Nicosia, V., Tang, J., Mascolo, C., Musolesi, M., Russo, G., Latora, V.: Graph metrics for temporal networks. In: Temporal Networks, pp. 15–40 (2013)
164. Praprotnik, S., Batagelj, V.: Spectral centrality measures in temporal networks. Ars. Mathematica Contemporanea **11**, 11–33 (2015)
165. Taylor, D., Myers, S.A., Clauset, A., Porter, M., Mucha, P.J.: Eigenvector-based centrality measures for temporal networks. Multiscale Modeling Simul. **15**, 537–574 (2017)
166. Bauer, F., Lizier, J.T.: Identifying influential spreaders and efficiently estimating infection numbers in epidemic models: a walk counting approach. EPL **99**, 68007 (2012)
167. Lawyer, G.: Understanding the influence of all nodes in a network. Sci. Rep. **5**, 8665 (2015)
168. Vázquez, A., Moreno, Y.: Resilience to damage of graphs with degree correlations. Phys. Rev. E **67**, 015101 (2003)

169. Restrepo, J.G., Ott, E., Hunt, B.R.: Characterizing the dynamical importance of network nodes and links. Phys. Rev. Lett. **97**, 094102 (2006)
170. Estrada, E., Rodríguez-Velázquez, J.A.: Subgraph centrality in complex networks. Phys. Rev. E **71**, 056103 (2005)
171. Brin, S., Page, L.: The anatomy of a large-scale hypertextual web search engine. Comput. Netw. ISDN Syst. **30**, 107–117 (1998)
172. Page, L., Brin, S., Motwani, R., Winograd, T.: The PageRank citation ranking: Bringing order to the web, Stanford InfoLab (1999)
173. Kleinberg, J.: Authoritative sources in a hyperlinked environment. J. ACM **46**, 604–632 (1998)
174. Gibbs, S.: Google has 'outgrown' its 14-year-old mission statement, says Larry Page. The Guardian (3 Nov 2014)
175. Rosvall, M., Bergstrom, C.T.: Maps of random walks on complex networks reveal community structure. Proc. Nat. Acad. Sci. U.S.A. **105**, 1118–1123 (2008)
176. Delvenne, J.-C., Yaliraki, S.N., Barahona, M.: Stability of graph communities across time scales. Proc. Nat. Acad. Sci. U.S.A. **107**, 12755–12760 (2010)
177. Durrett, R.: Some features of the spread of epidemics and information on a random graph. Proc. Nat. Acad. Sci. U.S.A. **107**, 4491–4498 (2010)
178. Masuda, N., Porter, M.A., Lambiotte, R.: Random walks and diffusion on networks. Phys. Rep. **716–717**, 1–58 (2017)
179. Ermann, L., Frahm, K.M., Shepelyansky, D.L.: Google matrix analysis of directed networks. Rev. Mod. Phys. **87**, 1261 (2015)
180. Ding, Y., Yan, E., Frazho, A., Caverlee, J.: PageRank for ranking authors in co-citation networks. J. Assoc. Inf. Sci. Technol. **60**, 2229–2243 (2009)
181. Jing, Y., Baluja, S.: Visualrank: applying PageRank to large-scale image search. IEEE Trans. Pattern Anal. Mach. Intell. **30**, 1877–1890 (2008)
182. Iván, G., Grolmusz, V.: When the Web meets the cell: using personalized PageRank for analyzing protein interaction networks. Bioinformatics **27**, 405–407 (2010)
183. Perra, N., Fortunato, S.: Spectral centrality measures in complex networks. Phys. Rev. E **78**, 36107 (2008)
184. Adamic, L., Glance, N.: The political blogosphere and the 2004 US election: divided they blog. In: Proceedings of the 3rd International Workshop on Link Discovery, pp. 36–43 (2005)
185. Lü, L., Zhang, Y.-C., Yeung, C.H., Zhou, T.: Leaders in social networks, the delicious case. PLoS ONE **6**, e21202 (2011)
186. Travençolo, B.A.N., da Costa, L.F.: Accessibility in complex networks. Phys. Lett. A **373**, 89–95 (2008)
187. Arrigo, F., Grindrod, P., Higham, D.J., Noferini, V.: Non-backtracking walk centrality for directed networks. J. Complex Netw. **6**, 54–78 (2018)
188. Arrigo, F., Higham, D.J., Noferini, V.: Beyond non-backtracking: non-cycling network centrality measures. Proc. Roy. Soc. A: Math. Phys. Eng. Sci. **476**, 20190653 (2020)
189. Lin, Y., Zhang, Z.: Non-backtracking centrality based random walk on networks. Comput. J. **62**, 63–80 (2019)
190. Krzakala, F., Moore, C., Mossel, E., Neeman, J., Sly, A., Zdeborová, L., Zhang, P.: Spectral redemption in clustering sparse networks. Proc. Nat. Acad. Sci. U.S.A. **110**, 20935–20940 (2013)
191. Angel, O., Friedman, J., Hoory, S.: The non-backtracking spectrum of the universal cover of a graph. Trans. Am. Math. Soc. **367**, 4387–4388 (2015)
192. Aleja, D., Criado, R., García del Amo, A.J., Pérez, Á., Romance, M.: Non-backtracking PageRank: from the classic model to Hashimoto matrices. Chaos, Solitons Fractals **126**, 283–291 (2019)
193. Barucca, P., Tantari, D., Lillo, F.: Centrality metrics and localization in core-periphery networks. J. Stat. Mech. P023401 (2016)
194. Pastor-Satorras, R., Castellano, C.: The localization of non-backtracking centrality in networks and its physical consequences. Sci. Rep. **10**, 1–12 (2020)

195. Lü, L., Chen, D., Ren, X.-L., Zhang, Q.-M., Zhang, Y.-C., Zhou, T.: Vital nodes identification in complex networks. Phys. Rep. **650**, 1–63 (2016)
196. Liu, X., Li, D., Ma, M., Szymanski, B.K., Stanley, H.E., Gao, J.: Network resilience. Phys. Rep. **971**, 1–108 (2022)
197. Li, M., Liu, R.-R., Lü, L., Hu, M.-B., Xu, S., Zhang, Y.-C.: Percolation on complex networks: theory and application. Phys. Rep. **907**, 1–68 (2021)
198. Pei, S., Makse, H.A.: Spreading dynamics in complex networks. J. Stat. Mech. P12002 (2013)
199. Pei, S., Wang, J., Morone, F., Makse, H.A.: Influencer identification in dynamical complex systems. J. Complex Netw. **8**, cnz029 (2020)
200. Piraveenan, M., Thedchanamoorthy, G., Uddin, S., Chung, K.S.K.: Quantifying topological robustness of networks under sustained targeted attacks. Soc. Netw. Anal. Min. **3**, 939–952 (2013)
201. Wandelt, S., Sun, X., Feng, D., Zanin, M., Havlin, S.: A comparative analysis of approaches to network-dismantling. Sci. Rep. **8**, 13513 (2018)
202. Grassia, M., De Domenico, M., Mangioni, G.: Machine learning dismantling and early-warning signals of disintegration in complex systems. Nat. Commun. **12**, 5190 (2021)
203. Rogers, E.M.: Diffusion of Innovations. Simon and Schuster, New York (2010)
204. Centola, D.: How Behavior Spreads: The Science of Complex Contagions. Princeton University Press (2018)
205. Pastor-Satorras, R., Castellano, C., Van Mieghem, P., Vespignani, A.: Epidemic processes in complex networks. Rev. Mod. Phys. **87**, 925 (2015)
206. Granovetter, M.: Threshold models of collective behavior. Am. J. Sociol. **83**, 1420–1443 (1978)
207. Baxter, G.J., Dorogovtsev, S.N., Goltsev, A.V., Mendes, J.F.F.: Heterogeneous k-core versus bootstrap percolation on complex networks. Phys. Rev. E **83**, 051134 (2011)
208. Watts, D.J.: A simple model of global cascades on random networks. Proc. Natl. Acad. Sci. U.S.A. **99**, 5766–5771 (2002)
209. Dodds, P.S., Watts, D.J.: Universal behavior in a generalized model of contagion. Phys. Rev. Lett. **92**, 218701 (2004)
210. Boccaletti, S., Latora, V., Moreno, Y., Chavez, M., Hwang, D.-U.: Complex networks: Structure and dynamics. Phys. Rep. **424**, 175–308 (2006)
211. Newman, M.E.J.: Spread of epidemic disease on networks. Phys. Rev. E **66**, 016128 (2012)
212. Cornuejols, G., Fisher, M.L., Nemhauser, G.L.: George L, Location of bank accounts to optimize float: an analytic study of exact and approximate algorithms. Manag. Sci. **23**, 789–810 (1977)
213. Nemhauser, G.L., Wolsey, L.A., Fisher, M.L.: An analysis of approximations for maximizing submodular set functions-I. Math. Program. **14**, 265–294 (1978)
214. Leskovec, J., Krause, A., Guestrin, C., Faloutsos, C., VanBriesen, J., Glance, N.S.: Cost-effective outbreak detection in networks. In: Proceedings 13th ACM SIGKDD Conference on Knowledge Discovery and Data Mining, pp. 420–429 (2007)
215. Goyal, A., Lu, W., Lakshmanan, L.V.: SIMPATH: an efficient algorithm for influence maximization under the linear threshold model. In: 2011 IEEE 11th International Conference on Data Mining (ICDM), pp. 211–220. IEEE (2011)
216. Chen, W., Wang, Y., Yang, S.: Efficient influence maximization in social networks. In: Proceedings of the 15th ACM SIGKDD International Conference on Knowledge Discovery and Data Mining, pp. 199–208 (2009)
217. Hu, Y., Ji, S., Jin, Y., Feng, L., Stanley, H.E., Havlin, S.: Local structure can identify and quantify influential global spreaders in large scale social networks. Proc. Natl. Acad. Sci. U.S.A. **115**, 7468–7472 (2018)
218. Chen, W., Wang, C., Wang, Y.: Scalable influence maximization for prevalent viral marketing in large-scale social networks. In: Proceedings of the 16th ACM SIGKDD International Conference on Knowledge Discovery and Data Mining, pp. 1029–1038. ACM (2010)
219. Dijkstra, E.W.: A note on two problems in connexion with graphs. Numer. Math. **1**, 269–271 (1959)

220. Nematzadeh, A., Ferrara, E., Flammini, A., Ahn, Y.-Y.: Optimal network modularity for information diffusion. Phys. Rev. Lett. **113**, 088701 (2014)
221. Curato, G., Lillo, F.: Optimal information diffusion in stochastic block models. Phys. Rev. E **94**, 032310 (2016)
222. Yan, S., Tang, S., Fang, W., Pei, S., Zheng, Z.: Global and local targeted immunization in networks with community structure. J. Stat. Mech. P08010 (2015)
223. Wang, Y., Cong, G., Song, G., Xie, K.: Community-based greedy algorithm for mining top-k influential nodes in mobile social networks. In: Proceedings of the 16th ACM SIGKDD International Conference on Knowledge Discovery and Data Mining, pp. 1039–1048 (2010)
224. Pei, S., Teng, X., Shaman, J., Morone, F., Makse, H.A.: Efficient collective influence maximization in cascading processes with first-order transitions. Sci. Rep. **7**, 45240 (2017)
225. Bunde, A., Havlin, S.: Fractals and Disordered Systems, 2nd edn. Springer, Heidelberg (1996)
226. Cohen, R., Havlin, S.: Complex Networks: Structure, Robustness and Function. Cambridge University Press, Cambridge (2010)
227. Erdös, P., Rényi, A.: On random graphs, I Publicationes Mathematicae (Debrecen) **6**, 290–297 (1959)
228. Callaway, D.S., Newman, M.E.J., Strogatz, S.H., Watts, D.J.: Network robustness and fragility: percolation on random graphs. Phys. Rev. Lett. **85**, 5468 (2000)
229. Bollobás, B.: Random Graphs in Modern Graph Theory. Springer, New York (1998)
230. Newman, M.E.J., Strogatz, S.H., Watts, D.J.: Random graphs with arbitrary degree distributions and their applications. Phys. Rev. E **64**, 026118 (2001)
231. Clauset, A., Shalizi, C.R., Newman, M.E.J.: Power-law distributions in empirical data. SIAM Rev. **51**, 661–703 (2009)
232. Radicchi, F.: Predicting percolation thresholds in networks. Phys. Rev. E **91**, 010801 (2015)
233. Karrer, B., Newman, M.E.J., Zdeborová, L.: Percolation on sparse networks. Phys. Rev. Lett. **113**, 208702 (2014)
234. Radicchi, F., Castellano, C.: Beyond the locally treelike approximation for percolation on real networks. Phys. Rev. E **93**, 030302 (2016)
235. Marinari, E., Monasson, R.: Circuits in random graphs: from local trees to global loops. J. Stat. Mech. P09004 (2004)
236. Marinari, E., Monasson, R., Semerjian, G.: An algorithm for counting circuits: application to real-world and random graphs. EPL **73**, 8 (2005)
237. Bianconi, G., Marsili, M.: Loops of any size and Hamilton cycles in random scale-free networks. J. Stat. Mech. P06005 (2005)
238. Wormald, N.C.: The asymptotic connectivity of labeled regular graphs. J. Comb. Theory Ser. B **31**, 156–167 (1981)
239. Bhatia, N.P., Szegö, G.P.: Stability Theory of Dynamical Systems. Springer, Berlin Heidelberg (2002)
240. Newman, M.E.J.: Spectral methods for community detection and graph partitioning. Phys. Rev. E **88**, 042822 (2013)
241. Coja-Oghlan, A., Mossel, E., Vilenchik, D.: A spectral approach to analyzing Belief Propagation for 3-coloring. Comb. Probab. Comput. **8**, 881–912 (2009)
242. Morone, F., Min, B., Bo, L., Mari, R., Makse, H.A.: Collective influence algorithm to find influencers via optimal percolation in massively large social media. Sci. Rep. **6**, 30062 (2016)
243. Luo, S., Morone, F., Sarraute, C., Travizano, M., Makse, H.A.: Inferring personal economic status from social network location. Nat. Commun. **8**, 15227 (2017)
244. Bau, S., Wormald, N.C., Zhou, S.: Decycling numbers of random regular graphs. Random Struct. Algorithms **21**, 397–413 (2002)
245. Kobayashi, T., Masuda, N.: Fragmenting networks by targeting collective influencers at a mesoscopic level. Sci. Rep. **6**, 37778 (2016)
246. Osat, S., Faqeeh, A., Radicchi, F.: Optimal percolation on multiplex networks. Nat. Commun. **8**, 1540 (2017)
247. Santos, F.C., Pacheco, J.M.: Scale-free networks provide a unifying framework for the emergence of cooperation. Phys. Rev. Lett. **95**, 098104 (2005)

248. Pacheco, J.M., Santos, F.C., Chalub, F.A.C.: Stern-judging: a simple, successful norm which promotes cooperation under indirect reciprocity. PLoS Comput. Biol. **2**, 178 (2006)
249. Teng, X., Yan, S., Tang, S., Pei, S., Li, W., Zheng, Z.: Individual behavior and social wealth in the spatial public goods game. Phys. A **402**, 141–149 (2014)
250. Szolnoki, A., Perc, M.: Collective influence in evolutionary social dilemmas. EPL **113**, 58004 (2016)
251. Szolnoki, A., Perc, M.: Leaders should not be conformists in evolutionary social dilemmas. Sci. Rep. **6**, 23633 (2016)
252. Newman, M.E.J.: Networks: An Introduction. Oxford University Press, Oxford (2010)
253. Dorogovtsev, S.N., Mendes, J.F.F., Samukhin, A.N.: Metric structure of random networks. Nucl. Phys. B **653**, 307–338 (2003)
254. Baxter, G.J., Dorogovtsev, S.N., Goltsev, A.V., Mendes, J.F.F.: Bootstrap percolation on complex networks. Phys. Rev. E **82**, 011103 (2010)
255. Singh, P., Sreenivasan, S., Szymanski, B.K., Korniss, G.: Threshold-limited spreading in social networks with multiple initiators. Sci. Rep. **3**, 2330 (2013)
256. Radicchi, F., Castellano, C.: Fundamental difference between superblockers and superspreaders in networks. Phys. Rev. E **95**, 012318 (2017)
257. Ghavasieh, A., Stella, M., Biamonte, J., De Domenico, M.: Unraveling the effects of multiscale network entanglement on empirical systems. Commun. Phys. **4**, 129 (2021)
258. De Domenico, M., Biamonte, J.: Spectral entropies as information-theoretic tools for complex network comparison. Phys. Rev. X **6**, 041062 (2016)
259. Karp, R.M.: Reducibility among combinatorial problems. In: Miller, R.E., Thatcher, J.W. (eds.) Complexity of Computer Computations, pp. 85–103. Plenum Press, New York (1972)
260. Kirchhoff, G.: Über die Auflösung der Gleichungen, auf welche man bei der Untersuchung der linearen Verteilung galvanischer Ströme geführt wird. Ann. Phys. Chem. **72**, 497–508 (2847)
261. Beineke, L.W., Vandell, R.C.: Decycling graphs. J. Graph Theory **25**, 59–77 (1997)
262. Zhou, H.-J.: Spin glass approach to the feedback vertex set problem. Eur. Phys. J. B **86**, 455 (2013)
263. Im, Y.S., Kahng, B.: Dismantling efficiency and network fractality. Phys. Rev. E **98**, 012316 (2018)
264. Goh, K.-I., Salvi, G., Kahng, B., Kim, D.: Skeleton and fractal scaling in complex networks. Phys. Rev. Lett. **96**, 018701 (2016)
265. Gallos, L.K., Song, C., Havlin, S., Makse, H.A.: Scaling theory of transport in complex biological networks. Proc. Nat. Acad. Sci. U.S.A. **104**, 7746–7751 (2007)
266. Mézard, M., Parisi, G.: The Bethe lattice spin glass revisited. Eur. Phys. J. B **20**, 217–233 (2001)
267. Bayati, M., Borgs, C., Braunstein, A., Chayes, J., Ramezanpour, A., Zecchina, R.: Statistical mechanics of Steiner trees. Phys. Rev. Lett. **101**, 037208 (2008)
268. Guggiola, A., Semerjian, G.: Minimal contagious sets in random regular graphs. J. Stat. Phys. **158**, 300–358 (2015)
269. Schmidt, C., Pfister, H.D., Zdeborová, L.: Minimal sets to destroy the k-core in random networks. Phys. Rev. E **99**, 022310 (2019)
270. Chen, Y., Paul, G., Havlin, S., Liljeros, F., Stanley, H.E.: Finding a better immunization strategy. Phys. Rev. Lett. **101**, 058701 (2008)
271. Hadidjojo, J., Cheong, S.A.: Equal Graph Partitioning on estimated infection network as an effective epidemic mitigation measure. PLoS ONE **6**, e22124 (2011)
272. Schneider, C.M., Mihaljev, T., Havlin, S., Herrmann, H.J.: Suppressing epidemics with a limited amount of immunization units. Phys. Rev. E **84**, 061911 (2011)
273. Holme, P., Kim, B.J., Yoon, C.N., Han, S.K.: Attack vulnerability of complex networks. Phys. Rev. E **65**, 056109 (2002)
274. Holme, P.: Efficient local strategies for vaccination and network attack. EPL **68**, 908 (2004)
275. Cohen, R., Havlin, S., Ben-Avraham, D.: Efficient immunization strategies for computer networks and populations. Phys. Rev. Lett. **91**, 247901 (2003)

276. Aspnes, J., Chang, K., Yampolskiy, A.: Inoculation strategies for victims of viruses and the sum-of-squares partition problem. J. Comput. Syst. Sci. **72**, 1077 (2006)
277. Preciado, V.M., Zargham, M., Enyioha, C., Jadbabaie, A., Pappas, G.: Optimal vaccine allocation to control epidemic outbreaks in arbitrary networks. In: 52nd IEEE Conference on Decision and Control, pp. 7486–7491 (2013)
278. Prakash, B.A., Adamic, L., Iwashyna, H., Tong, H., Faloutsos, C.: Fractional Immunization on Networks. Austin, Texas (2013)
279. Hethcote, H.W.: The mathematics of infectious diseases. SIAM Rev. **42**, 599–653 (2000)
280. Diekmann, O., Heesterbeek, J.A.P.: Mathematical epidemiology of infectious diseases: model building, analysis and interpretation, vol. 5. John Wiley & Sons, West Sussex, England (2000)
281. Moreno, Y., Nekovee, M., Pacheco, A.F.: Dynamics of rumor spreading in complex networks. Phys. Rev. E **69**, 066130 (2004)
282. Altarelli, F., Braunstein, A., Dall'Asta, L., Lage-Castellanos, A., Zecchina, R.: Bayesian inference of epidemics on networks via belief propagation. Phys. Rev. Lett. **112**, 118701 (2014)
283. Lokhov, A.Y., Mézard, M., Ohta, H., Zdeborová, L.: Inferring the origin of an epidemic with a dynamic message passing algorithm. Phys. Rev. E **90**, 012801 (2014)
284. Shah, D., Zaman, T.: Detecting sources of computer viruses in networks: theory and experiment. In: ACM SIGMETRICS Performance Evaluation Review, pp. 203–214. ACM (2010)
285. Shah, D., Zaman, T.: Rumors in a network: Who's the culprit? IEEE Trans. Inf. Theory **57**, 5163–5181 (2011)
286. Comin, C.H., da Fontoura Costa, L.: Identifying the starting point of a spreading process in complex networks. Phys. Rev. E **84**, 056105 (2011)
287. Pei, S., Kandula, S., Yang, W., Shaman, J.: Forecasting the spatial transmission of influenza in the United States. Proc. Natl. Acad. Sci. U.S.A. **115**, 2752–2757 (2018)
288. Scarpino, S.V., Petri, G.: On the predictability of infectious disease outbreaks. Nat. Commun. **10**, 898 (2019)
289. Pei, S., Shaman, J.: Counteracting structural errors in ensemble forecast of influenza outbreaks. Nat. Commun. **8**, 925 (2017)
290. Kandula, S., Yamana, T., Pei, S., Yang, W., Morita, H., Shaman, J.: Evaluation of mechanistic and statistical methods in forecasting influenza-like illness. J. R. Soc. Interface **15**, 20180174 (2018)
291. Keeling, M.J., Rohani, P.: Modeling Infectious Diseases in Humans and Animals. Princeton University Press, Princeton (2011)
292. Dall'Asta, L.: Inhomogeneous percolation models for spreading phenomena in random graphs. J. Stat. Mech. P08011 (2005)
293. Karrer, B., Newman, M.E.J.: Message passing approach for general epidemic models. Phys. Rev. E **82**, 016101 (2010)
294. Ferguson, N.M., Cummings, D.A.T., Cauchemez, S., Fraser, C., Riley, S., Meeyai, A., Iamsirithaworn, S., Burke, D.S.: Strategies for containing an emerging influenza pandemic in Southeast Asia. Nature **437**, 209–214 (2005)
295. The New York Times: Outbreak in Sierra Leone Is Tied to Single Funeral Where 14 Women Were Infected, 29 Aug 2014, p. A7 (2014)
296. Auerbach, D.M., Darrow, W.W., Jaffe, H.W., Curran, J.W.: Cluster of cases of the acquired immune deficiency syndrome. Patients linked by sexual contact. Am. J. Med. **76**, 487–492 (1984)
297. Qin, S.-M., Zhou, H.-J.: Solving the undirected feedback vertex set problem by local search. Eur. Phys. J. B **87**, 273 (2014)
298. Min, B.: Identifying an influential spreader from a single seed in complex networks via a message passing approach. Eur. Phys. J. B **91**, 18 (2018)
299. Lipton, R.J., Rose, D.J., Tarjan, R.E.: Generalized nested dissection. SIAM J. Numer. Anal. **16**, 346–358 (1979)
300. Buldyrev, S.V., Parshani, R., Paul, G., Stanley, H.E., Havlin, S.: Catastrophic cascade of failures in interdependent networks. Nature **464**, 1025–1028 (2010)

301. Centola, D.: The spread of behavior in an online social network experiment. Science **329**, 1194–1197 (2010)
302. Scheffer, M., Bascompte, J., Brock, W.A., Brovkin, V., Carpenter, S.R., Dakos, V., Held, H., van Nes, E.H., Rietkerk, M., Sugihara, G.: Early-warning signals for critical transitions. Nature **461**, 53–59 (2009)
303. Scheffer, M.: Anticipating critical transitions. Science **338**, 344–348 (2012)
304. Haldane, A.G., May, R.M.: Systemic risk in banking ecosystems. Nature **469**, 351–355 (2011)
305. Yang, Y., Nishikawa, T., Motter, A.E.: Small vulnerable sets determine large network cascades in power grids. Science **358**, eaan3184 (2017)
306. Chalupa, J., Leath, P.L., Reich, G.R.: Bootstrap percolation on a Bethe lattice. J. Phys. C **12**, L31 (1979)
307. von Neumann, J.: Theory of Self-Reproducing Automata. University Illinois Press, Urbana (1966)
308. Balogh, J., Bollobś, B., Duminil-Copin, H., Morris, R.: The sharp threshold for bootstrap percolation in all dimensions. Trans. Am. Math. Soc. **364**, 2667–2701 (2012)
309. Gottschau, M.: Bootstrap percolation on degenerate graphs (2016). arXiv:1605.07002
310. Dorogovtsev, S.N., Goltsev, A.V., Mendes, J.F.F.: Critical phenomena in complex networks. Rev. Mod. Phys. **80**, 1275–1335 (2008)
311. Goltsev, A.V., Dorogovtsev, S.N., Mendes, J.F.F.: k-core (bootstrap) percolation on complex networks: Critical phenomena and nonlocal effects. Phys. Rev. E **73**, 056101 (2006)
312. Cellai, D., Lawlor, A., Dawson, K.A., Gleeson, J.P.: Tricritical point in heterogeneous k-core percolation. Phys. Rev. Lett. **107**, 175703 (2011)
313. Cellai, D., Lawlor, A., Dawson, K.A., Gleeson, J.P.: Critical phenomena in heterogeneous k-core percolation. Phys. Rev. E **87**, 022134 (2013)
314. Watts, D., Dodds, P.: Influentials, networks, and public opinion formation. J. Consum. Res. **34**, 441–458 (2007)
315. Szymanski, B., Lizardo, O., Doyle, C., Karampourniotis, P., Singh, P., Korniss, G., Bakdash, J.: The spread of opinions in societies. In: Modeling Sociocultural Influences on Decision Making, pp. 91–114 (2016)
316. Dhar, D., Shukla, P., Sethna, J.P.: Zero-temperature hysteresis in the random-field Ising model on a Bethe lattice. J. Phys. A **30**, 5259–5267 (1997)
317. Ohta, H., Sasa, S.-I.: A universal form of slow dynamics in zero-temperature random-field Ising model. EPL **90**, 27008 (2010)
318. Balogh, J., Pittel, B.: Bootstrap percolation on the random regular graph. Random Struct. Algor. **30**, 257–286 (2007)
319. Ramos, M., Shao, J., Reis, S.D.S., Anteneodo, C., Andrade, J.S., Jr., Havlin, S., Makse, H.A.: How does public opinion become extreme? Sci. Rep. **5**, 10032 (2015)
320. Parshani, R., Buldyrev, S.V., Havlin, S.: Interdependent networks: reducing the coupling strength leads to a change from a first to second order percolation transition. Phys. Rev. Lett. **105**, 048701 (2010)
321. Gao, J., Buldyrev, S.V., Havlin, S., Stanley, H.E.: Robustness of a network of networks. Phys. Rev. Lett. **107**, 195701 (2011)
322. Cohen, E.: Size-estimation framework with applications to transitive closure and reachability. J. Comput. Syst. Sci. **55**, 441–453 (1997)
323. Karampourniotis, P.D., Szymanski, B.K., Korniss, G.: Influence maximization for fixed heterogeneous thresholds. Sci. Rep. **9**, 5573 (2019)
324. Mézard, M., Zecchina, R.: Random k-satisfiability problem: from an analytic solution to an efficient algorithm. Phys. Rev. E **66**, 056126 (2002)
325. Ackerman, E., Ben-Zwi, O., Wolfovitz, G.: Combinatorial model and bounds for target set selection. Theor. Comput. Sci. **411**, 4017–4022 (2010)
326. Dreyer, P.A., Jr., Roberts, F.S.: Irreversible k-threshold processes: graph-theoretical threshold models of the spread of disease and of opinion. Discrete Appl. Math. **157**, 1615–1627 (2009)
327. Reichman, D.: New bounds for contagious sets. Discrete Math. **312**, 1812–1814 (2012)

328. Feige, U., Krivelevich, M., Reichman, D.: Contagious sets in random graphs. Ann. Appl. Probab. **27**, 2675–2697 (2017)
329. Coja-Oghlan, A., Feige, U., Krivelevich, M., Reichman, D.: Contagious sets in expanders. In: Proceedings of the 26th Annual ACM-SIAM Symposium on Discrete Algorithms, pp. 1953–1987. SIAM (2015)
330. Janson, S., Thomason, A.: Dismantling sparse random graphs. Comb. Probab. Comput. **17**, 259–264 (2008)
331. Sun, J., Liu, R., Fan, Z., Xie, J., Ma, X., Hu, Y.: Lower bound of network dismantling problem. Chaos **28**, 063128 (2018)
332. Wormald, N.C.: Differential equations for random processes and random graphs. Ann. Appl. Probab. **5**, 1217–1235 (1995)
333. Wormald, N.C.: The differential equation method for random graph processes and greedy algorithms. In: Lectures on Approximation and Randomized Algorithms, vol. 73, p. 155 (1999)
334. Achlioptas, D., D'souza, R.M., Spencer, J.: Explosive percolation in random networks. Science **323**, 1453–1455 (2009)
335. da Costa, R.A., Dorogovtsev, S.N., Goltsev, A.V., Mendes, J.F.F.: Explosive percolation transition is actually continuous. Phys. Rev. Lett. **105**, 255701 (2010)
336. Riordan, O., Warnke, L.: Explosive percolation is continuous. Science **333**, 322–324 (2011)
337. Grassberger, P., Christensen, C., Bizhani, G., Son, S.-W., Paczuski, M.: Explosive percolation is continuous, but with unusual finite size behavior. Phys. Rev. Lett. **106**, 225701 (2011)
338. Friedman, E.J., Landsberg, A.S.: Construction and analysis of random networks with explosive percolation. Phys. Rev. Lett. **103**, 255701 (2009)
339. Newman, M.E.J., Ziff, R.M.: Fast Monte Carlo algorithm for site or bond percolation. Phys. Rev. E **64**, 016706 (2001)
340. Ma, M., Korniss, G., Szymanski, B.K.: Learning parameters for balanced index influence maximization. In: Proceedings the 9th Conference on Complex Networks and their Applications, Madrid, Spain, 1–3 Dec 2020. Studies in Computational Intelligence, vol. 449, 5 Feb, pp. 167–177. Springer (2021)
341. Gao, L., Wang, H., Zhang, Z., Zhuang, H., Zhou, B.: HetInf: social influence prediction with heterogeneous Graph Neural Network. Front. Phys. **9**, 787185 (2022)
342. Ren, X.-L., Gleinig, N., Helbing, D., Antulov-Fantulin, N.: Generalized network dismantling. Proc. Natl. Acad. Sci. U.S.A. **116**, 6554–6559 (2019)
343. Ribeiro, H.V., Alves, L.G.A., Martins, A.F., Lenzi, E.K., Perc, M.: The dynamical structure of political corruption networks. J. Complex Netw. **6**, 989–1003 (2018)
344. Artime, O., De Domenico, M.: Percolation on feature-enriched interconnected systems. Nat. Comm. **12**, 2478 (2021)
345. Guimerà, R., Reichardt, I., Aguilar-Mogas, A., Massuci, F.A., Miranda, M., Pallarès, J., Sales-Pardo, M.: A Bayesian machine scientist to aid in the solution of challenging scientific problems. Sci. Adv. **6**, eaav6971 (2020)
346. Artime, O., Grassia, M., De Domenico, M., Gleeson, J.P., Makse, H.A., Mangioni, G., Perc, M., Radicchi, F.: Robustness and resilience of complex networks. Nat. Rev. Phys. **6**, 114–131 (2024)
347. Ren, X.-L., Nino, A.-F.: Ensemble approach for generalized network dismantling. In: Cherifi, H., Gaito, S., Mendes, J. F., Moro, E., Rocha, L.M. (eds.) Complex Networks and their Applications VIII, p. 783–793. Springer International Publishing (2020)
348. Grassia, M., Mangioni, G.: CoreGDM: geometric Deep learning network decycling and dismantling. In: Teixeira, A.S., Botta, F., Mendes, J.F., Menezes, R., Mangioni, G. (eds.) Complex networks XIV, pp. 86–94. Springer Nature Switzerland (2023)
349. Fan, C., Zeng, L., Sun, Y., Liu, Y.-Y.: Finding key players in complex networks through deep reinforcement learning. Nat. Mach. Intell. **2**, 317–324 (2020)
350. Ribeiro, H.V., Alves, L.G.A., Martins, A.F., Lenzi, E.K., Perc, M.: The dynamical structure of political corruption networks. J. Compl. Netw. **6**, 989–1003 (2018)
351. Holme, P., Saramäki, J.: Temporal networks. Phys. Rep. **519**, 97–125 (2012)

352. Boccaletti, S., Bianconi, G., Criado, R., Del Genio, C.I., Gómez-Gardenes, J., Romance, M., Sendina-Nadal, I., Wang, Z., Zanin, M.: The structure and dynamics of multilayer networks. Phys. Rep. **544**, 1–122 (2014)
353. Kivelä, M., Arenas, A., Barthelemy, M., Gleeson, J.P., Moreno, Y., Porter, M.A.: Multilayer networks. J. Complex Netw. **2**, 203–271 (2014)
354. Estrada, E., Ross, G.J.: Centralities in simplicial complexes, applications to protein interaction networks. J. Theor. Biol. **438**, 46–60 (2018)
355. Aksoy, S.G., Joslyn, C., Marrero, C.O., Praggastis, B., Purvine, E.: Hypernetwork science via high-order hypergraph walks. EPJ Data Sci. **9**, 16 (2020)
356. Serrano, D.H., Gómez, D.S.: Centrality measures in simplicial complexes: applications of topological data analysis to network science. Appl. Math. Comput. **382**, 125331 (2020)
357. Bianconi, G.: Higher-Order Networks: An Introduction to Simplicial Complexes. Cambridge University Press (2021)
358. Torres, J.J., Bianconi, G.: Simplicial complexes: higher-order spectral dimension and dynamics. J. Phys. Complex. **1**, 015002 (2020)
359. Flores, J., Romance, M.: On eigenvector-like centralities for temporal networks: discrete vs. continuous time scales. J. Comput. Appl. Math. **330** 1041–1051 (2018)
360. Lv, L., Zhang, K., Zhang, T., Bardou, D., Zhang, J., Cai, Y.: PageRank centrality for temporal networks. Phys. Lett. A **383**, 1215–1222 (2019)
361. De Domenico, M., Solé-Ribalta, A., Cozzo, E., Kivelä, M., Moreno, Y., Porter, M.A., Gómez, S., Arenas, A.: Mathematical formulation of multilayer networks. Phys. Rev. X **3**, 041022 (2013)
362. Wang, D., Zou, X.: A new centrality measure of nodes in multilayer networks under the framework of tensor computation. Appl. Math. Model. **54**, 46–63 (2018)
363. Wu, M., He, S., Zhang, Y., Chen, J., Sun, Y., Liu, Y.-Y., Zhang, J., Poor, H.V.: A tensor-based framework for studying eigenvector multicentrality in multilayer networks. Proc. Nat. Acad. Sci. U.S.A. **116**, 15407–15413 (2019)
364. De Domenico, M., Sole-Ribalta, A., Gomez, S., Arenas, A.: Navigability of interconnected networks under random failures. Proc. Nat. Acad. Sci. U.S.A. **11**, 8351 (2014)
365. Barthelemy, M.: Spatial Networks. Phys. Rep. **499**, 1–101 (2011)
366. De Arruda, G.F., Barbieri, A.L., Rodríguez, P.M., Rodrigues, F.A., Moreno, Y., da Fontoura Costa, L.: Role of centrality for the identification of influential spreaders in complex networks. Phys. Rev. E **90**, 032812 (2014)
367. Krioukov, D., Papadopoulos, F., Kitsak, M., Vahdat, A., Boguñá, M.: Hyperbolic geometry of complex networks. Phys. Rev. E **82**, 036106 (2010)
368. Liu, W.-C., Huang, L.-C., Liu, C.W.-J., Jordán, F.: A simple approach for quantifying node centrality in signed and directed social networks. Appl. Netw. Sci. **5**, 46 (2020)
369. Shapley, L.S.: A value for n-person games. In: Kuhn, H.W., Tucker, A.W. (eds.) Contributions to Game Theory, pp. 307–317. Princeton University Press (1953)
370. Gómez, D., González-Arangüena, E., Manuel, C., Owen, G., del Pozo, M., Tejada, J.: Centrality and power in social networks: a game theoretic approach. Math. Soc. Sci. **46**, 27–54 (2003)
371. Michalak, T.P., Aadithya, K.V., Szczepanski, P.L., Ravindran, B., Jennings, N.R.: Efficient computation of the Shapley value for game-theoretic network centrality. J. Artif. Intell. Res. **46**, 607–650 (2013)
372. Gutierrez, F., Barocio, E., Uribe, F., Zuniga, P.: Vulnerability analysis of power grids using modified centrality measures. Discrete Dyn. Nat. Soc. **2013**, 135731 (2013)
373. Tyloo, M., Pagnier, L., Jacquod, P.: The key player problem in complex oscillator networks and electric power grids: Resistance centralities identify local vulnerabilities. Sci. Adv. **5**, 084101 (2019)
374. Tyloo, M., Coletta, T., Jacquod, P.: Robustness of synchrony in complex networks and generalized Kirchhoff indices. Phys. Rev. Lett. **120**, 084101 (2018)
375. Zotenko, E., Mestre, J., O'Leary, D.P., Przytycka, T.M.: Why do hubs in the yeast protein interaction network tend to be essential: reexamining the connection between the network topology and essentiality. PLoS Comput. Biol. **4**, e1000140 (2008)

376. Missiuro, P.V., Liu, K., Zou, L., Ross, B.C., Zhao, G., Liu, J.S., Ge, H.: Information flow analysis of interactome networks. PLoS Comput. Biol. **5**, e1000350 (2009)
377. Kim, Y.-A., Wuchty, S., Przytycka, T.M.: Identifying causal genes and dysregulated pathways in complex diseases. PLoS Comput. Biol. **7**, e1001095 (2011)
378. Cowen, L., Ideker, T., Raphael, B.J., Sharan, R.: Network propagation: a universal amplifier of genetic associations. Nat. Rev. Genet. **18**, 551 (2017)
379. Teng, X., Pei, S., Morone, F., Makse, H.A.: Collective influence of multiple spreaders evaluated by tracing real information flow in large-scale social networks. Sci. Rep. **6**, 36043 (2016)
380. Burleson-Lesser, K., Morone, F., Tomassone, M.S., Makse, H.A.: K-core robustness in ecological and financial networks. Sci. Rep. **10**, 3357 (2020)

Chapter 2
Social Media Influencers and Politics

Abstract For ages, influencing the masses has been the privilege of the elites. New decentralized and interactive social media have changed all that by empowering people to produce their own content of information to spread influence. In this chapter we apply the theories of influencer identification of Chap. 1 to understand the spread of influence in social media. We focus on the arrival of new influencers in the political arena and study how these new influencers are driving opinion trends, how they contribute to polarization of the masses, and how they spread fake news and misinformation to influence society at large. We also discuss how they can determine nationwide elections. The theories of influencers described in Chapt. 1 have to be fed with enormous amounts of data to be effective. These data are generated continuously in the digital era. We first review the process of data generation and collection from social media. To measure opinion trends, we augment the theory of influencers with artificial intelligence, through machine learning and natural language processing, to learn and predict opinion trends including predicting nationwide elections. We process large amount of data with AI and classify people and influencers according to different topics, trends and political tendencies. At the same, AI can differentiate fake news and misinformation from genuine news. Combining the theory of influence with AI, we will understand the rise of political movements that can bring new political ideas to power, and the influence of misinformation in political campaigns and opinion trends.

2.1 The Emergence of New Influencers Versus the Traditional News Corporate Elites

The political discourse has been dominated by 'elites' for much of the history of humankind. The standard definition of elites in political sciences focuses on a small self-reinforcing group of political, economic, corporate, and communications leaders (opinion journalists, editorial boards, and newsrooms of traditional media). These elites can be traditional news corporate media organizations like CNN, The New York Times, Fox News, The Wall Street Journal and the like or the political parties machineries.

H. A. Makse and M. Zava, *The Science of Influencers and Superspreaders*, Understanding Complex Systems, https://doi.org/10.1007/978-3-031-78058-5_2

Elite corporate organizations are complemented by 'elite celebrities', i.e., highly influential and visible members of society like politicians, party machineries and poll watchers, but also celebrities in other spheres of influence beyond politics like arts and sports, executives of major corporations and other member of society who are politically active and educated people. These elite people and organizations may represent less than 1% of society, yet they directly affect many other voters and members of society.

While the elites have traditionally affected the direction of politics, are involved in the discussion and dissemination of views, and influence many others, their power to drive the political discourse is now being challenged by the rise of new organic influencers from social media. The emergence of social media has democratized access to publishing, and moved the focus from ability to publish to ability to influence. Social media influencers come from a variety of backgrounds and are increasingly important in shaping society and politics. They are diverse, young and are usually skeptical of the traditional media that has dominated the cycle news for long time.

Nowadays, social media provide a lower barrier of entry into celebrity status than before. It also amplifies views, creating the new generation of social media influencers who organically emerges from society and they do not necessarily respond to the old elites. The key question is whether this new generation of social media influencers are now or will be an important driver of opinions and trends, and whether they amplify existing centers of traditional influence or introduce new ones. By applying the algorithms developed in Chap. 1 and collecting big-data from social media, in this chapter we show evidence that the emergence of influencers in society is changing the status quo. These dynamics lead to the paradox of the digital era: increasing isolation in the age of social media sharing.

2.2 Models of Mass Communication

To understand the propagation of information and misinformation in the population we require a theoretical framework. Traditional models of information diffusion were developed mainly to explain the flow of information, and they presumed mainly trusted sources and did not take into account malicious actors. Thus, the first empirical question to answer is indeed whether the mechanisms for the spread of information and misinformation are indeed the same or different. Although there are strong reasons to suspect that the mechanisms will indeed not to be the same (e.g., misinformation may rely more heavily on one-step flows or herding dynamics) the question cannot be answered prior to doing the empirical work.

The first thing is to put the problem into a firm theoretical framework for the propagation of information. The classical way to understand this problem in communication theory is under the framework of the two-step flow of communication model proposed by Lazarsfeld and Katz [10, 11] in the 40's, and further elaborated in a series of papers [12, 13]. This framework defines an idealized model of diffusion of information, e.g., political opinions, ideas, beliefs and innovations [1].

But this model follows an outdated communication system of the pre-digital era. Nowadays, social networks have changed our way of communicating, and the reach and prominence of influencers have changed the way information diffusion is distributed. Because of this, we have to adapt the model to a multiple-step model of mass communication to consider a more appropriate framework that faithfully describes the current information flow.

Below, we examine this theory in the light of the transformative effects emerging from social media and the rise of new influencers, and show how a multiple-step model of mass communication provides a more accurate framework to describe the flow of information between the news media, opinion leaders, influencers and the masses of adopters.

2.2.1 Classical Two-Step Model of Mass Communication of Lazarsfeld and Katz

There are not many people who form their opinions by digging the news directly from the original source. Thus, the main assumption of the two-step model is that 'adopters' of opinions do not get the news from the 'sources of information', but they form their opinions following 'opinion leaders' [10, 14–16] who can interpret and put the information into context with the idea of clarifying what the adopters should think or take into account. In this case, the model looks like a two-step chain seen in Fig. 2.1:

$$\textbf{Two step model:}\ \text{source} \rightarrow \text{opinion leaders} \rightarrow \text{adopters}. \tag{2.1}$$

The sources of information are a small group of mainstream media outlets endowed with high credibility and legitimacy among information consumers. These groups have access to the original sources of information, as well as the political, social, and scientific elites, so they can verify the information and the characters that played the role of sources. Afterwards, they pass this information to opinion leaders who are able to summarize their opinions in the form of information for the masses. Opinion leaders usually were local news papers or other 'community leaders' that will reshape the news following the opinion of their community at the same time than influencing it. This community leaders have the credibility of their local community on the ground [10]. We can see these 'opinion leaders' as 'early adopters' of opinions and beliefs whom transfer the information from the source to the rest of the people who become the 'opinion followers'.

In the language of the theories of Chap. 1 they are called the influencers, but to distinguish them from the new social media influencers, we will refer to them as 'elite or traditional influencers'. The opinion leaders use their influence within the local social structures to communicate and influence the general public to adopt beliefs and opinions as their own [1, 15].

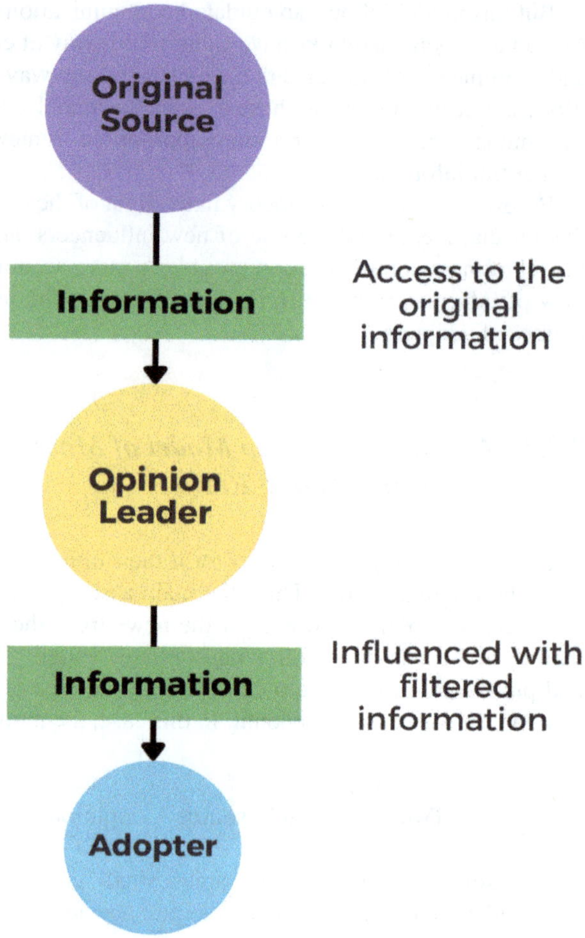

Fig. 2.1 Classical two-step model of communication from Lazarsfeld and Katz [10–13]. Figure courtesy of O. Lizardo, UCLA [28]

A clear example of this process occurred during COVID-19. During this pandemic many people wanted to gather information about the state of the situation. Only big mainstream media outlets like CNN, FOX News or others have real access to the original source of information (doctors, scientists, e.g., Dr. Anthony Fauci, director of the National Institute of Allergy and Infectious Diseases and Chief Medical Advisor to the US President) who provided the latest scientific advances on a relevant issue (e.g., mitigation measures to the COVID-19 pandemic). Then, this media share their information with opinion leaders, TV pundits and others, that receive the information, generate an opinion about it and share this information to their community. Opinion leaders leverage their legitimacy and influence to get people to adopt the guidelines for pandemic mitigation recommended by members of the scientific elite (the original source). The community has a trust bond on these opinion leaders so they will not doubt about the facts and the data and this is how influence occurs.

As we can see, the main character in this linear graph is the opinion leader, serving as the key intermediary in the diffusion chain and thus is endowed with significant amount of power [17]. As depicted in Fig. 2.1, the key social processes are thus access and influence mediating either via direct interpersonal relations, or locally mediated via community organizations and other information outlets.

2.2.2 Two-Step Model in the Modern Age of Social Media

Social media have completely changed the information channels and processes. Information is now democratized, everyone has access to millions of sources in real time, at the same time that new sources and channels are being created without any restriction. All this has disrupted the two-step model explained above. Many individual users with no resources or connection to high-level characters are now in control of the communication for millions of people, and the figure of the opinion leader appears now in the center of the phenomena.

We cannot talk anymore about a single chain-like dynamics, and the the two-step model is ripe for disruption. The control of the information spreading is now limited. Before news were easier to control and to know who has access to the information and how it is distributed. Now, adopters have access to the original source at the same time they can act as spreaders of the information, mixing their opinions with the opinion leaders. The new graph of information flow shows new edges as there are many agents that access the original sources, behind the opinion leaders.

At the same time this disruption is significant because of the lack of knowledge of the new adopters with access to the original source without fully understanding it. In this scenario, a new disruption appears: sources of misinformation. In this new graph, the opinion is not only controlled by the opinion leaders that have knowledge and access to the data. Adopters now connecting to sources may not perform a good judgment of the source. This could lead to the spread of misinformation. These new actors undercut the opinion leader's advantage and control of the popular opinion, generating a chaos that allows to spread whatever new a source wants, real or fake, in seconds.

In the social media era, the disruption of the simplistic two-step chain of influence has happened in two major ways. First, the increasing number of users that bypass the traditional news outlets to access directly the source on social media [19]. These new opinion leaders then become the new social media influencers with audiences that rival in number those from traditional news outlets. Here, rather than a few traditional or mainstream news sources, multiple sources, some of which of suspect credibility spreading misinformation, compete to provide people with a range of opinions and beliefs on a given subject. This disrupts the first step in the traditional two-step communication chain because now potential adopters can access sources of information without the mediation of conventional opinion leaders and can become influencers themself by spreading the information further in the social network. Thus, in the age of social media, anyone can become an influencer with even larger audiences than the traditional media. We will see many examples of these new influencers when we discuss the rise of influencers in American politics during the 2016 and 2020 presidential elections.

Second, according to many metrics, the usual mechanisms via which opinion leaders could exercise their influence have been declining over time. This includes declining rates of membership in civic institutions and religious attendance [20], the decline of local news media outlets [21], geographic isolation and segregation along political lines [22], and increasing polarization and withering of opinion diversity in interpersonal political networks [23].

2.2.3 Multistep Model of Mass Communication

As discussed by many authors [19, 24–27], the above considerations have naturally led to a generalization of the two-step model into a multistep model of mass communication applicable to the post-social media world. In particularly, Ref. [28, 29] has proposed a formulation of the multistep model depicted in Fig. 2.2, which we discussed next. The model combines classical opinion-leader mediated approaches in mass communication, influencer and network diffusion theory, and computational approaches to the spread of opinions that take into account the multiscale dynamics in social networks [30–32].

In this integrative model, there are, in principle, multiple sources competing for the attention of the general public. Some of these sources are the traditional media outlets of the two-step model, mainly attempting to diffuse what they see as accurate and credible information. However, there are new sources that try to diffuse their news, and, every so often, outlets with no credibility, yet with a massive reach. Also, adopters who were in the bottom of the food chain before, receiving the information from the opinion leaders, are now accessing directly these sources through transitive links bypassing the elite opinion leaders. This creates shorter chains than those modeled by the two-step flow, leading some to speak of a return to 'one-step' models of communication in which online media sources communicate directly to consumers and other potential adopters [19, 33]. These one-step formulations are similar to the 'hypodermic needle' and 'mass society' approaches to mass communication [34], against which the classical two-step model was first developed [11, 12].

This represents a democratization of information access by the general public. Nowadays, anyone can become an influencer by accessing information that is not filtered or manipulated by the old opinion leader elites. We will study these dynamics using examples from the 2016 and 2020 US presidential elections below.

In conclusion, the new multistep model is a combination one-step [19], two-step [10, 12], and multistep dynamics [25], all combined and co-existing. One-step dynamics exist as a special case in the multistep model (see Fig. 2.3).

Traditional opinion leaders do not have an exclusive role any more. While they still have access to the original sources and can be followed by adopters and influence them, there are some adopters that bypass them and go directly to the source evalu-

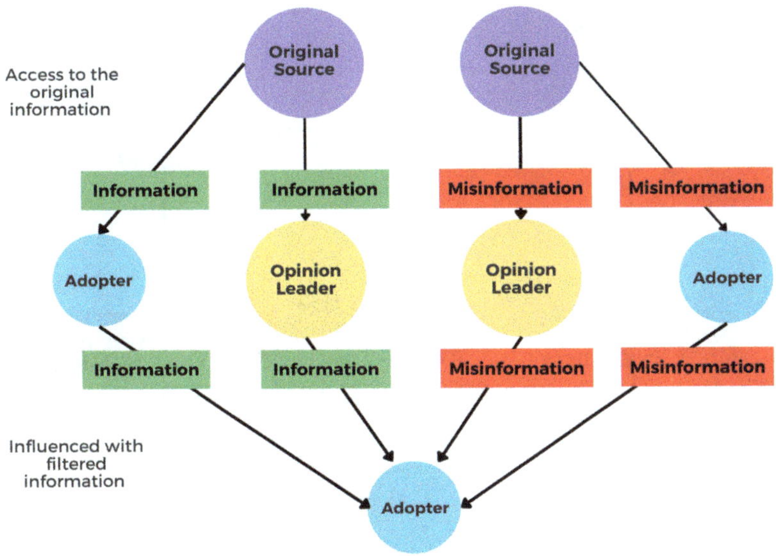

Fig. 2.2 A formulation of the multistep model of mass communication and information diffusion in a post-social-media world proposed by Lizardo [28]. The multistep dynamics encompasses not only an increasing number of information flow links, but also an increasing number of actors. A plethora of competing opinion leaders enter the scene, including new organically emerging social media influencers, some of whom are directly connected to sources spreading misinformation. This dynamic disrupts the traditional opinion leader's monopoly on cultural authority (e.g., every issue now has multiple sides and "alternative reality" interpretations [18]). Adopters are now exposed to multiple, divergent and frequently contrary opinions on the same subject, with multiple actors attempting to exercise influence for their interpretation of reality. Figure courtesy of O. Lizardo, UCLA [28]

ating it on their own terms (e.g., with skepticism). At the same time as the flow is not controlled by opinion leaders anymore, it has become more difficult for traditional social elites to set the agenda of the political discourse, while people are exposed to multiple and sometimes contradictory information on the same topic.

Bennett and Manheim [19] already noticed that the significant role that opinion leaders played before has diminished in the last years with the advent of social media. This capitulation has dramatic consequences for society. These opinion leaders are now more likely to reinforce latent opinions rather than to frame them. At the same time, the refined targeting of messages directly to individuals is increasingly fragmented and differentiated, contributing to individualizing processes, polarization, and news spin.

However, as said, this does not mean that the standard two-step chain has disappeared. They still exist (amid declining influence) in Twitter [35–39] or Facebook [27, 40]. Nevertheless, they are now embedded in a more complex structure, mixing different communication flows such as three-step or longer diffusion chains. One example depicted in Fig. 2.4 are the outsiders or marginal members of densely connected

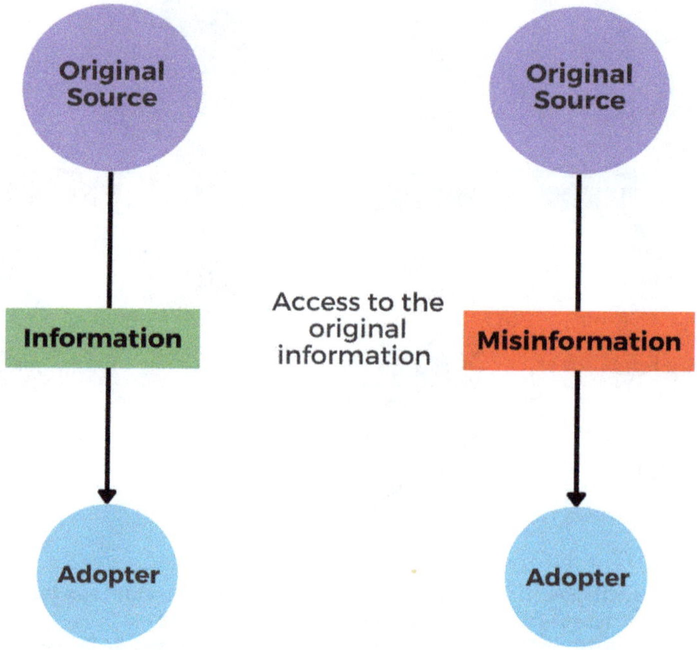

Fig. 2.3 One-step flow. Example of one-step flow as a special case of the multistep model of political communication and information diffusion. Figure courtesy of O. Lizardo, UCLA [28]

communities who serve as information brokers and influencers between groups [14]. Information is re-transmitted to more influential members of these communities [25], whereby they go through a process of diffusion via imitation and cascade dynamics [15, 30].

This structure also gives the opportunity for adopters to connect among themselves into densely connected clusters introducing new channels to increase polarization, novel cascading and herding via imitation dynamics into the process of opinion adoption [15]. Both information and misinformation can spread rapidly via these densely connected clusters. The lack of resources, supervision or education leads in these clusters to a higher rate of misinformation spread. The adoption of opinion and beliefs can now happen via mechanisms independent of source credibility or opinion-leader-based influence. Instead, opinion herding and global threshold dynamics, where the credibility and probability of adoption of an opinion increases adoption regardless of validity, become a significant part of the system.

Reference [29] analyzed this multistep model detecting the problem of misinformation showing that it can accommodate complex dynamics. For instance, some sources can start the spreading of fake news or misinformation. This source can be reached by an adopter (who may put their own idiosyncratic 'spin' on it). If this adopter has enough influence to spread this information to his community and reach a densely connected cluster with cascade effects and opinion herding processes

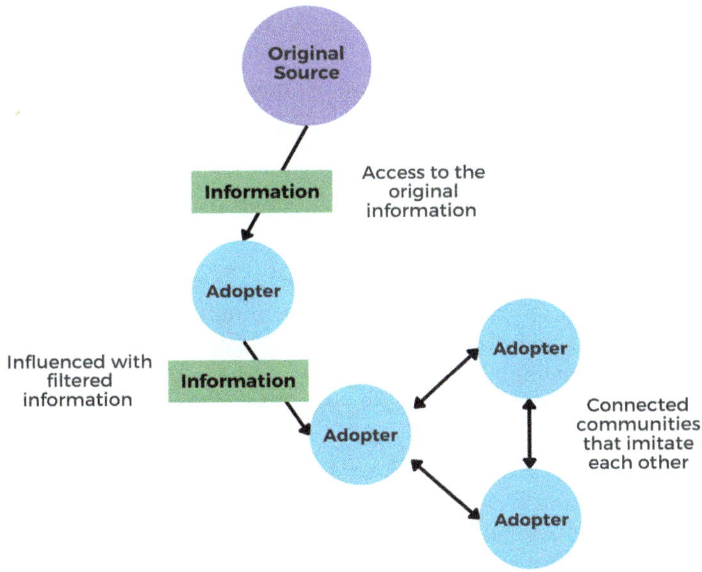

Fig. 2.4 Multistep flow. Example of multistep information diffusion chain. Figure courtesy of O. Lizardo, UCLA [28]

(source → adopter → community) [15], the effect of this original misinformation on the community can be massive and without control.

Of course, opinion leaders can be part of this process, despite that a new horizontal influence from one adopter to the other has raised, the opinion leaders still have a role in the communication model and can also, pass a piece of misinformation into densely connected communities (source → opinion leader → community) [25]. In a more complex scenario (Fig. 2.2), even the same community can be targeted with different information (information and misinformation) from several types of sources, traditional opinion leaders, other adopters, or directly consuming the information online, generating confusion and losing trust from the sources they end up not engaging with.

In conclusion, the rise of social media poses interesting theoretical questions as to how to properly model the flow of information in the modern digital age. The influencer theories of Chap. 1 augmented by machine learning analysis of big data collected from social networks can measure the different channels of information in the multistep dynamics. In the following sections, we analyze the differences between the traditional elites and new influencers and among their respective influence techniques used as political and diplomatic strategies. Influencers cause other people to think and act differently. We show evidence that influencers have changed the dynamics of political groupings and parties developed new identities and affiliations, and found different forms of self-reinforcing ideologies. By collecting large datasets from social media during the 2016 and 2020 US presidential elections, the evolution of new segments of influencers is measured in contrast to the old elites, and study their roles in the spreading of fake news and the rise of political polarization in the framework of the multistep model of communication.

2.3 Finding Influencers in Social Networks

Chapter 1 discusses different approaches to influencer identification. As already discussed, there is no way to known a priori which centrality or algorithm will perform better in each particular situation. Studies then have been devoted to compare the centralities and influencer finding algorithms using simulations of spreading behavior in both synthetic and real-world networks. We review some of these results next.

Beyond node locations in social networks, researchers have explored the role of human behavior in identifying influential spreaders of information in social networks. Aral and Walker [41, 42] performed an in vivo randomized experimentation involving 1.3 million Facebook users to elucidate how users' attributes (e.g., gender, age, marital status) are related to influence and susceptibility to opinion trends. The authors designed an experiment to analyze users' adoption of a commercial Facebook application after receiving influence-mediating messages from their peers. Using a continuous-time single-failure proportional hazards model, which explicitly incorporates the impact of homophily [41], they estimated the influence and susceptibility of the users. Personal attributes were found necessary in stratifying users' influence and susceptibility. In particular, influential people tend to be clustered with influential friends, which suggests a multiplier effect from infecting influencers. They further argued that influence and susceptibility are important in the peer-to-peer diffusion of products.

In a recent modeling work by Aral and Dhillon [43], it was found that neglecting the joint impact of influence and susceptibility causes an underestimation of influence propagation by an average of 21.7% in traditional models given the same seed sets. Moreover, influencers found in empirical influence models are less well-connected and less central, with only 19.8% overlap with the influencers identified in traditional models. These results call for the inclusion of user attributes and behaviors in finding more realistic influencers.

From another viewpoint, Min et al. [44] used the social mechanisms and activity patterns of connection formation to find influential spreaders in real-world online social networks. The authors surveyed 3 online social platforms to estimate the probabilistic tendency of connection formation for each individual, and considered four classes of social mechanisms for link creation proposed in sociology [45, 46]: exchange interaction of reciprocal links, balance interaction that closes a triangle, collective action linking to hubs, and structural hole interaction that connects two different communities. It was found that, among the above four mechanisms, the collective action, i.e., the tendency to connect to a hub is the best predictor for influencers. Meanwhile, people connecting different communities are more likely to be influential spreaders if a network has pronounced modular structure.

In evaluations that rely on simulations of a specific spreading model, contradictory or inconclusive results are obtained. For example, da Silva et al. found that degree and accessibility are efficient in SIR models [47]; Kitsak et al. found that k-core outperforms degree in SIR and SIS models [48], while Borge-Holthoefer and Moreno found that the influence of node does not depend on the k-core value in rumor-spreading models [49]. On the contrary, k-core is more relevant to the ability to prevent the spreading on a global scale. In addition, when it comes to treatment delivery administration, it was found that clustering matters more than centrality metrics in selecting seed nodes for large-scale treatment campaigns [50]. Gu et al. [51] claimed that ranking individual influencers itself is an ill-defined problem. They reported that the set of top k influencers does not necessarily have nodes in common with the top $k + 1$ influencers. These contradictory results are considered to originate from the simplification of human behaviors in spreading models, in which the complex features of diffusion processes are partly neglected. As a consequence, further validations of different metrics based on real-world information flow becomes necessary.

Using information diffusion data in four dissimilar real-world social media, including a blog-sharing community, LiveJournal, scientific journals of the American Physical Society, and two online social networks, Facebook and Twitter, Pei et al. [52, 53] compared the performance of several centralities in identifying influential spreaders. By collecting all directed diffusion links, directed diffusion graphs were constructed, on which the influence of each node can be determined by tracking the diffusion instances layer by layer in a breadth-first-search fashion. Meanwhile, centrality measures such as degree, k-core, and PageRank were calculated on the social networks constructed using social ties in these systems. In all platforms, nodes with high influence have a wide distribution of degrees. In contrast, they are consistently located in the high k-core region. This suggests that the influence of a given node depends more on its global location rather than its local connections.

Later, Teng et al. [54] further investigated the performance of several metrics in identifying multiple influencers. Because it is rare for a real-world spreading process to be simultaneously initiated by a given set of influencers, direct empirical validation for multiple influencer identification is difficult. To overcome this problem, a virtual multi-source spreading process is constructed by combining the behavioral patterns of individuals expressed in the data. Comparison results show that non-interacting metrics such as degree, PageRank, and k-core are unreliable in identifying multiple spreaders. Instead of considering only local properties, multiple influencers could be identified by metrics designed from a collective viewpoint, such as collective influence [55]. Influencer identification was also applied to locate influencers of fake news on Twitter during the 2016 U.S. presidential election [56–58]. This work will be discussed in detail below.

Influencer identification was also studied in multi-layer and multiplex social networks [59], where cascading effects are likely to occur as a result of the emergent collective effect of coupling layers [60]. According to Al-garadi et al. [61], the accuracy of influencer identification can be increased not only by improving algorithms, but also by concerning the multiplex network topology, which is a better representation of information diffusion in a range of social networks. In some applications, the prior knowledge about network structure is incomplete or even erroneous. As for influencer identification with the existence of such noise, Erkol et al. reported in [62] that the accuracy of identified influencers does not necessarily decrease as the noise level increases.

2.4 Artificial Intelligence to Predict Elections and Political Influencers

To discuss the application of the multistep model, we will next describe the rise of political influencers in recent presidential campaigns in the US. However, we must classify influencers and their followers according to their political ideologies to understand the political influencer movement. This is a problem for machine learning.

Figure 2.5 shows anecdotal evidence of new political influencers emerging during the rise of Donald Trump to the nomination as a presidential candidate in the Republican primaries for 2016 election. `@LindaSuhler` was one of the most influential pro-Trump Twitter users according to analyses of cascades of retweeting performed in [57]. Figure 2.5shows the retweet network during the Republican primaries between Donald Trump and Ted Cruz in 2016. We observe the polarization between both camps in separating the network into two distinct communities. More surprising is the prominent position in the Trump community of `@LindaSuhler`. The node's size in this network denotes the influence of the user in the Twitter community as calculated by the CI algorithm from Sect. 1.6.1. We can see that `@LindaSuhler` has more influence in the community than Trump. According to the account profile, the handler `@LindaSuhler` corresponds to 'Linda Suhler, PhD' (Fig. 2.5). The internet has no record of this person, and direct Twitter messages from a Science journalist, John Bohannon, who worked on a piece in Science on the advent of social media data in predicting elections [63], was never answered. Whether Linda Suhler is a troll or a real user is not known. However, this is a clear example of a new influencer unknown to the news media who has contributed to the rise of a new political power, an emergent opinion leader.

The situation is not just anecdotal. Using a dataset of 1 billion tweets collected throughout the months preceding 2016 and 2020 US presidential elections, a series of papers have investigated the rise of these influencers at a massive scale [56, 57, 64]. By applying influencer finder algorithms, the set of influencers that drove the attention during these presidential elections was uncovered. To classify these political influencers correctly and understand their influence in the political campaign, we first need to understand the political preferences of the social media users, which, in turn, can also forecast the elections accurately [56, 65]. Albeit this has not been done before, machine learning can help us predict the vote turnout using different variables.

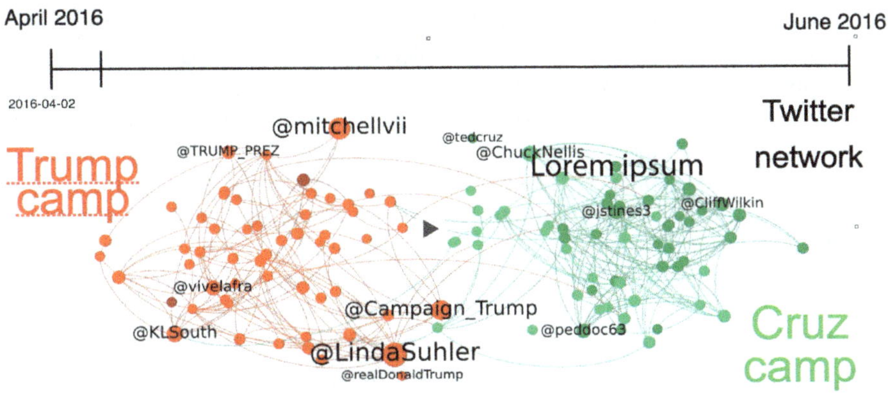

Fig. 2.5 **Who is Linda Suhler?** An organic political influencer supporter of Trump who represents the kind of new influencer with overreaching influence despite not belonging to any traditional news media organization. Retweet network during the 2016 US republican primary campaign. Size of the node is proportional to the influence

The analysis will tell us about the political ideology of the user. Such a classification adds another layer of information, as one is able to understand not only the spreading of information and misinformation among the users, but how the news influence the different political campaigns. This level of classification is achieved by machine learning (natural language processing, NLP), and Granger causality analysis [66], which are explained next. But first, we discuss the main competitor of AI in gathering people's opinions: the traditional pollsters and their failures in capturing societal trends.

2.5 Why Polls Are Failing to Predict Elections

Pollsters are in trouble. The victory of Donald Trump in the 2016 presidential election was a turning point for the polling industry as none of the pollsters and political pundits, including those in Trump's campaign could anticipate Trump's victory [67]. Traditional polling methods [68] are facing increasing challenges to predict election outcomes across the world [69–75]. The failure of traditional surveys has been widely discussed in the press and on the specialized literature [69]. Studies convey a complex image of the extent and causes of these challenges [69–73]. The use of AI and network algorithms can provide alternative methods by abandoning traditional pollsters in favor of using alternative data to provide election forecasts.

The key is the analysis of billions of data points from social media, like Twitter and Facebook, where users can express their opinions freely. These modern methods differ from traditional surveys, which are based on direct interactions with a small number of individuals. Traditional polling methods include live telephone surveys with respondents reached by random digit dial (RDD), online surveys using opt-in samples of internet users, and interactive voice response (IVR), to contact prospective respondents [68]. Traditional pollsters reach people mainly by dialing landline numbers so that they miss the majority of the population using cell phones [68, 76]. Online surveys with opt-in samples are surveys conducted in small online panels of social media users, typically a thousand users from Facebook that are made up of volunteers who are recruited through Facebook ads and who received some form of compensation for completing surveys, such as small amounts of money or frequent flyer miles. IVR performs automatic calls made by robots, again, to landline numbers.

Using these methods, the polling industry has recently faced increasing challenges in predicting elections. Firstly, it is becoming increasingly difficult to solicit people's opinions over the phone, and survey response rates have decreased over the years [71, 77]. Pools are plagued by ever-declining response rates, currently falling into the low single digits [77], believed to have reached 6% recently [77], and will continue to decline. Response rates are even lower for other methodologies, like internet polling and IVR. The Economist featured an article reflecting the challenges telephone pollsters are increasingly facing [78].

Compounded with declining response rates is the problem of misrepresentation of the survey samples, i.e., the sample surveyed by pollsters do not represent the demographic distributions of the voting population. Technically, this problem is solved by reweighting the survey samples according to the general demographic features of the population in a process called sample-balancing or raking [79, 80]. However, in countries where the vote is not obligatory, like the US, re-weighting a sample to the general population demographics (which is obtained, for instance, from the Census Bureau [68, 69]) fails since the general population demographics do not necessarily match the demographics of the voting population, i.e., the voter turnout or people who vote. Indeed, it is very difficult to predict which demographic groups will turn out at the voting station.

This represents the 'hidden vote' problem, which is believed to be one of the major reasons for the generalized failure of pollsters to predict the triumph of Trump in the 2016 US presidential election [69]. If a group that is underrepresented in the polls, like white men without college degrees from rural areas who overwhelmingly supported Trump in 2016 [81], decides to vote on election day, the re-weighting of the polls fails, leading to inaccurate predictions.

This group was heavily undersampled by all pollsters in 2016. Even with this historical information at hand, which supposedly allowed pollsters to resample their surveys more carefully in the subsequent election, pollsters again under-predicted the support for Trump in the 2020 election in some states or under-predicted the support for Biden in newly created battleground states like Georgia [82]. The inability of the pollsters to accurately predict the voter turnout to deal with this sampling misrepresentation renders the predictions of the pollsters questionable.

While nonresponse and misrepresentation bias present steep challenges to pollsters, they are not the only problems. Surveys are also affected by social-desirability biases, known as Bradley effect [83, 84]. This is the tendency of people to give socially desirable responses instead of choosing responses that are reflective of their true feelings. This bias appears when choosing controversial candidates. It is the tendency of respondents to not to tell the truth about their support for controversial candidates like Trump, which could otherwise open themself to social ostracism. Respondents may feel under pressure to provide 'politically correct' answers producing a bias towards candidates accepted by news media and the establishment (Clinton), in detriment of the controversial one (Trump).

This mechanism is believed to have been at place in the 2016 US election, contributing to the failure of pollsters to predict the winning of Trump, since many Trump supporters were undetected to pollsters. The tendency of people to lie to pollsters has been studied by Zhuo et al. [65] as a major contributor to the pollster's failure to predict the 2019 Argentina presidential election, as we will describe next. This election involved a controversial candidate (Cristina Fernández), unfavored by the media, who was heavily under-predicted by the polls.

2.5.1 A Case Study: The Failure of Pollsters to Predict the 2019 Argentina Election

The events leading up to the primary elections in Argentina 2019 are a telling example of the challenges presented to the polling industry [85–87], and how these challenges can be tackled with alternative data. On primary election day on August 11, 2019 (called PASO in Spanish: Primarias, Abiertas, Simultáneas y Obligatorias; in English: Open, Simultaneous, and Obligatory Primaries), none of the pollsters in the country predicted the wide +16% winning margin of the win of candidate Alberto Fernández (formula Alberto Fernández and Cristina Fernández, FF) over the incumbent president Macri (formula Macri-Pichetto, MP). The top five pollsters of the country, Real Time Data, Poliarquía, Isonomía, Giacobbe and Elypsis missed this result [87]. They all considered that Macri would win the election.

It was documented in the press [88] that these firms held several telephone conferences with foreign investors prior to the PASO election. One of the most active pollsters was the consulting firm Elypsis [89]. This was its message: "Macri wins by one point: 38 to 37%." Macri was right-leaning, favoring foreign investors, while Fernández was left-leaning. Elypsis had created a pool of investors that received ongoing election forecasts. Argentina's economy is volatile, with a bond market that is highly dependent on the political orientation of the government, either left (against foreign investment) or right-leaning (favoring foreign investment). Thus, foreign investors make their investing decisions based on election trends and outcomes. A single poll showing that left-leaning Fernández was ahead of the right-leaning Macri was able to move the bond market by -8% and increase the 'country risk' by several points [90].

As a result of the pollster's predictions favoring Macri, the bond market rose excessively in the days preceding the election. With Macri's loss, there was a historic collapse of the bond market by 40%, and the banks lost 1 billion dollars in the wager [91–93]. On the Monday after election day, this speculative stumble was coined the "Lunes negro" (Black Monday).

Robert Gibbins, founder of Autonomy Capital, a USD 6 billion hedge fund specializing in emerging markets, was one of the most prominent investors exposed to this market collapse, totaling losses of up to $1 billion as documented in the Wall Street Journal [91] and the Financial Times [92], making it one of the most prominent investors caught on the wrong side of market turmoil in that election. Bankers defended themselves for being on the wrong side of the trade by saying that they would not have bet so much if all the pollsters and the Macri administration had not endorsed the unrealistic pollster's predictions.

This story has been widely featured in the Argentinian press and social media [87, 88, 90–92]. Operatives of the political party of Fernández (Frente de Todos) had even proposed to institute legal proceedings against the pollsters under charges of 'market manipulation' [93].

One of the former directors at Elypsis, Luciano Cohen, has provided access to the raw data of the polls conducted before and after the elections by Elypsis, which were analyzed in [65]. For this election, Elypsis employed online panels from Facebook around 1,900 people assembled from volunteers who were recruited by promising frequent flyer miles using the ad tools provided by Facebook to target different specific demographics for online advertisement. Considering that pollsters do not share their raw data publicly, much less in the case of failed predictions, the analysis of the data provided by Elypsis represents a unique opportunity to understand the problems that traditional polls are facing and why they fail.

Figure 2.6a shows the analysis results done in [65] of the survey conducted by Elypsis immediately before the elections on 1,900 respondents. The data show the number of respondents for Fernández (blue), Macri (red), and Third Party (grey) separated by age and gender. Comparison with the actual distribution of age and gender of the entire population obtained from the Census Bureau is shown in Fig. 2.6b. It is clear that the ensemble of respondents to the Elypsis polls is biased toward people older than 51 years old, while the census population (Fig. 2.6b) shows the opposite; the majority of the population is younger than 50. The Elypsis survey does not have significant coverage among people younger than 30, even though it was done online on Facebook. The largest group surveyed by Elypsis are females between 51 and 65 years old who were overwhelmingly in favor of Macri. In fact, in all groups above 30, Macri is the clear favorite in the Elypsis poll. This clearly shows the sampling bias of these polls.

To deal with this sample bias, pollsters adjust their raw data to population benchmark distributions given by the Census Bureau [68, 69] (Fig. 2.6b) by re-weighting the raw data using sample-balancing or raking [79, 80]. The poll sample is weighted to match the population on a set of relevant demographic variables, for instance, age, gender, location and socio-economic variables, like education level or income (we will discuss these methods in more detail in Sect. 2.10). However, when the raw-data distribution is drastically undersampled and oversampled as the Elypsis case, a small error in the most representative groups could propagate to produce inaccurate results.

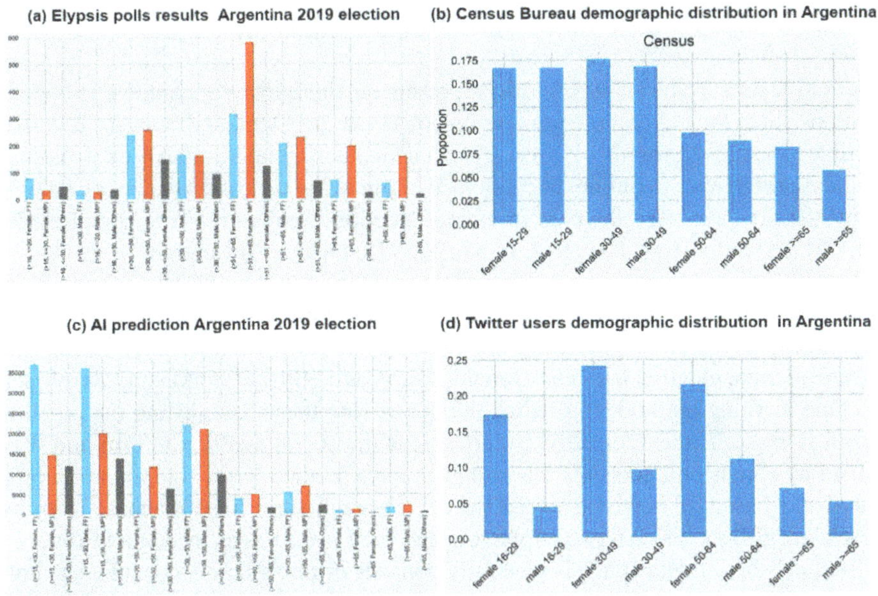

Fig. 2.6 Polling results in the Argentina 2019 election and AI predictions. **a** Elypsis polling results before PASO separated by age and gender (Macri = red, Fernández = blue, Others = grey). Results are highly biased to people older than 50 compared to the Census distribution shown in (**b**). **b** Argentina Census Bureau 2010 demographic distribution by age and gender. **c** Machine learning predictions by candidates, age, and gender from Twitter users obtained in [65] from the 5 million users used to analyze the Argentina 2019 election. **d** Demographics of Twitter users during the Argentina election by gender and age. Figure reproduced from [65].

Even after re-weighting, a significant sample bias produces inaccurate results since essential segments of society, like youngsters, are highly underrepresented.

Beyond the sampling problem, another problem in this survey is the tendency for the respondents to lie about their preference for Fernández, who was controversial and the underdog in the light of the established media [65]. This social-desirability bias was the main culprit for the wrong predictions.

The fact that people lied to the polls can be seen in the analysis of a unique longitudinal panel done by Elypsis where the same questions were asked to the same people before and after the election as analyzed in [65]. Specifically, when comparing 'Who are you going to vote in the PASO' with 'Who did you vote in the PASO' – using the same sampling and post-stratification methodology in both surveys – it was found that about 18% of the people did not disclose their true vote intention [65]: people lied and hid their true voting intentions to the pollsters.

Lack of anonymity is a common problem in a polarized environment, as respondents may not feel comfortable expressing their honest views [69]. A further concern with the type of polls conducted by Elypsis is that participants typically rush through the surveys to obtain their rewards and don't respond thoughtfully [69].

2.6 Can Alternative Data Replace the Polls?

Alternative data from social networks represent an alternative for capturing people's opinions since they overcome the low-response rate problem, and they are less susceptible to social-desirability biases. As social media users continuously express their political preferences in online discussions without being exposed to direct questions, many studies have been devoted to extract election predictions from social media [94, 95], specially from Twitter [56, 57, 96, 97] since its API is free.

Initial studies at capturing the electoral trends using social media had mixed results as well [98–100]. Later, Bovet et al. [56] analyzed 171 million tweets from 11 million users in the five months preceding the election day on November 2016 during the US presidential election between Donald Trump and Hillary Clinton and developed machine learning methods to predict elections using these data. Using the pipeline shown in Fig. 2.7, they collected tweets across the social media platform and then trained machines on a monthly basis to produce a prediction on vote intention on a daily basis based on newly acquired tweets. Through supervised machine learning these algorithms capture how people vote and their preferences throughout time.

Twitter's based data analysis generally consists of three main steps: (1) data collection, (2) data processing, and (3) data analysis.

2.6.1 *Data Collection*

Collection of tweets is based on Twitter's public API (Application Programming Interface). To be able to use the Twitter API, one needs to apply for a developer account to create a project and an application (app). These instructions are for the pre-Musk acquisition of Twitter era. Nowadays, the API is not free anymore, and significant funds would be needed to conduct this study, which was free before. The API is the software intermediary that allows the app to communicate and collect data from the Twitter app. Once the app is created, Consumer Keys (API Key and API Secret Key) and Authentication Tokens (Access Token and Access Token Secret) are provided.

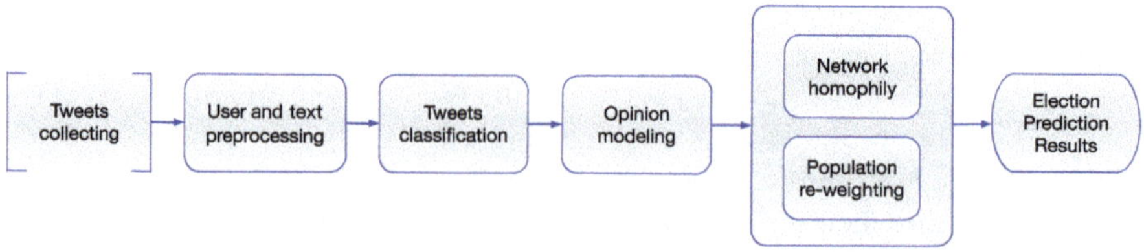

Fig. 2.7 Pipeline for the election prediction algorithm. From collecting information to forecasting elections and beyond. Figure reproduced from [65]. Copyright © 2021, The Authors

There are two APIs: the REST API and the Streaming API. Typically, one collects data using the Twitter REST API, which is called for a given query, for instance, 'Trump' and has a rate limit of 450 requests with a maximum of 100 tweets per 15 minutes, meaning that 45,000 tweets can be collected every 15 minutes. The Twitter Streaming API can also be used (using the Tweepy library from Python), which is not rate-limited, but it only yields around 1% of the total amount of tweets for a given query.

Appendix A lists freely available codes in Python and tutorials to perform data acquisition using the Twitter API and running the analytic pipeline shown in Fig. 2.7, including machine learning to predict vote intention. Collecting tweets using the REST API and the Streaming API for the same query is recommended. Using both APIs allows for a comparison of the results and biases of each collection method. API can be used to collect tweets from a given user. The REST API must be run periodically to collect the latest tweets. In contrast, the Streaming API will collect continuously (but may not provide all of the tweets if there is a lot of activity and many tracked users). The REST API is rate-limited per app, so anyone using it must create several apps to track different topics to the maximum extent. The Streaming API is limited per user, so anyone using it must create several Twitter accounts with several IP addresses. Using these APIs, one collects tweets according to specific queries, such as the names of the candidates in elections. This is a reminder that these APIs are significantly costly after the acquisition of Twitter by Elon Musk.

Ref. [56] collected tweets continuously using the Twitter Search API from June 1st, 2016 to November 8th, 2016, and gather a total of 171 million tweets in the English language, mentioning the two top candidates from the Republican Party (Donald Trump) and Democratic Party (Hillary Clinton) by using two different queries with the following keywords: *hillary OR clinton OR hillaryclinton* and *trump OR realdonaldtrump OR donaldtrump*. Zhou et al. [65] analyzed the 2019 Argentina election using the Twitter public API by collecting tweets from March 1, 2019 until October 27, 2019, filtered according to the following queries (corresponding to the candidates' name and handlers): *Alberto AND Fernandez*, *alferdez*, *CFK*, *CFKArgentina*, *Kirchner*, *mauriciomacri*, *Macri*, *Pichetto*, *MiguelPichetto*, and the main third candidate in the primaries *Lavagna*. Only tweets in Spanish were selected. It is important to define carefully the queries to achieve a uniform data collection across all candidates or topics to be analyzed or compared.

2.6.2 Data Pre-processing

Data pre-processing includes all those pre-processing techniques which aim to guarantee the credibility of the Twitter dataset or any other social media data. This is, for example, bot detection and removal. It is of fundamental importance to discern between bots or fake users from genuine users [56, 101, 102], the latter representing for instance the actual voters in an election. Botnets (a network of bots) and trolls are also identified at this pre-processing stage to guarantee the quality of the predictions.

Techniques for detection of bots as well as fake news posts are discussed in more detail in Sect. 2.11.

Data processing employs different national language processing (NLP) techniques to transform unstructured social media data into structured data that can be used in machine learning algorithms to produce predictions. The NLP pipeline is a process where text is transformed into structured data to exploit the richness of the dataset. After the removal of bots, typical techniques in data mining and NLP [103, 104] are applied such as text standardization performed by stop word removal and word tokenization. A pre-processing pipeline for sentiment analysis may include:

1. Tokenization of the raw text,
2. followed with a parts-of-speech (POS) tagging,
3. lemmatization, and
4. word/sentences embeddings like LASER (Language-Agnostic Sentence Representations) and MUSE (Multilingual Unsupervised and Supervised Embeddings) embeddings.

Several machine learning models are tested afterward to find the best model as explained in Sect. 2.4.

2.6.3 Data Analysis and Prediction

Finally, methods of data analysis and predictions can be classified into four main approaches: volume analysis, sentiment analysis, network analysis and AI [94].

The simplest way to predict an election is to use volume analysis using the number of mentions for a party or a candidate. For instance, a prediction for Trump versus Biden can be obtained by counting the number of tweets that mention each candidate. This method was used to forecast the result of the 2009 German parliament election [105] by examining the relative frequency of the mentions of political parties in Twitter messages posted during the election campaign. This method is inaccurate since it does not consider whether the mentions are positive or negative towards a candidate or party. Machine learning can then perform this classification.

Volumetric analyses have received further criticism in Ref. [106] since the results are contingent on arbitrary choices by the authors. However, this work inspired many researchers to improve upon and to use social media to predict elections. Later, Gaurav et al. [107] proposed a model, based on the number of times the name of a candidate is mentioned in tweets before elections, to predict the winner of three presidential elections held in Latin America (Venezuela, Paraguay, Ecuador) from February to April 2013. Based on volumetric analyses are also the works of Lui et al. [108] and Bermingham [109].

Ceron et al. and other authors [110–113] performed a sentiment analysis study on the tweets to check the popularity of political candidates in the Italian parliamentary election of 2011 and in the French presidential election of 2012. By going beyond sentiment analysis and by considering all the terms used in tweets, even the terms usually considered neutral, Refs. [100, 114] have shown that a more accurate measurement of Twitter opinion during the 2012 US election was possible. Moreover, Sylwester et al. [115] showed that it is possible to differentiate Republican and Democrat Twitter users based on their usage of words.

Ceron et al. [114, 116, 117] used a supervised machine learning approach based on a hand-labeled training set to estimate the proportion of tweets in favor of each candidate in the 2012 US election and the 2012 Italian center-left primaries. Beauchamp [100] extracted significant textual features from Twitter by fitting a model to existing polls and showed that these features improved state-level polls prediction. Caldarelli et al. [118] used the derivative of the volume to forecast the results of the Italian elections. Singh et al. [119] employed sentiment analysis to predict the victory of Trump in the election of 2016, and Ref. [120] proposed a method based on sentiment analyses and machine learning on historical data to predict the number of seats that contesting parties were likely to win in the Punjab election of 2017. Other works [121, 122] used social network analyses to identify the position of a party in the online community by measuring its centrality. The parties that are more supported are, in general, those with a higher centrality.

The number of papers attempting to use social media as a sensor of political opinion and public preferences is significant, since it is becoming accepted that alternative data will, at some point outdated traditional approaches to polls. The current state of the art on the topic is provided by many contributions in social science journals in particular in social computing conferences where this topic is mainstream, e.g., International AAAI Conference on Web and Social Media (ICWSM), ACM Conference On Computer-Supported Cooperative, Social Computing (CSCW), and The Web Conference (WWW). There are many works that deal with public opinion (and political processes) in a time-resolved way, some of them purely with a machine learning approach (NLP), some of them purely from a network perspective and some of them combine both.

Section 2.7 will review work by Bovet et al. [56, 57] and Zhou et al. [65] that combines both, networks and machine learning. Their machine learning models use an in-domain training set produced by hashtag co-occurrence networks from Twitter supporters to develop a predictor of election outcomes wherever social media is prevalent. Machine learning is used to predict the opinions of millions of users who discuss politics and share their preferences towards the candidates via social networks like Twitter and Facebook. This information is then used to study the influencers of each political party and to understand the process of polarization in society.

Analyses of social media data also anticipate opinion trends. This is called in the literature 'nowcasting' in contrast to 'forecasting' [117]. For example, in the case of the 2012 US Presidential elections, it has been shown that Twitter was able to anticipate by roughly eight days the surveys on the vote intention for the presidential candidates, Obama and Romney [117]; a result that was also found in the analysis of Clinton versus Trump in [56].

In the statistical literature [116, 123], it has been shown that in particular when analyzing social media, individual classification error through any machine learning approach remains high and does not vanish due to aggregation because of the large variance in estimates. In contrast, it can easily propagate up to the extent that, in many applications with thousands or millions of posts, one could see the error increasing to 15%-20%. This is problematic if one is mainly interested in estimating some aggregate measure through social media analysis, as with studies that want to map written opinions in tweets to votes. In this respect, instead of performing an individual classification of each post and then aggregating the predicted values, algorithms that directly estimate the aggregated distribution of opinions can be more robust. By using an algorithm that directly estimates the aggregated distribution of views. However, no individual classification of the voting preference of the users is possible. So, there is a trade-off involved between a more accurate aggregate level vs. a loss of individual information.

Despite the large amount of literature, the debate about whether Twitter or other social media outlets can be used to infer political opinions are still open. Online social networks are continuously filled by false, erroneous data through trolls, bots, and misinformation campaigns to a level that distinguishing between what is genuine and what is not is generally difficult [102]. By virtue of this, the great challenge of algorithms is to discover and interpret real data from 'junk data' that could lead to accurate predictions of electoral or opinion trends.

An essential limitation of social media-based methods is the misrepresentation bias. While social media solves the low response rates of polls by 'surveying' millions of users with non-intrusive methods, these large numbers of respondents might not represent the demographics of the voting population. Thus, the opinions of Twitter users may not be representative of the entire population [124], and re-sampling methods need to be used, importing along with them the same problems that plagued the traditional polls. Solutions to these problems are discussed in Sect. 2.10.

2.7 Machine Learning to Predict Elections

We describe methods to infer political opinions and vote intention from Twitter and Facebook users by using machine learning and network theory developed in [56, 57, 65]. At the core of the model is the capability to predict the intention of vote of a social media user for a given candidate and to identify demographic categories related to geolocation, age, gender, county/state, rural-nonrural, and income of users in social media.

The first step of the method is data collection, as discussed in Sect. 2.6.1. This kind of data acquisition needs approval from Institutional Review Boards (IRB) at the institutions where the research is conducted. To preserve privacy, data need to be fully anonymized at collection. Using the Twitter Search API and an 'ad hoc' throttling strategy, data can be uniformly collected for each query as a function of time. The uniformity of the collection is essential to guarantee the significance of any analysis. Depending on the project's target, a query can be the name of the candidate, a particular keyword, or a hashtag. Employing this methodology, Ref. [65] has collected 70 million tweets from 5 million unique users (approximately

20% of the electorate) from March 2019 until election day for the 2019 Argentina election. Ref. [56, 57] collected 171 million tweets from 11 million unique users during the six months preceding the 2016 US elections, and Ref. [64] collected 702 million tweets sent by 20 million users during the 2020 US election.

The analysis of the collected data consists of several layers. First, an in-domain training set is constructed. A labeled training set of tweets are built with explicit opinions about the two presidential candidates by taking advantage of the fact that a large number of Twitter users label their own tweets by using hashtags. The use of a hashtag that explicitly expresses an opinion in a tweet represents a 'cost' in terms of self-exposition by Twitter users [114] and therefore allows one to select tweets that clearly state support or opposition to each candidates.

A study of the co-occurrence hashtag network is performed as a first analysis. Given their self-explanatory meaning, the most significant hashtags allow classifying tweets. The first task is therefore to classify the hashtags as expressing support or opposition to each of the candidates. For this purpose, we start by identifying the most important hashtags in term of their total number of occurrences and then use the relations between hashtags co-occurring in tweets to discover new hashtags. For this, a network of hashtag co-occurrence $H(V, E)$ is built, where the set of vertices $v_i \in V$ represents hashtags, and an edge e_{ij} is drawn between v_i and v_j if they significantly appear together in many tweets, according to p-value statistics.

Figure 2.8 shows a typical co-occurrence network in the US 2016 election. Two main clusters are visible. One cluster corresponding to the Pro-Trump: #MAGA,

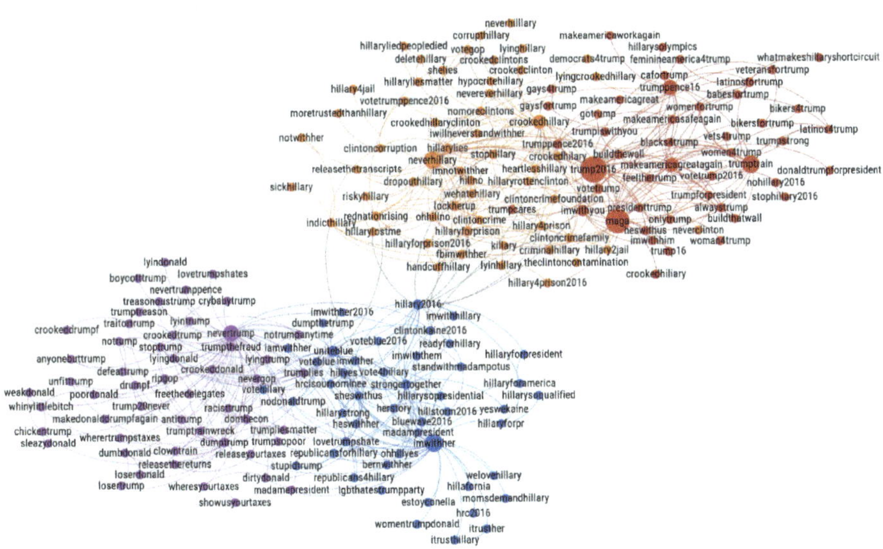

Fig. 2.8 Example of hashtag classification via network of co-occurrence. Network of hashtags obtained during the 2016 US election with data acquisition from June 1 to September 1, 2016. The node's size is proportional to the total number of hashtag occurrences. Figure reproduced from [56]. Copyright © 2018, The Authors

#MakeAmericaGreatAgain and Anti-Clinton: #NeverHillary, #CrookedHillary. The other cluster corresponding to Pro-Clinton: #IAmWithHer, #MadamPresident, and Anti-Trump: #NeverTrump, #TrumpTheFraud. Inside of these two clusters, the separation between Pro-Trump (red) and Anti-Clinton (orange), or Pro-Clinton (blue) and Anti-Trump (purple), is also visible. The coloring corresponds to clusters found by community detection algorithm from [125, 126]. This approach was used in [56] to automatically build a training set of the order of one million tweets (out of 171 million total) which is two order of magnitude larger than previous methods based on hand labeling [114, 123]. The training set of labeled tweets contained two classes: (1) pro-Clinton or anti-Trump and, (2) pro-Trump or anti-Clinton, and tweets belonging to both classes simultaneously are discarded to avoid ambiguous tweets. Using this in-domain training set, particularly labeled for a given election, not only helps to capture the informalities of Twitter language but also permits capturing the rich context of every election.

After discarding bots and pre-processing the data with methods explained in Sect. 2.6, a supervised machine learning approach is then used to predict the opinion of the remaining Twitter users beyond the labeled ones. Despite the astonishing progress reached by NLP and, more in general, by ML techniques, algorithms applied for election prediction suffer from two main problems [127]: **(a)**, they lack generalization, which means that they do not perform well when applied to different elections, and **(b)** they lack real-time prediction capabilities. General multi-platform and multi-lingual NLP algorithms are needed. Several machine learning models can be trained to find the best model to fit the data. These include the following non-exhausting list of machine learning methods. The list might already look outdated at the time of publication due to the frenetic pace of innovation in AI. It should be taken as more of an indication of the historical importance of the main NLP methods. A more current list at publication appears in Appendix A.

- Models based on techniques such as support vector machine, logistic regression, Naïve Bayes method, random forest, and decision Tree can be used for a simple and quick assessment of the data.
- Recurrent neural networks (RNN), Long Short-Term Memory (LSTM), GRU, and variants can be used to predict the current voting intention of each user from the time series of acquired data. In RNNs, nodes are connected sequentially to form a directed graph. An example is the LSTM network, particularly practical for modeling sequential data, such as human language. It has been used to predict elections in [130]. These techniques have been rendered obsolete by Large Language Models (LLM).
- Convolutional Neural Networks have been also used to understand election trends.
- The word2vec embedding algorithm learns word associations from the training set. Different flavors of this algorithm exist and have been used for political classifications. Transformers have rendered this technique obsolete.
- The BERT model (Bidirectional Encoder Representations from Transformers) is a transformer-based ML technique for pre-training developed by Google. Pre-trained models are available in different languages. BERT [128] is designed to

pre-train deep bidirectional representations from unlabeled text. The pre-trained BERT model is fine-tuned with one additional output layer to create state-of-the-art models for various tasks such as question answering, language inference, and sentiment analysis. It has been used to model the US 2020 election in [129]. It has many variants. Transformers are the core of current LLMs.

- Graph Neural Networks (GNN) [131] in all its variants are quite popular [132] to understand trends from social media through its network structure of information spreading.
- Large Language Models (LLMs) are the latest advances in generative AI and have practically rendered all the algorithms above obsolete, especially for sentiment analysis and opinion classification. They are discussed in Sect. 2.8.

Of relevance to the problem of predicting elections from time series are recurrent neural networks, RNNs and Large Language Models. Considering elections as a stand-alone event is not the best strategy since it leads to non-generalizable solutions; this limits the researcher in understanding the common characteristics among different elections. An example of these characteristics is the citizens' response to a dire economic situation. Likely, this response is election-independent. In other cases, identifying some variables' patterns could also help define a general algorithm. It may be the case of national demographic characteristics or voter turnout. These features are the substratum of an algorithm capable of predicting political elections and, more generally, people's opinions. Instead of focusing on a single election, alternative data from social media allows the use of time series from past elections to extract those futures and patterns and use them to predict new elections.

RNNs (like LSTM) and Transformers offer the possibility to achieve this result. LSTM is a neural network that makes predictions according to data from previous times, i.e., it has a concept of 'memory' explicitly built into the model structure. The time series under investigation is a time series of states. A set of variables characterizes each state (defined as a state for each candidate for every day of the political campaign). Macro data (e.g., turnover, demographic distribution), and volumetric social data (e.g., number of mentions, reactions, hashtags) can be used (see Sect. 2.10) to process candidate approbation computed using NLP (e.g., Transformers) analysis on tweets, posts, and comments. Further homophily analysis on the retweets/posts networks can be performed with Graphical neural networks. Once trained on past elections, the LSTM network can optimally assign a weight to a new state (future elections) for each candidate. At this point, a further layer can combine each day the states of the competing candidates in a sort of zero-sum game, returning the percentage approbation of each candidate.

This model reveals features and patterns that different political campaigns have in common. The model goes through new states in every election and improves its predictions. Because the model relies on national demographics and voter turnout (Sect. 2.10), the sample bias problem will be automatically considered.

For each of the models considered above, hyperparameter optimization can be performed with a 10-fold cross-validation optimizing F1 score. The performance of the classifiers is tested with different regularization methods (Ridge Regression,

Lasso, Elastic net, etc.), and the model that better performs in terms of predictions and runtime is selected as the final predictor.

In the study of US and Argentina elections performed in [56, 65], one of the best scores was obtained, surprisingly, by a simple logistic regression classifier (with L_2 penalization, ridge regression). In this study, the logistic regression performs better than CNNs, RNNs, and SVM with an average group accuracy equal to 83% and recall and F1-score equal to 83%. Support vector machine is the second best with an average accuracy of 81%. LLMs did not exist at the time of these studies.

Logistic regression is a very simple classifier. These results do not imply that it should be considered the best performer overall. While the logistic regression performed well in the Argentina study, the recently introduced LLMs represent a quantum leap in accuracy and versatility, particularly in sentiment analysis of text data and in understanding opinion classification to track election trends. We briefly discuss these remarkable advancements in the next section. The simplicity of the best architectures found in [56, 65] seems to suggest that partisan users of social media are 'easy' to be identified from the words they use on social media.

A final layer of analysis is used to move from tweet classification to user classification. After classifying every tweet into favorable to one candidate or another, the user classification is performed according to the candidate for which the majority of their tweets are classified. This simple user classification can be improved by taking into account the time series of the tweets for every user.

2.8 Generative AI and Large Language Models: The Revolution Will Not Be Televised

Very recently, the introduction of Large Language Models (LLMs) has revolutionized the field of generative AI. LLMs such as OpenAI's ChatGPT—https://openai.com/index/hello-gpt-4o/, Meta's Llama—https://en.wikipedia.org/wiki/Llama_(language_model), Claude AI—https://en.wikipedia.org/wiki/Claude_(language_model), and others have shown astonishing accuracy in capturing entire domains of knowledge. We could not write a book on AI without mentioning these remarkable developments. Elaborating on LLMs would require writing another book altogether, so we leave the discussion of these astonishing developments for a later edition and briefly mention the latest results below.

These models, trained on massive collections of text data, utilize transformers as the basic architecture. The primary function of LLMs is to understand and generate human-like text, adapting to various language tasks with considerable flexibility and accuracy. Since the release of OpenAI's GPT-3.5 and, subsequently, the more advanced versions leading up to ChatGPT-4o, the field of LLMs has witnessed transformative changes. The initial versions, like GPT-2 featured 1.5 billion parameters (https://en.wikipedia.org/wiki/GPT-2), while GPT-3 expanded to 175 billion parameters (https://en.wikipedia.org/wiki/GPT-3), and the latest GPT-

4 boasts around 1.76 trillion parameters. This exponential growth in parameters has enabled LLMs to exhibit emergent properties, complex behaviors, and capabilities that resemble aspects of human intelligence, such as advanced reasoning and a complex understanding of context.

Recent developments have focused on enhancing the models' abilities to grasp the subtle nuances of language and context, which is essential for applications ranging from customer service to creative writing. Innovations include fine-tuning techniques that allow these models to become more specialized in specific tasks and training methodologies that aim to mitigate inherent biases in the training data. These improvements affect LLMs' functional capabilities and strive to align their outputs with ethical standards and social norms, enhancing their reliability and trustworthiness as tools.

2.8.1 LLMs in Sentiment Analysis: Predicting Social Trends and Elections

LLMs are particularly effective in sentiment analysis in identifying and categorizing opinions expressed in written text. This application is especially useful in monitoring social media to gauge public sentiment on various topics, such as political elections or consumer products. For example, by analyzing the sentiment of online social media posts and comments, LLMs can classify these inputs into emotional categories like positive, negative, or neutral towards an election candidate or product. This ability allows us to track overall sentiment trends and potentially get a good estimation of the public opinion of certain events, announcements, or people.

To implement sentiment analysis, an LLM is first trained on a dataset where pieces of text are labeled according to their sentiment. The model learns to associate specific words and phrases with particular emotions. For instance, words like "happy" or "excellent" might be associated with positive sentiments, whereas "disappointed" or "terrible" might indicate negative sentiments. As the model processes more data, it becomes more adept at detecting these patterns, even in nuanced or complex sentences where the main opinion is hidden behind sarcasm or a joke.

In practical terms, using an LLM for sentiment analysis on social media involves collecting relevant posts and feeding them into the model, which then assesses and classifies the sentiment of each post. This analysis can be scaled to handle large volumes of data in real-time, providing immediate insights into public opinion. Such tools are invaluable for political analysts, marketers, and social researchers who need to keep a pulse on public attitudes and reactions without the delay of traditional survey methods. For political campaigns and analysts, this means gaining insights into which issues resonate with voters and how sentiments are shifting in real-time.

2.8.2 Llama Model from Meta

Due to its open-source nature, Llama is our favorite architecture for performing test analyses. The Llama models are a family of large language models developed by Meta, with Llama 3 being the most advanced model to the date of publication (June 2024). Llama 3 is a 70 billion parameter model that has been fine-tuned for various natural language processing tasks, including sentiment analysis. While running predictions on the 70B model would be hard on common-sized GPU clusters, Llama 3 provides an 8 billion parameter version, ideal for running on small local servers.

We can access Llama 3 through the Hugging Face Transformers library to download and use it. Llama 3 can be run on a single GPU machine to classify the sentiment of social media posts on topics like elections by feeding the text data into the model and generating predictions. By analyzing the sentiment of social media posts, Llama 3 can help predict public opinion on different topics, such as elections. This can be done by classifying posts as positive, negative, or neutral and then aggregating the results to gauge overall sentiment. This can provide valuable insights for political analysts, researchers, and campaigns.

The beauty of Llama is that Meta provides the model already trained; that is, all the 8 billion parameters, which can be used to test applications in great detail. Running this model does not require a large investment in computer power; a USD $5,000 server with an Nvidia GPU (price as of June 2024) suffices to run big data applications at the level of a full election.

The server for training deep learning models and running locally hosted LLMs would mostly rely on the GPU, which has a lot of weaker processing units (compared to processing cores in a CPU) that can work in parallel. A very important feature for training models and hosting local LLMs is the VRAM of the GPU, which determines how large of a model the GPU can hold and serve. Typical specifications of the machine are as follows (as of June 2024): (a) A AMD Ryzen 9 7950X Hexadeca-core (16 Core) 4.50 GHz Processor, (b) Nvidia GeForce RTX 4090 Windforce V2 PCI-E 4.0 24GB Graphics Card (which has the largest VRAM in consumer-level GPUs), and (c) 2× 48 GB DDR5 Memory.

This system can train large deep-learning models on different data problems and, for this use case, host the Llama 3-8B model for inference.

2.8.3 Using OpenAI's ChatGPT and Llama 3 for Sentiment Analysis and Opinion Classification

Classifying social media posts into sentiments and opinions can be done with any LLM. OpenAI, as the most famous and pioneer LLM developer right now, provides their LLM as a ready-to-use website and API for a fee, without the need to host and maintain it locally. Using OpenAI's latest model (GPT-4 & GPT-4o) is expensive for typical academic use cases ($5 per 1M input and $15 per 1M output tokens, as of June 2024). GPT-3.5 is available for a more reasonable price ($0.50 per 1M input and $1.50 per 1M output tokens).

Using the GPT-3.5 model, we were able to classify 375K Facebook posts from the 2020 US elections into pro-Trump / pro-Biden / neutral classes. This costs around $30 per experiment (we had to do several experiments using different prompts) and close to $100 in total. This would not be the end of the operating costs since we might want to re-run the experiment with different prompts and classify more posts from larger datasets (such as Twitter) and also the comments on posts with longer prompts.

Although it varies case-by-case and in different domains, our benchmark indicates that GPT-3.5's performance appears to be lower than the latest Llama 3-8B in various benchmarks. Given the API costs for OpenAI, even using the cheap GPT-3.5 model might not be worth it if compared to hosting local Llama and other open-source models for research purposes. While all these results are preliminary, these exciting developments indicate that the revolution has just started. We now go back to more pedestrian pre-chatGPT methods.

2.9 Machine Learning Predictions

The machine learning model can monitor the opinion of every user over time and predict their voting intentions using all the cumulative data. For instance, assigning a user as a Trump supporter only when the majority (with a given statistical significance) of her/his tweets are classified as supporting Trump. Figure 2.9 shows that considering the opinions of all Twitter users during several months preceding the election, the results of several elections are captured by the machine learning model. Figure 2.9a show the predictions in Argentina 2019. The model predicted the landslide defeat of Macri by Fernández at the primary election on August 11, 2019 and the general elections on October 27, 2019; a result not anticipated by pollsters [65]. The final machine learning prediction in that election was accurate down to 0.6%. The model predicted for the PASO primary elections: Fernández = 50.2%, Macri = 36.9% and Third Party = 12.9%, while the official result of the PASO was: Fernández = 49.49%, Macri = 32.93% and Third Party = 17.58%. For the general election on October 27, the model predicted: Fernández = 48.9%, Macri = 39.6% and Third Party = 11.4%, while the official final result of the election was: Fernández = 48.24%, Macri = 40.26%, and Third Party = 11.5% [65].

In the 2016 US presidential election, Fig. 2.9b shows the instantaneous snapshot predictions of the opinion trend in a time window of 14 days. This measurement does not correspond to the cumulative averages discussed above. However, it provides an accurate fitting to the results of the New York Times Aggregator of Polls at 'The Upshot' [133] ($r = 0.93$ and RMSE = 0.31%). The Upshot Aggregator assembles a large number of polls, of the order of thousands, and weights them with proprietary information to produce a weighted average of the most trustable pollsters in US. While the results of machine learning follows the instantaneous aggregate of the

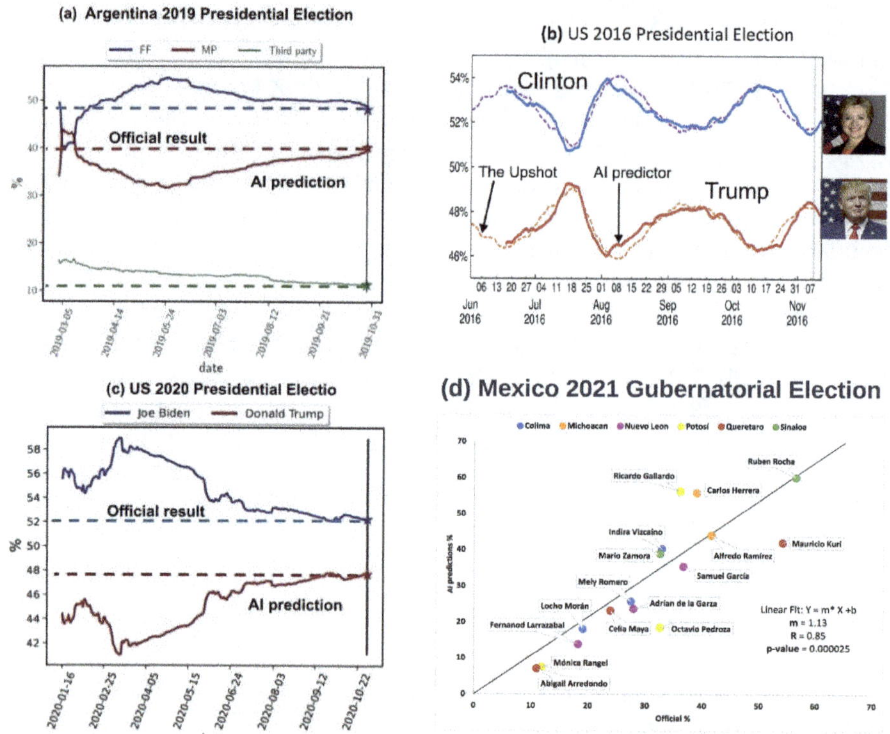

Fig. 2.9 AI for elections. **a** AI prediction in the Argentina 2019 election captures the election result on Oct 27, 2019. Figure reproduced from [65]. Copyright © 2021, The Authors. **b** Instantaneous snapshot of opinion trends in US 2016 campaign follows closely (r = 0.93, RMSE = 0.31%) the Aggregate of US polls done by The Upshot at the New York Times. Figure reproduced from [56]. Copyright © 2018, The Authors. **c** Results of the machine learning model for the US 2020 election capture the victory of Biden against Trump. **d** Results of the machine learning model in six of the gubernatorial elections celebrated on June 6, 2021 in Mexico using Facebook data. Each dot is a candidate for different colored states, as shown in the legend. We observe a significant correlation (r = 0.85) between the AI predictions and the official results

pools at the national level, they did not reproduce the results of the electoral college. The cumulative average and the improvements discussed in Sect. 2.10 are needed to capture the vote intention at the county and state level by taking into account the biases of the online data to reproduce the electoral college results.

The result of the US 2020 election is captured by the cumulative method as shown in Fig. 2.9c. Finally the same method applied to Facebook data produced accurate predictions in the gubernatorial elections in Mexico 2021 (Fig. 2.9d). In summary, the cumulative prediction where the activity of the user is followed over time can be used as a predictor of the user opinion and a good election predictor.

There are further generalizations of the machine learning method, such as multi-partite political systems and implementations for public opinion outside an election

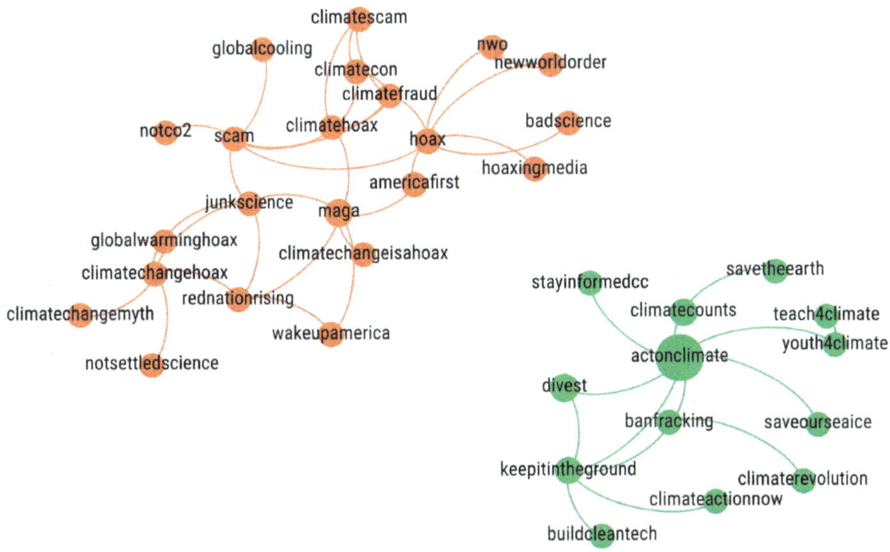

Fig. 2.10 Twitter hashtag co-occurrence networks for climate change. The network shows the strong polarization on the topic separating those believers from non-believers, a result that can be used to generalize the machine learning method from elections to this topic. Figure reproduced from [56]. Copyright © 2018, The Authors

setting. For example, Fig. 2.10 shows the co-occurrence network of hashtags related to climate change. The network splits into two groups, one with hashtags supporting action toward climate change #ActOnClimate (green) and the other with hashtags depicting climate change as a hoax #GlobalWarmingHoax and #ClimateHoax (orange). This community is intertwined with hashtags representing the Republican Party like #MAGA and #AmericaFirst. This result shows the clear polarization on the topic and can be used to define an in-domain training set to predict people's opinion on climate change, as done with the elections. This result indicates that machine learning and co-occurrence hashtag network methods can be generalized to topics beyond the election setting. The technique could also be applied to events where we have witnessed dramatic failures of traditional polls to predict critical voting processes, like in 2016 Brexit and Colombia's referendum on the peace process. This brings the fundamental question of how the results can be generalized to different scenarios.

Despite this progress, methods based on alternative data are still plagued by sampling bias and mis-representations, in similar fashion as the standard polls [63]. A further concern is coming from the Twitter's API, either the streaming API or the firehose, because how Twitter samples its contents is unknown [134, 135]. Solving these biases is critical to develop reliable predictors of human behavior using AI and alternative data. They represent unsolved open problems in the application of AI to predict human behavior. The AI predictions will not be reliable until these sampling problems are properly addressed. Possible solutions are discussed next.

2.10 Sampling Biases of Social Media

The difficulties of pollsters represent an opportunity for emerging new techniques to capture the opinion trends using AI and network theory. Network theoretical analytics have the ability required to filter, interpret and aggregate data to produce predictions by obtaining opinions directly from voters who express themselves on social media like Twitter and Facebook without the need to solicit responses. If declining response rates continue, predictions can only get more challenging for traditional pollsters, while for AI, predictions will get better and better as more training data are incorporated to the models.

However, enormous challenges remains to successfully apply alternative data to predict human behaviour. The critical issue is that Twitter or Facebook, as any social media platform, are naturally biased [136–139]. Even in countries with a strong penetration of Twitter, the adoption is still below 25% (Pew Research Center 2021 [138]). Twitter over-represents young people [138, 140]. Social media users are not fully representative of the population [137], being younger and more Democratic (in the US) than the rest of the population (Pew Research Center [140]). And only a minority of users produce political content in social media [139].

Also, social media does not represent the opinions of millions of people who live in more rural areas than city dwellers. Figure 2.11a shows the coverage of Twitter users per county in the US 2016 election studied in [56]. Each county is colored according to the AI predictions for 100% Democratic Twitter in blue to 100% Republican Twitter in red, and compared with the official results in Fig. 2.11b. The differences between the Twitter prediction and the results are apparent when plotted in terms of counties. We see that the majority of the large urban counties are Democratic, according to the Twitter prediction and coverage, while the predicted Republican counties are primarily in rural areas.

Twitter coverage is highly concentrated on urban populations, while the large areas of rural counties are not covered at all by Twitter. The blank counties in Fig. 2.11a have no Twitter coverage at all. While these areas are sparsely populated, when taken together, they represent a large portion of each state vote that contributes to the final count in the state and the electoral college. Twitter does not capture these rural areas. It was argued that the polls did not capture these rural areas either. This explains why the instantaneous voting population from Twitter in Fig. 2.9b agrees so well with the aggregate of the pools of The New York Times. In contrast, when the data is aggregated at the county level to predict the state elections, both results failed to predict the electoral college and the victory of Trump.

Indeed, rural voters were a critical aspect that allowed Donald Trump to win the 2016 election [69], see [81]. More rural counties tend to be more Republican. Most rural counties went for Trump by 70% overall. Rural America is not a monolith,

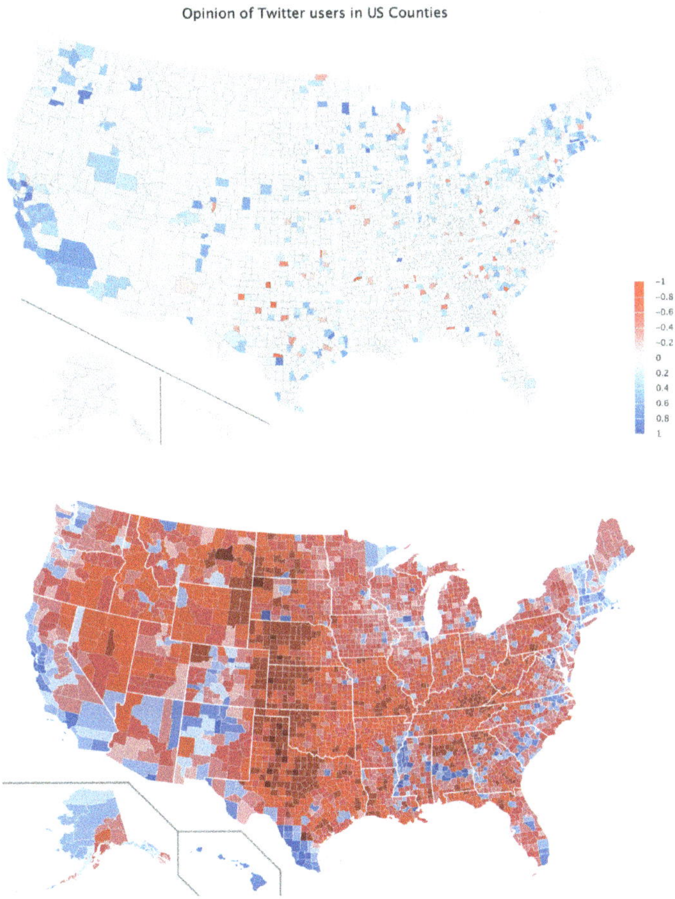

(a)

(b)

Fig. 2.11 Coverage of Twitter users in the 2016 US election. a The coverage of Twitter is clearly biased towards the large urban areas where Clinton was largely a favorite. The rural areas are sparsely covered by Twitter explaining the difficulties to capture the election results at the county/state level which determined the electoral college results. As shown in the official results per county in (**b**), they were largely determined by the overwhelming victory of Trump in the rural areas that are not covered by Twitter. The color bar goes from blue for a Democratic majority to red for a Republican one. Figure reproduced from Wikipedia [141]

but most rural counties have demographics that have favored Trump. While most Americans live in cities, the US Electoral College gives rural areas outsize voting power [142].

Furthermore, studies have shown that adults with lower educational levels, located in rural areas, and with anti-government views are less likely to take part in traditional surveys [143–145]. Rural versus urban areas, as well as income level, are the most difficult variables to control in any sample. Correcting for these biases in the polling samples is an important aspect that has led to wrong predictions in traditional polls [69]. Controlling for these areas becomes crucial to predict elections in the US. While pollsters find mounting difficulties in addressing these areas, detailed geolocalized data from social media might be able to control for this bias, see Sect. 2.10.2.

The use of alternative data to understand trends in society requires to resolve these key biases so that the method works not only in the the particular context of one election but is indeed feasible to predict all outcomes. AI methods require the following:

- predict voter turnout, in particular in the US,
- adequate sample of rural areas with weaker online coverage,
- filter of bots and fake news outlets (Sect. 2.11),
- re-weight the sample on a state-by-state basis to match the electoral college in the case of the US,
- address the opinion of voters who are not on social media,
- address the opinion of voters who are on social media but do not express their voting preferences directly, and
- address the social desirability bias [73], i.e., the tendency of respondents to hide their intention to vote for controversial candidates.

These problems can be addressed as follows.

2.10.1 Demographic Categories

First, relevant demographic categories should be extracted from social media. These include features of social media users like age and gender, which can be obtained from public profiles of Twitter users. Location is also provided in the position field of Twitter data and can geolocalize a user in a given city, county, or state depending on the information provided in the public profile. These locations can be matched on Google Maps to extract the latitude and longitude. These data can be completed by using homophily (the majority of their ego-centered network) to assign a probable location to those profiles that do not display this field. The geolocalization data allows the detection of rural and non-rural users and assigns to the users several demographic categories such as income, education, or religious level that can be identified from $j = 1$ to $N = 7$: $j = \{1 =$ age, $2 =$ gender, $3 =$ rural/non-rural user, $4 =$ income level, $5 =$ geolocalization to city/state/county of the user, $6 =$ education level, $7 =$ religion $\}$. The voter preference for the candidates is obtained with machine learning from Sect. 2.7, adding another index, for instance, in 2020: $i = \{1 = \text{Biden}, 2 = \text{Trump}\}$. These demographic categories and classifications provide the necessary information to treat the biases inherent to social media, as discussed next.

2.10.2 The Coverage Problem: Rural-Nonrural Divide

To predict the opinion in society, which includes the online and real world, we first correct for the notorious Twitter bias for having a demography concentrated mainly in the larger urban areas and younger voters [140]. It is important to cover correctly the sparsely populated rural areas. To do so, the geolocalization of the user to each county in the US should be established such that the results can be rescaled to represent all counties in the US according to its population.

Each county is classified by its degree of urbanization using the USDA Rural-Urban Continuum codes [146] into 9 codes spanning from 'Counties in metro areas of 1 million population or more' (code 1) to 'Non-metro—Completely rural or less than 2,500 urban population, not adjacent to a metro area' (code 9). Ideally, uniform coverage of every county in the US should be obtained to re-weight each of the 3,243 US counties with the re-balancing method (explained in more detail in Sect. 2.10.3). However, the social media coverage is not uniform, as shown in Fig. 2.11a, especially in the most rural areas (codes 4–9) where typically Twitter does not have any coverage. To re-weight these areas, counties can be agglomerated by rural areas using the city clustering algorithm used in studies of city growth [147–149]. Counties are agglomerated according to their county-rural code to define rural super-counties until one obtains a proper Twitter coverage of each super-county. Likewise, each county in the US has an average income, education, and religious classification. The same method is used to rescale the data using these demographic classes.

2.10.3 Predicting Voter Turnout and Addressing Sampling Bias

Next, we explain how the problem of sampling bias in social media and voter turnout can be solved with a regression model. We use county-level demographic data in US from the Census Bureau and the social media categories described in Sects. 2.10.1 and 2.10.2. The core model predictor takes into account the following demographic categories:

- n_{ij} is the direct Twitter user count in support of candidate i in demographic category $j = 1, \ldots, N$, with $N = 7$, the categories from Sect. 2.10.1.
- p_{jk} is the distribution of demographic category j in (super) county k obtained from Twitter,
- q_{jk} same as p_{ik} obtained from the US Census Bureau,

- $r_{jk}(t)$ is the predicted turnout rate a time t, i.e. today's election.

Without taking turnout into account, the expected number of voters for candidate i in county k can be re-scaled by raking [79, 80] as follows:

$$\text{Model (1)}: \ \hat{n}_{ik} = \sum_j n_{ij} \cdot \frac{q_{jk}}{p_{jk}} \tag{2.2}$$

With voter turnout included, we obtain:

$$\text{Model (2)}: \ \hat{n}_{ik} = \sum_j r_{jk}(t) \cdot n_{ij} \cdot \frac{q_{jk}}{p_{jk}} \tag{2.3}$$

A zero-order model for turnout assumes that the turnout rate is the same as the last presidential election $t - 1$ (available from Census Bureau). This is what most traditional pollsters assume. Then:

$$\text{Model (2')}: \ r_j(t) = r_{jk}(t-1)$$

To improve on this model, we train a linear regression model to predict the change in turnout at time t: $dr_{jk}(t)$

$$\text{Model (3)}: \ r_{jk}(t) = r_{jk}(t-1) + dr_{jk}(t) = r_{jk}(t-1) + f_j(\text{macro}_t, \text{micro}_k) \tag{2.4}$$

We use micro variables specific to the county k, which do not change much with time t. These are the demographic distributions. The macro variables are socio-economic variables like unemployment, economic outputs, revenues, taxes, etc. and, crucially, the opinion variables that change in time t even on a weekly basis. The socio-economic variables are obtained from Census Bureaus and Department of Labor, while the opinion trends are obtained from the Twitter analysis. We can train this model by looking back a few election cycles and across all counties. The simplest model is a linear regression for this:

$$f_j(\text{macro}_t, \text{micro}_k) = \alpha_j \cdot \text{macro}_t + \beta_j \cdot \text{micro}_k + \gamma_j, \tag{2.5}$$

and then one can train a more elaborated Model (2.4), by first testing for instance for the US 2020 if Model (2.3) is any better than the simpler Model (2.2) without turnout.

The next step in sophistication if a logistic regression. In this case, turnout-rate is modeled directly, instead of a delta-rate, as:

$$\text{Model (5)}: \ r_{jk}(t) = \text{logistic}([\text{macro}_t, \text{micro}_k, \log(r_{jk}(t-1))]), \tag{2.6}$$

where we enter the previous year log-rate as an additional regressor variable. The more variables we include the less motivation we have to use anything other than linear. Further nonlinear models can include kernel logistic regression.

2.10.4 How to Take Into Account Voters Who Are Not on Social Media

The social media sample requires re-weighting to compensate for voters who are not on social media. The strategy is to stratify the sample by demographic categories as explained above and then use Model 2.3 and Model (2.6) to perform sample-balancing [79, 80]. Perhaps the most important demographic is the rural-urban divide since Twitter is underrepresented in rural areas. As explained above, the predictor takes into account the county-level demographic (divided into rural-urban) by n_{ij}, p_{jk} obtained from Twitter, and q_{jk} and p_{ik} obtained from the Census Bureau, and r_{jkt} predicted turnout rate. Model (2.3) then considers those people not in social media by reweighting the demographic categories.

While traditional pollsters also do this reweighting to deal with misrepresentation, applying it to social media offers some advantages: (a) samples are in the 10's millions and not in the thousands, (b) most importantly, machine learning can predict the voter turnout r_{jk} based on different micro and macro variables, see Eqs. (2.4)–(2.6), while pollsters use $r_{jk} = 1$ and cannot take into account the rise of new groups from one election to the other, like the 'silent' rural vote for Trump in the US in 2016.

2.10.5 How to Capture Voters on Social Media Who Do Not Express Voting Preferences

The capability of social media to perform longitudinal opinion tracking over extended periods and transversal network analysis and homophily can address this problem. First, a given user can be followed from election to election to infer his/her current voting preferences. For those users who do not express their opinions in social media, we could build the network of social media interactions (using retweets, mentions and interactions). We then use network features such as homophily (looking at the majority in the ego-centric network) and eco-chambers to infer groups with the same ideology.

This method is typically used in biological networks to infer functionality of proteins by 'guilty by association', and it is effective at characterizing groups that do not necessarily express their political opinions explicitly. Community detection of the network obtained by defining an association when a user retweet or mention another user on a specific topic (climate change, economics, alt-right, QAnon, liberal issues, etc.) can capture the voter intention by homophily even if the classifier is unable to predict the user. As society gets more polarized, the polarization will only increase the accuracy of these network tools in predicting the vote intention even for those users who do not explicitly say for whom they will vote.

2.10.6 *Desirability Bias: The Disconnects Between How Social People Say They Will Vote and How They Actually Vote*

Likewise, social-desirability biases [73], i.e., the tendency of respondents to hide their intention to vote for controversial candidates, can be addressed and investigated by tracking the same people before and after the election, a unique capability of social media analysis not available to pollsters. Another way to test for the true preferences of voters is to check the prediction of machine learning (inferred from what they say) with the homophily in the social network (inferred from what their friends say). Inconsistencies between the two indicators is a measure of the desirability bias and allows to correct for this tendency in the electorate by tagging users with high desirability bias in future elections.

2.10.7 *Predicting the Undecided*

Machine learning provide an identification of undecided voters as those users who tweet equal (or similar number according to a given threshold) number of tweets for one or the other candidate. There are also certain users who speak about the candidates but the classifier is unable to clearly classify them. This is the undecided group. Network theory metrics can then predict these undecided. First, the concept of homophily and the ego-centric network (looking at the majority of their ego-centered network) is able to predict the undecided. Second, community detection [126] is able to test this classification as well. The network of contacts is built from the activity of users like re-tweets, mentions, etc. These methods were applied by Zhou et al. [65].

2.11 Detecting the Malicious: Bots, Botnets, Trolls, Fake News, Misinformation, Disinformation, and Conspiracy Theories (Anything Else?)

A fundamental problem that originated with the rise of social media is the proliferation of misinformation, disinformation and fake news. Not surprisingly, intense research efforts have been devoted to better comprehend the phenomenon of fake news as well as to the development of algorithms for detection of fake news. Despite the existence and application of well-established ML and NLP techniques, a fully automated approach to the fake news problem is still a work in progress. In order to understand political influence, it is necessary to classify the information according to the political orientation with machine learning as discussed in Sect. 2.7. Likewise, due to the spread of fake news and misinformation in political campaigns, it is essential to understand how this misinformation spreads in political networks.

But what is fake news? This is fabricated information that propagates '*false and deceptive content, often sensational, disseminated under the guise of news reporting, shared on social media platforms*' (Collins Dictionary). There are so many examples of fake news, that is difficult to find a single representative one. But as of late, we could cite a recent News Nature article [150] in the scientific literature that explains how fake news spreaders and misinformation campaigns are targeting prominent COVID-19 scientists. In the study, misinformation is defined as '*Verifiably false or misleading information with the potential to cause public harm, for example by undermining democracy or public health or encouraging discrimination or hate speech.*' [151].

This study analyzed online misinformation targeting Dr. Fauci, head of the US National Institute of Allergy and Infectious Diseases, and Chief Medical Advisor to the President, as well as other prominent scientists. A widely circulated misinformation story targeted Fauci, making false claims that Fauci lied about issues such as the effectiveness of masks and hydroxychloroquine as a treatment for COVID-19 [151]. According to this study, in a highly circulated video posted on Facebook [152], Tucker Carlson, the host of the popular political talk show on Fox News, claimed that '*Fauci admitted that surgical masks don't really work. They are, in short, a form of make-believe. Yet once again, while under oath, Tony Fauci claimed exactly the opposite.*' The video had 1.8M views, 11k comments, 65k likes, and 15k angry reactions as of January 2022. Fact-checkers have debunked this form of misinformation [153, 154], and the science is clear that masks do reduce the spread of COVID-19 and that there is no evidence that hydroxychloroquine is effective as a COVID-19 treatment. Despite this, Facebook and Twitter did not take action to stop the spread of this misinformation campaign from their platforms [150].

Another classical case of misinformation viral spreading event is the debunked '*Pizzagate*' fake story which then transformed into a conspiracy theory that went viral during the 2016 US presidential election. According to Wikipedia [155], a single post from a Twitter account posting antisemitic and white supremacist posts (which have been taken down) shown in Fig. 2.12, falsely claimed that the New York Police Department, NYPD, had discovered a pedophilia ring run by members of the Democratic Party, including Hillary Clinton, in the basement of a pizzeria in Washington DC. The post went viral ending in an incident where a conspiracy theory believer, armed with an assault-like AR-15 rifle, entered the pizzeria to investigate, and discharged the weapon. The story was debunked by fact-checker Snopes [156], among many others. This event was the precursor of the QAnon conspiracy theory [157] spreading among the American far-right political sphere false claims that Hollywood celebrities and Democratic politicians, among others, are running a global child sex trafficking ring.

Fig. 2.12 Example of misinformation and fake news that have been debunked by fact-checking organizations. Post on Twitter that initiated the *Pizzagate* conspiracy theory

Fakes news are intertwined with conspiracies theories. For instance, Alex Jones, director of the website www.InfoWars.com which is a well-known fake news website according to fact-checker Politifact [158, 159] and many others [160], and, as handler `@realAlexJones` (now permanently banned from Twitter), top influencer of fake news and conspiracy theories in the 2016 US elections (see Table 2.1 and Fig. 2.17), has advanced conspiracy theories claiming that the 2012 Sandy Hook Elementary School shooting was fake and "no one died" in Sandy Hook and that survivors were "crisis actors" [160]. All these claims have been debunked [161], and InfoWars has filed for bankruptcy in U.S. court due to defamation suits related to the spreading of this conspiracy theory [160].

The problem of fake news is compounded by the proliferation of bots, botnets and trolls designed to influence people, and to artificially inflate the traffic supporting a given idea in social media. A bot (short for *software robot*) is a piece of software or app that automatically injects information in the Twitter system or any other social media platform. A bot injects information automatically through an algorithm without human intervention. A botnet is a network of bots that works at unison to spread a given message or ideology. They are typically designed to artificially increase the social media engagement of targeted account or topics and hashtags making the targeted topic to look more engaged.

Unlike a machine-run bot, an internet troll is a real person who intentionally spread a given ideology or idea by engaging into conflicts with other social media users by posting inflammatory and digressive posts in online communities with the intention to provoke emotional responses or manipulate discussions. Case studies of fake news influencers and Russian trolls meddling in the US elections will be discussed in Sects. 2.12 and 2.13, respectively.

Table 2.1 Top five most mentioned websites in Twitter in the US 2016 election. The outlets are classified according to media categories, from fake news to traditional media, according to ideology. We show the number of tweets with a URL pointing toward each hostname. Tweets with several URLs are counted multiple times. Table reproduced from [57]. Copyright © 2019, The Authors

	Fake news		Extreme bias (right) news		Right news	
	Websites	# tweets	Websites	# tweets	Websites	# tweets
1	thegatewaypundit.com	761756	breitbart.com	1854920	foxnews.com	1122732
2	truthfeed.com	554955	dailycaller.com	759504	dailymail.co.uk	474846
3	infowars.com	478872	americanthinker.com	179696	washingtonexaminer.com	462769
4	therealstrategy.com	241354	wnd.com	141336	nypost.com	441648
5	conservativetribune.com	212273	freebeacon.com	129077	wsj.com	310416
	Center news		Left news		Extreme bias (left) news	
	Websites	# tweets	Websites	# tweets	Websites	# tweets
1	cnn.com	2291736	nytimes.com	1811627	dailynewsbin.com	189257
2	thehill.com	1200123	washingtonpost.com	1640088	bipartisanreport.com	119857
3	politico.com	1173717	huffingtonpost.com	1057518	bluenationreview.com	75455
4	usatoday.com	326198	nbcnews.com	512056	crooksandliars.com	73615
5	reuters.com	283962	abcnews.go.com	467533	occupydemocrats.com	73143

The lines of research currently active to detect malicious and deceiving activities from bots to fake news in social media involve classification problems using machine learning and network theory techniques. In the former, the most used approach focuses on using supervised/unsupervised machine learning algorithms to identify patterns in fake news, and detect them before they are able to spread among the population. The underline pipeline of these algorithms generally consists of two fundamental steps: feature selection/extraction and model selection [162].

Studies uncovered the existence of large numbers of online social networks automated accounts (bots) meant to distribute biased or fake information to leverage the public opinion [163]. Bots play a central role in the spreading of misinformation in social networks. They are the epicenter of all sort of problems afflicting modern social media, but they have existed way before the appearance of social media platforms. They trace their history to the origins of AI with the introduction of ELIZA in 1966, a *chatbot* algorithm designed to maintain human conversations that could pass the Turing's test; which was proposed by Alan Turing as a criterion of intelligence [164]. So far these conversational programs have never passed such a test [165]. Social media bots can target various audiences by creating fake trends [166]. Bots are also used as unsolicited spam, advertise events, and products of doubtful legality. Bots detection is a field by itself, and complex models have been proposed for their identification [166, 167].

Bots are more prevalent in Twitter than other social platforms due to the facility to access the Twitter API that allows any app to obtain tweets freely and reply to them automatically via bots. Not all bots are malicious, though. It is a question why Twitter allows this kind of automated traffic to proliferate in its platform. In fact, it would be easy for Twitter to stop the proliferation of bots, but this might affect their bottom line as the stock value of the company is valuated by the traffic it generates. As of June 2022, Elon Musk is acquiring Twitter and a point of heated disagreement with the company is the amount of fake accounts currently existing in the platform. It is also a question how much of the Twitter traffic is due to non-human means, although some estimates put at 15% of traffic from automated bot accounts [101].

Both bots and fake news detection concern a classification problem. In the first case, the aim is to distinguish between human-ran accounts from automated non-human accounts. In the second case, the algorithm identifies misleading information spreading over the social network. With appropriate modifications, the algorithm developed for one case can be used to tackle the other problem and vice versa [101].

Despite the enormous progress made in the field of ML and network theory, the lack of extensive social media datasets (primarily because of privacy concerns) represents one of the main barriers to the development of detection algorithms for bots and fake news. The army of bots, botnets, and trolls that artificially manipulate traffic in social media, together with the concomitant rise of fake news will eventually destroy the credibility of social media platforms that allow for their proliferation. Thus, detecting bots and differentiating what is fake from genuine news is existential for social media, as well as society. In what follows, we review attempts to identify these malicious elements by first focusing on the features used by the most common algorithms for fake news/bots detection, and then concentrate on model selection.

2.11.1 Feature Selection/Extraction

The choice of meaningful features characterizing either fake news (content) or bots (sources) is the first step toward an exhaustive AI algorithm for detection. Features can be roughly categorized as follows [168]:

- Features extracted from news content,
- features extracted from news sources,
- features extracted from the environment (e.g., social network structure).

Fig. 2.13, reproduced from Bondielli and Marcelloni [168], summarizes the taxonomy of different types of features used in the literature.

Features extracted from news content can be either textual or not-textual (headline, body text, images, and video). According to Conroy *et al.* [169], textual features consist of the information extracted from the news text and syntactic features are frequency of function words and phrases (n-grams and bag-of-words), or punctuation and parts-of-speech (POS) tagging. According to Shu et al. [170] lexical features are total words, characters per word, frequency of large words, unique words, hashtags, punctuation counts, etc.

NLP is used to extract semantic features. Opinion mining and sentiment analysis [171–173] use features based on emotions expressed in the text. Latent Dirichlet Allocation (LDA) then extract topics from social media posts. Non-textual content

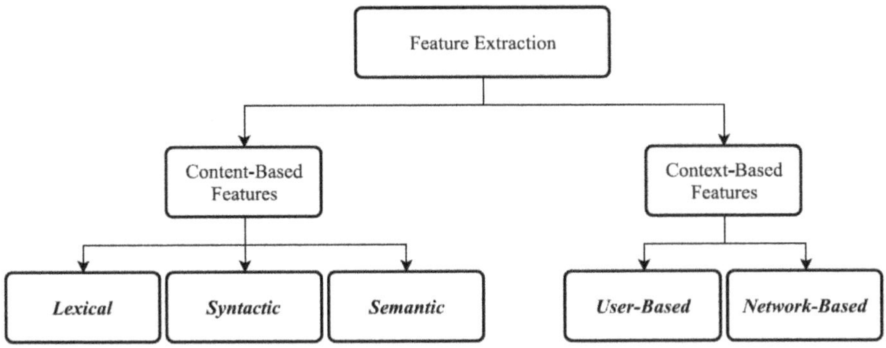

Fig. 2.13 Fake news detection. Different types of features used in the literature for fake news detection. Figure reproduced from [168]. Copyright © 2019, Elsevier Inc

also plays a role in differentiating fake from genuine news. Mangal et al. [174] uses a VGGNet to extract features from an article's image and textual features.

News source features consist of information about the publisher of the news article. One of the most common ways to extract the source's trustworthiness is through fact-checking lists, which require human labeling and are not scalable. Recently studies highlighted the possibility of assessing the reliability of a source using lexical and host-based features. Examples of hostname features are the domain location and the IP address, which could be extracted from an URL [175]. Tokens that appear in the hostname, punctuation frequency, and the lengths of the hostname are shown to be valuable traits [166, 176, 177].

Vosoughi et al. [178] uses social context features from user engagements of news consumption. They are referred to as user-based features and consist of statistics of user engagement and temporal patterns from social media, such as reactions, comments, shares, intervals from publication time, the average number of followers, type of the account (verified, not-verified), the age of the account on the platform or the presence of a clear description for the profile and of URLs linking to external resources [57, 179].

Network-based features are used to model information concerning the properties of the network where the news is shared. Examples are the users polarization or the centrality of the users explained in Chap. 1, and propagation structures, diffusion patterns, and properties of the sub-graph in which the news spread, such as density and clustering coefficient.

Ma et al. use temporal level features such as the variations of post-level feature values [180]. RNNs are utilized to capture the temporal changes in posts. Based on the time series for various metrics of relevant posts (e.g., number of posts), mathematical features can be computed, such as SpikeM parameters [181].

Identifying meaningful features for classification problems involves state-of-the-art machine learning techniques, from the most rudimentary, although most used, TF-IDF analysis up to more advanced embedding techniques, such as Word2Vec.

Features based on data from social media seem to lead to exciting results. However, as stated in [182], "*the analysis of fake news content is not sufficient to establish an effective and reliable detection system.*" Similarly to the problem of fake news identification, the characterization of bots can not be based only on features extracted from news sources, and a complete picture, including both content and context features, is needed to reach the goal.

2.11.2 Detection of Bots

According to [56] bots are easy to identify, in principle. In practice, they may not. The simplest strategy to identify a tweet that comes from a bot is to extract the name of the Twitter app used by the bot to post the tweet from their *source* field and keep only those tweets originating from an 'official' Twitter client. When a bot uses programmatic access to Twitter's API to send tweets, the value of the *source* field of these automated tweets corresponds to the name of the app used to create the bot. This name is unique to the app, given by Twitter to the app during the creation of access tokens. Since this name is unique to the app used by the bot, it can be used to identify the bots. A real person sending a tweet or retweeting would use a client's name that is common to many users, such as those using an iPhone or an Android-based smartphone. These client names are considered as 'official' clients.

This fact can be used to differentiate a tweet coming from a bot or a person. Although, it will classify as a bot also a real person who partially automates his/her tweets, while also sending real tweets. Supplementary Table 14 in Ref. [56] provides a typical list of clients that are identified as 'official'. For data collected during the US 2016 presidential election, there are 18 such official clients with names like 'Twitter for iPhone', 'Twitter Web Client', 'Twitter for Android', 'Google' 'Facebook', 'Mobile Web', etc. The rest of the clients are considered 'non-official' (not belonging to this official list) and emanating from bots. These non-official clients represents a variety of applications, from apps sometimes used by professional for automating some tasks (e.g. www.sprinklr.com or www.dlvrit.com) to manually programmed bots with a variety of unique client's names. The study of Bovet et al. [57] have found that political bots post around 8% of the total number of tweets in a typical election cycle.

This simple method allows to identify tweets that have not been automated and scales very easily to large datasets. Typically, one removes all tweets sent from non-official clients when computing the activity of people and predicting elections, but bots can be kept when building the retweet networks, as one may want to include automated accounts that play a role in the diffusion of news, since these bots can influence people in the network. This method is well suited to discriminate bots at the tweet level, can be applied to historical data and scales well to large systems. However, it may reject some tweets written from real users, since real users may use non-official clients to manage their tweets and automate their activity to a certain degree. The client-based method could lead to: (1) false positives: real people using third party applications (typically professionals), and, (2) false negatives: the most sophisticated social bots could use official clients (to be more difficult to detect) and use another software, like Selenium, on top that automatizes the official client, see for example [183].

In order to have a better bot detection, methods such as Botometer [101, 184] can be used. Botometer is a machine learning framework that classifies Twitter bots from real users. It uses more than thousand features extracted from public data and meta-data about users (friends, tweet content and sentiment, network patterns, and activity time series). Contrary to the classification done at the tweet level based on official clients, Botometer classifies bots at the user level. Advanced bots might not be detected by the client-based method, but this is also a problem for the more advanced Botometer since it relies on a training set of known bots [101].

Botometer does not scale well to large systems, and it is not well suited for historical data as it requires several tweets per user (up to 200) to classify a user as a bot/non-bot, and also requires results of a Twitter search of tweets (up to 100) mentioning each user, which cannot be done retroactively. Reference [56] has performed a comparison between the client-based method and the results of Botometer and found that the former has good accuracy but suffer from a relatively high number of false positives compared to Botometer.

Knauth [185] used content features (tweets and users' metadata) to train an Adaptive Boosting machine called *AdaBoost*. Using the public bot dataset Cresci-2017 [166, 186] from the Botometer project [101, 184], containing 6708 users for training and 1677 users for testing, the author obtains 99% for both precision and accuracy. Haidermota et al. [187] experimented with classical classifiers, such as Naïve Bayes, random forest, and logistic regression. It is reported that all classifiers performed poorly, except for random forest with 95% precision for the test data. The authors showed that the AdaBoost method improved the overall performance of all detectors. In [188], a Naïve Bayesian classifier achieved an accuracy of 90.9%, while in [189] a Naïve Bayes classifier had the best performance with 91.7% precision, recall, and F-measure; performing better than the other models like support vector machines, neural network, k-nearest neighbors, and decision tree. In [190], the authors used four supervised machine learning techniques: SVM, Bayesian network, decision tree, and multi-layer artificial neural network. The Bayesian network and random forest outperform other classifiers with an accuracy of 90.90% and 91.39%, respectively. Semi-supervised techniques have also been used. For example, Alharthi et al. [191] developed a semi-supervised method to label their dataset and classify Twitter accounts into spam/genuine. They applied label propagation and labeled spreading algorithms and had a good performance with an accuracy of 91%.

Deep Learning classifiers have risen in popularity for bot detection, given their ability to produce discriminant features. Reference [102] presents a machine learning method based on deep neural networks to detect bots. In [192] the authors proposed a contextual LSTM that considers both content and context features to improve bot detection. Using the Cresci-2017 dataset [186], the model showed 96% precision and accuracy. Wei and Nguyen [193] used a bidirectional RNN to identify bot accounts in Twitter with LSTM. Their model results in precision higher than 90%. In [194]

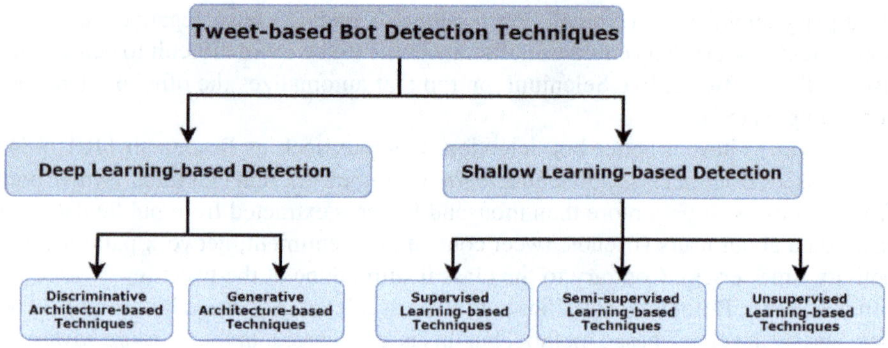

Fig. 2.14 Taxonomy for tweet-based bot detection techniques proposed by Derhab et al. Figure reproduced from [195]

the authors perform a simple CNN architecture to classify textual tweets into bots or humans. They tried several word embedding models resulting in 85% accuracy and 97% precision of bot detection.

In general, these different methods should be used concurrently on the same dataset to cross-validate the performance of the algorithms against each other. For a more detailed reading and review of further work, we refer to [195] whom proposed the taxonomy of bot detection algorithms shown in Fig. 2.14.

2.11.3 Detection of Fake News

In the modern digital era, information is power, and we are drowned by it. However, with the increased availability of information, there also comes the increasing prevalence of misinformation, disinformation, and fake news. This is a major issue, straining the information economy of advanced democracies and leading to suboptimal outcomes across various institutions (e.g., health, education, politics, elections).

There is a difference between misinformation and disinformation. The former consists of false information that is spread regardless of an intent to harm or deceive. The latter is created with a deliberate purpose of deceiving or harming the population. Thus, while someone can spread misinformation without knowing that is is false, and can share it widely, misinformation turns into disinformation when someone shares it knowingly that it is false to cause divisiveness and cast doubts in the population. While this distinction is essential, it is also quite challenging to differentiate between both forms of false information since it requires knowing the purpose of the spreader. In Sect. 2.13 we will analyze the disinformation campaign performed by the IRA during the 2016 US presidential election. In this case, the motives of the IRA were clear, and this campaign can be classified as disinformation. In general, we will not distinguish between these two forms of false news spreading.

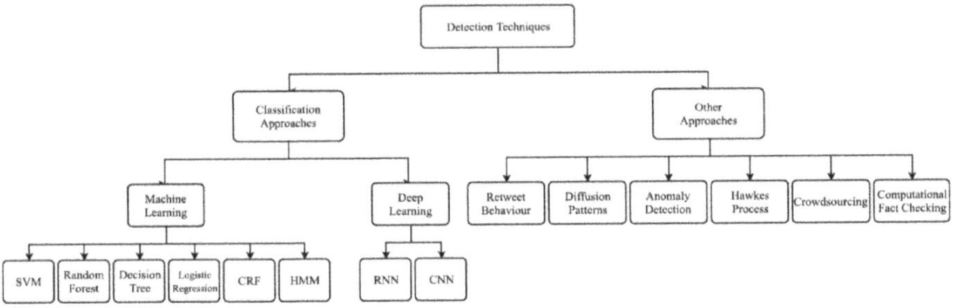

Fig. 2.15 Taxonomy of the different approaches to fake news detection proposed in the literature. Figure reproduced from [168]

Tracing how misinformation spreads via social media channels is a crucial task in contemporary social science, requiring collaborative research teams cutting across the disciplinary divides separating the social and computational sciences. Possible approaches proposed in the literature to detect false information consist of:

- Machine learning, including deep learning approaches, face the task as a classification problem; they associate labels such as genuine or not genuine to a particular piece of text.
- Time series analysis (data mining techniques), diffusion patterns in social networks, anomaly detection, and knowledge graphs try to spot the characteristic traits of genuine/not-genuine news and assess their credibility based on these traits.
- Monitoring and classification of fake news and their potential spreaders based on curated lists from fact-checking organizations like https://snopes.com and https://factcheck.org.

These strategies are achieving promising results, although combining them seems to be the best solution [168].

Figure 2.15, reproduced from Bondielli and Marcelloni [168], summarizes the different approaches. Among the many classical ML models, Support Vector Machines (SVMs) are widely used methods for classification, by maximizing the distance among the classes. The first attempts at fake news and deception detection employed content-based features [196, 197]. Later approaches exploit a mixture of content-and context-based features to train the classifiers [198, 199]. Logistic regression models have also been used in several studies [200, 201], as well as decision tree algorithms. The decision tree performs a recursive split on feature values to determine the class. Despite their relative simplicity, these algorithms are highly resource-demanding and, therefore, not scalable.

Deep Learning classifiers are very popular because they speed up the process of feature selection, a task that is time-consuming and may lead to biased features. Therefore, the problem is shifted from modeling relevant input features to model the network itself to solve the task efficiently. Recent works on fake news detection have exploited RNNs and CNNs in their models [202–207]. Other studies focus on computational-oriented fact-checking algorithms. Ciampaglia et al. [208, 209] proposed the use of Wikipedia info-boxes to generate a Knowledge Graph representing genuine information. A Knowledge graph represents human knowledge by considering entities as nodes connected by the relationship that relates them [209]. It is a directed multigraph, where a concept can be an abstract idea such as a thought, a real-world entity such as a person, e.g., Biden, or an object, like a vaccine. An example of two nodes linked by a relation is the sentence: Biden (entity) is the president (relation) of US (entity), which is represented as a link from Biden $\rightarrow$ US, and the link $\rightarrow$ represents 'president'.

This approach assumes that Wikipedia is the ground truth of genuine information. The authors define a measure of semantic proximity using a transitive closure algorithm to check claims of genuine or fakeness against the knowledge graph. The lowest the proximity of the text with the knowledge graph, the higher the probability the text does not correspond to the truth.

Other works explore the possibility of combining ML and network theory analysis so that the outputs of the latter results are the input of the former or vice-versa. Despite promising results, the lack of data and the difficulty represented by the different languages/cultures are the most significant barriers to solving this problem [210].

Different frameworks for evaluating tweets and their authors as possible sources of misinformation have been proposed in [57, 211]. These works assess the credibility of users based on their retweet behavior. In [211] the authors combine the Gini coefficient and PageRank to compute a score of acceptability of a given tweet.

In [57], Bovet et al. tackle this problem by combining a fact-checking algorithm and the CI centrality measure from Sect. 1.6.1 to find the top fake news spreaders in the retweet network, evaluating their impact on the 2016 and 2020 US presidential elections. Results will be discussed in Sect. 2.12.2.

Overall, the current state of the art in fake news detection is quite advanced, and some effective algorithms have been developed. Many groups are working to develop fake news detectors using different databases. A compilation and benchmark datasets with a repository for the most implemented solutions and the most relevant papers on the topic including libraries to find Fake News Detection models and implementations are offered at Papers with Code, a free resource from Meta [212]. A research group at Simon Fraser University in British Columbia, Canada, developed an online fake news detector that identifies bias and misinformation in news articles by pasting a piece of text in an online form available at [213]. By collecting news articles with veracity labels from fact-checking websites, they train text classification systems to detect fake from real news. The following list was compiled by [214] represents a non-exhaustive list of baseline models used to compare the performance of the latest algorithms:

- FAKEDETECTOR: inference based on explicit and latent textual features and relationship connections based on the gated diffusive unit (GDU) model [214].
- Hybrid CNN: classify textual news articles based on a CNN model [202].

- LIWC: news article classification based on Linguistic Inquiry and Word Count language feature extraction and textual content information modeled by LSTM [215].
- TRIFN: exploits the user, news and publisher relationships for fake news detection [216].
- DEEPWALK: network embedding model [217] into a latent feature space of news articles, creators, and subjects with an added SVM model to determine the classes of news [214].
- LINE: scalable network embedding model that optimizes an objective function that preserves both the local and global network structure [218] to which a classifier model is added for news articles, creators, and subjects [214].
- PROPAGATION: label-propagation based model [219], which is cast into labels according to a label-score mapping [214].
- RNN: textual content based RNN model [220] to learn latent representations of the textual input that is fused to predict credibility labels [214].
- SVM: a model based on raw text inputs to extract a set of explicit features which are used as input for SVM [221] based classification model [214].

Reference [214] tested these algorithms using a dataset of tweets compiled by the fact-checking organization PolitiFact at its official Twitter account `@PolitiFact`, as well as the fact-checked articles were written regarding these statements in the PolitiFact website [158]. PolitiFact is a fact-checking organization operated by the Tampa Bay Times and the Poynter Institute and it has been identified as one of the top 10 influencers in the left-leaning news class of outlets during the 2016 US presidential election in the study of [57], as displayed in Table 2.3. At PolitiFact journalists fact-check statements and news articles made by members of the US Congress, the White House, lobbyists, and political groups collected from speeches, news articles, and social media. The evaluation produces a credibility rating, ranging from "True" for completely accurate statements to "Pants on Fire!" for totally false claims.

Zhang et al. [214] formulate the fake news detection problem in the following way:

- Define News Articles as a set $\mathcal{N} = \{n_1, n_2, \ldots, n_m\}$, where $n_i = (n_i^t, n_i^c)$ represents its *textual content* and *credibility label*, respectively.
- Define News Subject as a set of news subjects $\mathcal{S} = \{s_1, s_2, \ldots, s_k\}$, where $s_i = (s_i^t, s_i^c)$ represents its *textual description* and *credibility label*, respectively.
- Define News Creator as a set $\mathcal{U} = \{u_1, u_2, \ldots, u_n\}$, where $u_i = (u_i^p, u_i^s)$ represents the *profile information* and *credibility label* of the creator, respectively.
- Define a News Augmented Heterogeneous Social Network (News-HSN) $G = (V, E)$, where $V = \mathcal{U} \cup \mathcal{N} \cup \mathcal{S}$ covers the sets of news articles, creators, and subjects, and $E = \mathcal{E}_{u,n} \cup \mathcal{E}_{n,s}$ involves the authorship links between news articles and news creators, and the topic indication links between news articles and news subjects.

The Fake News Detection problem is formulated as [214]:

- Given a News-HSN network $G = (V, E)$, the problem aims at learning an inference function $f : \mathcal{U} \cup \mathcal{N} \cup \mathcal{S} \rightarrow \mathcal{Y}$ to predict the credibility labels of the set $\mathcal{U} \cup \mathcal{N} \cup \mathcal{S}$. The learning function f incorporates various kinds of heterogeneous information of G, including both the textual content/profile/description information, as well as the connections among them.

Based on this model, Ref. [214] shows that FAKEDETECTOR outperforms the above list of algorithms with a 0.63 accuracy score for inferring articles in the benchmark database of the PolitiFact network is made of 14,055 articles, 3,634 creators, and 152 subjects. While the accuracy is 14.5% higher than the score obtained by the rest of the state-of-the-art models, still, there is a lot of room for improvement, given the overall low accuracy reached.

Despite this astonishing progress reached by NLP techniques, the development of a universal algorithm for detecting fake news still faces some challenges:

- Current machine learning models using text classification exist [222–226], yet, these models consider only text features to identify fake news, which are not all the features that can be considered.
- Other approaches combine text classification with metadata embedded with the news (URLs, mentions, hashtags, authors, publication time, and interactions from social media) and curated lists from fact-checked organizations [57]. However, features characterizing fake news are changing all the time. For instance, it is impossible to keep up with new URLs and websites spreading fake news as curated by fact-checking organizations. Furthermore, fake news about COVID-19 may be structurally different from fake news on politics, impeding transfer learning from topic to topic.
- Works exploring the possibility of assessing the legitimacy of a text by comparing it with a ground truth Knowledge Graph built from a trustful source of information like Wikipedia [208, 209] has its limitations since it is difficult to define trustable sources of information. Moreover, Knowledge Graphs face scalability problems since they can easily reach order of billions of nodes.
- The absence of training sets covering a vast set of domains (ranging from politics, economics, medicine, etc.) and languages and the lack of NLP models capable of understanding and replicating the language features of fake news (i.e., exclamation points, capital letters, time patterns, etc.) hinder the development of fake news detectors across all topics.
- Finally, NLP algorithms mostly focus on a unique language (English) and on a unique social platform, Twitter, because of its free API service, while the problem of fake news is multi-platform and multi-lingual.

Considering these multiple hurdles to overcome, it is understandable that the problem of fake news detection has not yet been solved, even with the enormous research efforts of thousands of scientists working at Twitter and Facebook. This is surprising since these are the very same platforms that serve as the substrate of the

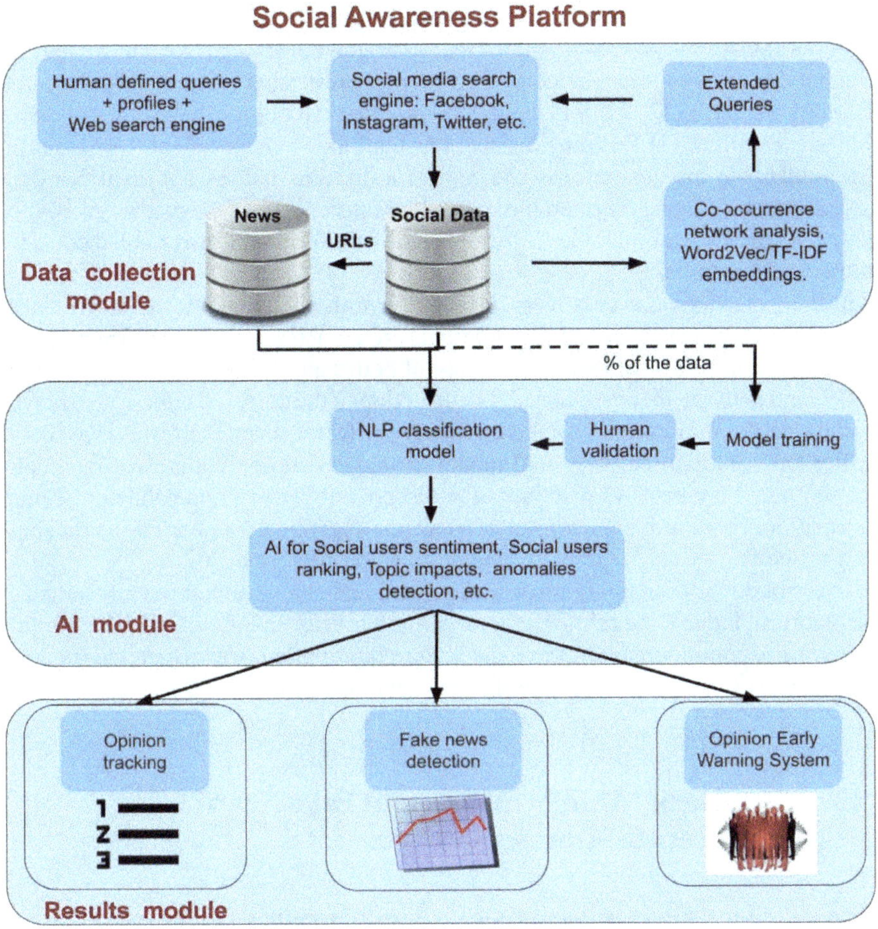

Fig. 2.16 General flow of fake news detection pipeline. From collection to fake news detection

problem, and they have all the data available and all the resources to support the best AI scientists in the world. Still, the problem remains open.

The crux of the matter is that the problem requires the development of a multistep approach integrating a variety of (mis)information flow mechanisms to examine how information and misinformation spread in social media as shown in Fig. 2.2. Building a universal misinformation detector requires a pipeline depicted in Fig. 2.16 to first, spot and predict opinion trends and identify users' polarization on a topic of broad interest to society (e.g., climate change, COVID-19, etc.), and then to track misinformation propagation to understand the role it plays in shaping opinion trends, and identify the underlying pathways.

To deal with the limitations of using hand-labeled training sets to teach the model how to differentiate between fake and genuine news, an adversarial machine learning model capable of generating fake content can be used. While Generative adversarial networks or GANs have revolutionized the paradigm of learning from counterfeit data as applied to images, the applications to NLP are restricted because of the discrete nature of the text [227, 228]. From the methodological point of view, the following steps are required. **(1)** Defining an operator that allows the gradient flow from the discriminator to the generator in the case of a discrete dataset. **(2)** Identifying the most important (high-order combinations of) features of fake news that go beyond the text itself, for example, the IP from which the information was published or the protocol of the domain.

In this perspective, access to social data from the major platforms is crucial to obtain a wide range of metadata to be analyzed, such as profile information and post interactions. This step requires the use of both state-of-the-art machine learning models and network theory analysis. The selection of the correct database also plays a crucial role. A multi-domain database covering different social platforms (Facebook, Twitter, etc.) is required. The final model should reveal key features of fake news and, through the fake news discriminator, discern between genuine and not genuine information. From a dynamic point of view, the classified data enrich the database from which features are extracted, leading to better performances.

This strategy is domain-agnostic and general and can be applied to any language and platform. Indeed, the reinforcement learning process at the base of the algorithm optimization continuously improve the feature distribution over which the model is optimized. This could solve in one shot the challenges mentioned above.

2.12 The Rise of Misinformation and Fake News Influencers in American Politics

Using the AI machinery and the network theory of influencers, the emergence of influencers as replacements of the old elites can be studied. In turn, the influence of the new cradle of influencers across the political spectrum can be assessed. The multistep model of diffusion of information outlined in Sect. 2.2.3 provides a general guideline to understand this process based on decades of social-scientific research. It provides a framework to understand how the various processes outlined in the model (e.g., opinion-leader intermediation, direct media access, cascading and herding dynamics in cohesive communication) operate, and which mechanisms are most important in driving the spread of both information and misinformation in social media platforms. Network analytics tools, together with AI analysis, allow us to predict the adoption of specific opinions and beliefs, as well as outlining the conditions that either increase or decrease polarization along a given opinion dimension.

The next step consists of identifying opinion leaders, i.e., influencers in the language of the theories of Chap. 1. It is clear that these individuals play a central role in the diffusion of the (mis)information. This observation leads to an interesting conclusion. To find early adopters and opinion leaders, we need to find the most central individuals in the system. Chapter 1 reviewed the algorithms that allow us to identify these drivers of opinion: the actors who play the role of opinion leaders in the multistep model of information diffusion chains.

Social media have changed how information spreads from source to adopters. On the one hand, multiple sources, some of which of suspect credibility, compete to provide people with a range of opinions and beliefs on a given subject. Individuals with insufficient knowledge in a given field are seen as opinion leaders (influencers). The combination of these phenomena leads to the segregation of social users according to their beliefs.

References [57, 64] identified emerging influential actors of fake and genuine news and compared them with the influence of traditional elites within the Twitter platform in the context of the US elections of 2016 and 2020. These works analyzed how these actors change between these two major events. Several research questions can be answered:

- How and when the new influencers have emerged and changed across both elections?
- Have they become more influential than the traditional elites?
- How have their audiences changed?
- Are the influential actors of the 2016 election central to 2020, or have new actors taken their place?
- Are the networks in 2020 more polarized when compared to 2016?
- Is it the case that people can rise to significant influence and affect polarization, positively or negatively, based on their use of modern technologies?
- Does this mostly apply to people who are already influential, or have relative outsiders had a major impact?
- Do the interactions matter (e.g., does the shouting affect votes)?
- Finally, has the rise of social media new influencers contributed to the increased polarization in American society?

These questions can be tackled by the theoretical machinery explained so far. There has been a lot of attention given to these topics, but most work has been anecdotal or journalistic. Specific stories have been heavily reported in the media - Russian interference in the US elections by manipulating social media groups [229], Wall Street Journal investigations of internal documents leaks on the harmful societal effects of Facebook (The Facebook Files, WSJ [230]), Donald Trump's tweeting to 85M+ followers, etc. It is essential to move from specific anecdotes of journalistic reporting to scientifically study broader data that can systematically address these questions over time.

Analysis of the importance of influential actors versus elite media corporations are performed by comparing their evolution in US social media politics. Analyzing the Twitter networks where these actors reside in, Bovet and Makse [57] and Flamino et al. [64] observe how the network topology has changed between consecutive elections from 2016 and 2020, and how this contributes to the social dynamics observed in the respective elections. These observations, paired with qualitative data from both elections, help set the stage for understanding the power of new social media influencers as it continues to integrate into the elite US political environment.

The 2016 investigation in Ref. [57] revealed that in terms of tweet volume, fake news represented 10% and extremely biased news 15% of the volume of the tweets linking to a news outlet media. Although, it is found that approximately the same ratio of users using automated Twitter clients in each media category, it is found that computerized accounts (bots) diffusing fake news are much more active than the automated accounts diffusing other types of news. These results confirmed the role of bots in the diffusion of fake news and also showed that automated accounts play a role in the diffusion of traditional news. Yet, this role is smaller than the role of human-controlled accounts.

It is interesting to characterize the spread of news on Twitter to measure the influence of new emerging influencers on political polarization. The influence of emerging influencers is measured versus the influence exerted by the old traditional news media. The comparison follows on analyses presented in [56] for the 2016 election and [64] for 2020.

2.12.1 Building the Network of Information Flow

We search for influencers who set the political agenda and sway masses. They are found using the methodology set out in Chap. 1 that consists of, first, constructing the network representing the flow of information concerning each type of news and then identifying the most influential nodes in these networks. Building the network is the critical issue here because it determines who are the influencers. All the theories in Chap. 1, including the IMP, assume that the network of transmission of information is well defined, but leave the question of how to define this network open.

One may even ask why we need the network at all to find the influencers. For instance, a plausible way to define influencers in social media is by the number of retweets that they get, for instance in Twitter, or the number of interactions, likes, or engagement in general. The total downstream follower count of the influencers either directly or via those who retweet their content is a way to measure influence. This is the simplest way to identify an influencer, and this method does not need the knowledge of the network.

Furthermore, one could use the follower counts to corroborate that the users labeled as influencers are producing widely read content. However, since we identify influencers by their ability to spread information which occurs via retweeting, not passive following, it is not advisable to use the followers count to identify influencers.

The followers count would miss the lion's share of influence, which is indirect, via the users who retweet their content. However, if no one is following the retweeters, then those labeled as influencers would be more likely called diffusers not influencers. In order to justify the label of influencer, the user needs to have a large direct and indirect follower counts and those who retweet them. Thus, the importance of influencers in the spread of information over online networks is demonstrated by their downstream follower count. The downstream count includes both the number of users who follow the influencer, and more importantly, the number of users who follow those who retweet influencer content. Retweeting, but not passive following, is a way to identify influencers.

While influencers can be measured by the reach of their tweets, like the number of retweets, the network approach measures influence by different metrics attacking directly the Influence Maximization Problem [232] described in Sect. 1.3.2. This problem assumes that information travels in a network, and this network determines the influencers which are characterized by centralities like CI and others.

Twitter users are embedded in relatively stable communication networks created by the exchange of retweets. A 2015 study by Metaxas *et al.* [231] found that '*retweeting indicates not only interest in a message, but also trust in the message and the originator, and agreement with the message contents.*' Using retweet data, we can identify influencers who are the users with the greatest ability to effectively propagate new information over the retweet network.

In general, any ranking of influencers based on the solution of IMP [232] should represent a more optimal solution for the minimal set of influencers than just counting retweets. For instance, CI is a global optimization of influence over the entire network and it is a plausible approximate solution of the IMP. Being a global maximization of influence, it takes into account not only the first layer of retweets but the second, and the third, and so on. The parameter ℓ in $\text{CI}(_\ell)$, Eq. (1.84), controls how many layers of influence one wants to consider. By consider ℓ = diameter of the graph, then we find the influencers that influence the network.

Finding influencers is a collective process. We need to find them all at once, and not individually one by one and independently of each other. For instance, if we identify influencers by their retweets isolated from the network, we could end up with two influencers with high retweet count, but who are both influencing the same part of the network, i.e., these influencers could be very close to each other, even retweeting each other.

Instead, when we search for the minimal set of collective influencers, it is crucial to consider this kind of influence overlap. Two highly retweeted influencers that overlap cannot be counted as two influencers: only one will suffice to be added to the list of collective influencers; they cover the same area of the network and one of them is redundant. When considering many influencers as the minimal set to influence the whole community, this kind of overlapping behaviour readily determines who is an influencer and who is not. Thus, being an influence is a collective behavior and the influence depend on the location in the network respective to the other influencers: your influence is not yours but depends on the collective network of interactions. This is a property of the network that cannot be captured by any local property of the users, like the retweet count or the follower counts considered in isolation from the network.

Thus, finding influencers is a global optimization problem and any local quantity will be only an approximation to the optimal set. However, both definitions (retweet-based and CI-based) should be related since we build the network where CI is calculated with the retweets, so a large CI node has a large number of people retweeting. So, in a sense, CI also captures the retweet-based definition. To confirm these ideas, one may want to plot the ranking of any metric, like CI, versus the retweet count of the influencers and also the retweet count times retweet of the retweet to check if the influencers are widely spreading messages or not. Such benchmarking has been performed in detail by Teng et al. [54].

The network of information flow in Twitter is built as follows. We consider an edge going from node v to node u in the network when user u retweets the tweet of user v. The direction of the edge represents the direction of the information from $v \to u$. We do not consider multiple links with the same direction between the same two users, and neither do we consider self-links, i.e., when a user retweets their own tweet. The out-degree of a node is its number of outgoing links and is equal to the number of different users that have retweeted at least one of their tweets. A node in-degree is its number of ongoing links and represents the number of different users they retweeted. Different extensions are possible. One may want to consider a weighted network where the edges are weighted by the number of retweets rather than a binary network. This is an important consideration since it is not the same as retweeting one tweet that consistently retweets many. Likewise, one can consider any interactions, like shares, likes, or mentions between users to define the links.

In other social media networks, like Facebook, the idea is the same. Any interaction or engagement between users can be considered as a transmission of information and can, in principle, be used to construct the network over which the information flows. It also serves as the substrate where the influencer metrics are calculated. We note that the network thus defined depends strongly on the type of topic that generates the traffic of information. The information flow network is completely dynamic since links are being created all the time. It also depends on the type of topic chosen to define it. This allows one to find the influencer for a given thematic community and for a given time scale. In contrast, a network of followers is static and does not change much with time unless new followers appear in the system, and it does not capture the theme of the conversation of the influencers. By using the retweet network instead of the follower network, we capture the effective dynamical network of information diffusion.

2.12.2 Fake News Influencers in the 2016 US Presidential Election

Using the dataset of 171 million tweets posted during the 2016 US presidential election, Ref. [57] reconstructed the retweet network of users. They also classified each tweet into fake and extremely biased, right and right-leaning, left and left leaning and center according to the URLs contained in the tweets using fact-checking organizations as detailed next.

Fake news identification starts by first identifying all the tweets in the dataset that contain at least one URL linking to a website outside Twitter. These URLs are then separated into two main categories based on the websites to which they link. The first category includes websites containing misinformation, fake news, and other malicious categories and the second website contains traditional, fact-based news outlets. Here, the term traditional is used in the sense that news outlets in this category follow the traditional rules of fact-based journalism.

A finer classification is also considered for news outlets spreading misinformation in two sub-categories: *fake news* and *extremely biased news*. Fake news websites are those that have been flagged as consistently spreading fabricated news or conspiracy theories by fact-checking groups. Extremely biased websites include more controversial websites that do not necessarily publish fabricated information but distort facts and may rely on propaganda, decontextualized information, or opinions distorted as facts.

The classification of misinformation websites in [57] is based on a curated list of websites which, in the judgment of a media and communication research team headed by Melissa Zimdars at Merrimack College, are either 'fake, false, conspiratorial, or regularly misleading websites' according to the authors [233, 234]. The original list of misinformation websites have been posted at http://www.opensources.co. This site has been taken down by the authors, after Dr. Zimdars reported 'a significant amount of harassment ... almost entirely from the readers of a handful of sites on the list' [235], occurring after the fake-news list appeared in the Los Angeles Times and went viral. A copy of the list is kept by the authors of [57] at www.github.com/alexbovet/opensources. The list includes sites well known for spreading misinformation across the political spectrum, like *Info Wars, Occupy Democrats* and *Breitbart News*; although the great majority of sites are from the alt-right movement.

The group at Merrimack College classified websites by analyzing several of their aspects, such as whether they tried to imitate existing reliable websites, whether they were flagged by fact-checking groups, or by analyzing the sources cited in articles. The main fact-checking organizations used are known organizations such as https://snopes.com, https://hoax-slayer.com (closed down on May 2021) and https://factcheck.org. The websites in the extremely biased category are classified according to their political orientation, Democrats or Republicans, by manually checking the bias report of each website using www.allsides.com and www.mediabiasfactcheck.com (MBFC).

To validate the classification of misinformation outlets, the website www.mediabiasfactcheck.com is used to verify the level of factual reporting. All the websites classified in the extremely biased category have a level of factual reporting which is *mixed* or worse. The curated list of outlets from 2016 at www.github.com/alexbovet/opensources has been augmented with new outlets appearing on Twitter in 2020 that are classified as "questionable" in www.mediabiasfactcheck.com in the misinformation category according to the method presented in [236]. This site is maintained by a small independent team of researchers and journalists and is widely used for labeling bias and factualness of news sources in research studies (see, e.g. [236–238]).

MBFC was found to offer the largest set of biased and low-factual news sources among five fact-checking datasets [239]. Outlets in the questionable category have all factual reporting graded as 'mixed', 'low' or 'very low' in www.mediabiasfactcheck.com. By comparison with the classification from [57], the outlets with 'mixed' factual reporting are placed in the *extremely biased* category and outlets with 'low' or 'very low' in the *fake news* category.

To compare the influence of fake news outlets with genuine traditional media outlets in political campaigns, a classification for traditional news websites based on their political orientation is used [64]. The most important conventional news outlets are classified by manually inspecting the list of top 250 URL's hostnames, representing 79% of all URLs shared on Twitter during the 2016 US election. The news outlets are classified by the right, right-leaning, center, left-leaning, or left based on their reported bias on www.allsides.com and www.mediabiasfactcheck.com. The news outlets in the right-leaning, center, and left-leaning categories are more likely to follow the traditional rules of fact-based journalism. As we move toward more biased categories, websites are more likely to have mixed factual reporting. As for misinformation websites, insignificant outlets are discarded by keeping only websites that accumulate more than one percent of the total number of tweets of their respective category.

This kind of classification based on fact-checking organizations and curated lists of fake and genuine websites are subject to subjectivity, and are embodied by a certain level of controversy. Therefore, they require a higher level of validation, for instance, by comparing these lists to the domain-level ideological alignment scores of news outlets presented in [240] and based on the average self-declared ideological alignment of Facebook users sharing URLs directed to news outlets.

The classification of news used in the work of [57, 64] includes the following eight categories: fake news, extremely biased (right) news, right-leaning news, right, center, left-leaning, left, and extremely biased (left) news. In some analyses, these categories are grouped for analytic convenience as fake/extremely biased (right) news, right-leaning/right, center, left-leaning/left, and extremely biased (left) news.

In the work of [57], the authors have identified 30.7 million tweets with URLs directing to a news outlet website, sent by 2.3 million users during the period from 1st of June to election day of the US 2016 election. It is interesting to look at the top fake news and extremely biased websites with the largest traffic during this US presidential election and how they compare to the established news outlets. Table 2.1 shows the number of tweets with a URL pointing toward a given website that is spreading fake news, extreme bias news, or genuine news separated by ideology [57, 64]. The website with the largest traffic on Twitter is www.cnn.com, with 2.2 million mentions. Almost with the same traffic, www.Breitbart.com, the American alt-right site supporting Trump appears in second place overall and as the top spreader of extremely biased news, with a surprising 1.8 million mentions. This misinformation site is even more important than the www.NYTimes.com, which received slightly less traffic.

Breitbart is the most shared source on Twitter by right-wing supporters, even more important than Fox News, a result that was also found in [245, 246]. This is remarkable, considering that Breitbart was founded in 2007, and 9 years later, it has become more influential than the almighty New York Times. Breitbart executive

Table 2.2 Top five most mentioned websites in Twitter in the US 2020 election. The outlets are classified according to media categories, spreading information classified as fake news to traditional media according to ideology from left to right. We show the number of tweets with a URL pointing toward each hostname. Tweets with several URLs are counted multiple times. Table reproduced from [64]

	Fake news		Extreme bias (right) news		Right news	
	Websites	# tweets	Websites	# tweets	Websites	# tweets
1	thegatewaypundit.com	1883852	breitbart.com	2192997	foxnews.com	3136578
2	hannity.com	428483	dailymail.co.uk	600523	nypost.com	1701531
3	waynedupree.com	258838	bongino.com	346103	wsj.com	887537
4	judicialwatch.org	233085	thenationalpulse.com	215017	dailycaller.com	771765
5	truepundit.com	176647	freebeacon.com	197092	forbes.com	748636
	Center news		Left news		Extreme bias (left) news	
	Websites	# tweets	Websites	# tweets	Websites	# tweets
1	thehill.com	2256888	nytimes.com	6775402	occupydemocrats.com	18151
2	apnews.com	1182504	washingtonpost.com	6438506	lancastercourier.com	5815
3	usatoday.com	993957	cnn.com	5577352	deepleftfield.info	5753
4	businessinsider.com	773328	politico.com	2290755	tplnews.com	402
5	newsweek.com	756820	nbcnews.com	2231564	bipartisanreport.com	3243

chairman was Steve Bannon, later appointed as the White House Chief Strategist in the Trump administration. This represents a remarkable explosive growth of a new influential news organization which has displaced the old elites and whose growth was based entirely on spreading stories considered racist and misleading by liberals and traditional conservatives alike, see [247]. Another site spreading fake news with enormous influence was www.thegatewaypundit.com, a far-right fake news site whose associated Twitter account has been permanently suspended by Twitter in 2021, and who's motto is '*We report the truth—and leave the Russia Collusion fairy tale to the Conspiracy media.*'

Later, Flamino et al. [64] analyzed 35.9 million tweets with URLs directing to a news outlet website during the five months period preceding the 2020 US presidential election. The same analysis of tweets pointing to news organizations' websites in 2016 was done for this election. Table 2.2 shows the analogous results to Table 2.1 in the 2020 election. Between 2016 and 2020, the traffic decreased the most in the center category and increased most in the left-leaning category, with a smaller increase in the right-leaning category. The shift away from the center indicates the increasing polarization, both among users and media outlets. However, most of the decrease in the fraction of center media outlets reflects the shift of CNN.com, which was categorized by www.allsides.com as center in 2016 and as left in 2020, combined with CNN accumulating more than twice the number of tweets in 2020 than the top outlet of the center category (www.thehill.com).

Compared with the 2016 data, www.breitbart.com is still the top fake news site in terms of traffic. But this time, there is a noticeable growth of the left-leaning sites, with the most popular one being The New York Times with three times the traffic as the most popular fake news Breitbart. All top 5 left-leaning news sites, The New

York Times, Washington Post, CNN, Politico, and NBC, have surpassed the top fake news spreader, Breitbart. May this be an indication that the fake news movement is dying? To answer this question, we need to look beyond these numbers and use the theory of influencers and Granger causality analyses to understand whether or not the fake movement still has a grasp on society. We treat this problem next.

In order to detect the influencers for each category of public opinion, the retweet network is constructed for each community. Using the collective influence algorithm from Sect. 1.6.1 [55], the influencers can be identified in each camp of public opinion. The out-degree of each node is considered when applying the collective influence theory to a directed network. A directed version of the algorithm is discussed in Sect. 5.2.2. A Twitter user highly ranked by the directed CI algorithm is a super-source of information. Yet, it does not necessarily need to be directly retweeted by many users. It is sufficient for the user to be surrounded by highly retweeted users.

Figure 2.17 shows the networks formed by the top 100 influencers of each retweet graph in the 2016 election by including those nodes that make the networks connected. A clear difference is apparent between the graphs representing the flow of fake and extremely biased news and the graphs for left, left-leaning and center news. The left, left-leaning, and center news outlets correspond to larger networks in terms of number of nodes and edges, revealing their larger reach and influence on Twitter. The retweet networks corresponding to fake and extremely biased news outlets are the most dense with an average degree $\langle k \rangle$ between 6 and 7, more than twice the number for left, left-leaning, and center news. The retweet network for right and right-leaning news has characteristics in between those two graphs with a slightly larger size than the networks for fake and extremely biased news and a larger average degree than left, left-leaning, and center news. These results show that users spreading fake and extremely biased news, although in smaller numbers, are not only more active on average, but also more connected, indicating that the audience of fake and extremely biased news forms tighter community structures, which may facilitate the spread of misinformation among themselves. The heterogeneity of the degree distribution plays an important role in spreading processes on networks as a larger degree heterogeneity indicates the presence of hubs that accelerate the diffusion in the entire network. This suggests that different mechanisms of information diffusion could be at play in the center and left-leaning news networks, where high degree nodes may play a more important role than in the fake and extremely biased news networks.

The investigation of the 2016 US election [57] using the CI algorithm produced the ranking of influencers seen in Table 2.3 for the top 10 influencers and their echo-chamber networks in Fig. 2.17. These data reveal very different profiles of the top spreaders of fake and extremely biased news compared to traditional news spreaders. Influencers in traditional news mainly consist of famous journalists or media verified by Twitter, whatever their political orientation. In contrast, influencers in fake news and extremely biased news propagation are mostly unverified or later-deleted accounts with deceiving user profiles that have been created recently. That is, a large number of influencers spreading fake and extremely biased news are new emerging influencers who did not exist a few years back in the 2016 election, i.e., they do not belong to the traditional elite media. Some of these influencers are unknown,

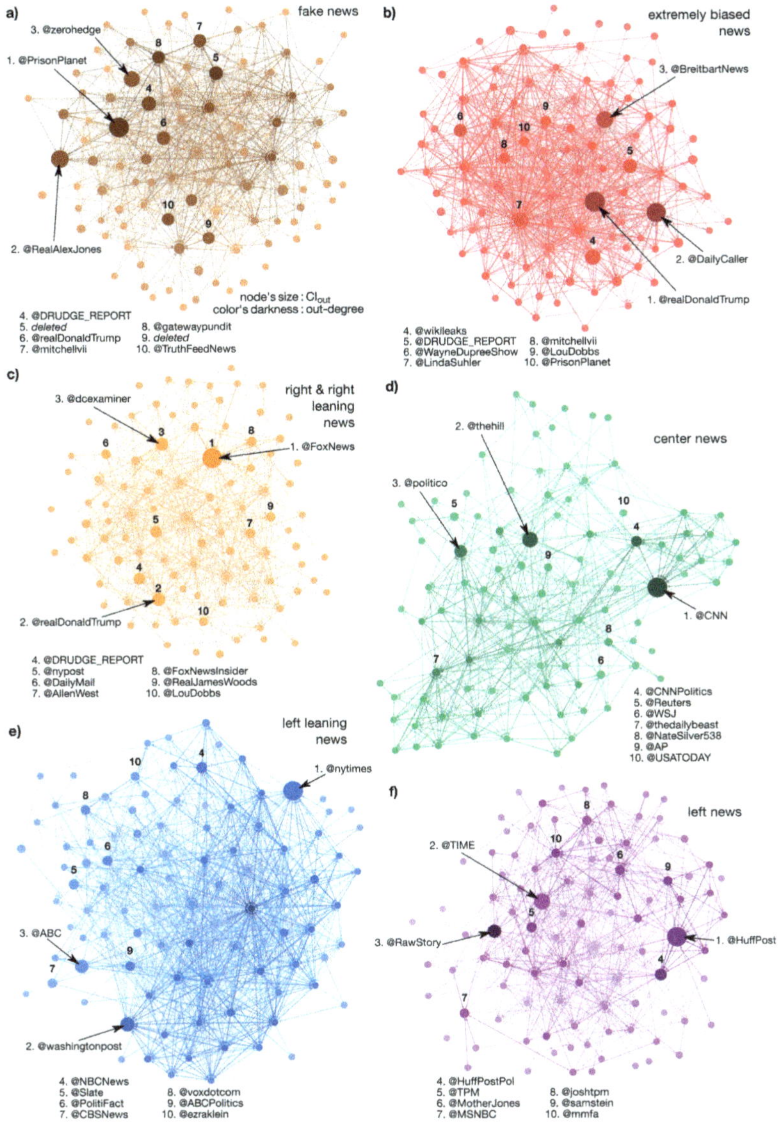

Fig. 2.17 Retweet networks formed by the top 100 influencers of each media category during the 2016 US election. Retweet networks for users spreading: **a** fake news, **b** extremely biased news, **c** right and right-leaning news, **d** center news, **e** left leaning news, and **f** left news. Each link in each network is classified in one of these categories of news. Each network is built using each type of link among the top 100 influencers ranked according to their collective influence. The direction of the links represents the flow of information between users. The size of the nodes is proportional to their CI_{out} values, and the shade of the nodes' color represents their out-degree from dark (high out-degree) to light (low out-degree). The legends state the top influencers in each network. The network of fake in (**a**) and extremely biased news in (**b**) are characterized by a connectivity that is larger than average and less heterogeneous than that of the networks of traditional news from left to right. Figure reproduced from [57]. Copyright © 2019, The Authors

Table 2.3 Top 10 CI news spreaders of the retweet networks corresponding to each media category in 2016. Verified users have a checkmark next to their username. Verifying its accounts is a feature offered by Twitter that "lets people know that an account of public interest is authentic" (https://www.help.twitter.com/en/managing-your-account/about-twitter-verified-accounts). Unverified accounts do not have a checkmark, and accounts marked as *deleted* have been deleted, either by Twitter or by the users themselves. Table reproduced from [57]. Copyright © 2019, The Authors

Rank	Fake news (7 verified, 2 deleted, 19 unverified)	Extreme bias (right) news (15 verified, 1 deleted, 9 unverified)	Right news (22 verified, 0 deleted, 2 unverified)	Right-leaning news (20 verified, 1 deleted 4 unverified)
1	@PrisonPlanet ✓	@realDonaldTrump ✓	@FoxNews ✓	@WSJ ✓
2	@RealAlexJones ✓	@DailyCaller ✓	@realDonaldTrump ✓	@WashTimes ✓
3	@zerohedge	@BreitbartNews ✓	@dcexaminer ✓	@RT_com ✓
4	@DRUDGE_REPORT	@wikileaks ✓	@DRUDGE_REPORT	@realDonaldTrump ✓
5	@realDonaldTrump ✓	@DRUDGE_REPORT	@nypost ✓	@RT_America ✓
6	@mitchellvii ✓	@seanhannity ✓	@FoxNewsInsider ✓	@WSJPolitics ✓
7	deleted	@WayneDupreeShow ✓	@DailyMail ✓	@DRUDGE_REPORT
8	@TruthFeedNews	@LindaSuhler	@AllenWest ✓	@KellyannePolls ✓
9	@RickRWells	@mitchellvii ✓	@RealJamesWoods ✓	@TeamTrump ✓
10	deleted	@LouDobbs ✓	@foxandfriends ✓	@LouDobbs ✓
Rank	Center news (24 verified, 0 deleted, 1 unverified)	Left-leaning news (25 verified, 0 deleted 0 unverified)	Left news (21 verified, 0 deleted, 0 unverified)	Extreme bias (left) news (7 verified, 1 deleted, 17 unverified)
1	@CNN ✓	@nytimes ✓	@HuffPost ✓	@Bipartisanism ✓
2	@thehill ✓	@washingtonpost ✓	@TIME ✓	@PalmerReport ✓
3	@politico ✓	@ABC ✓	@thedailybeast ✓	@peterdaou ✓
4	@CNNPolitics ✓	@NBCNews ✓	@RawStory ✓	@crooksandliars ✓
5	@Reuters ✓	@Slate ✓	@HuffPostPol ✓	@BoldBlueWave
6	@NateSilver538 ✓	@PolitiFact ✓	@NewYorker ✓	@Shareblue ✓
7	@AP ✓	@CBSNews ✓	@MotherJones ✓	@Karoli
8	@business ✓	@voxdotcom ✓	@TPM ✓	@RealMuckmaker
9	@USATODAY ✓	@ABCPolitics ✓	@Salon ✓	@GinsburgJobs
10	@AP_Politics ✓	@ezraklein ✓	@thinkprogress ✓	@AdamsFlaFan

non-public figures, yet they play a major role in the diffusion of disinformation and information on Twitter.

Out of the tweets that link to a website, 29% of these tweets are linked to sites that disseminate fake or extremely biased news. Outside these two categories, influencers are mainly journalists or news outlets with verified Twitter accounts, e.g. `@NYTimes` and `@CNN`. However, the majority of fake news influencers have unverified or deleted accounts. In particular, accounts with deceiving profiles are found among the top fake and extremely biased influencers.

It is interesting to observe the top influencers in each category in Table 2.3. The top influencer spreading extremely biased news in the 2016 campaign is `@realDonaldTrump`, the official account belonging to Donald Trump, who later become the president of the USA, and his account was permanently suspended by

Twitter on January 8, 2021 “due to the risk of further incitement of violence” [248] after the events leading to the violent storm of the US Capitol on January 6, 2021 (later restored by Elon Musk after he acquired Twitter). Donald Trump is seconded by `@DailyCaller` and `@BreitbartNews`; the Twitter account of the Breitbart News and the most trafficked website in the extreme bias category (Table 2.1). `@LindaSuhler`, who was already identified as an influencer in the Republican primary campaign in Fig. 2.5 appears as number 7 in spreading extremely biased news. Wikileaks appears in a surprising fourth place. The top 3 influencers spreading and mentioning fake news websites are `@PrisonPlanet`, `@RealAlexJones` and `@zerohedge`, all well-known popular accounts in the alt-right movement, while president `@realDonalTrump` appears in sixth place. In fifth and ninth place of fake news, ‘deleted’ accounts appear, pointing to the phenomenon that many accounts in the fake news movement become highly influential very rapidly with a posterior deletion of the account. This is a remarkable feat, considering how difficult it is to rise to the top ranks in the political discussion.

2.12.3 The Conundrum of Social Media: Good Data Versus Bad Data

The categories and classifications of news outlets utilized in the previous section and in other sections of this book are a matter of opinion, rather than a statement of fact. In fact, they are a matter of subjectivity. The categorizations and labels assigned to the corresponding classes of news as reproduced in this book are those used in the papers of [57, 64] and have originated in publicly available datasets from fact-checking organizations credited above. In Appendix A we provide links to the datasets used in these studies, so anyone can reproduce the results. Furthermore, anyone in the world can download the data and develop a different machine learning classifier or any other method that could come up with any other classification of the posts in different categories. Thus, the classification is subject to the kind of classifier used. The views and conclusions contained in these articles and this book, as well as the classification of news media and individuals into fake news, extremely biased news, influencers, and categories of political orientation from left to right should not be interpreted as representing the opinions of the authors of this book nor the funders of the works.

While the results presented in this book are the accepted view among the majority of scholars on what represents fake news, this view is not universally accepted by all. As an example, let’s recall that the machinery described in this book together with the fact-checking classifications from established organizations of scholars uncover Donald Trump as the top spreader of extremely biased news and a top 5 spreader of fake news in 2016, Table 2.3. This is consistent with academic consensus posing that Donald Trump and his presidency have ushered the digital era of fake news and misinformation in political elections, see for instance Ref. [241].

However, not everybody would agree with this classification. Trump, and supporting media, would most probably disagree, as can be seen from a tweet that `@realDonaldTrump` posted on January 17, 2018 at 12:35 PM announcing the 'Fake News Award' to Nobel Laureate in Economics, CUNY professor and contributor to The New York Times Paul Krugman for predicting an economic recession after Trump victory in 2016, see [242]. Quoting from the tweet: "*"The Fake News Awards, those going to the most corrupt & biased of the Mainstream Media, will be presented to the losers on Wednesday, January 17th, rather than this coming Monday. The interest in, and importance of, these awards is far greater than anyone could have anticipated!*". It received 135,308 likes shortly after.

Fake news and misinformation identification is a contentious issue with no easy solution.

There is a certain a level of subjectivity in any fake news classification which should be addressed as well. While no one would disagree that Alex Jones and Infowars denying that "no one died" in the 2012 Sandy Hook Elementary School shooting is a self-evident paradigmatic case of fake news, disinformation, and conspiracy theory (a jury ordered him to pay $45.2 million to the parents in punitive damages for spreading the lie [243]), any fake news classification offers an spectrum of possibilities. A Wall Street Journal editorial [244] argued that sometimes disagreement is confused with disinformation, giving an excuse for censorship.

In the 21st century, we live in a conundrum: good data versus bad data. In social media, the outcome of expanding the use of technology is to have positive influencers, which, in principle, outweigh the negative aspects of bots, trolls, and influencers driving negative, divisive, and contradictory opinions. However, when social media came up over a decade ago, little did we know that these platforms would become the medium for extreme polarization, hate speech, bots, trolls, and the spreading of fake news and misinformation. There will be those who think that social media is doomed and there is nothing we can do to solve it (remember MySpace?). And there will be those who insist there will always be a way to respond to crises.

Unfortunately, the bottom line of the tech companies that created these platforms (read: Twitter, Facebook, etc.) provides little incentive to solve the problem of misinformation. The conundrum of fake news is exacerbated by the AI algorithms that run social media platforms. These algorithms have been designed by humans and trained by data ultimately created by humans. And these humans respond to the corporate goals of the companies they work for. And these companies respond to the Milton Friedman's philosophy of 'profits at any cost'. For social media companies, profits are measured by user engagement (number of likes, hates, etc.). It has also been shown that there is no more efficient way to increase engagement than by sparring controversy. And this has contributed to the extreme polarization shown in Fig. 2.21. In conclusion, these companies have little economic incentive to decrease the amount of controversies running in their platforms.

The tech companies that may survive are the ones that strike the right balance between freedom of speech and the goodness of AI that the new information era deserves. This will be the 'Google' of the future. While Google may have become a mere advertisement company that uses AI to profit from the 'user as a product', the next Google will be a company that uses AI for the benefit of humankind to solve the current pressing problems.

2.12.4 *Granger Causality for Influence and the Multistep Model of Mass Communication in Social Media*

To understand how fake news influence the opinion of people in the 2016 presidential election, Bovet *et al.* [56] performed a Granger-causality test between the time series of activity of influencers and the supporters of each presidential candidate: Trump and Clinton.

The Twitter activity time series are studied by counting the number of tweets in a 15 minute interval. In addition to the activity related to each media group, the time series of the activity of the supporters of each presidential candidate are measured. Supporters are classified using machine learning explained in Sect. 2.7.

Figure 2.18a shows the Pearson correlation coefficients between activity time series related to the different types of news outlets, Trump supporters and Clinton supporters. It was found that the activity of Trump supporters is equally correlated with all kinds of news outlets, while the activity of Clinton supporters is mainly correlated with the media clusters consisting of center, left and left-leaning news.

The cross-correlation between activity time series informs us about their temporal relation, however they do not allow us to draw conclusions about causation

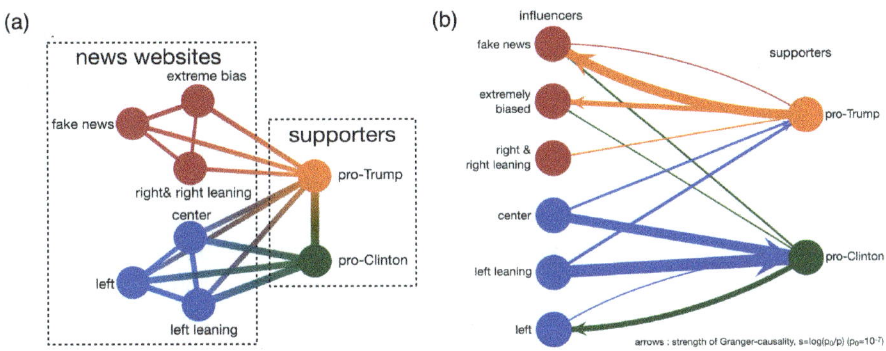

Fig. 2.18 Activity correlation and Granger causality relations between news outlets and supporters in the 2016 US presidential election. a Correlation between different types of news websites and supporters. Edges indicate correlations larger than $r = 0.5$. Two clusters of news websites are marked by red (fake news, extreme bias and right & right leaning) and blue (center, left leaning and left) respectively. (**b**) The Granger causality relations between the activity of the top 100 influencers in each media category and the activity of supporters. Arrows represent the direction of Granger-causation ($p < p_0 = 10^{-7}$) between two activity time series, The width of arrows is proportional to the strength of the causation: $s = \log_{10} p_0/p$. Figure adapted from [57]. Copyright © 2019, The Authors

between time series. In order to investigate the causal relations between news media sources and Twitter dynamics, the statistical test of Granger-causality [66] between the activity of news influencers and supporters of the presidential candidates is used.

A stationary stochastic process, represented by a time series Y_t, is said to Granger-cause another stochastic process X_t when a regression using the past values of Y_t and X_t is better able to predict X_t than the same regression model using only the past values of X_t [66]. Using a linear autoregressive model for the time series:

$$X_t = \sum_{j=1}^{m} a_j X_{t-j} + \epsilon_t, \tag{2.7}$$

where ϵ_t is a white noise level, and m is the order of the model. The null-hypothesis of the test is that the predictive error of this model is not significantly decreased when adding the lagged values of Y_t up to lag l in the regression:

$$X_t = \sum_{j=1}^{m} a_j X_{t-j} + \sum_{j=1}^{l} b_j Y_{t-j} + \epsilon_t, \tag{2.8}$$

In other words, the coefficients b_j for $i = 1, ..., l$ of the regression are equal to zero. As for the cross-correlation analysis, the residuals of the signal of the 15-minute tweet volume time series are used to test for Granger causality. Removal of trends and circadian patterns in the time series is performed, resulting in stationary time series tested by an augmented Dickey-Fuller test [249].

Figure 2.19 shows examples of the residual component of activity time series for the top 100 left-leaning influencers, Clinton supporters, the top 100 fake news influencers, and Trump supporters. It is observed that the peaks in the activity of the Clinton supporters (Fig. 2.19a) appear a few minutes after a similar peak in the activity of the top 100 influencers in the left-leaning category. This indicates that the source of information in the left are the influencers, who, in this case, correspond to the traditional news organization, and then the Clinton supporters just follow by retweeting and commenting on the news generated by the media. This dynamics follows the classical two-step model of political communication and information diffusion from Sect. 2.2.1 where the news is generated by the original source, then this information is transmitted to the opinion leaders represented by the traditional left-leaning and center media, New York Times, CNN, et al., and only then the information reaches the adopters who then follow the news by commenting and retweeting among themself a few minutes later.

The classical model does not apply to the Trump supporters and their fake news channels. As shown in Fig. 2.19b, the situation is reversed when we look at the activity of Trump supporters and the top 100 influencers of fake news. For instance, a large peak of activity in the Trump supporters around July 6 at 18:00 hs is followed a few minutes later with a peak in activity by the influencers of fake news. Similarly, another peak in activity of the Trump supporters around July 7 at 12:00 hs is followed

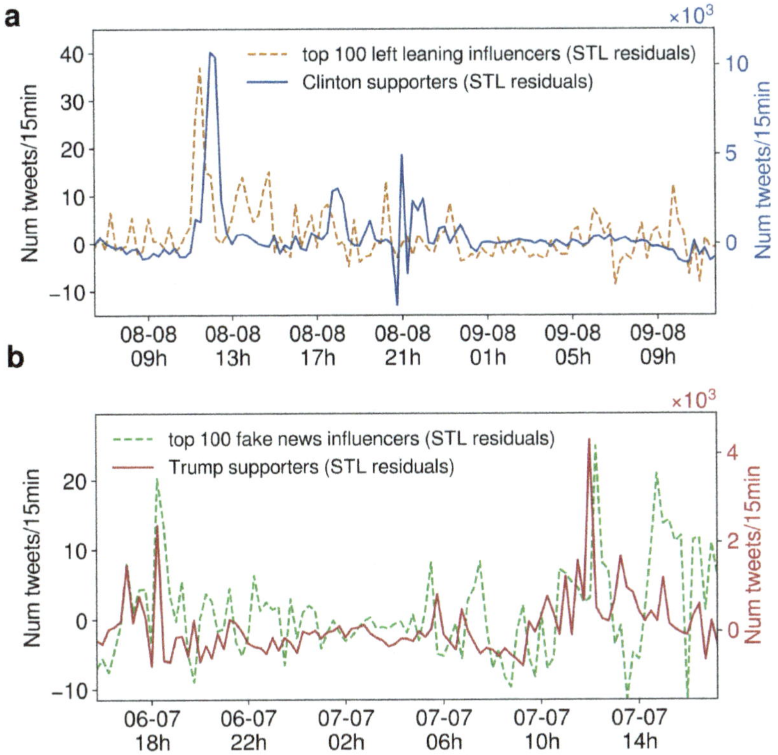

Fig. 2.19 Example of Granger-causality relation between the activity of influencers and supporters. a Activity time series corresponding to the top 100 left-leaning influencers (black) and the Clinton supporters (blue, right vertical axis). **b** Activity time series of the top 100 fake news influencers (black) and the Trump supporters (red, right vertical axis). We show the residuals of the STL filtering after the removal of the seasonal (daily) and trend components. A Granger-causality relation is apparent from the top 100 left-leaning influencers to the Clinton supporters (**a**), i.e., the information of the influencer's time series helps predict the Clinton activity time series. This is apparent as peaks in the left-leaning influencers activity (black) tend to precede peaks in the activity of Clinton supporters (blue). A Granger-causality relation from the Trump supporters to the top 100 fake news influencers (**b**) is also apparent. Figure reproduced from [57]. Copyright © 2019, The Authors

by a peak in activity a few minutes later, again, by the fake news influencers. This is remarkable. These examples indicate that the fake news are first generated organically by the people, Trump supporters, and then these news are picked by the fake news influencers and the extremely biased influencers, for which Trump is the top one, whom then retweet back the fake news to their supporters to amplify the fake news echo-chamber.

This anecdotical evidence in Fig. 2.19 is rigorously corroborated by statistical analysis of Granger causality shown in Fig. 2.18b. This figure shows the Granger causality relations between the activity of the top 100 influencers of each news source category and the supporters of Clinton and Trump. The p-values of the null hypothesis obtained with an F-test between the activity of the top 100 influencers of each news source category and the activity of the supporters are reported in Table 2.4. The activity of only the top influencers are considered since they are the most important sources of information, and it is assumed that they are the ones triggering the activity of the rest of the population. The Granger-causality relations in-between influencers groups is not reported as the groups do not form disjointed sets.

The value of the lag in the Granger analysis is chosen by an information criterion by fitting a bi-variate autoregressive model to the pairs of time series. The Bayesian information criterion (BIC) gives the smallest lag (22), the Akaike information criterion (AIC) gives the largest (48) and the Hannan–Quinn information criterion (HQC) gives a lag of 37 [57]. It is chosen to include up to 37 lagged values, i.e. up to 9:25 hs, since including more values does not significantly changes the results. This value is consistent with the observation of the retweet delay distribution since approximately 75% of the retweets happen during this lag.

Small values of p-values indicate a stronger rejection of the null hypothesis, and therefore a stronger evidence of Granger-causality between the two time series. Only p-values smaller than $p_0 = 10^{-7}$ in Fig. 2.18b are considered. This small p-value represents a highly significant criterium. It is observed that this condition is fulfilled for several pairs of time series revealing the causality between different groups. However, a broad range of p-values smaller than p_0 is observed which allows to compare the strength of the causal relations and reveals interesting patterns of mass communication. These patterns address directly the validity in this setting of the communication theories of Sect. 2.1, from the one-step model to the multistep model of information diffusion. On a related topic, it is interesting to understand the advantages and pitfalls of p-value statistical analysis. A discussion appears in Benjamin et al. [250] where it has been proposed to change the default p-value threshold for statistical significance for claims of new discoveries from 0.05 to 0.005.

The center- and left-leaning influencers have the strongest causation on the supporters' activity, with a stronger effect on the Clinton supporters than on the Trump supporters. Since the Clinton supporters dominate Twitter activity, they are also the main drivers of global activity. The influencers of right and right-leaning and extremely biased news have no significant effect on the supporter's activity. The influencers of left news and fake news have a smaller impact on the supporter's activity.

The reverse Granger causality relations are also shown in Fig. 2.18b and in Table 2.4b revealing the presence of feedback loops, when two time series have a significant Granger causality relation in both directions, between the fake news influencers and the Trump supporters and between the left news influencers and the Clinton supporters. No significant causation is revealed from the Trump or Clinton supporters activity toward the center and left leaning influencers. This indicate that left leaning and center news influencers are the ones driving Twitter activity. All the other influencers are more strongly influenced by supporters activity than are influencing it. This suggest that they are in fact following Twitter activity rather than driving it. In particular, the activity of fake and extremely biased news influencers is influenced by the activity of Trump supporters.

Table 2.4 Test of the multistep model of mass communication. p-values of the Granger causality analysis between the activity of the top 100 influencers of each media category and the activity of the presidential candidate supporters. **(a)** p-values obtained with an F-test for the direction influencers Granger-cause supporters activity. **(b)** p-values obtained with a F-test for the direction supporters activity Granger-cause influencers activity. Smaller values of the p-values indicate a stronger rejection of the hypothesis and, therefore, stronger evidence of Granger-causality between the two time series. Table reproduced from [57]. Copyright © 2019, The Authors

a)	*influencers*					
↙	fake news	extreme bias	right	center	left leaning	left
supporters						
pro-Clinton	1.8×10^{-3}	8.1×10^{-1}	8.8×10^{-2}	7.9×10^{-37}	1.4×10^{-45}	2.2×10^{-9}
pro-Trump	7.4×10^{-9}	4.9×10^{-1}	1.1×10^{-2}	3.3×10^{-13}	2.2×10^{-15}	2.0×10^{-7}

b)	*supporters*	
↙	pro-Clinton	pro-Trump
influencers		
fake news	2.1×10^{-11}	8.6×10^{-36}
extremely biased	2.2×10^{-9}	8.9×10^{-22}
right & right leaning	2.6×10^{-6}	5.6×10^{-10}
center	3.4×10^{-6}	9.4×10^{-2}
left leaning	1.3×10^{-6}	2.8×10^{-2}
left	1.4×10^{-22}	1.9×10^{-7}

These results reveal two very different dynamics of news diffusion for traditional, center, and left-leaning news and misinformation:

- The influencers spreading traditional center and left-leaning news largely determine (Granger-cause) the opinion of the Clinton supporters by following a one-step classical model of information diffusion. Thus, the traditional news dynamics on the left is governed by journalists and newsrooms, as it has always been in accordance with the classical two-step model of diffusion of information.
- Remarkably, this causality is reversed for the fake news sphere: the opinion of Trump supporters largely Granger-causes the dynamics of influencers spreading fake and extremely biased news. Fake and extremely biased news dynamics seem to be governed by the ensemble of Trump supporters rather than by a small group of influencers. This result supports a multistep model of communication between Trump supporters, although in this case, reversing the directionality of flow of information from adopters to influencers and opinion leaders.

In the fake news movement of Trump supporters, the opinion leaders are not the classical journalists, nor politicians. This movement seems to be organically generated by the people, whom first choose the type of fake news and misinformation that they want to consume, read, comment, retweet and amplify in social media, and then the influencers and opinion leaders like `@realDonaldTrump`, `@PrisonPlanet` and `@BreitbartNews` just follow on what these Trump supporters have generated organically. Thus, it is as though the rise to power of Trump has been engendered by the people and not the other way around.

In summary, the Granger causality test reveals that the opinion of the Clinton supporters are mostly influenced by center and left-leaning influencers, while the opinion of Trump supporters influences the fake news influencers. This is a remarkable turn on the classical view of mass communication of Sect. 2.2.1, and could explain how and why the fake news movement has brought Donald Trump to power from the grassroots.

2.12.5 Fake News Influencers in the 2020 US Presidential Election

Identifying the top news spreader of each news category during the 2020 US election allows to understand how the influencer landscape evolved compared to 2016 and to understand how their role in the diffusion of disinformation and traditional news evolved and how they affected the increased polarization in society.

Figure 2.20 shows the evolution of the average number of followers of the top 5 influencers of 2016 and 2020 for each news media category. The average follower count of influencers of the extremely biased right news and of the right news is shown to have rapidly increased from 2016 to 2020 and to now be comparable with the average follower counts of the traditional media of the center of left leaning categories. The inset in Fig. 2.20 reveals that the increase in follower count in the extremely biased right news and in the right news categories is mainly due to the presence of `@realDonaldTrump`.

The list of influencers in 2020 for each news categories is shown in Table 2.5 and displayed in Fig. 2.21 bottom panel with their echo-chamber networks. In addition to changes in account types and verified users, a significant shuffle is observed in the elite rankings from 2016 to 2020. Figure 2.22, which maps the change in rankings of the top 10 users per considered bias category, illustrates several interesting changes between 2016 and 2020. First, we see the rise of highly influential users from obscurity. Across all categories we see a set of new users entering the top 25 ranking having been completely unranked, or very lowly ranked, in 2016. These users include `@TeaPainUSA`, `@svdate`, `@kylegriffin1`, `@marklevinshow`, `@DavidHarrisJr`. Some of these users are well connected to well known news outlets, but a significant portion of them are not, which demonstrates the ability for non-elite users to garner a large influence over the political discourse overnight. Notably, while Donald Trump was the top spreader of extremely biased right news and the top 5 of fake news in the campaign preceding the 2016 election, his influence has diminished during the 2020 campaign being replaced by his son, `@DonaldJTrumpJr`, who became the top influential spreader in the right

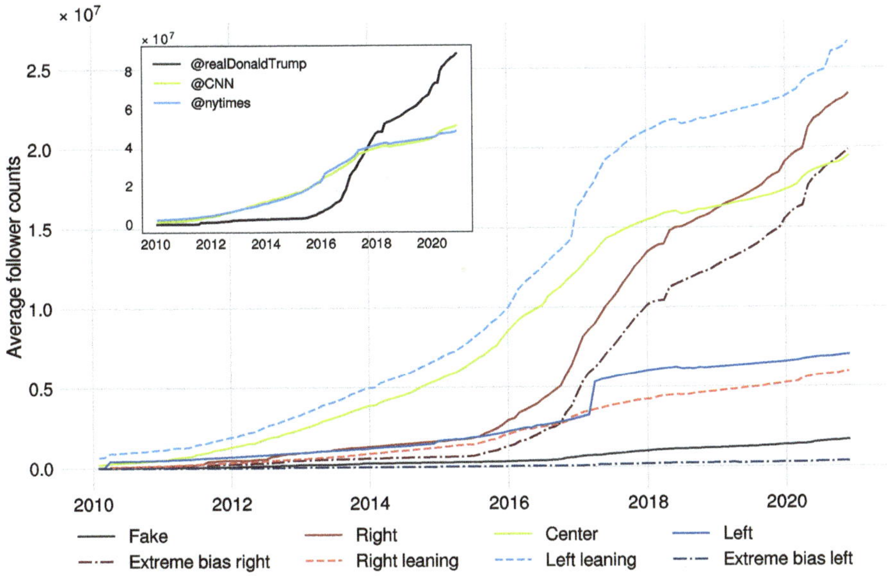

Fig. 2.20 The evolution of the average follower counts of the top 5 influencers, across 2016 and 2020, in each news media category. For the fake news, the top 5 are: @seanhannity, @PrisonPlanet, @gatewaypundit, @RealAlexJones and @DavidJHarrisJr; for the extreme bias right: @realDonaldTrump, @DonaldJTrumpJr, @BreitbartNews, @DailyCaller and @dbongino; for the right: @realDonaldTrump, @DonaldJTrumpJr, @FoxNews, @marklevinshow and @jsolomonReports; for the right leaning: @nypost, @WSJ, @DonaldJTrumpJr, @WashTimes and @EricTrump; for the center: @thehill, @CNN, @AP, @Reuters and @politico; for the left leaning: @CNN, @nytimes, @kylegriffin1, @washingtonpost, @ABC; for the left: @MSNBC, @HuffPost, @thedailybeast, @TIME and @kylegriffin1. Inset: Followers count evolution for @realDonaldTrump, @CNN and @nytimes

and the extreme bias right news. The tweeting rate for bots is four times larger than for traditional media, suggesting a larger role of bots in the diffusion of fake news.

We see similar patterns in the volatility of the extreme bias/fake news categories. For both extreme bias categories there has been a complete shift in the top 5 users. The previous top 5 all drastically dropped in rank, being replaced by a group of users who were largely unranked in 2016. While we see similar rises and falls in the other news categories, the magnitude of the shift in these extreme networks is remarkable. The extreme bias right/fake news plot also shows a distinct pattern of lines rising from the lower ranked/unranked users in 2016 to the top 10 of 2020 and a similar dropping of the high ranked of 2016. We also observe a number of news category shifts. In particular we see a shift in @CNN and @politico, previously the first and third highest ranked influencers in the center category in 2016, from center to left leaning. This shift of such large and influential traditional media accounts across news categories is indicative of the increased content polarization on Twitter.

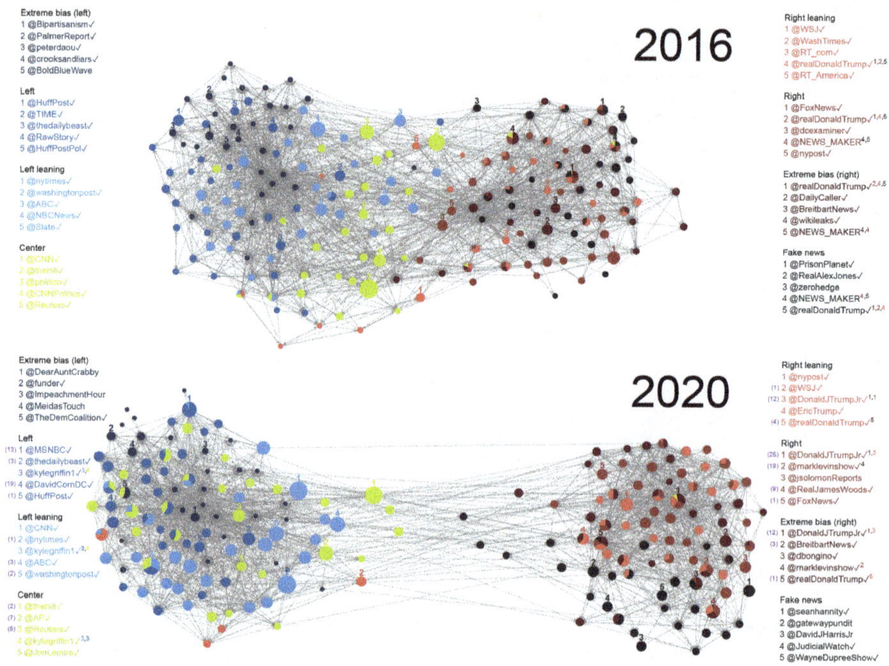

Fig. 2.21 Retweet network formed by the top 30 influencers of each media category during the presidential election in 2016 (top) and 2020 (bottom). The 2016 network is reproduced from [57] and the 2020 network from [64]. The direction of the links represents the flow of information between users. The size of the nodes is proportional to their out-degree in the complete combined network, i.e. the number of different users that have retweeted at least one of her/his tweets with a URL directing to a news outlet, and the color of the nodes indicates to which news category they belong. Nodes that belong to several news categories are represented by pie charts where the size of each slice is proportional to their CI_{out} ranking, taking into accounts only their rank among the top 30. Both networks layout have been computed using a force-directed graph layout. The tables on either side of the networks show the top 5 users in each news category. Users ranked in the top 5 of multiple news categories have superscripts, in the form of a colored number, indicating the rank and news classification of their other top 5 positions. Additionally, verified users are indicated by a checkmark and for users in the 2020 tables their rank from 2016 is given by the purple number to the left of their 2020 rank. It is important to note that users may have changed their user handle since 2016. Those users will have their new handle displayed in the 2016 tables for consistency. Three usernames in the top 10 changed between 2016 and 2020: `@DRUDGE REPORT` became `@NEWS MAKER`, `@HuffingtonPost` became `@HuffPost`, and `@TruthFeedNews` became `@TAftermath2020`. Figure reproduced from [64]

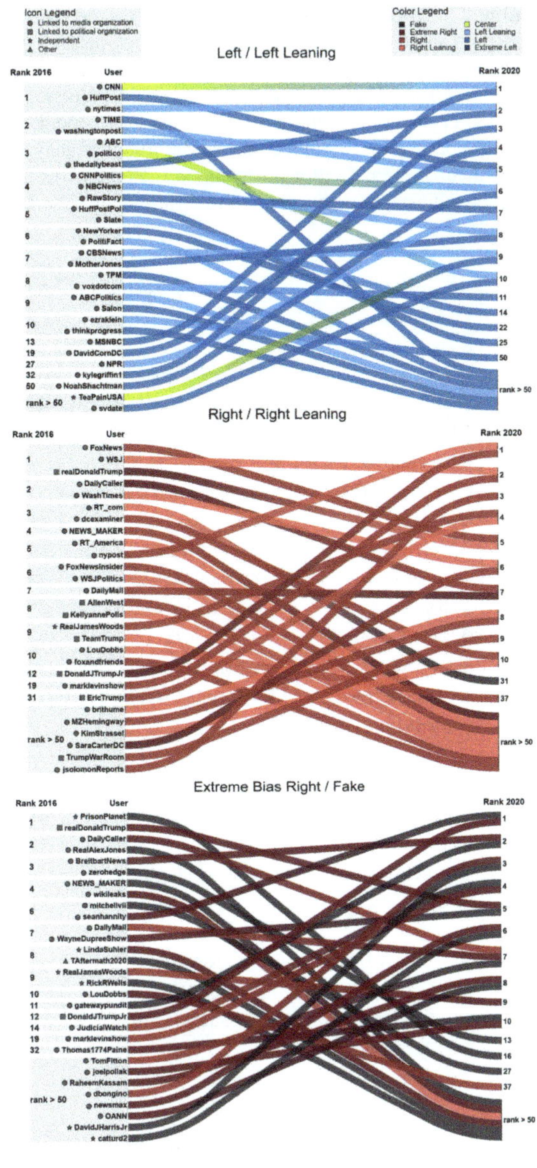

Fig. 2.22 Change in rankings from 2016 to 2020. All users in this figure were ranked in the top 10, by CI influence, in at least one media category, in 2016 or in 2020. This figure maps the user's rank in 2016, displayed to the left of the user's username, to their rank in 2020 on the right. Each user only has one flow, whose starting color represents their highest ranked media bias in 2016 and becomes the color of their highest ranked media bias in 2020, if a user has the same rank in multiple bias categories the color most suiting the subfigure is chosen. Each subfigure shows two media biases and the corresponding top ranked users. Figure adapted from [64]

Table 2.5 Top 10 CI news spreaders of the retweet networks corresponding to each media category in 2020. Verified users have a checkmark next to their username. Unverified accounts do not have a checkmark and accounts marked as *deleted* have been deleted, either by Twitter or by the users themselves. If a user held a position in the top 25 in 2016 as well, we mark that position for reference in red parentheses next to the username. Despite `@realDonaldTrump` having their account permanently banned, due to the role they played in the 2020 Election, we have chosen to keep their original Twitter username in the table. However, we count this account as deleted, and marked their name with a star (⋆) to differentiate their account from other accounts. Table adapted from [64]

rank	fake news (10 verified, 8 deleted, 7 unverified)	extreme bias (right) news (23 verified, 2 deleted, 0 unverified)	right news (23 verified, 1 deleted, 1 unverified)	right-leaning news (23 verified, 2 deleted 0 unverified)
1	@seanhannity✓	(12) @DonaldJTrumpJr✓	(25) @DonaldJTrumpJr✓	@nypost✓
2	deleted	(3) @BreitbartNews✓	(19) @marklevinshow✓	(1) @WSJ✓
3	@DavidJHarrisJr	@dbongino✓	@jsolomonReports	@DonaldJTrumpJr✓
4	@JudicialWatch✓	@marklevinshow✓	(9) @RealJamesWoods✓	@EricTrump✓
5	@WayneDupreeShow✓	(1) @realDonaldTrump⋆	(1) @FoxNews✓	(4) @realDonaldTrump⋆
6	@catturd2	@newsmax✓	@SaraCarterDC✓	(2) @WashTimes✓
7	@TomFitton✓	@DailyMail✓	@DailyCaller✓	@marklevinshow✓
8	@OANN✓	@RaheemKassam✓	@MZHemingway✓	@brithume✓
9	@dbongino✓	@RealJamesWoods✓	@TrumpWarRoom✓	@RealJamesWoods✓
10	@Thomas1774Paine	@joelpollak✓	(3) @dcexaminer✓	@KimStrassel✓
rank	center news (24 verified, 0 deleted, 1 unverified)	left-leaning news (24 verified, 0 deleted 1 unverified)	left news (23 verified, 0 deleted, 2 unverified)	extreme bias (left) news (3 verified, 1 deleted, 21 unverified)
1	(2) @thehill✓	@CNN✓	(13) @MSNBC✓	@DearAuntCrabby
2	(7) @AP✓	(1) @nytimes✓	(3) @thedailybeast✓	@funder✓
3	(5) @Reuters✓	@kylegriffin1✓	@kylegriffin1✓	@ImpeachmentHour
4	@kylegriffin1✓	(3) @ABC✓	(19) @DavidCornDC✓	@MeidasTouch
5	@JonLemire✓	(2) @washingtonpost✓	(1) @HuffPost✓	@TheDemCoalition✓
6	@Newsweek✓	@CNNPolitics✓	@NoahShachtman✓	@grantstern✓
7	@yarotrof✓	@NPR✓	(4) @RawStory✓	(15) @OccupyDemocrats
8	(9) @USATODAY✓	(4) @NBCNews✓	(7) @MotherJones✓	@Stop_Trump20
9	@ProjectLincoln	(7) @CBSNews✓	@TeaPainUSA	@InSpiteOfTrump
10	@JoeBiden✓	@politico✓	@svdate✓	@froggneal

2.12.6 *Latent Ideology Analysis to Understand the Increasing Polarization in American Politics*

To extend the understanding of the dynamics between influencers and Twitter users, Flamino et al. [64] quantified the amount of polarization of the Twitter users and of the news influencers using positions on a latent ideology scale independent of the news classification based on fact-checking organizations. This analysis serves to validate the manual fact-checked classification of fake news and misinformation. To do this, they follow the method developed by Barberá [251] to infer the ideology of

Twitter users based on the political actors they follow. The method, further developed in [252], considers a set of influential political actors, journalists and news media accounts (the elite) and searches for all the Twitter users that follow them (masses). This bipartite network of followers is then projected on a one dimensional scale using correspondence analysis [253, 254], which applies a singular value decomposition (SVD) of the adjacency matrix standardized to take into account the difference of popularity of the elites and the difference of activity of the masses.

The latent ideology analysis of polarization in Twitter follows the method of [251, 252] adapted to using retweet interactions instead of following relations. Like in [252], correspondence analysis [253] (CA) is used to infer the ideological positions of Twitter users. The adjacency matrix, $\mathbf{A}$, of the retweet network between the influencers and their retweeters is the matrix with element A_{ij} equal to the number of times user i retweeted influencer j. Bots are removed with the methods of Sect. 2.11.2. Users that show a low interest in the US election are also removed by removing users that retweeted less than three different influencers. For the 2016 data, the matrix $\mathbf{A}$ has 751,311 rows corresponding to distinct users, 593 columns corresponding to influencers, and the total number of retweets equal to 39,385,772. For the 2020 data, the matrix $\mathbf{A}$ has 2,034,970 rows corresponding to distinct users, 591 columns corresponding to influencers and the total number of retweets equal to 153,463,788.

The CA method consists on the following steps [64, 254]: The matrix of standardized residuals of the adjacency matrix is computed as

$$\mathbf{S} = \mathbf{D}_r^{-1/2}(\mathbf{P} - \mathbf{rc})\mathbf{D}_\mathbf{c}^{-1/2}, \tag{2.9}$$

where $\mathbf{P} = \mathbf{A}(\sum_{ij} A_{ij})^{-1}$ is the adjacency matrix normalized by the total number of retweets, $\mathbf{r} = \mathbf{P1}$ is the vector of row sums, $\mathbf{c} = \mathbf{1}^T\mathbf{P}$ is the vector of column sums, $\mathbf{D}_r = \text{diag}(\mathbf{r})$ and $\mathbf{D}_c = \text{diag}(\mathbf{c})$.

Using the standardized residuals allows the inference to take into account the variation of popularity and activity of the influencers and the users, respectively [252]. A SVD is then computed such that

$$\mathbf{S} = \mathbf{U}\mathbf{D}_\alpha\mathbf{V}^T, \tag{2.10}$$

with $\mathbf{U}\mathbf{U}^T = \mathbf{V}\mathbf{V}^T = \mathbf{I}$ and $\mathbf{D}_\alpha$ being a diagonal matrix with the singular values on its diagonal. The positions of the users are then given by the standard row coordinates: $\mathbf{X} = \mathbf{D}_r^{-1/2}\mathbf{U}$ where we only consider the first dimension, corresponding to the largest singular value. Finally, the ideological positions of the users are found by standardizing the standard row coordinates to have a mean of zero and a standard deviation of one. The ideological position of the influencers is given by the median of the weighted positions of their retweeters.

Two users are close on the resulting latent ideology scale if they follow similar elites. This method has been shown to produce ideological estimates of the members of US Congress that was highly correlated with ideological estimates based on roll call voting similarity such as DW-NOMINATE [251]. For 2016 and 2020 elections, Ref. [64] considers the union of the top 100 influencers of each type of news as the sets of elites and use the sets of users that retweeted at least 3 different influencers as the masses. Then, they extract the coordinates of each user on the first dimension of the results of the correspondence analysis applied to the weighted network of retweets between the users and the influencers.

Users are more likely to retweet influencers that have an ideology similar to their own. The act of retweeting is a form of endorsement of the content being retweeted [231], in particular since Twitter offers other types of interactions allowing an user to comment on the content such as quote tweets and replies. Finally, for 2016 and 2020, the coordinates of all users are standardized in order to have a mean of zero and a standard deviation of one. Two users are close together on the latent ideology scale if they tend to retweet similar influencers.

Figure 2.23 shows the result of this analysis. The latent ideology of the top 5 influencers of each category is shown as a box plot representing the distribution of the ideology of the users having retweeted them. The distribution of ideology positions of the users and of the influencers shows increased polarization in 2020 compared to 2016. This is confirmed by a Hartigans' dip test for unimodality [255]. Increasing

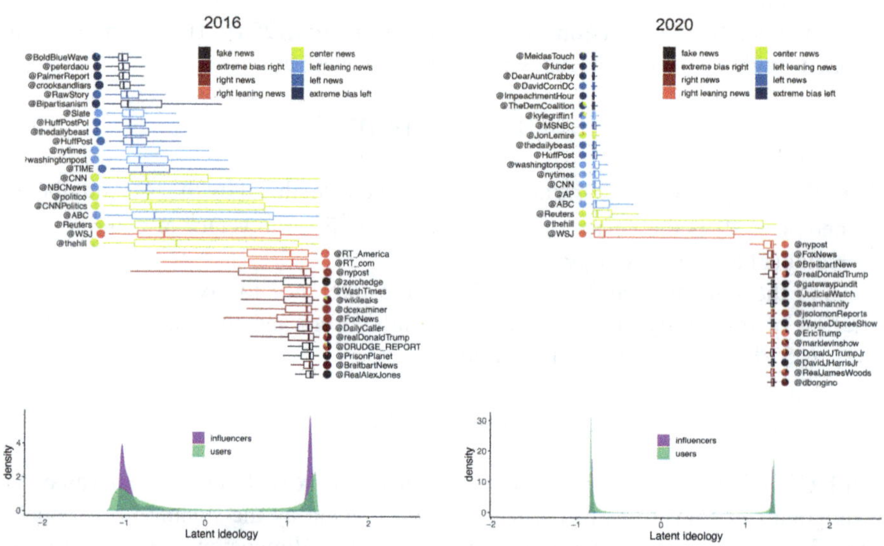

Fig. 2.23 Latent ideology scale of influencers and their retweeters in 2016 (left) and 2020 (right). The latent ideology of the top 5 influencers of each category is shown as a box plot representing the distribution of the ideology of the users having retweeted them. The distribution of the ideology estimate of the users is shown in green and the distribution of the ideology estimate of the top 100 influencers of each news category (computed as the median of the ideology of their retweeters) is displayed in purple. Pie charts next to the influencers' names represent the news categories they belong to (weighted by their respective CI ranks in each category). Figure reproduced from [64]

values of the test statistic indicates distributions that increasingly deviate from a unimodal distribution, corroborating the growing division found in the similarity networks of Fig. 2.21. Furthermore, Fig. 2.23 reveals a clear increase in polarization of the users and influencers in 2020 compared to 2016 and an alignment of their latent ideologies in two distinct groups, mirroring the news classifications groupings seen in Fig. 2.21. This represents a confirmation of the ideological and fake news categories obtained from the fact-checking organization, this time obtained with the independent statistical method of latent ideology.

The retweet networks shown in Fig. 2.21 illustrate the increased polarization found by the latent analysis. The two force-directed network layouts are computed using the same parameters and show two main clusters, with the right-biased and fake news influencers in one cluster and the left-biased influencers in the other. The increased separation evidenced in 2020 is notable. In 2016 the center influencers are mostly between the two clusters; in 2020 the separation between the two clusters increases, and only a few influencers remain within a central position (e.g. `@thehill`). Fake and extremely biased news are sent mostly by influencers whose accounts are unverified or deleted, with fake news seeing a significant increase in deleted influencer accounts from two in the top 25 in 2016 to eight in 2020. The reasons adopters decide to believe in these influencers are completely different. Reference [256] demonstrated that sparse connectivity of social networks leads to social learning drawing the wrong conclusions from limited hypothesis testing, and humans rely on previous interactions to bypass the scrutiny of news verification. References [257, 258] show how this leads to the polarization of opinions, which results in strong competition between parties and little collaboration [259].

While in 2016 several influencers were retweeted by users from both ideological sides, in 2020 only a few remain with a wider ideological base (e.g. `@thehill` and `@WSJ`). These results confirm the shift in bias from the center to the right and the left observed using the news classification by solely relying on the retweet patterns between influencers and their retweeters. Moreover, we find a high correlation (above 0.90 for 2016 and 2020) between the users' latent ideology distribution and their average leaning distribution computed using the news categories they posted in. This high correlation indicates that the bias shift observed at the level of the news outlets is also present at the level of the users' retweeting pattern, and serves as an independent validation of the news outlet classification employed in the study.

Future research should be done in every election to monitor the dynamics of polarization and fake news spreading and to further tease apart the political shifts. Researchers should also monitor interactions on other social media and during non-election periods to permit finer grained analysis of the new entrants. Eventually, these analyses might be able to identify mechanisms that might close the chasms between a highly polarized American society and improve the quality of political discussion and knowledge. For the moment, the present analytics seem to indicate that this may not happen anytime soon.

2.13 Influence of Twitter Russian Trolls During the 2016 US Presidential Election

According to the Special Counsel's report by Mueller [229], Russian trolls have leveraged social networks to wage a propaganda war during the 2016 presidential election. This campaign was done through the Internet Research Agency (IRA), also known as Glavset or the Trolls from Olgino [260], which is a Russian troll farm company that conducted online influence operations using fake news and disinformation dissemination to promote the Kremlin's interests by provoking and amplifying political and social discord in the United States.

At the beginning of 2018, Twitter committed to the US Congress to provide to the public information regarding foreign interference on its platform. On October 2018, Twitter released all the accounts and related content associated the operations from the IRA since 2016. This large dataset consists of more than three thousand accounts affiliated with the IRA, originating in Russia including more than 9 million tweets dating back to 2009. Meanwhile, Twitter estimates that 9% of the tweets from IRA accounts were election-related.

Many works have been devoted to characterizing these Russian trolls on Twitter [261–264]. Researchers have analyzed behavioral differences between the troll farm of the IRA spreading fake news and disinformation and regular American users who absorb this information. Analyzing the reactions of targeted users, and gauging the impact of repeated exposure to malicious propaganda, are the key points for understanding the manipulation campaigns [265].

Zhou et al. [266] combined the IRA-Twitter-provided dataset with the collected dataset of 171 million tweets covering the 2016 election between Trump and Clinton from [56]. The IRA dataset was jointed with this large dataset to obtain a new merged dataset, which helps to investigate the behavioral differences and interactions between IRA users and regular users (non-IRA). In-depth analysis was performed of both IRA trolls and non-IRA users from the large-scale merged dataset to understand the influence of Russian trolls activities during the 2016 US president election. Of primary interest are three largely unanswered questions:

- How much influence do Russian trolls have during the US President election?
- How Russian trolls organize themselves and interact with US citizens?
- What topics have been promoted by the Russian trolls to infiltrate the American online conversation.

Firstly, the authors investigate Russian trolls' tweets on each type of media categories to understand what are their relative importance and how much influence they have on different media types. IRA has a strong influence on the networks spreading fake news, extremely biased right news, right news and local news, but shows quite weak influence on mainstream media of left-leaning and right news during the election.

To investigate how IRA influenced US citizens in the Twitter political discussion, community detection is performed on the largest connected component of the

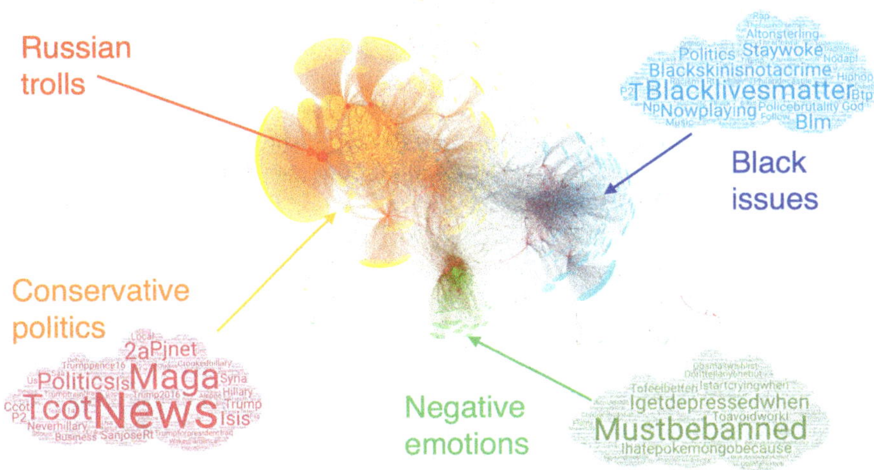

Fig. 2.24 Interaction network between IRA trolls and their echo chamber of non-IRA followers and second layer followers. The colors of nodes represent the community to which every node belongs, and red nodes represent IRA trolls. World clouds of top hashtags from three identified IRA communities are shown. Three communities are identified: C_1: Conservative Politics, C_2: Black Issues, and C_3: Negative emotions

aggregated directed interaction network of IRA trolls and non-IRA followers and influenciable users. This network contains 497 IRA and 179,000 non-IRA users. This component represents the one-layer interaction network where non-IRA users directly connect to IRAs. The hashtags with IRA's tweets are then used to find the second spreading layer surrounding the IRA troll echo-chamber to then construct the two-layer interaction network which is shown in Fig. 2.24. Through generalized Louvain community detection analysis [267], it is found that this connected component of IRA followers is composed of three communities C_1, C_2, and C_3 shown in Fig. 2.24. The Russian trolls mainly amplify misinformation and manipulate opinions through:

- C_1: Conservative Politics, with associated hashtags: #Tcot, #MAGA,
- C_2: Black Issues, with associated hashtags: #BlackLivesMatter, #BlackSkinIsNotaCrime,
- C_3: Negative Emotions, with associated hashtags: #IGetDepressedWhen, #IStartCryingWhen.

The IRA regularly played hashtag "games", attempting to define meaning of hashtags. From the IRA dataset, 20,649 unique hashtags are observed and in average 0.8 hashtags per tweet are posted. Of all IRA's tweets during the election, 58% have one or more hashtags. From the hashtag clouds showed in Fig. 2.24, the three communities identified are discussing about political news and election in support of the Republican party in C_1, diffusing topics referring to African Americans issues with #BlackLivesMatter and #BlackSkinIsNotACrime in C_2, and C_3 where the IRA leveraged to persuade people to discuss social issues or diffuse depressed emotions on a broad scale, which might stir controversy and conflict, e.g. #MustBeBanned, #IGetDepressedWhen and #IStartCryingWhen.

Overall, IRA organizes itself by following two distinct strategies. In C_1, which is the largest community, Russian trolls are the super-spreaders diffusing fake and political right news to support Donald Trump during the election. In the two other communities, C_2 black issues and C_3 negative emotions, IRA spread memes regarding African Americans and depressed emotions by actively connecting to non-IRA users.

The Twitter dynamics between IRA and non-IRA reveals the existence of a two-layer interaction networks obtained by Granger causality analysis. Employing the opinion classifier of Sect. 2.7 and the CI algorithm (Sect. 1.6.1), non-IRA followers of IRA are classified into four categories, including Donald Trump influencers and supporters, Hillary Clinton influencers and supporters. It is found that Trump supporters have a strong causation on the IRA activity in the interaction network regarding political news. Clinton supporters also have a causation on IRA, but the significance is not so strong as that of the Trump supporters. More importantly, IRA Granger causes Trump influencers activities during the election. This finding suggests that IRA was able to infiltrate the political discourse by positioning itself as intermediary between supporters and Trump influencers on political news, especially fake and extremely biased news. In the community referring to African Americans, Russian trolls have a causal relation on Trump supporters activity. Furthermore, IRA drives activities of Trump and Clinton influencers and Clinton supporters in the communities where IRA accounts spread depressed emotions and discuss social issues.

The volume of tweets in different languages posted by IRA trolls is shown in Fig. 2.25. Since the creation of their first Twitter account on April 14, 2009, a cumulative of 3.8 thousand Russian trolls continuously try to manipulate the public opinions via Twitter across all languages. 3,586 Russian trolls have ever sent tweets in Russian or English, of whom 57% trolls sent tweets in English and never in Russian. Starting from 2016, IRA mainly targeted English-speaking audiences on Twitter by posting more tweets in English than in Russian. From June to November in 2016, 556 IRA troll accounts published 391,680 tweets that amplify political and social discord in the US, especially in the 2016 president election. IRA also operates a network of bot Twitter accounts that enabled the IRA to amplify existing content on Twitter. 18% of these tweets are from the unofficial clients described in Sect. 2.11.2, which helps to perform automatic posting for IRA.

Besides spreading disinformation, IRA trolls promote discord by posting thousand of tweets about local news websites without obvious political ideologies. Surprisingly, KSNT, an NBC-affiliated television station licensed to Topeka, Kansas, is the most dominant media outlet in term of number of tweets with more than 3,000. The following most important local news outlets targeted by IRA trolls are KRON4, KSN and NOLA, serving San Francisco, Kansas and New Orleans.

In order to characterize IRA's campaign during the election on Twitter, the same methods explained in Sect. 2.11.3 are used, augmented by a curated list of fake news outlets by Grinberg et al. [58]. The following categories are investigated: fake news,

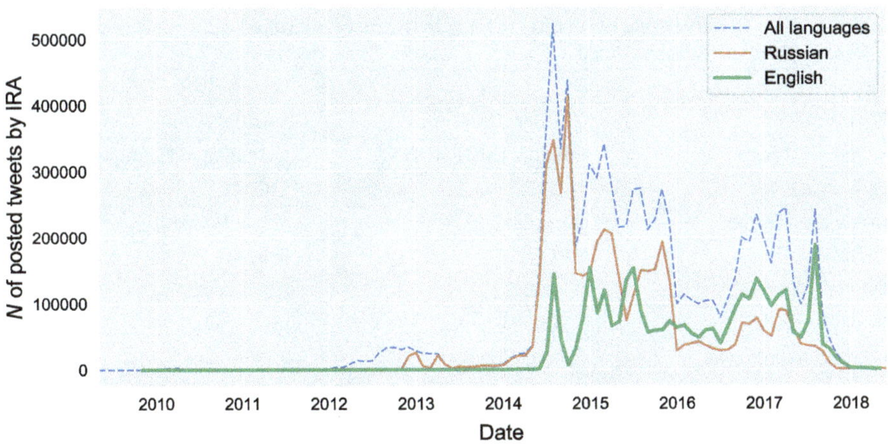

Fig. 2.25 Activity of IRA trolls over time. Volume of tweets posted by IRA trolls in different languages since their inception in 2009 until dissaperance around 2018

extremely biased right news and extremely biased left news, left, left-leaning, center, right-leaning and right.

Employing the CI algorithm from Sect. 1.6.1 adapted to consider only the out-degree, the top 100 influencers are found for each of the three communities. Table 2.6 shows top 10 influencers in each community. In C_1, the top 3 are Russian trolls (`@TEN_GOP`, `@Pamela_Moore13` and `@America_1st_`), which are the three large red nodes clearly shown in the yellow community in Fig. 2.24. The trolls `@TEN_GOP` and `@Pamela_Moore13` are also found in retweet networks of fake and extremely biased right news. These accounts were also identified in the Mueller report [229]: `@TEN_GOP` as claiming to be related to the Tennessee Republican Party, and `@Pamela_Moore13` claiming to be a Trump supporter with 70,000 followers. The other one, `@America_1st_` is an anti-immigration persona according to the Mueller report [229]. The rest of the influencers in this community are mainly Trump supporters, like `@LindaSuhler` of Fig. 2.5 fame.

Surprisingly, we also see that `@HillaryClinton` inadvertently got entangled in this IRA community as one of the top 10 influencers. We also see that important news media, like `@CNN` and `@FoxNews`, are involved in the communities surrounding the IRA trolls, specially C_2 and C_3 talking about black issues and emotions. Democrats like Bernie Sanders and again Hillary Clinton are also influencers of C_2 and C_3. This indicates how the Russian trolls have succeeded to infiltrate the mainstream American political discussion in Twitter by meddling with regular conversations in social communities. Russian trolls have meddled with the election process with such a great success that even main figures like Hillary Clinton and Bernie Sander and elite news media like CNN have retweeted the information propagated by the IRA becoming important IRA followers and influencers.

Table 2.6 Top 10 influencers according to CI_{out} in the three identified communities originating in IRA trolls. IRA trolls are in red

C_1	C_2	C_3
@TEN_GOP	@HillaryClinton	@HillaryClinton
@Pamela_Moore13	@realDonaldTrump	@realDonaldTrump
@America_1st_	@CNN	@CNN
@Always_Trump	@wikileaks	@JackPolakoff
@bfraser747	@POTUS	deleted
@wikileaks	deleted	@wikileaks
@SandraTXAS	@FoxNews	@mmpadellan
@RealJamesWoods	@BernieSanders	@peterdaou
@LindaSuhler	@CNNPolitics	@POTUS
@HillaryClinton	@thehill	@Always_Trump

This anecdotical evidence of IRA influence at every level of American politics can be rigorously corroborated by Granger causality analysis as done in Sect. 2.12.4 by analyzing the time series of Twitter activity. The two-layer interaction network in Fig. 2.24 contains IRA and non-IRA users which can be Trump or Clinton influencers/supporters. By using the residuals of the STL filtering of the 15 min tweet volume time series, the influence direction between the IRA trolls and users can be obtained. Figure 2.26 depicts the Granger causality of the residuals of the time series between the different groups in C_1.

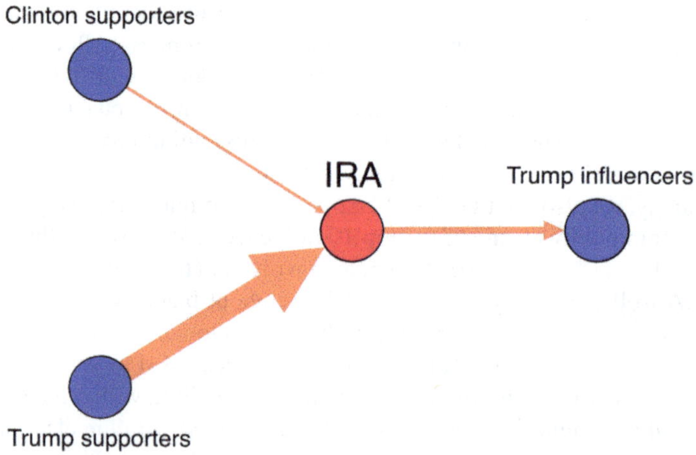

Fig. 2.26 How the IRA influenced American politics. Granger causal network reconstruction between IRA and non-IRA users classified as Trump and Clinton supporters in the C_1 community of conservative news including fake and right news. The IRA trolls do not create the news but they amplify it by receiving influence from Clinton and Trump grassroots supporters (mainly the latter) and then influence the Trump influencers that then transmit the deceptive news broadly

Table 2.7 Granger causality test for the IRA trolls. Results of Granger causality test of Twitter activities between IRA and non-IRA for each community. On the second column, '→ IRA' represents non-IRA users from the first row Granger-causing IRA trolls; 'IRA →' represents IRA trolls Granger-causing non-IRA users from the first row. Only significant results are listed. p-value < 0.05:*, p-value < 0.01:**, p-value < 0.001***

		Trump influencers	Clinton influencers	Trump supporters	Clinton supporters
C_1	→ IRA			0.0027**	0.0165*
	IRA →	0.0101*			
C_2	→ IRA				
	IRA →			0.0324*	
C_3	→ IRA				
	IRA →	0.0257*	0.0251*		<0.001***

Tables 2.7 and Fig. 2.26 show the Granger-causality results between IRA and non-IRA in the interaction network C_1. Here, Trump supporters have a strong causation on the IRA activity. Clinton supporters also have a causation on IRA, but the significance is not so strong as Trump supporters. The IRA trolls then influence Trump influencers activities. This finding suggests that IRA role is intermediary between supporters and influencers on conservative political news, especially fake and right news.

The Granger causal graph in Fig. 2.26 indicates that the news are generated by the grassroots, specially the Trumpian camp, which are then amplified by the IRA trolls who later influence the influencers in the Trump camp. Thus, the IRA is the bottleneck of this information transmission process showing that they have exerted a large influence in the political debate. This shows the remarkable mechanism achieved by the Russian trolls to effectively meddle with the electoral process. It is another example of the multistep model of mass communication, this time positioning the IRA as an important node in the spreading network.

In C_2 on African Americans issues, Russian trolls have quite a causal relation on Trump supporters activity and no other causal relation could be found in this community. In C_3, IRA drives activities of Trump and Clinton influencers and Clinton supporters by diffusing depressed social issues.

The fact that Clinton and Sanders have ended up as superspreaders of news generated by the IRA may not be a surprise after all. The present analysis clearly indicates that the IRA trolls did not generate direct conflicts with the left. Instead, they tried to infiltrate the already existing communities of interest to the different parties, like the leftist Black Lives Matter community, or the Trumpian MAGA community with enormous success. Thus, it is not surprising that the presidential candidates got entangled in IRA driven discussions.

What is more worrisome is that nobody has realized the extend of this influence during the actual electoral campaign and the IRA trolls have been left to roam freely to exert their influence across vast swaths of the ideological debate. The present analysis shows that the tools presented in this book can be readily used to identify this kind of meddling with electoral processes and the normal democratic debate. The machinery presented can understand in real-time the development of emergent social media communities such as those deployed by troll farms that could be easily detected in real-time before they become entrenched at the very heart of the political debate.

References

1. Rogers, E.M.: Diffusion of Innovations. Simon and Schuster, New York (2010)
2. Ledford, H.: How Facebook, Twitter and other data troves are revolutionizing social science. Nature **582**, 328–330 (2020)
3. Eric Schmidt at Techonomy, Techonomy Conference in Lake Tahoe, Calif., Youtube video, 4 Aug. 2010. https://www.youtube.com/watch?v=UAcCIsrAq70. Accessed Apr. 2022
4. Zoder, S.: How Much Is Your Data Worth? Forbes, 6 Aug. 2019. https://www.forbes.com/sites/stephanzoder/2019/08/06/how-much-is-your-data-worth/?sh=503ca14970fc. Accessed Jun. 2022
5. Wibson, How Much Is Your Data Worth? At Least $240 per Year. Likely Much More, 19 Jan. 2018. https://medium.com/wibson/how-much-is-your-data-worth-at-least-240-per-year-likely-much-more-984e250c2ffa. Accessed Jun. 2022
6. Goodson, S.: If You're Not Paying For It, You Become The Product, 5 Mar. 2012. Forbes. https://www.forbes.com/sites/marketshare/2012/03/05/if-youre-not-paying-for-it-you-become-the-product/?sh=226970b75d6e. Accessed Apr. 2022
7. Foot, P.: Abortion and the Doctrine of Double Effect. Oxford Review, Number 5 (1967)
8. Awad, E., Dsouza, S., Shariff, A.: Universals and variations in moral decisions made in 42 countries by 70,000 participants. Proc. Nat. Acad. Sci. USA **117**, 2332–2337 (2020)
9. Grind, K.: Tech's first big plan to tackle Covid-19 stumbles: "An App Is Not Going to Fix This". Wall Street J. https://www.wsj.com/articles/techs-first-big-plan-to-tackle-covid-19-stumbles-an-app-is-not-going-to-fix-this-11590757102. Accessed Apr. 2022
10. Lazarsfeld, P., Berelson, B., Gaudet, H.: The People's Choice: How the Voter Makes Up His Mind in a Presidential Campaign. Columbia University Press (1944)
11. Katz, E., Lazarsfeld, P.: Personal Influence: The Part Played by People in the Flow of Mass Communications. Free Press (1955)
12. Berelson, B., Lazarsfeld, P., McPhee, W., Voting, W.: A Study of Opinion Formation in a Presidential Campaign. University of Chicago Press (1954)
13. Katz, E.: The two-step flow of communication: an up-to-date report on an hypothesis. Public Opin. Q. **21**, 61–78 (1957)
14. Burt, R.: The social capital of opinion leaders. Ann. Am. Acad. Pol. Soc. Sci. **566**, 37–54 (1999)
15. Watts, D., Dodds, P.: Influentials, networks, and public opinion formation. J. Consum. Res. **34**, 441–458 (2007)
16. Rogers, E., Cartano, D.: Methods of measuring opinion leadership. Public Opin. Q. **26**, 435–441 (1962)
17. Marsden, P.: Restricted access in networks and models of power. Am. J. Soc. **88**, 686–717 (1983)
18. Alternative facts: Wikipedia, Wikimedia Foundation, 25 Mar. 2022. https://en.wikipedia.org/wiki/Alternative_facts. Accessed Apr. 2022
19. Bennett, W., Manheim, J.: The one-step flow of communication. Ann. Am. Acad. Political Soc. Sci. **608**, 213–232 (2006)
20. Putnam, R.: Bowling Alone: America's declining social capital. J. Democr. **6**, 65–78 (1995)

21. Hayes, D., Lawless, L.: The decline of local news and its effects: new evidence from longitudinal data. J. Polit. **80**, 332–336 (2018)
22. Brown, J., Enos, R.: The measurement of partisan sorting for 180 million voters. Nat. Hum. Behav. **5**, 998–1008 (2021)
23. Lee, B., Bearman, P.: Political isolation in America. Netw. Sci. **8**, 333–355 (2020)
24. Robinson, J.: Interpersonal influence in election campaigns: two step-flow hypotheses. Public Opin. Q. **40**, 304–319 (1976)
25. Weimann, G.: On the importance of marginality: one more step into the two-step flow of communication. Am. Sociol. Rev. **47**, 764–773 (1982)
26. Hilbert, M., Vásquez, J., Halpern, D., Valenzuela, S., Arriagada, E.: One step, two step, network step? Complementary perspectives on communication flows in Twittered citizen protests. Soc. Sci. Comput. Rev. **35**, 444–461 (2017)
27. Karlsen, R.: Followers are opinion leaders: the role of people in the flow of political communication on and beyond social networking sites. Eur. J. Commun. **30**, 301–318 (2015)
28. Lizardo, O.: Unpublished (2022)
29. Serafino, M., Clemente, G.V., Flamino, J., Szymanski, B.K., Lizardo, O., Makse, H.A.: Analysis of flows in social media uncovers a new multi-step model of information spread. J. Stat. (2025)
30. Singh, P., Sreenivasan, S., Szymanski, B., Korniss, G.: Threshold-limited spreading in social networks with multiple initiators. Sci. Rep. **3**, 1–7 (2013)
31. Szymanski, B., Lizardo, O., Doyle, C., Karampourniotis, P., Singh, P., Korniss, G., Bakdash, J.: The spread of opinions in societies. In: Modeling Sociocultural Influences On Decision Making, pp. 91–114 (2016)
32. Karampourniotis, P., Sreenivasan, S., Szymanski, B., Korniss, G.: The impact of heterogeneous thresholds on social contagion with multiple initiators. PloS One **10**, e0143020 (2015)
33. Park, C.: Revisiting the two-step flow model on Twitter: Interconnection of self-identified South Korean Twitter opinion leadership, news consumption, news links, and news curation. Electron. News **13**, 63–77 (2019)
34. Lasswell, H.D.: The structure and function of communication in society. In: Schramm, W., Roberts, D.F. (eds.), The Process and Effects of Mass Communication, pp. 84–99. University of Illinois Press, Urbana (1971). Original published in Bryson L. (ed.), The Communication of Ideas, pp. 37–51. Harper and Row, New York (1948)
35. Wu, S., Hofman, J., Mason, W., Watts, D.: Who says what to whom on Twitter. In: Proceedings of the 20th International Conference on World Wide Web, pp. 705–714 (2011)
36. Weeks, B., Ardèvol-Abreu, A., Zúñiga, H.: Online influence? Social media use, opinion leadership, and political persuasion. Int. J. Publ. Opin. Res. **29**, 214–239 (2017)
37. Winter, S., Neubaum, G., Stieglitz, S., Ross, B.: Opinion leaders: a comparison of self-reported and observable influence of Twitter users. Inf. Commun. Soc. **24**, 1533–1550 (2021)
38. Choi, S.: The two-step flow of communication in Twitter-based public forums. Soc. Sci. Comput. Rev. **33**, 696–711 (2015)
39. Dubois, E., Gaffney, D.: The Multiple facets of influence: Identifying political influentials and opinion leaders on Twitter. Am. Behav. Sci. **58**, 1260–1277 (2014)
40. Turcotte, J., York, C., Irving, J., Scholl, R., Pingree, R.: News recommendations from social media opinion leaders: effects on media trust and information seeking. J. Comput. Mediat. Commun. **20**, 520–535 (2015)
41. Aral, S., Walker, D.: Identifying influential and susceptible members of social networks. Science **337**, 337–341 (2012)
42. Aral, S., Walker, D.: Identifying social influence in networks using randomized experiments. IEEE Intell. Syst. **26**, 91–96 (2011)
43. Aral, S., Dhillon, P.S.: Social influence maximization under empirical influence models. Nat. Hum. Behav. **2**, 375–382 (2018)
44. Min, B., Liljeros, F., Makse, H.A.: Finding influential spreaders from human activity beyond network location. PLoS One **10**, e0136831 (2015)

45. Monge, P.R., Contractor, P.S., Contractor, N.S.: Theories of Communication Networks. Oxford University Press, US (2003)
46. Gallos, L.K., Rybski, D., Liljeros, F., Havlin, S., Makse, H.A.: How people interact in evolving online affiliation networks. Phys. Rev. X **2**, 031014 (2012)
47. Da Silva, R.A.P., Viana, M.P., da Fontoura Costa, L.: Predicting epidemic outbreak from individual features of the spreaders. J. Stat. Mech. 07005 (2012)
48. Kitsak, M., Gallos, L.K., Havlin, S., Liljeros, F., Muchnik, L., Stanley, H.E., Makse, H.A.: Identifying influential spreaders in complex networks. Nat. Phys. **6**, 888–893 (2010)
49. Borge-Holthoefer, J., Moreno, Y.: Absence of influential spreaders in rumor dynamics. Phys. Rev. E **85**, 026116 (2012)
50. Chami, G.F., Kontoleon, A.A., Bulte, E., Fenwick, A., Kabatereine, N.B., Tukahebwa, E.M., Dunne, D.W.: Diffusion of treatment in social networks and mass drug administration. Nat. Commun. **8**, 1929 (2017)
51. Gu, J., Lee, S., Saramäki, J., Holme, P.: Ranking influential spreaders is an ill-defined problem. EPL **118**, 68002 (2017)
52. Pei, S., Muchnik, L., Andrade, J.S., Jr., Zheng, Z., Makse, H.A.: Searching for superspreaders of information in real-world social media. Sci. Rep. **4**, 5547 (2014)
53. Pei, S., Muchnik, L., Tang, S., Zheng, Z., Makse, H.A.: Exploring the complex pattern of information spreading in online blog communities. PLoS One **10**, e0126894 (2015)
54. Teng, X., Pei, S., Morone, F., Makse, H.A.: Collective influence of multiple spreaders evaluated by tracing real information flow in large-scale social networks. Sci. Rep. **6**, 36043 (2016)
55. Morone, F., Makse, H.A.: Influence maximization in complex networks through optimal percolation. Nature **524**, 65–68 (2015)
56. Bovet, A., Morone, F., Makse, H.A.: Validation of Twitter opinion trends with national polling aggregates: Hillary Clinton vs Donald Trump. Sci. Rep. **8**, 8673 (2018)
57. Bovet, A., Makse, H.A.: Influence of fake news in Twitter during the 2016 US presidential election. Nat. Commun. **10**, 7 (2019)
58. Grinberg, N., Joseph, K., Friedland, L., Swire-Thompson, B., Lazer, D.: Fake news on Twitter during the 2016 US presidential election. Science **363**, 374–378 (2019)
59. Osat, S., Faqeeh, A., Radicchi, F.: Optimal percolation on multiplex networks. Nat. Commun. **8**, 1540 (2017)
60. Lee, K.-M., Goh, K.-I.: Strength of weak layers in cascading failures on multiplex networks: case of the international trade network. Sci Rep. **6**, 26346 (2016)
61. Al-Garadi, M.A., Varathan, K.D., Ravana, S.D., Ahmed, E., Chang, V.: Identifying the influential spreaders in multilayer interactions of online social networks. J. Int. Fuzzy Syst. **31**, 2721–2735 (2016)
62. Erkol, S., Faqeeh, A., Radicchi, F.: Identifying influential spreaders in noisy networks. EPL **123**, 58007 (2018)
63. Bohannon, J.: The pulse of the people. Science **355**, 470–472 (2017)
64. Flamino, J., Galeazzi, A., Feldman, S., Macy, M.W., Cross, B., Zhou, Z., Serafino, M., Bovet, A., Makse, H.A., Szymanski, B.K.: Political polarization of news media and influencers on Twitter in the 2016 and 2020 US presidential elections. Nat. Hum. Behav. **7**, 904–916 (2023)
65. Zhou, Z., Serafino, M., Cohan, L., Caldarelli, G., Makse, H.A.: Why polls fail to predict elections. J. Big Data **8**, 137 (2021)
66. Granger, C.W.J.: Investigating causal relations by econometric models and cross-spectral methods. Econometrica **37**, 424 (1969)
67. Jacobs, J., House, B.: Bloomberg.com, Trump Says He Expected to Lose Election Because of Poll Results, 13 Dec 2016. https://www.bloomberg.com/politics/articles/2016-12-14/trump-says-he-expected-to-lose-election-because-of-poll-results. Accessed Jul. 2021
68. Tourangeau, R., Conrad, F.G., Couper, M.P.: The Science of Web Surveys. Oxford University Press, New York (2013)

69. Kennedy, C., et al.: Ad Hoc committee on 2016 election polling, an evaluation of 2016 election polls in the US American Association for Public Opinion Research, AAPOR report (2016). https://www.aapor.org/Education-Resources/Reports/An-Evaluation-of-2016-Election-Polls-in-the-U-S.aspx. Accessed Apr. 2022
70. Prosser, C., Mellon, J.: The twilight of the polls? A review of trends in polling accuracy and the causes of polling misses. Gov. Oppos. **53**, 757–790 (2018)
71. Groves, R.M.: Nonresponse rates and nonresponse bias in household surveys. Publ. Opin. Q. **70**, 646–675 (2006)
72. Coppock, A.: Did shy Trump supporters bias the 2016 polls? Evidence from a nationally-representative list experiment. Stat. Polit. Pol. **8**, 29–40 (2017)
73. Brownback, A., Novotny, A.: Social desirability bias and polling errors in the 2016 presidential election. J. Behav. Exp. Econ. **74**, 38–56 (2018)
74. Durand, C., Blais, A.: Quebec 2018: a failure of the polls? Canadian J. Polit. Sci./Revue Canadienne de Sci. Politique **53**, 133–150 (2020)
75. Bekele, A.: Are faulty opinion polls to blame for Brexit?. Spokane Intercollegiate Research Conference (2018). https://digitalcommons.whitworth.edu/sirc/2018/ie1/6/. Accessed Apr. 2022
76. National Public Radio, Special Series Election 2010. In: Halloran, L. (ed.), Pollsters Need Cell Phone Users More Than Ever, 19 June 2010. https://www.npr.org/templates/story/story.php?storyId=127937110. Accessed Jan. 2022
77. Kennedy, C., Hartig, H.: Response rates in telephone surveys have resumed their decline. Pew Res. Cent. (2019). https://www.pewresearch.org/fact-tank/2019/02/27/response-rates-in-telephone-surveys-have-resumed-their-decline. Accessed Apr. 2022
78. The Economist, A new iPhone feature poses a threat to opinion pollsters, 18 Sep. 2019. https://www.economist.com/united-states/2019/09/26/a-new-iphone-feature-poses-a-threat-to-opinion-pollsters. Accessed Apr. 2022
79. Battaglia, M., Izrael, D., Hoaglin, D., Frankel, M.: Tips and Tricks for Raking Survey Data (A.K.A. Sample Balancing) American Association for Public Opinion Research, JSM, pp. 4740–4745 (2004)
80. Izrael, D., Hoaglin, D., Battaglia, M.P.: A SAS Macro for Balancing a Weighted Sample, Statistics and Data Analysis, Paper 258-25 (2000)
81. National Public Radio, Rural Voters Played A Big Part In Helping Trump Defeat Clinton. In: Kurtzleben, D. (ed.), 14 Nov. 2016. https://www.npr.org/2016/11/14/501737150/rural-voters-played-a-big-part-in-helping-trump-defeat-clinton. Accessed Jan. 2022
82. Leonhardt, D.: 'A Black Eye': why political polling missed the mark. Again. https://www.nytimes.com/2020/11/12/us/politics/election-polls-trump-biden.html. Accessed July 2021
83. Payne, J.G.: The Bradley effect: mediated reality of race and politics in the 2008 US presidential election. Am. Behav. Sci. **54**, 417–435 (2010)
84. Krumpal, I.: Determinants of social desirability bias in sensitive surveys: a literature review. Qual. Quant. **47**, 2025–2047 (2013)
85. Argentina election, Wikipedia, Wikimedia Foundation, 18 Apr. 2022. https://en.wikipedia.org/wiki/2019_Argentine_general_election#Opinion_polls. Accessed July 2021
86. Encuestas de intención de voto para las elecciones presidenciales de Argentina de 2019, Wikipedia, Wikimedia Foundation, 28 Aug. 2021. https://es.wikipedia.org/wiki/Anexo:Encuestas_de_intenci'on_de_voto_para_las_elecciones_presidenciales_de_Argentina_de_2019. Accessed Apr. 2022
87. Clarín, E.P.: Encuestadoras bajo fuego: por qué erraron en las PASO y qué dicen para Octubre 27 Sep. 2019. (in Spanish): https://www.clarin.com/politica/encuestadoras-fuego-erraron-paso-dicen-octubre_0_T72H9hdl.html. Accessed Apr. 2022. English translation at: https://www.kcore-analytics.com/kcore-references. Accessed Apr. 2022
88. Clarín, M.B.: Intrigas en la Casa Rosada y pases de factura en la City por el 'lunes negro', 15 Aug. 2019. (in Spanish): https://www.clarin.com/opinion/intrigas-casa-rosada-pases-factura-city-lunes-negro_0_jnggAIsh5.html. Accessed Apr. 2022. English translation at: https://www.kcore-analytics.com/kcore-references. Accessed Apr. 2022

89. Elypsis Twitter account. https://twitter.com/elypsisweb. Accessed Apr. 2022
90. Clarín, D., Canedo, F.: Inquietud en los mercados Revuelo político por una encuesta que ayudó a subir el riesgo país, 19 Apr. 2019. (in Spanish) https://www.clarin.com/economia/revuelo-encuesta-ayudo-subir-riesgo-pais_0_qsecQJtQc.amp.html. Accessed Apr. 2022. English translation at: https://www.kcore-analytics.com/kcore-references. Accessed Apr. 2021
91. Levy, R.: Hedge Fund Loses $1 Billion in One Month on Argentina Bet. Wall Street J. (2019). https://www.wsj.com/articles/hedge-fund-loses-1-billion-in-one-month-on-argentina-bet-11567696547. Accessed Apr. 2022
92. Financial Times, Autonomy Capital lost 16% in Argentina market rout. https://www.ft.com/content/29764546-c821-11e9-a1f4-3669401ba76f. Accessed Apr. 2022
93. Impulso Baires: Bomba: Inversores demandarían a Elypsis por la encuesta del viernes, y los K cargarían con denuncia penal para investigar a operadores, 15 Aug. 2019. Retrieved from (in Spanish): https://www.impulsobaires.com.ar/nota/275336/bomba-inversores-demandarian-a-elypsis-por-la-encuesta-del-viernes-y-los-k-cargarian-con-denuncia-penal-para-investigar-a-operadores. Accessed Apr. 2022. English translation at: https://www.kcore-analytics.com/kcore-references. Accessed Apr. 2022
94. Zolghadr, M., Niaki, S.A.A., Niaki, S.T.A.: Modeling and forecasting US presidential election using learning algorithms. J. Ind. Eng. Int. **14**, 491–500 (2018)
95. Ravi, K., Ravi, V.: A survey on opinion mining and sentiment analysis: tasks, approaches and applications. Knowl. Based Syst. **89**, 14–46 (2015)
96. Jaidka, K., Ahmed, S., Skoric, M., Hilbert, M.: Predicting elections from social media: a three-country, three-method comparative study. Asian J. Commun. **29**, 252–273 (2019)
97. Jungherr, A.: Twitter use in election campaigns: a systematic literature review. J. Inf. Technol. Polit. **13**, 72–91 (2016)
98. Gayo-Avello, D.: No, you cannot predict elections with Twitter. IEEE Int. Comput. **16**, 91–94 (2012)
99. Gayo-Avello, D.: A meta-analysis of state-of-the-art electoral prediction from Twitter data. Soc. Sci. Comput. Rev. **31**, 649–679 (2013)
100. Beauchamp, N.: Predicting and interpolating state-level polls using Twitter textual data. Am. J. Pol. Sci. **61**, 490–503 (2017)
101. Varol, O., Ferrara, E., Davis, C.A., Menczer, F., Flammini, A.: Online human-bot interactions: detection, estimation, and characterization. In: Proceedings of the 11th International AAAI Conferences Weblogs Social Media, pp. 280–289 (2017)
102. Wu, Y., Fang, Y., Shang, S., Jin, J., Wei, L., Wang, H.: A novel framework for detecting social bots with deep neural networks and active learning. Knowl. Based Syst. **211**, 106525 (2021)
103. Manning, C., Schutze, H.: Foundations of Statistical Natural Language Processing. MIT Press (1999)
104. Li, D., Liu, Y.: Deep Learning in Natural Language Processing. Springer (2018)
105. Tumasjan, A., Sprenger, T.O., Sander, P.G., Welpe, I.M.: Predicting elections with twitter: what 140 characters reveal about political sentiment. In: Fourth international AAAI conference on weblogs and social media (2010)
106. Jungherr, A., Jürgens, P., Schoen, H.: Why the pirate party won the German election of 2009 or the trouble with predictions: a response to Tumasjan, A., Sprenger, T.O., Sander, P.G., Welpe, I.M.: Predicting elections with twitter: what 140 characters reveal about political sentiment. Soc. Sci. Comput. Rev. **30**, 329–234 (2012)
107. Gaurav, M., Srivastava, A., Kumar, A., Miller, S.: Leveraging candidate popularity on Twitter to predict election outcome. In: Proceedings of the 7th workshop on social network mining and analysis '13, 11 Aug. 2013, pp. 7–16 (2013)
108. Lui, C., Metaxas, P.T., Mustafaraj, E.: On the predictability of the US elections through search volume activity (Report no. 23) (2011). http://repository.wellesley.edu/scholarship/23/
109. Bermingham, A., Smeaton, A.F.: On using Twitter to monitor political sentiment and predict election results. In: Sentiment Analysis where AI Meets Psychology (SAAIP), pp. 2–10 (2011)
110. Ceron, A., Curini, L., Iacus, S.M., Porro, G.: Every tweet counts? How sentiment analysis of social media can improve our knowledge of citizens' political preferences with an application to Italy and France. New Media Soc. **6**, 340–358 (2014)

111. Ceron, A., D'Adda, G.: Enlightening the voters: the effectiveness of alternative electoral strategies in the 2013 Italian election monitored through (sentiment) analysis of Twitter posts. In: Proceedings of the ECPR Conference, pp. 4–7. ECPR, Bordeaux (2013)
112. González-Bailón, S., Banchs, R.E., Kaltenbrunner, A.: Emotions, public opinion, and US presidential approval rates: a 5-year analysis of online political discussions. Hum. Commun. Res. **38**, 121–143 (2012)
113. Monti, C., Rozza, A., Zappella, G., Zignani, M., Arvidsson, A., Colleoni, E.: Modelling political disaffection from Twitter data. In: Proceedings of the Second International Workshop on Issues of Sentiment Discovery and Opinion Mining, pp. 3–11. ACM, Chicago, IL (2013)
114. Ceron, A., Curini, L., Iacus, S.M.: Using sentiment analysis to monitor electoral campaigns: method matters-evidence from the United States and Italy. Soc. Sci. Comput. Rev. **33**, 3–20 (2015)
115. Sylwester, K., Purver, M.: Twitter language use reflects psychological differences between democrats and republicans. PLoS One **10**, 1–18 (2015)
116. Ceron, A., Curini, L., Iacus, S.M.: ISA: a fast, scalable and accurate algorithm for sentiment analysis of social media content. Inf. Sci. **367–368**, 105–124 (2016)
117. Ceron, A., Curini, L., Iacus, S.: Politics and Big Data: Nowcasting and Forecasting Elections with Social Media. Routledge, Taylor and Francis (2017)
118. Caldarelli, G., Chessa, A., Pammolli, F., Pompa, G., Puliga, M., Riccaboni, M., Riotta, G.: A multi-level geographical study of Italian political elections from Twitter data. PloS one **9**, e95809 (2014)
119. Singh, P., Sawhney, R., Kahlon, K.: Forecasting the 2016 US presidential elections using sentiment analysis. In: 16th Conference on e-Business, e-Services and e-Society (I3e), Nov 2017, Delhi, India, pp. 412–423 (2017)
120. Singh, P., Dwivedi, Y.K., Kahlon, K.S., Pathania, A., Sawhney, R.S.: Can twitter analytics predict election outcome? An insight from 2017 Punjab assembly elections. Gov. Inf. Q. **37**, 101444 (2020)
121. Newman, M.: Networks: An Introduction. Oxford University Press (2010)
122. Cuzzocrea, A., Papadimitriou, A., Katsaros, D., Manolopoulos, Y.: Edge betweenness centrality: a novel algorithm for QoS-based topology control over wireless sensor networks. J. Netw. Comput. Appl. **35**, 1210–1217 (2012)
123. Hopkins, D.J., King, G.: A method of automated nonparametric content analysis for social science. Am. J. Pol. Sci. **54**, 229–247 (2010)
124. Bode, L., Dalrymple, K.: Politics in 140 characters or less: campaign communication, network interaction, and political participation on Twitter. J. Polit. Mark. **15**, 311–332 (2016)
125. Raghavan, U.N., Albert, R., Kumara, S.: Near linear time algorithm to detect community structures in large-scale networks. Phys. Rev. E **76**, 1–11 (2007)
126. Blondel, V.D., Guillaume, J.-L., Lambiotte, R., Lefebvre, E.: Fast unfolding of communities in large networks. J. Stat. Mech. 10008 (2008)
127. Brito, K.D.S., Filho, R.L.C.S., Adeodato, P.J.L.: A systematic review of predicting elections based on social media data: research challenges and future directions. In: IEEE Transactions on Computational Social Systems, vol. 8, no. 4, pp. 819–843 (2021)
128. Devlin, J., Chang, M.-W., Lee, K., Toutanova, K.: BERT: pre-training of deep bidirectional transformers for language understanding (2018). arXiv:1810.04805
129. Chandra, R., Saini, R.: Biden versus Trump: modeling US general elections using BERT language model. IEEE Access **9**, 128494–128505 (2021)
130. Gaikar, D., Sapare, G., Vishwakarma, A., Parkar, A.: Twitter sentimental analysis for predicting election result using LSTM neural network. IRJET **6**, 3665–3670 (2019)
131. Vijayan, R., Mohler, G.: Forecasting retweet count during elections using graph convolution neural networks. In: 2018 IEEE 5th International Conference on Data Science and Advanced Analytics (DSAA), pp. 256–262 (2018)
132. Li, M., Perrier, E., Xu, C.: Deep hierarchical graph convolution for election prediction from geospatial census data. Proc. AAAI Conf. Artif. Intell. **33**, 647–654 (2019)

133. The Upshot at The New York Times, New York Times National Polling Average (2016). http://www.nytimes.com/interactive/2016/us/elections/polls.html. Accessed Jan. 2022
134. Morstatter, F., Pfeffer, J., Liu, H., Carley, K.M.: Is the sample good enough? Comparing data from Twitter's Streaming API with Twitter's Firehose. In: Proceedings of the Seventh International AAAI Conference on Weblogs and Social Media, ICWSM, Boston, MA (2013)
135. González-Bailón, S., Wang, N., Rivero, A., Borge-Holthoefer, J., Moreno, Y.: Assessing the bias in communication networks sampled from Twitter. Soc. Netw. **38**, 16–27 (2014)
136. Barberá, P., Rivero, G.: Understanding the political representativeness of Twitter users. Soc. Sci. Comput. Rev. **33**, 712–729 (2015)
137. Mellon, J., Prosser, C.: Twitter and Facebook are not representative of the general population: political attitudes and demographics of British social media users. Res. Polit. **4**(3) (2017)
138. Brooke, A., Anderson, M.: Pew Research Center, Social Media Use in 2021, 7 Apr. 2021. https://www.pewresearch.org/internet/2021/04/07/social-media-use-in-2021. Accessed Jan. 2022
139. Pew Research Center, National Politics on Twitter: Small Share of U.S. Adults Produce Majority of Tweets, 23 Oct. 2019. https://www.pewresearch.org/politics/2019/10/23/national-politics-on-twitter-small-share-of-u-s-adults-produce-majority-of-tweets. Accessed Jan. 2022
140. Wojcik, S., Hughes, A.: Pew Research Center, Sizing Up Twitter Users, 24 Apr. 2019. https://www.pewinternet.org/2019/04/24/sizing-up-twitter-users/. Accessed Jan. 2022
141. 2016 United States presidential election, Wikipedia, Wikimedia Foundation, 3 June 2022. https://en.wikipedia.org/wiki/2016_United_States_presidential_election. Accessed Jun. 2022
142. The New York Times, In the land of self-defeat, 4 Oct. 2019. https://www.nytimes.com/2019/10/04/opinion/sunday/trump-arkansas.html?action=click&module=Opinion&pgtype=Homepage. Accessed Jan. 2022
143. Chang, L., Kosnick, J.A.: National surveys via RDD telephone interviewing versus the Internet: comparing sample representativeness and response quality. Public Opin. Q. **73**, 641–678 (2009)
144. Battaglia, M.P., Frankel, M.R., Link, M.W.: Improving standard poststratification techniques for random-digit-dialing telephone surveys. Sur. Res. Methods **2**, 11–19 (2008)
145. Link, M.W., Battaglia, M.P., Frankel, M.R., Osborn, L., Mokdad, A.H.: A comparison of address-based sampling (ABS) versus random-digit dialing (RDD) for general population surveys. Public Opin. Q. **72**, 6–27 (2008)
146. Rural-Urban Continuum Codes, Economic Research Service, U.S. Department of Agriculture, Dec 10, 2020. https://www.ers.usda.gov/data-products/rural-urban-continuum-codes.aspx#.UYJuVEpZRvY. Accessed Apr. 2022
147. Makse, H.A., Havlin, S., Stanley, H.E.: Modelling urban growth patterns. Nature **377**, 608–612 (1995)
148. Rozenfeld, H.D., Rybski, D., Andrade, J.S., Jr., Batty, M., Stanley, H.E., Makse, H.A.: Laws of population growth. Proc. Nat. Acad. Sci. USA **105**, 18702–18707 (2008)
149. Rozenfeld, H.D., Rybski, D., Gabaix, X., Makse, H.A.: The area and population of cities: new insights from a different perspective on cities. Am. Econ. Rev. **101**, 560–580 (2011)
150. Owens, B.: Social-media platforms failing to tackle abuse of scientists. Nature News **601**, 28 (2022). Jan
151. Avaaz report, Scientists under Attack: how health experts became targets on Facebook and Co, 19 Jan. 2022. https://secure.avaaz.org/campaign/en/scientists_under_attack. Accessed Apr. 2022
152. Tucker Carlson Tonight, Youtube video, Lord Fauci Exposed: Hundred of emails published, 2 June 2021. https://www.facebook.com/TuckerCarlsonTonight/videos/lord-fauci-exposed-hundreds-of-emails-published/831537877742034. Accessed Jan. 2022
153. Teoh, F.: Health Feedback, Fauci's emails don't support viral claims on masks, hydroxychloroquine and virus engineering, 1 June 2021. https://healthfeedback.org/claimreview/faucis-emails-dont-support-viral-claims-on-masks-hydroxychloroquine-and-virus-engineering. Accessed Jan. 2022

154. Spencer, S.H., Robertson, L., Fichera, A.: Viral Posts, Pundits Distort Fauci Emails, 4 June 2021. https://www.factcheck.org/2021/06/scicheck-viral-posts-pundits-distort-fauci-emails. Accessed Jan. 2022
155. Pizzagate conspiracy theory, Wikipedia, Wikimedia Foundation, 28 Jan. 2022. https://en.wikipedia.org/wiki/Pizzagate_conspiracy_theory. Accessed Apr. 2022
156. LaCapria, K.: Snopes. Is Comet Ping Pong Pizzeria Home to a Child Abuse Ring Led by Hillary Clinton? 21 Nov. 2016. https://www.snopes.com/fact-check/pizzagate-conspiracy. Accessed Jan. 2022
157. QAnon, Wikipedia, Wikimedia Foundation, 29 Mar. 2022. https://en.wikipedia.org/wiki/QAnon. Accessed Jan. 2022
158. PolitiFact, Fact-checking website operated by Tampa Bay Times (2007–2018), Poynter Institute (2018-present). https://www.politifact.com. Accessed Apr. 2022
159. PolitiFact, Infowars is a conspiracy oriented website run by Alex Jones. https://www.politifact.com/personalities/infowars. Accessed Apr. 2022
160. Alex Jones, Wikipedia, Wikimedia Foundation, 20 Apr. 2022. https://en.wikipedia.org/wiki/Alex_Jones. Accessed Apr. 2022
161. Garcia, A.: Far Right Blogs, Conspiracy Theorists Attack Parkland Mass Shooting Survivor, Snopes, 21 Feb. 2018. https://www.snopes.com/news/2018/02/20/right-wing-media-david-hogg. Accessed Apr. 2022
162. Sharma, K., Qian, F., Jiang, H., Ruchansky, N., Zhang, M., Liu, Y.: Combating fake news: a survey on identification and mitigation techniques. ACM Trans. Intell. Syst. Technol. **10**, 4 (2019)
163. Ferrara, E., Varol, O., Davis, C., Menczer, F., Flammini, A.: The rise of social bots. Commun. ACM **59**, 96–104 (2016)
164. Turing, A.M.: Computing machinery and intelligence. Mind **49**, 433–460 (1950)
165. Bradeško, L., Mladenić, D.: A Survey of Chabot Systems through a Loebner Prize Competition, S2CID 39745939 (2012)
166. Cresci, S., Pietro, R.D., Petrocchi, M., Spognardi, A., Tesconi, M.: The paradigm-shift of social spambots: Evidence, theories, and tools for the arms race. In: Proceedings of the 26th International Conference on World Wide Web Companion, Perth, Australia, 3–7 Apr. 2017, pp. 963–972 (2017)
167. Yang, K.-C., Varol, O., Davis, C.A., Ferrara, E., Flammini, A., Menczer, F.: Arming the public with artificial intelligence to counter social bots. Hum. Behav. Emerg. Technol. **1**, 48–61 (2019)
168. Bondielli, A., Marcelloni, F.: A survey on fake news and Rumour detection techniques. Inf. Sci. **497**, 38–55 (2019)
169. Conroy, N.J., Rubin, V.L., Chen, Y.: Automatic deception detection: Methods for finding fake news. Proc. Annu. Meeting Assoc. Inf. Sci. Technol. 1–4 (2015)
170. Shu, K., Sliva, A., Wang, S., Tang, J., Liu, H.: Fake news detection on social media: a data mining perspective. ACM SIGKDD Explor. Newslett **19**, 22–36 (2017)
171. Volkova, S., Shaffer, K., Yea, J.J., Hodas, N.: Separating facts from fiction: Linguistic models to classify suspicious and trusted news posts on Twitter. Proc. 55th Annu. Meeting Assoc. Comput. Linguistics, pp. 647–653 (2017)
172. Cambria, E., Poria, S., Gelbukh, A., Thelwall, M.: Sentiment analysis is a big suitcase. IEEE Intell. Syst. **32**(6), 74–80 (2017)
173. M. Heidari, J. H. Jr Jones, O. Uzuner, An empirical study of machine learning algorithms for social media bot detection. In 2021 IEEE International IOT, Electronics and Mechatronics Conference (IEMTRONICS). IEEE (2021)
174. Mangal, D., Sharma, D.K.: Fake news detection with integration of embedded text cues and image features. In 8th International Conference on Reliability, Infocom Technologies and Optimization (Trends and Future Directions) (ICRITO), pp. 68–72 (2020)
175. Reis, J.C.S., Correia, A., Murai, F., Veloso, A., Benevenuto, F.: Supervised learning for fake news detection. In: IEEE Intelligent Systems, vol. 34, no. 2, pp. 76–81, Mar.–Apr. 2019

176. Mazzeo, V., Rapisarda, A., Giuffrida, G.: Detection of fake news on CoViD-19 on Web Search Engines. Front. Phys. **9**, 685730 (2021)
177. Daffa, W., Bamasag, O., AlMansour, A.: A survey on spam URLs detection in Twitter. In: 1st International Conference on Computer Applications and Information Security (ICCAIS), pp. 1–6, Apr. 2018
178. Vosoughi, S., Roy, D., Aral, S.: The spread of true and false news online. Science **359**, 1146–1151 (2018)
179. Kolomeets, M., Chechulin, A., Kotenko, I.: Bot detection by friends graph in social networks. JoWUA **12**, 141–159 (2021)
180. Ma, J., Gao, W., Wei, Z., Lu, Y., Wong, K.-F.: Detect rumors using time series of social context information on microblogging websites. In: CIKM'15: Proceedings of the 24th ACM International on Conference on Information and Knowledge Management, Oct. 2015, pp. 1751–1754
181. Kwon, S., Cha, M., Jung, K., Chen, W., Wang, Y.: Prominent features of rumor propagation in online social media. In: IEEE ICDM'13, pp. 1103–1108 (2013)
182. Manzoor, S.I., Nikita, J.S.: Fake news detection using machine learning approaches: a systematic review. In: 2019 3rd International Conference on Trends in Electronics and Informatics (ICOEI), pp. 230–234 (2019)
183. V. Sharma, How I built a Twitter Bot using Python and Selenium? Towards Data Science, May 31, 2020, https://towardsdatascience.com/how-i-built-a-twitter-bot-using-python-and-selenium-c036bfff6af8 [Accessed Aug 2022]
184. Botometer: An OSoMe project, https://botometer.osome.iu.edu [Accessed Apr 2022]
185. J. Knauth, Language-agnostic Twitter-bot detection. In Proc. Natural Lang. Process. Deep Learn. World, Oct. 2019, p. 550-558
186. Bot Repository Datasets, https://botometer.osome.iu.edu/bot-repository/datasets.html [Accessed Apr 2022]
187. Haidermota, M., Pansare, A., Mitra, D.: Classifying Twitter user as a bot or not and comparing different classification algorithms. Int. J. Adv. Res. Comput. Sci. **9**, 29 (2018)
188. B. Erşahin, Ö. Aktaş, D. Kılınç, C. Akyol, Twitter fake account detection, in Proc. Int. Conf. Comput. Sci. Eng. (UBMK), Oct. 2017, p. 388-392
189. A. H. Wang, Detecting spam bots in online social networking sites: A machine learning approach. In Proc. IFIP Annu. Conf. Data Appl. Secur. Privacy. Berlin, Germany: Springer, 2010, p. 335-342
190. Alarifi, A., Alsaleh, M., Al-Salman, A.: Twitter Turing test: identifying social machines. Inf. Sci. **372**, 332–346 (2016)
191. Alharthi, R., Alhothali, A., Moria, K.: Detecting and characterizing Arab spammers campaigns in Twitter. Procedia Comput. Sci. **163**, 248–256 (2019)
192. Kudugunta, S., Ferrara, E.: Deep neural networks for bot detection. Inf. Sci. **467**, 312–322 (2018)
193. Wei, F., Nguyen, U.T.: Twitter bot detection using bidirectional long short-term memory neural networks and word embeddings. In: Proceedings of the 1st IEEE International Conference on Trust, Privacy and Security in Intelligent Systems and Applications (TPS-ISA), pp. 101–109, Dec. 2019
194. Farber, M., Qurdina, A., Ahmedi, L.: Identifying Twitter bots using a convolutional neural network. In Proc, CLEF Working Notes (2019)
195. Derhab, A., Alawwad, R., Dehwah, K., Tariq, N., Khan, F.A., Al-Muhtadi, J.: Tweet-based bot detection using big data analytics. IEEE Access **9**, 65988–66005 (2021)
196. Zhang, H., Fan, Z., Zheng, J.-H., Liu, Q.: An improving deception detection method in computer-mediated communication. J. Netw. **7**, 1811–1816 (2012)
197. Pérez-Rosas, V., Mihalcea, R.: Experiments in open domain deception detection. In: Proceedings of the Conference on Empirical Methods in Natural Language Processing (EMNLP), Association for Computational Linguistics, pp. 1120–1125 (2015)
198. Qin, Y., Wurzer, D., Lavrenko, V., Tang, C.: Spotting rumors via novelty detection. CoRR (2016)

199. Yang, F., Liu, Y., Yu, X., Yang, M.: Automatic detection of rumor on Sina Weibo. In: Proceedings of the ACM SIGKDD Workshop on Mining Data Semantics. ACM, p. 13 (2012)
200. Chua, A.Y.K., Banerjee, S.: Linguistic predictors of rumor veracity on the internet. In: Proceedings of the International Multi Conference of Engineers and Computer Scientists (IMECS), pp. 387–391 (2016)
201. Silverman, C.: Lies, damn lies, and viral content. How news websites spread (and debunk) online rumors, unverified claims, and misinformation. Tow Center for Digit. J. **168** (2015)
202. Wang, W.Y.: 'Liar, liar pants on fire': a new benchmark dataset for fake news detection. In: Proceedings of the Fifty Fifth Annual Meeting of the Association for Computational Linguistics (Volume 2: Short Papers). Association for Computational Linguistics, pp. 422–426 (2017)
203. Song, C., Tu, C., Yang, C., Liu, Z., Sun, M.: CED: credible early detection of social media rumors. IEEE Trans. Knowl. Data Eng. **33**, 3035–3047 (2021)
204. Ruchansky, S., Seo, Y., Liu, Y.: CSI: a hybrid deep model for fake news detection. In: Proceedings of the ACM on Conference on Information and Knowledge Management, pp. 797–806. ACM (2017)
205. Cho, K., van Merrienboer, B., Gulcehre, C., Bahdanau, D., Bougares, F., Schwenk, H., Bengio, Y.: Learning phrase representations using RNN encoder-decoder for statistical machine translation. In: Proceedings of the Conference on Empirical Methods in Natural Language Processing (EMNLP). Association for Computational Linguistics, pp. 1724–1734 (2014)
206. Chen, Y.-C., Liu, Z.-Y., Kao, H.-Y.: Ikm at semeval-2017 task 8: convolutional neural networks for stance detection and rumor verification. In: Proceedings of the Eleventh International Workshop on Semantic Evaluation (SemEval-2017), pp. 465–469 (2017)
207. Ma, J., Gao, W., Mitra, P., Kwon, S., Jansen, B.J., Wong, K.-F., Cha, M.: Detecting rumors from microblogs with recurrent neural networks. In: Proceedings of the Twenty-Fifth International Joint Conference on Artificial Intelligence, IJCAI, address, pp. 3818–3824 (2016)
208. Ciampaglia, G.L., Shiralkar, P., Rocha, L.M., Bollen, J., Menczer, F., Flammini, A.: Computational fact checking from knowledge networks. PloS One **10**, e0128193 (2015)
209. Koloski, B., Perdih, T.S., Robnik-Šikonja, M., Pollak, S., Škrlj, B.: Knowledge graph informed fake news classification via heterogeneous representation ensembles. Neurocomputing. In Press (2022)
210. Hoy, N., Koulouri, T.: A systematic review on the detection of fake news articles (2021). arXiv:2110.11240
211. Kumar, K.P.K., Geethakumari, G.: Detecting misinformation in online social networks using cognitive psychology. Human-centric Comput. Inf. Sci. **4**, 14 (2014)
212. Papers with Code, Meta repository. https://paperswithcode.com/task/fake-news-detection. Accessed Apr. 2022
213. Fake News Detection at Simon Fraser University. http://fakenews.research.sfu.ca. Accessed Apr. 2022
214. Zhang, J., Dong, B., Yu, P.S.: FakeDetector: effective fake news detection with deep diffusive neural network. In: 2020 IEEE 36th International Conference on Data Engineering (ICDE), 20–24 Apr. 2020, pp. 1826–1829 (2020)
215. Rashkin, H., Choi, E., Jang, J., Volkova, S., Choi, Y.: Truth of varying shades: analyzing language in fake news and political fact-checking. In: Proceedings of the 2017 Conference on Empirical Methods in Natural Language Processing, pp. 2931–2937 (2017)
216. Shu, K., Wang, S., Liu, H.: Exploiting tri-relationship for fake news detection (2017). arXiv:1712.07709
217. Perozzi, B., Al-Rfou, R., Skiena, S.: Deepwalk: online learning of social representations. In: KDD '14: Proceedings of the 20th ACM SIGKDD International Conference on Knowledge Discovery and Data Mining, pp. 701–710 (2014)
218. Tang, J., Qu, M., Wang, M., Zhang, M., Yan, J., Mei, Q.: Line: large-scale information network embedding. In: WWW '15: Proceedings of the 24th International Conference on World Wide Web, pp. 1067–1077 (2015)

219. Hu, Q., Xie, S., Zhang, J., Zhu, Q., Guo, S., Yu, P.: HeteroSales: utilizing heterogeneous social networks to identify the next enterprise customer. In: WWW '16: Proceedings of the 25th International Conference on World Wide Web, Apr. 2016 pp. 41–50 (2016)
220. Mikolov, T., Karafiat, M., Burget, L., Cernocky, J., Khudanpur, S.: Recurrent neural network based language model. In: INTERSPEECH (2010)
221. Chang, C.-C., Lin, C.-J.: LIBSVM : a library for support vector machines. ACM Trans. Intell. Syst. Technol. **2**, 1–27 (2011). http://www.csie.ntu.edu.tw/~cjlin/libsvm
222. Li, X., Xia, Y., Long, X., Li, Z., Li, S.: Exploring text-transformers in AAAI 2021 shared task: Covid-19 fake news detection in English. In: International Workshop on Combating On line Hostile Posts in Regional Languages during Emergency Situation, pp. 106–115. Springer (2021)
223. Bhattarai, B., Granmo, O.-C., Jiao, L.: Explainable Tsetlin Machine Framework for Fake News Detection with Credibility Score Assessment (2021). arXiv:2105.09114
224. Yoon, S., Park, K., Shin, J., Lim, H., Won, S., Cha, M., Jung, K.: Detecting incongruity between news headline and body text via a deep hierarchical encoder. Proc. AAAI Conf. Artif. Intell. **33**, 791–800 (2019)
225. Karimi, H., Tang, J.: Learning hierarchical discourse-level structure for fake news detection (2019). arXiv:1903.07389
226. Shang, J., Shen, J., Sun, T., Liu, X., Gruenheid, A., Korn, F., Lelkes, A.D., Yu, C., Han, J.: Investigating rumor news using agreement-aware search. In: Proceedings of the 27th ACM International Conference on Information and Knowledge Management, pp. 2117–2125 (2018)
227. Li, J., Monroe, W., Shi, T., Jean, S., Ritter, A., Jurafsky, D.: Adversarial learning for neural dialogue generation. In: Proceedings of the 2017 Conference on Empirical Methods in Natural Language Processing, Copenhagen, Denmark, pp. 2157–2169 (2017)
228. Haidar, Md.A., Rezagholizadeh, M.: Textkd-gan: text generation using knowledge distillation and generative adversarial networks. In: Advances in Artificial Intelligence, pp. 107–118. Springer International Publishing (2019)
229. Mueller, R.S., III: Report On The Investigation Into Russian Interference in the 2016 Presidential Election, U.S. Department of Justice, Mar. 2019. https://www.justice.gov/archives/sco/file/1373816/download. Accessed Apr. 2022
230. The Facebook files, A Wall Street Journal investigation. https://www.wsj.com/articles/the-facebook-files-11631713039. Accessed Apr. 2022
231. Metaxas, P., Mustafaraj, E., Wong, K., Zeng, L., O'Keefe, M., Finn, S.: What do retweets indicate? Results from user survey and meta-review of research. In: Proceedings of the International AAAI Conference on Web Social Media, vol. 9 (2015)
232. Kempe, D., Kleinberg, J., Tardos, E.: Maximizing the spread of influence through a social network. In: Proceedings of the 9th ACM SIGKDD International Conference on Knowledge Discovery and Data Mining, pp. 137–143 (2003)
233. Zimdars, M.: Washington Post, My 'fake news list' went viral. But made-up stories are only part of the problem, 18 Nov. 2016. https://www.washingtonpost.com/posteverything/wp/2016/11/18/my-fake-news-list-went-viral-but-made-up-stories-are-only-part-of-the-problem. Accessed Jan. 2022
234. Zimdars, M.: False, Misleading, Clickbait-y, and/or Satirical 'News' Sources. https://docs.google.com/document/d/10eA5-mCZLSS4MQY5QGb5ewC3VAL6pLkT53V_81ZyitM/preview. Accessed Jan. 2022
235. Sandhu, S: I, Dr Melissa Zimdars removes fake news list and claims she was 'harassed', 19 Nov. 2016. https://inews.co.uk/news/technology/melissa-zimdars-removes-fake-news-list-claiming-harassed-doxed-32234. Accessed Jan. 2022
236. Cinelli, M., Quattrociocchi, W., Galeazzi, A., Valensise, C.M., Brugnoli, E., Schmidt, A.L., Zola, P., Zollo, F., Scala, A.: The COVID-19 social media infodemic. Sci. Rep. **10**, 16598 (2020)
237. Main, T.J.: The Rise of the Alt-Right. Brookings Institution Press (2018)
238. Stefanov, P., Darwish, K., Atanasov, A., Nakov, P.: Predicting the topical stance and political leaning of media using tweets. In: Proceedings of the 58th Annual Meeting of the Association for Computational Linguistics, pp. 527–537 (2020)

239. Bozarth, L., Saraf, A., Budak, C.: Higher ground? How ground truth labeling impacts our understanding of fake news about the 2016 US presidential nominees. Proc. Int. AAAI Conf. Web Soc. Media **14**, 48–59 (2020)
240. Bakshy, E., Messing, S., Adamic, L.A.: Exposure to ideologically diverse news and opinion on Facebook. Science **348**, 1130–1132 (2015)
241. Mould, T.: Introduction to the special issue on fake news: definitions and approaches. J. Am. Folklore **131**(522) (Fall 2018), 371–378. (University of Illinois Press)
242. Krugman, P.: The Economic Fallout, The New York Times, 9 Nov. 2016. https://www.nytimes.com/interactive/projects/cp/opinion/election-night-2016/paul-krugman-the-economic-fallout. Accessed Jan. 2022
243. Williamson, E., Hsu, T., Levenson, M.: Jury Orders Alex Jones to Pay $45.2 Million in Sandy Hook Case, The New York times, 5 Aug 2022. https://www.nytimes.com/2022/08/05/us/politics/alex-jones-verdict.html. Accessed Aug. 2022
244. Swaim, B.: How Disagreement Became 'Disinformation'. Wall Street J. (2022). https://www.wsj.com/articles/how-disagreement-became-disinformation-misinformation-nina-jankowicz-governance-board-czar-11652450632?st=gzk94hwgls7qrrz&reflink=article_email_share. Accessed May 2022
245. Blake, A.: Analysis: Trump backers' alarming reliance on hoax and conspiracy theory websites, in 1 chart. Washington Post, 22 Aug. 2017. https://www.washingtonpost.com/news/the-fix/wp/2017/08/22/trump-backers-disturbing-reliance-on-hoax-and-conspiracy-theory-websites-in-1-chart/. Accessed Apr. 2022
246. Faris, R., Roberts, H., Etling, B.: Partisanship, Propaganda, and Disinformation: Online Media and the 2016 U.S. Presidential Election. Berkman Center for Internet and Society, p. 72, 8 Aug. 2017
247. Grynbaum, M.M., Herrman, J.: Breitbart Rises From Outlier to Potent Voice in Campaign, The New York Times, 26 Aug. 2016. https://www.nytimes.com/2016/08/27/business/media/breitbart-news-presidential-race.html. Accessed Apr. 2022
248. Twitter, Permanent suspension of`@realDonaldTrump`, 8 Jan 2021. https://blog.twitter.com/en_us/topics/company/2020/suspension. Accessed June 2022
249. MacKinnon, J.G.: Approximate asymptotic distribution functions for unit-root and cointegration tests. J. Bus. Econ. Stat. **12**, 167–176 (1994)
250. Benjamin, D.J., et al.: Redefine statistical significance. Nat. Hum. Behav. **2**, 6–10 (2018)
251. Barberá, P.: Birds of the same feather tweet together: bayesian ideal point estimation using Twitter data. Polit. Anal. **23**, 76–91 (2015)
252. Barberá, P., Jost, J.T., Nagler, J., Tucker, J.A., Bonneau, R.: Tweeting from left to right. Psychol. Sci. **26**, 1531–1542 (2015)
253. Benzécri, J.-P., et al.: L'analyse des données, vol. 2. Dunod Paris (1973)
254. Nenadic, O., Greenacre, M.: Correspondence analysis in R, with two- and three-dimensional graphics: the CA package. J. Stat. Softw. **20**, 1–13 (2007)
255. Maechler, M.: Package 'diptest', R Package Version 0.75-5. R: a language and environment for statistical computing. R Foundation for Statistical Computing, Vienna, Austria (2013)
256. Michalski, R., Serwata, D., Nurek, M., Szymanski, B., Kazienko, P., Jia, T.: Temporal network epistemology: on reaching consensus in real world setting (2021). arXiv:2203.05368
257. Abdelzaher, T., Ji, H., Li, J., Yang, C., Dellaverson, J., Zhang, L., Xu, C., Szymanski, B.K.: The paradox of information access: growing isolation in the age of sharing (2020). arXiv:2004.01967
258. Xu, C., Li, J., Sun, D., Wang, R., Abdelzaher, T., Graham, J., Szymanski, B.: On polarization dynamics in the age of information overload, workshop. In: Proceedings of the 15th International AAAI Conference on Web and Social Media, 7 June 2021
259. Lu, X., Gao, J., Szymanski, B.K.: The evolution of polarization in the legislative branch of government. J. Roy. Soc. Int. **16**, 20190010 (2019)
260. Internet Research Agency, Wikipedia, Wikimedia Foundation, 14 Apr. 2022. https://en.wikipedia.org/wiki/Internet_Research_Agency. Accessed Apr. 2022

261. Howard, P., Ganesh, B., Liotsiou, D., Kelly, J., François, C.: The IRA, Social Media and Political Polarization in the United States, 2012–2018. University of Oxford (2018)
262. Badawy, A., Ferrara, E., Lerman, K.: Analyzing the digital traces of political manipulation: the 2016 Russian interference Twitter campaign. In: 2018 IEEE/ACM International Conference On Advances In Social Networks Analysis And Mining (ASONAM), pp. 258–265 (2018)
263. Stewart, L., Arif, A., Starbird, K.: Examining trolls and polarization with a retweet network. In: Proceedings of the ACM WSDM, Workshop on Misinformation and Misbehavior Mining on the Web (2018)
264. Zannettou, S., Caulfield, T., De Cristofaro, E., Sirivianos, M., Stringhini, G., Blackburn, J.: Disinformation warfare: Understanding state-sponsored trolls on Twitter and their influence on the web. In: Companion Proceedings of the 2019 World Wide Web Conference, pp. 218–226 (2019)
265. DiResta, R., Shaffer, K., Ruppel, B., Sullivan, D., Matney, R., Fox, R., Albright, J., Johnson, B.: The tactics and tropes of the internet research agency. New Knowledge (2018). https://disinformationreport.blob.core.windows.net/disinformation-report/NewKnowledge-Disinformation-Report-Whitepaper.pdf. Accessed Apr. 2022
266. Serafino, M., Zhou, Z., Bovet, A., Andrade, J.S., Jr., Makse, H.A.: Suspended accounts align with the Internet Research Agency misinformation campaign to influence the 2016 US election. Eur. Phys. J. Data Sci. **13**, 29 (2024)
267. De Meo, P., Ferrara, E., Fiumara, G., Provetti, A.: Generalized Louvain method for community detection in large networks. In: 2011 the 11th International Conference on Intelligent Systems Design and Applications, pp. 88–93. IEEE (2011)

Chapter 3
Influencers for Marketing

Abstract In this chapter, we show how the machinery developed in Chap. 1 on influencer identification can work in a social media marketing campaign. The network analysis also allows us to infer economic status through a measure of the location and influence in the social network. We show how this information can be used to increase the response rates in marketing campaigns by threefold.

3.1 Organic Influencers Versus Celebrity Influencers

In today's digital era, it is becoming ever more popular to market consumers on social media through influencers. Influencer marketing involves the endorsement and product placement from the influencer to the consumer through social media channels. In this kind of marketing, influencers are either common people with knowledge about the product or celebrity influencers with a large social following in the field of the product or elsewhere [1].

Instagram is characterized by the presence of content creators with a vast number of followers: from nano-influencers (<10 k) to macro-influencers (> 1 million) and celebrities. The Instagram influencer tiers are defined in terms of number of followers:

- Nano-influencers: 1,000–10,000 followers.
- Micro-influencers: 10,000–50,000 followers.
- Mid-tier influencers: 50,000–500,000 followers.
- Macro-influencers: 500,000–1,000,000 followers.
- Mega-influencers: 1,000,000+ followers.

Table 3.1 shows the list of the top 5 most followed accounts on Instagram as of January 2022, according to Wikipedia [2]. The most-followed individual is Portuguese footballer Cristiano Ronaldo with almost 400 million followers. Argentinian footballer Lionel Messi appears in fourth place with 300 million followers.

Despite this heterogeneity in the number of followers, the users with the biggest audience (e.g., celebrities or mega-influencers in Table 3.1) are taking almost all the money. According to a report on The Walt Street Journal [3], the top 1% of Twitch paid streamer influencers made over half of all revenue paid out in 2021 ($889

H. A. Makse and M. Zava, *The Science of Influencers and Superspreaders*,
Understanding Complex Systems, https://doi.org/10.1007/978-3-031-78058-5_3

Table 3.1 Instagram influencers. List of top 5 accounts with the largest number of followers in Instagram as of January 2022. Adapted from Wikipedia [2]

Rank	Username	Owner	Followers (millions)	Profession	Country
1	@instagram	Instagram	465	Social media	US
2	@cristiano	Cristiano Ronaldo	396	Footballer	Portugal
3	@kyliejenner	Kylie Jenner	306	TV personality	US
4	@leomessi	Lionel Messi	304	Footballer	Argentina
5	@selenagomez	Selena Gomez	293	Musician	US

million total). The same situation may apply to influencers in Instagram and TikTok. While these influencers take nearly all the action, consumers are more likely to trust *organic influencers*, who are regular people genuinely sharing their experience with the products they love, e.g., nano and micro-influencers, and not the celebrities mega-influencers. Ironically, creators of authentic contents are more difficult to be reached than other influencers because, in some cases, they are not even monetizing their activity on Instagram.

Recent studies show that people are more likely to purchase after seeing a peer's post rather than that of a mega-influencer (e.g., celebrity). In particular, Instagram users are more willing to identify themself with influencers who present themselves as ordinary people. In this context, recognizing these organic influencers is essential to establishing a trustful relationship with potential customers.

The methods to identify influencers explained in Chap. 1 can be readily used to find these authentic content creators on Instagram or other platforms that can influence their followers in an optimal way, thus performing marketing campaigns with minimal costs. This can be done for a specific product or, more in general, for a brand. Publicly available contents on Instagram can be used to detect consumers that are talking about a given product that they are using or consuming everyday. The methodologies can be used to both have a coverage of the active users discussing a given brand and product, and in addition, to spot genuine advocates for product or brands.

The method can be used to find the influencers from the communities that are organically generated. These influencers are people interested in a given product, and are not chosen by their number of followers, which is a static quantity that may not correlate with the exerted influence on the particular community. Again, 'connections are not everything' and brands could be better served finding nano and micro-influencers from the genuine discussions of their online communities.

Using big data collected from social media and machine learning, organic influencers can be found for brands and products. These influencers are dynamic; they constantly change according to the trends in social media. Billions of reactions, posts, and comments from social media platforms (Instagram, Facebook, TikTok, Twitter) can be collected on a particular topic. Then, NLP algorithms can be applied to classify the issues of interest, and machine learning can be used to classify social media users' positive and negative sentiments. Machine learning can detect consumers/

influencers talking positively about the products they are using or consuming daily and find genuine advocates for products and brands.

Finally, the network of users for a given brand, product and topic can be reconstructed from the interactions among users (posting, likes, retweets, etc.) to find the most suitable influencers tailored to a particular brand/product. These influencers are organic and genuine and enlarge the database of self-reported influencers used by brands and agencies. These methods can be used to find the missing influencers from self-reported influencer databases (from nano to mega), and thus tap into the 99% of the influencer market that is not being explored at the moment.

3.1.1 *Influencer's Study Case*

Next, we show a test case for the detection of Instagram influencers based on a particular topic/brand and geographical position/spoken language. We use as a test case Instagram with geographical location in Switzerland with a sample of users talking about Samsung during September 2021.

Instagram has lot of publicly available data that can be collected and processed. As a practical example of that we consider the Samsung study case. By tapping on the Instagram search bar #Samsung, the recent posts containing that hashtag can be obtained. Such posts contain lot of textual information that can be collected. A systematic collection can be done for other Samsung related hashtags, e.g., #samsungphotgraphy.

During September 2021, 25,000 posts were posted with 1M reactions from 15,000 users talking about Samsung products in Instagram. Then NLP can be used to classify the information provided by the users and then to geo-localize the content's creators, discover the product they are using, and more importantly, whether they create favorable content about Samsung's products or they express negative views about the brand.

The next step is to find the IG users who created the most viral and meaningful Samsung content. Machine learning can characterize the users by preferences, interests, and sentiment towards the brand or product (positive/negative/neutral). Table 3.2 shows a list of top influencers who are creating contents about Samsung's products with a significant following and engagement rates. The total number of

Table 3.2 Data analysis. List of top 5 IG influencers for Samsung according to their followers

User ID	Type	Followers	Following	Followers/Following
297506964	Micro	31879	2042	15.6
7166883433	Micro	22024	268	82.2
23548046	Micro	12886	1157	11.1
4510376375	Micro	12576	942	13.3
11708703018	Nano	9658	1187	8.1

influencers with >1000 followers that appear in one month of collecting data is around 100. These influencers are dynamic and change all the time.

Advanced analytics are helpful to select the best contents' creators by collecting and analysing the contents of the influencers and followers published during a marketing campaign. Then advanced key performance indicators (KPI) can be used to study the performance of the influencer in the campaign to find the best organic influencer. For example, we may look for an influencer having a number of posts about Samsung that is high and with high engagement over an extended period of time.

3.2 Finding the Influencers from the Network of Interactions

The above analysis provides a simple detection and assessment of influencers. However, it misses the network analysis that is important to situate the influencer in its proper sphere of influence. We further elaborate on a method that utilizes the network of interactions of influencers and followers to design better marketing campaign at the minimal cost by targeting the right influencers for any given brand/product.

By constructing the network of users talking about a given brand/product, the methods of Chap. 1 can be applied to improve the marketing campaign by identifying authentic influencers. The collected billions of data points from publicly available contents on Instagram and other platforms can be used to detect consumer/influencers who are talking positively about the products they are using or consuming every day.

Data-driven marketing can capture the direct and indirect effects of marketing campaigns on Instagram and other platforms. The theory of influencers from Chap. 1 can identify the leading influencers in particular topics (e.g., arts) and further evaluate the conversion rate between a user watching a story on Instagram and buying a product on an e-commerce site. Using machine learning to analyze Instagram data (e.g., comments on a post), and it is possible to provide critical insights into the sentiment of potential clients of a given brand and the competition on specific topics/products.

A network-driven influencer marketing campaign consists of the following three stages:

- **Stage 1**: Identification of leading influencers from network analysis interactions, top publications, and user sentiment.
- **Stage 2**: Evaluation of the direct/indirect traffic generated by influencers' stories and posts.
- **Stage 3**: Tracking the traffic due to the brand's Instagram page, assessing the conversion rate between users interacting with the Instagram page and a visit/purchase on the brand's e-commerce website.

Below, we elaborate on how the influencer theory can be applied to each of the stages mentioned above, the data we need to access, and how we use it to extract valuable information. The flow of information in the marketing campaign is depicted in Fig. 3.1 for a typical brand.

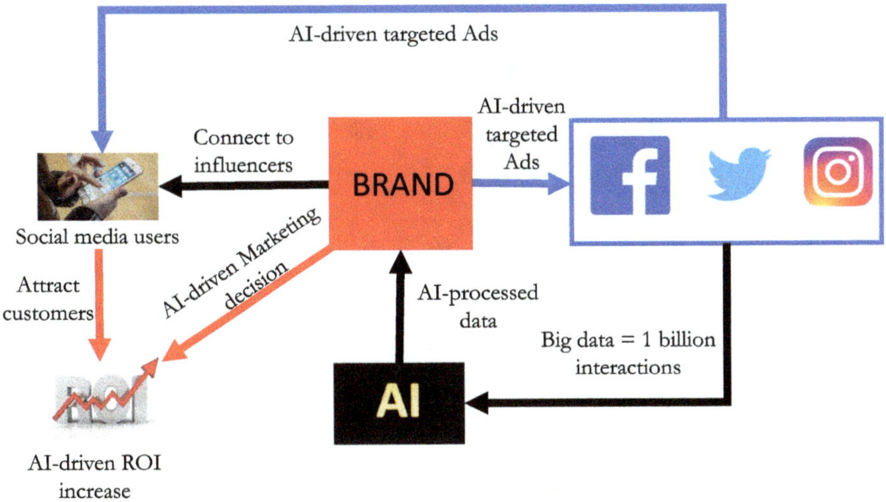

Fig. 3.1 **Pipeline for an influencer IG campaign**. From influencer identification to increase ROI

3.2.1 Influencers Detection from the Network

The influencer network uncovers the relevance of users on a specific topic. An influencer network is an easy-to-read representation of the interaction between users (e.g., posts or comments) about a specific topic (represented by one or more hashtags) or with a specific Instagram page.

An example of an influencer network can be seen in Fig. 3.2, which shows a study of a brand's Instagram page. The network is constructed by first collecting public posts in Instagram according to the topic of interest. For instance, here the topic of interest revolves around a given brand and the type of product. In the present case study, the brand sells art products, so we choose #arts, etc. The page of the brand is also followed to then build the ecosystems associated with the given brand and product. This forms the social network that can be then used to find the relevant influencers.

The CI algorithm can be used to capture weak links in this network who connect communities of followers of the brand as shown in Fig. 3.2. One of the communities in the network is the echo-chamber of the brand itself. This is expected since the collection of data is done using the hashtags associated with the company. The central node is the brand's account in Instagram, and it has a set of users directed connected with the central node; they are the followers. However, they are not the best influencers. Neither is the brand itself, since the brand will influence only its echo-chamber. That is, being a central hub in the network is not translated directed into the top influencer.

The network analysis then discovers another community. This community is dominated by an influencer as shown in Fig. 3.2 and Table 3.3 showing the top 5 influencers in the brand's network identified by CI.

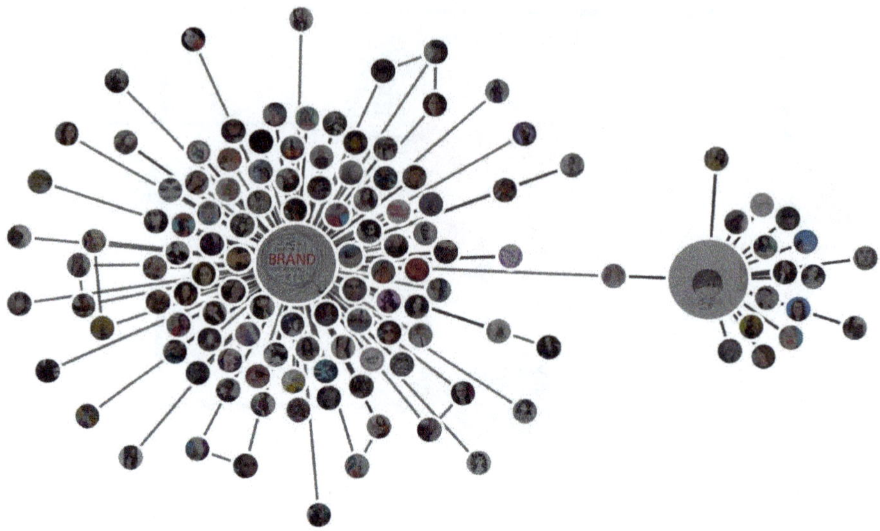

Fig. 3.2 **Finding influencers**. The brand network and its organic influencers

Table 3.3 CI values of top 5 influencers of the brand obtained from the network of interactions

Username	CI	Followers	Publications	Following	Age	Gender
edgar_is_drawing	22977	1036	91	1979	Null	Null
papeleria	22972	60063	1760	659	null	Null
ammy_ximemex	656	274	108	318	19	Female
rosek_91	656	367	140	850	36	Male
pamee_mapa	652	968	82	292	Null	Null

This methodology identifies typical organic influencers. They are not directly associated to the brand, but are using the products of the brand and organically advertising then via the hashtags and comments on the brand's products. However, that is not all. If we look at the network again, we clearly see that there is a 'weak-link' influencer who is connecting the two communities. The CI algorithm uncovers this weak link: a user that is weakly connected, since it has degree two, but it is strategically located bridging the gap between the two large communities. This is the quintessential Granovetter 'weak link' and its influence is a vivid example of the 'strength of weak ties'. This influencer can be readily obtained when we run the CI algorithms for $\ell = 2$, since the two communities are connected by a shortest path of length $\ell = 2$. This analysis also highlights why one should try different algorithms or different parameters to identify different aspects of the influencers.

3.2.2 *Sentiment Analysis Towards the Brand*

Recognizing the most successful posts in terms of interactions (e.g., comments or likes) and average sentiment of users' comments is another feature to understand what is thriving in a marketing campaign and the critical elements of its success (e.g., talking about specific products).

As a solution to the need for having feedback about a social media marketing campaign and position the brand with respect to the competition, machine learning can extract the sentiment of people reacting to posts about the brand's products/offers with the methods explained in Sect. 2.7. This information includes top posts, positive/negative comments, and words used by users of not only the brand under study but also the competition.

A influencer marketing campaign should have scientific metrics behind to assess its level of success. When brands do campaigns on social platforms like Instagram, direct links to the brand's e-commerce website are among the key sources of traffic and potential customers. Instagram also authorizes direct links in the specific user's biography field and stories.

Google Analytics data and a proper setting of the links used in a social campaign, the number of users coming from each influencer involved in a marketing campaign can be directly tracked in order to develop metrics of success of the campaign. This is the key performance indicator that evaluates the success of an activity in which the company engages, a particular case of KPI is the conversion rate.

IG allows to add tags and links that will enable the posts of the influencers to be tracked to measure their engagement and effectiveness in brand recognition via the conversion rate. Moreover, further KPIs are given by searching for correlations between posts about the brand (e.g., tagging the brand's Instagram page) published by an influencer and clicks on the links in stories.

3.3 Socioeconomic Influencers

Big data analysis can also be used to infer an individual's economic status for marketing purposes. Apart from helping marketers create optimized campaigns, it could help decision makers maximize the effects of large-scale economic stimulus policies. Back in 2017, Luo et al. [4] was granted access to two massively large data sets: a social network comprised by mobile phone calls and text messaging data as well as residential communication in Mexico for 122 days along with financial banking data from credit card transactions.

They ask the following question: Does a person's role in the national mobile phone network reveal anything about their economic status? The group discovered that the optimality of an individual's location in the network, which is measured by the collective influence CI metric, is highly correlated with the individual's economic status at the population level: the larger the CI, the higher the socioeconomic level.

The wealthiest 1-percenters have higher diversity in mobile contacts and are centrally located, surrounded by other highly connected people (network hubs). On the other hand, the poorest individuals have low contact diversity and are weakly connected to fewer hubs. This analysis was used to carry out a marketing campaign that showed a threefold increase in response rate by targeting individuals identified by their social network metrics as compared to random targeting.

Both in marketing as well as policy analysis, algorithms to infer economic analysis can be used. Yet, for more comprehensive analysis, it is ideal to include the identity-based 'consumer behavior' [5]. Identity can be defined as any category label with which a consumer self-associates that is amenable to a clear picture of what a person looks like, thinks, feels and does [6]. Building on this definition, networks are useful to define the identity of a person and therefore infer tastes, thoughts, feelings as well as the economic well-being of individuals to identify what a normal person is likely to purchase. For a more robust analysis, marketers also include social networks such as Facebook, Twitter or Instagram to run an analysis that could potentially help them identify and target the individuals for which a product/campaign was made.

3.3.1 Targeted Advertisement Campaign Using Influencer Theory

The relation between individuals' economic status and their position in social networks has long been a concern of studies in socioeconomics. It is commonly believed that patterns of social ties affect individuals' economic status: colloquially speaking, it is all about networking. Broad social connections usually lead to more economic benefits including job promotions, higher salary and increased power in negotiations [7–9]. In order to explore the specific correlation between social network structure and economic well-being, comprehensive datasets of both social ties and economic development are required. In reality, this requirement becomes a major obstacle in performing such research.

Progress made by Eagle et al. [10] shows that such correlation is revealed by exploiting the aggregated data of mobile communication network and the economic development at the community level. In this study, the communication network is constructed using records of phone calls during August 2005 in the UK, which consists of 65×10^6 nodes and 368×10^6 reciprocated social ties, covering more than 90% of the total mobile phone user population. The economic development was measured by UK government's Index of Multiple Deprivation (IMD), which is a composite measure of relative prosperity of 32,482 communities encompassing the entire country. Based on these two datasets, it was found that the economic prosperity is strongly correlated ($r = 0.73$) with the social network diversity measured by the Shannon entropy:

$$H(i) = -\sum_{j=1}^{k} p_{ij} \log p_{ij}. \tag{3.1}$$

Here, k is the degree of the ith community and p_{ij} is the proportion that j is included in i's total call volume: $p_{ij} = V_{ij} / \sum_{h=1}^{k} V_{ih}$.

Compared with other metrics such as degree ($r = 0.44$) and the volume of total communication ($r = -0.33$), the network diversity is a better indicator of economic prosperity at the community level. This strong correlation still holds when another metric of network diversity of 'structural holes' is adopted ($r = 0.72$). Unlike the clustering coefficient, the structure holes consider the missing link between any two of a node's neighbors [11]. Using the principal component analysis, a composite of Shannon entropy and structural holes performs even better ($r = 0.78$).

To further investigate such correlation at the individual level, an individual-level communication network in Mexico is exploited in Luo et al. [4], combined with data of credit limits of the individuals obtained from banking institutions. They analyzed two large-scale sources - the telecommunications and financial data of the whole country's population of México. Results showed that an individual's location in the network, measured as the optimal collective influence to the structural integrity of the social network, is highly correlated with personal economic status.

Constructed from mobile and residential phone communications over a period of 122 days, the communication network contains 110 million users connected with 246 million ties. A subset of 500,000 people contained financial credit card information. All data, including phone calls and bank clients, was anonymous with no names attached. As a synthetical consideration of financial status including income, debt and credit history, the credit limits of these users are extracted from an encrypted bank database and identified by the clients' phone numbers registered in the bank. The result shows that the richest and poorest people exhibit entirely different patterns of communication.

The social network patterns of influence observed mimicked the patterns of economic inequality in the Mexican society. As is illustrated in Fig. 3.3, individuals at

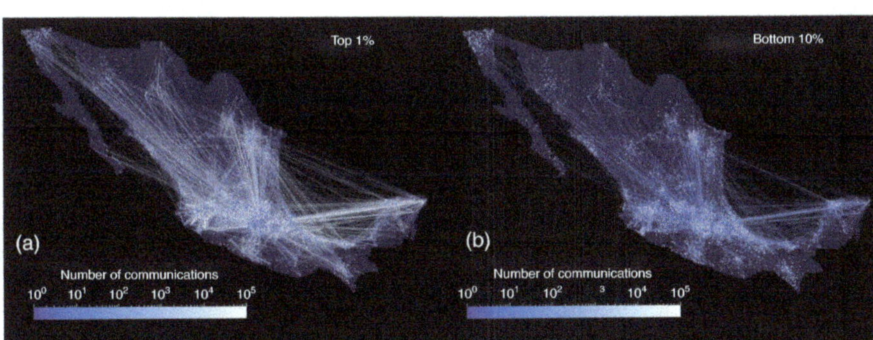

Fig. 3.3 Communication patterns for populations with different economic status. Comparison of communication activity of the population in the top 1% **(a)**, and bottom 10% **(b)** of total credit limit classes. Nodes are communities of bank clients who have registered zip codes. Node size indicates the number of bank clients inside each community, and the color and thickness of edges reflect the number of communication events (phone calls) between different communities. Figure reproduced from [4]. Copyright © 2017, The Authors

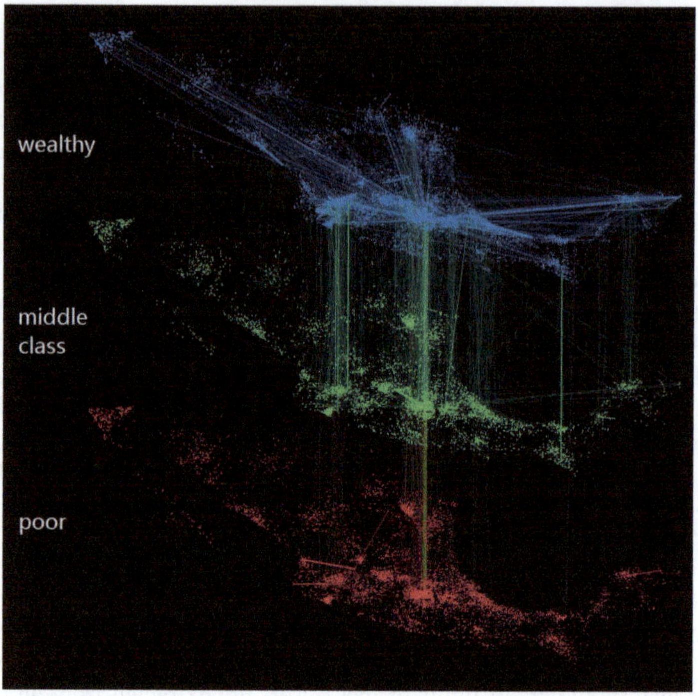

Fig. 3.4 Mobile phone network. The income inequality in México is reflected in the different communication patterns between economic classes, from poor, middle class, to wealthy as defined by the banking information of 500,000 people embedded in the mobile phone communication network of 110 million people at the country level as studied in [4]

the top 1% largest credit limit in their credit cards exhibit far more diversity in social connections than the ones at the bottom 10% credit limit. In general, the wealthiest 1-percenters show higher diversity in mobile contacts, while the poorest individuals tend to make more local communications. The patterns of communication in the mobile phone networks of three layers of economic status, poor, middle class and wealthy, are shown in Fig. 3.4. It is visually clear that the poorer classes connect only within their immediate geographic locations and do not directly interact with the middle classes and much less with the wealthy classes. On the contrary, the higher the economic status the longer the geographical ties but only with members of the same class. This result shows that income inequality has a correlate in segregated communication patterns between the economic classes.

To identify the affluent individuals according to their location in the social network, several metrics that are scalable to large-scale networks are examined, including degree centrality, PageRank, k-core index and collective influence (see Chap. 1). Taken age as an independent factor that affects the financial wellness, the k-core index ($R^2 = 0.96$) and CI ($R^2 = 0.93$) capture the strongest correlation between financial status and network measures. However, CI provides a better resolution of the economic heterogeneity since its value spans over several orders of magnitude.

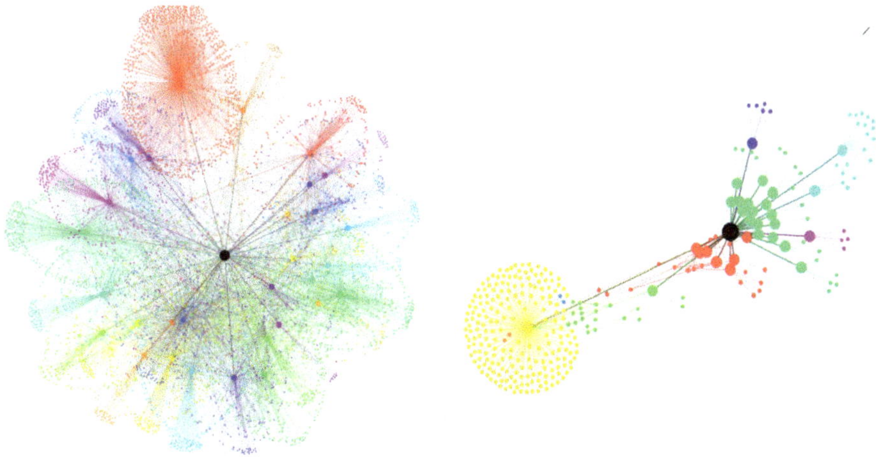

Fig. 3.5 Patterns of network influence by CI mimic patterns of income inequality. Left: Example of the ego-centric network (extended to two layers) for an individual (black node) in the top 1% wealthy class. Right: Ego-centric network of an individual in the bottom 10% class. The networks show two distinct patterns of social ties according to high and low economic status: the former is characterized by large CI, the latter by low CI. Figure reproduced from [4]. Copyright © 2017, The Authors

A top CI node is surrounded by hubs that are hierarchically placed at different distances, which is exactly the pattern of connections shown by a typical wealthy individual in Fig. 3.5, left. A node with low CI has a lower concentration of surrounding hubs, and this is the pattern of connection for a typical poor individual as depicted in the example of Fig. 3.5, right. Furthermore, the performance of CI is investigated independently for people in different age groups. It was found that CI is a more reliable metric for older people, whose economic levels are more mature and stable. By combining CI and age into a composite index with $\alpha = 0.5$:

$$\text{ANC} = \alpha\text{Age} + (1 - \alpha)\text{CI}, \tag{3.2}$$

a strong correlation can be achieved at $R^2 = 0.99$.

For pragmatic use and validation, Luo et al. [4] carried out a marketing campaign that showed a three-fold increase in response rate by targeting individuals identified by their social network metrics as compared to random targeting. This strategy can also be useful in maximizing the effects of large-scale economic stimulus policies.

The strategy consists on performing a social marketing campaign whose objective is the acquisition of new credit card clients, by sending messages to affluent individuals (as identified by their CI values) and inviting the recipients to initiate a product request. The marketing campaign was conducted by Grandata Inc, https://grandata.com and was reported in [4].

In the marketing campaign, clients were approached by SMS messages offering a benefit. The text sent during the campaign did not provide a specific product. Instead, the only information provided was to notify the client that he/she was eligible for an offer from the bank. This somehow eliminated the bias caused by the nature of a product which may have a different appeal to wealthy or poor people. The marketing campaign sent the following messages:

- Request your credit card with benefits from (Bank_name) by calling at (Bank_phone_number). Fees and requirements at (Bank_url).
- (Bank_name) has a special offer for you. If you're interested call at (Bank_phone_number). Fees and requirements at (Bank_url).
- (Bank_name) has a credit card fit for you. Request it by calling at (Bank_phone_number). Fees and requirements at (Bank_url).
- (Bank_name) has a credit card with benefits. Request it at (Bank_phone_number). Fees and requirements at (Bank_url).
- (Bank_name) offers you a credit card with benefits. Request it by calling at (Bank_phone_number). Fees and requirements at (Bank_url).
- (Bank_name) has an exclusive offer for you, call at (Bank_phone_number). Fees and requirements at (Bank_url).

The CI values extracted from the communication network were used to classify the targeted people. Specifically, the communication network resulting from the calls and SMS exchanged between users over a period of 91 days was used from December 2015 to February 2016. The resulting social network contained 7.19 $\times 10^7$ people and 3.51 $\times 10^8$ links. The campaign was conducted on a total of 656,944 people who were targeted by the above SMS messages offering the product according to their CI values in the social network. Messages were also sent to a control group of 48,000 people, chosen randomly.

The success of the campaign is evaluate by the response rate, i.e., the number of recipients who requested the product divided by the number of targeted people. In the control group, the response rate to the messages was 0.331%. The results show that groups of increasing CI show an increase in their response rate, with a sound three-fold gain in the rate of response of the top influencers (as identified by top CI values) when compared to the random case, reaching a remarkable 1% response rate for high CI. The results of the experiment are summarized in Table 3.4 and in Fig. 3.6.

The 3-fold improvement in the response rate obtained when using network-based targeting advertisement in comparison with random targeting highlights the possibility of predicting both financial status and benefits of socially-targeted policies based on network metrics, leading to tangible improvements in social marketing campaigns. The high performance of CI among network metrics also suggests the possible role of accessing and mediating information in financial opportunity and wellbeing [12]. This has an immediate impact in designing optimal marketing campaigns by identifying the responsive targets based on their influential position in the social network. This finding may be also raised to the level of a principle, which would explain the emergence of the phenomenon of collective influence itself as the result of the local optimization of socio-economic interactions.

Table 3.4 A marketing campaign driven by influencer identification. Results of the real-life marketing campaign conducted by Grandata Inc using targeting advertisement guided by CI. Individuals ("Count") were targeted according to their quantile CI ranking in the social network obtained from phone communications activity. The response to the campaign ("Answered Yes") was computed to calculate the Response Rate. Table reproduced from [4]. Copyright © 2017, The Authors

CI range	Count	Quantile	Answered yes	Response rate (%)
[0,48]	66495	0.1	170	0.26
(48, 246]	65164	0.2	218	0.33
(246, 600]	65961	0.3	316	0.48
(600, 1144]	65376	0.4	332	0.51
(1144, 1992]	65477	0.5	363	0.55
(1992, 3408]	65477	0.6	458	0.70
(3408, 6032]	65736	0.7	493	0.75
(6032, 11772]	65641	0.8	555	0.8
(11772, 28740]	65683	0.9	657	1.0
(28740, 2719354]	65683	1.0	573	0.87

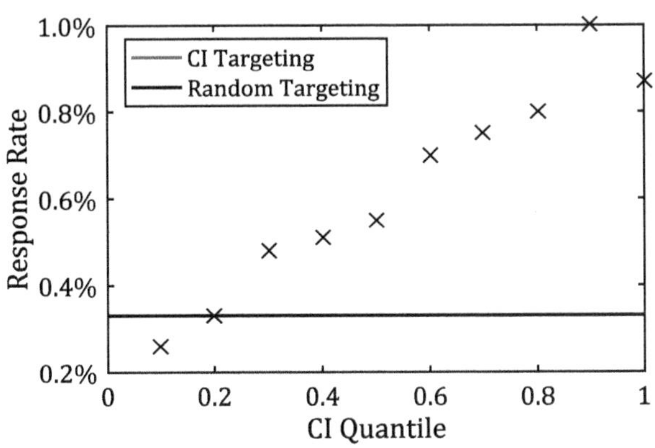

Fig. 3.6 Response rate versus CI quantile in the real-life CI-targeted marketing campaign. The response rate increases approximately linearly with CI ranking. The CI-targeted campaign shows a three-fold gain for the top influencers with high CI, as compared with a campaign targeting a randomized control group, demonstrating the use of network-based methods of influencers in marketing. Figure reproduced from [4]. Copyright © 2017, The Authors

References

1. Influencer marketing, Wikipedia, Wikimedia Foundation, 14 Dec. 2021, https://en.wikipedia.org/wiki/Influencer_marketing. Accessed Apr. 2022
2. List of most-followed Instagram accounts, Wikipedia, Wikimedia Foundation, 29 Jan. 2022. https://en.wikipedia.org/wiki/List_of_most-followed_Instagram_accounts. Accessed Apr. 2022

3. Bentley, E., Kim, K.: Twitch streamer earnings increase for top gamers, data from hack shows. Wall Street J. https://www.wsj.com/articles/twitch-streamer-earnings-increase-for-top-gamers-data-from-hack-shows-11633802185. Accessed Jan. 2022
4. Luo, S., Morone, F., Sarraute, C., Travizano, M., Makse, H.A.: Inferring personal economic status from social network location. Nat. Commun. **8**, 15227 (2017)
5. Golebiewski, J.: Evolution of the theory of marketing: a micro- and macro-approach. Ann. Mark. Manage. Econ. **1**, 13–21 (2015)
6. Reed, A., Forehand, M.R., Puntoni, S., Warlop, L.: Identity-based consumer behavior. Int. J. Res. Mark. **29**, 310–321 (2012)
7. Granovetter, M.S.: The strength of weak ties. Am. J. Soc. **78**, 1360–1380 (1973)
8. Seibert, S.E., Kraimer, M.L., Liden, R.C.: A social capital theory of career success. Acad. Manag. J. **44**, 219–237 (2001)
9. Brass, D.J., Burkhardt, M.E.: Potential power and power use: an investigation of structure and behavior. Acad. Manag. J. **36**, 441–470 (1993)
10. Eagle, N., Macy, M., Claxton, R.: Network diversity and economic development. Science **328**, 1029–1031 (2010)
11. Burt, R.S.: Structural Holes: The Social Structure of Competition. Harvard University Press, Cambridge, MA (2009)
12. Granovetter, M.: The impact of social structure on economic outcomes. J. Econ. Perspect. **19**, 33–50 (2005)

Chapter 4
Influencers as Superspreaders of Disease

Abstract On December 2019 the COVID-19 pandemic stroke the world from a cluster of initial patients in Wuhan, China. In the absence of vaccines or treatments, governments worldwide implemented lockdowns to halt the spread of the pandemic, triggering large social and economic disruptions. Unfortunately, the massive lockdowns did not stop the spread of the disease. In this chapter we elaborate on the possibility to implement intelligent quarantines. Intelligent quarantines are isolation strategies that use available digital technologies—contact tracing, AI and influencer network theory—to identify and isolate superspreaders of disease. Intelligent quarantines leave all the work to intelligent machines rather than humans, who have had a hard time implementing successful strategies against the disease. Influencer theory shows that intelligent targeted quarantines are possible to stop efficiently the spread of the disease, in conjunction with global quarantines. They key is to identify the set of superspreaders using the algorithms of Chap. 1.

4.1 How Digital Contact Tracing Can Stop the Spreading of COVID-19

Two years after the initial COVID-19 outbreak and with nearly 500 million cases worldwide, striking differences are observed in how the coronavirus spread across different countries. For instance, China flattened the curve in 3 months and counts 100,000 total cases as of January 2022. On the other side of the spectrum, in the United States, with less than a quarter of the population of China, cases have reached 75 million and still increasing after two years into the pandemic.

The reasons behind these striking differences in the extend of COVID-19 spreading in the two most powerful economies in the world could be many, and may never be scientifically determined. However, there is general consensus that the 'Zero Covid' policy applied in China has served the country remarkably well in containing the spread of the disease [1]. Countries like New Zealand and Australia also embraced at a given point zero-Covid policies with similar results. These countries have gone to extraordinary—sometimes draconian, measures to prevent spreading.

H. A. Makse and M. Zava, *The Science of Influencers and Superspreaders*,
Understanding Complex Systems, https://doi.org/10.1007/978-3-031-78058-5_4

It is still a question how sustainable these policies are in the long run, in light of the appearance of highly transmissible new variants like Omicron and the winding down of people's tolerance after two years of pandemic [2, 3]. It is not the purpose of this chapter to discern whether these policies are acceptable, sustainable in democratic societies or not. In this chapter we want to decipher whether it is possible to use current digital technologies to optimally dismantle the transmission chain of the disease, and, at the same time, propose intelligent quarantines that minimize disruptions to the population.

The Zero Covid policy implemented in China refers to 'Find, Test, Trace, Isolate and Support'. It relies on stringent border controls, mass testing, massive concentrated lockdowns and extensive digital contact tracing supported by mitigation software to extinguish new outbreaks [1]. The kind of software developed for COVID-19 mitigation includes mobile apps for contact tracing, digital passports for vaccination status, and software to improve social distancing and lockdowns.

Out of all these measures, the discussion in this chapter only pertains to a scientific assessment of the efficacy of digital contact tracing strategies and optimized targeted isolation strategies to stop the spread of disease. As we will see below, this is a network theoretical problem whose solution can be achieved by the influencer theoretical tools developed in Chap. 1, adapted to treat epidemic outbreaks.

Vaccines may take months or years to reach all the population or at least the herd immunity threshold, specially in countries like USA where the anti-vaxxer movement is very strong. In the meantime, contact tracing could result crucial to contain the spread of COVID-19. Why? To answer this question we need to look at the epidemiological profile of the disease.

The way to understand an epidemiological profile is through modelling the disease transmission with the type of models from Sect. 1.6.5.1. These models are based on a series of events: susceptible (S), exposure (E), infectious (I), and recovered (R) [4–6]. A typical model, for instance, is the SIR model, where the series of disease transmission events starts at being susceptible (S), then turns into exposed and infectious at the same time (I), to then turn to recovered (R), when the individual reach immunity or die. This is not a good model for COVID-19 since there are many other events that can occur in between those events. For instance, the events of exposure and being infectious occur at different times. One first is exposed to the virus, gets infected, but not immediately become infectious, thus, splitting the (I) event into two: exposed (E) and infectious (I).

We then arrive to the SEIR model of infectious disease transmission which takes into account the susceptible state (S), the exposure (E), the onset of infectiousness (I), and finally the transition to recovery (R). One may argue that, still, this is a very simple model of COVID-19 transmission, since many other states can occur between (S) and (R). For instance, the subject may enter into isolation after the onset of symptoms, or into quarantine after exposure, or the subject may be even asymptomatic, yet infectious. And after a few months after recovery, the individual can get infected again. Thus, many other states or event can be considered for a more realistic model of COVID-19 transmission. However, for simplicity, we analyze

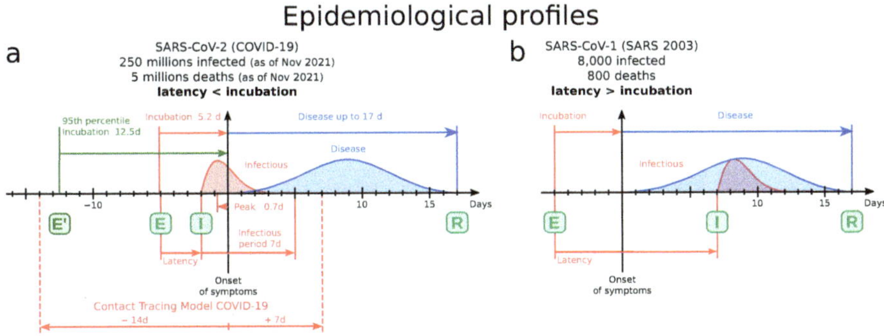

Fig. 4.1 Infectiousness profiles of SARS-CoV-1 and SARS-CoV-2. a Infectiousness profile of coronavirus SARS-CoV-2 responsible for COVID-19 obtained during the early days of the pandemic in [7]. The numbers correspond to the early variant of the virus and may not correspond to the latest variants like Delta and Omicron. The COVID-19 pandemic is modeled by a SEIR model. From exposure **E** the virus is incubated in average for 5.2 days (12.5 days 95th percentile), starting the symptoms two days after infectiousness **I** and lasting the disease up to 17 days to recover (R). According to this number, a window of −14/+7 days from the first symptoms is used to detect infectious and exposures. **b** Infectiousness profile of coronavirus SARS-CoV-1 responsible for SARS-2003. Data obtained from [7]. As opposed to COVID-19, we note that in this case the latency is longer than the incubation period, and the peak of infectiousness then appears after the onset of symptoms. Thus, when the patient presents its first symptoms, upon isolation, the transmission of disease is interrupted. In this case, isolating the patients after the symptoms is an effective way to control the pandemic. On the contrary, COVID-19 in (**a**) is characterized by a latency shorter than incubation, and, even more troublesome, with a pre-symptomatic peak of transmission appearing before the onset of symptoms. Thus, in this case, even if the patient isolates after the symptoms appear, most of its infections have occurred already. This indicates that the only way to stop the chain of transmission of COVID-19 is by going into the past, before symptoms, and perform contact tracing to capture and isolate the contacts of the infected person. This crucial difference in the epidemiological profiles of these two coronaviruses explain why SARS was contained successfully in 2003 producing around 8,000 infections and 800 deaths, while COVID-19 kept spreading reaching a much larger worldwide population. Figure reproduced from [11]. Copyright © 2022, Serafino et al.

the epidemiological profile using a SEIR model, which captures the essence of the transmission chain of events.

An epidemiological profile under the SEIR model is characterized by the following time intervals (Fig. 4.1):

- Incubation time: time from exposure (E) to onset of symptoms.
- Latency: time from exposure to onset of infectiousness (I).
- The infectious period: the interval of time starting at (I) covering the time when the patient is contagious.
- The extend of the disease: from the onset of symptoms until recovery or death (R).

The comparative values of these intervals determine the epidemiological profile and the virulence of the disease. In the beginning of the pandemic, a study done by He et al. [7] has determined these parameters for the early variant of COVID-19. Many other studies followed [8, 9] due to the importance of establishing the parameters defining the disease spreading. The values of the corresponding parameters of SARS-CoV-2 are depicted in Fig. 4.1a as obtained from the literature [7–9] and are compared with the profile for SARS-CoV-1 responsible for the SARS pandemic in 2003 in Fig. 4.1b. The values correspond to the early variant of the virus and might be different to those of the latest variants like Delta and Omicron [10], although the general discussion is still valid.

Serafino et al. [11] discusses the relevant information in this profile: how the period in which a person transmits the virus compares with the period of manifestation of symptoms of the disease (fever, coughing, headaches and so on). As oppose to the regular flu, in the novel coronavirus, the infected person spreads the disease before the symptoms appear. In other words, the latency is shorter than the incubation period, and the patient becomes contagious before she/he starts to feel the symptoms of the disease. In fact, the peak of infectiousness, that is, when the patient is at its most contagious stage, occurs, in average, about a day or two before the onset of symptoms, according to the study of [7].

These numbers are crucial to understand the rapid spread of COVID-19. They imply that when the patient starts with the symptoms of the disease, she/he has already transmitted the virus to the majority of the secondary infections during the previous two days. In other words, by the time a person has fever provoked by COVID-19, even if she/he quarantines, it is too late already to stop the spread to the next layer of infections. What is worse: the patient has already infected most of them, without even knowing it. Remarkably, at the peak of the disease, when the symptoms are the strongest, the patient is not contagious anymore as the infectious period has ended, see Fig. 4.1a.

This epidemiologic profile of the new coronavirus implies that, even if the patient enters into isolation after the onset of symptoms, this is mainly of no consequence to stop the transmission chain. This is because, at the onset of symptoms, the infectious period has ended (see Fig. 4.1a) and the patient is not contagious anymore.

Thus, to stop the chain of transmission (in the absence of vaccines) one needs to perform contact tracing to track the past exposures of the patient from the first time of reporting symptoms, and then isolate those contacts, not necessarily the patient (although it is always advisable to do so as well; the numbers in the epidemiological profile are averages and may not correspond to a given individual). We need to go back in time to identify the contacts that have already occurred before the patient reports the symptoms to the health authorities or hospital. Without contact tracing, the chain of transmission cannot be broken. The situation is exacerbated due to the existence of fully asymptomatic cases, i.e., infected people who do not feel any symptoms, yet they transmit the disease without knowing it.

The relation between latency and incubation of SARS-CoV-2 is inverted in the SARS-CoV-1 coronavirus that was responsible for the SARS pandemic in 2003. As we see in the epidemiological profile of SARS-CoV-1 in Fig. 4.1b, in this case, the latency is longer than the incubation period, according to studies reported in [7]. Patients of SARS 2003 become contagious a few days after the appearance of symptoms. In this case, if the patient enters into isolation after reporting symptoms, then the chain of transmission is effectively broken. Thus there is no need to perform contact tracing back in time before the onset of symptoms since all contagious exposures could only happen during the manifestation of the disease, when the patient is already in isolation.

This situation explains why SARS in 2003 was contained successfully without spreading worldwide. This coronavirus infected 'only' about 8,000 people with around 800 deaths worldwide. On the other hand, the new coronavirus spread to all continents infecting 500 million people and causing 6 million deaths as of early 2022 across the world.

Countries that implemented effective earlier contact tracing protocols, like South Korea and China [12, 13], were able to contain the pandemic more successfully than counties who did not implement contact tracing protocols. Inspired by this evidence, Serafino et al. [11] have shown scientifically how contact tracing works in a real setting.

4.2 Contact Tracing Protocols

Thus, the only way to stop the spreading of the disease is by going backwards in time to perform a contact tracing model looking at all the contacts of the infected person before their onset of symptoms. Contact tracing is a method to identify persons who may have come into contact with an infected person to let them know that they may have been exposed to the disease. Contact tracing is commonly performed to slow the spread of the disease and it has been used with tuberculosis, measles, sexually transmitted infections (including HIV), Ebola, and other diseases. There are two ways of performing contact tracing:

- Old fashion phone calls: A health department officer calls the infected person to track who they have been in contact with, particularly anyone who may have been less than 6 feet away of an infected person for a total of 15 minutes according to the CDC [14].
- Modern digital contact tracing: By crossmatching the infected patients with their anonymized geolocalization from a digital medium, like an app or smartphone, digital contact tracing tracks the contacts for a time period in the past. An app sends an alert to the possible infected people who can then enter into isolation.

Back in March 2020, when the epidemiology profile shown in Fig. 4.1a was published by the group in China [7], we realized that without contact tracing (and vaccines) the pandemic was heading towards a catastrophic consequence for humanity. Comparing the epidemiological profile of COVID-19 to SARS-CoV-1 it was clear that, with no vaccines or treatment, only contact tracing could stop the pandemic. We, along with a group of scientists from the Universidade Federal de Ceará, Brazil led by Prof. José Soares de Andrade, health authorities in the city of Fortaleza, Brazil, as well as IT leaders like Grandata Inc, decided to develop, test and validate a contact tracing protocol using big-data from geolocalization and network theoretical tools. In a paper eventually published in [11], we mathematically proved that the pandemic can be effectively stopped with a minimal number of isolations identified by influencer algorithms fed with GPS localization data from users of smartphones, in conjunction with a global quarantine. Considering that the peak of infectiousness as well as 44% (95% confidence interval, 25-69%) of infected cases occur during the pre-symptomatic stage [7] (Fig. 4.1a), performing contact tracing becomes essential to stop the spreading disease. The study drew upon the algorithms from Chap. 1, including k-cores and other centralities to identify superspreaders, and found the following:

The main driver that prolong the pandemic after the lockdowns is the existence of superspreading k-core structures in the transmission network that persist even after the application of massive lockdowns. Once the k-cores in the contact network are identified, the optimal strategy to break the chain of transmission (GCC) is to isolate/quarantine a minimal number of 'weak links' with high betweenness centrality connecting the large k-cores.

The analysis deployed a contact tracing algorithm to track the chain of disease transmission across society. The algorithm sets forth intelligent quarantine protocols to halt the epidemic spreading with minimal socio and economic disruptions. This is possibly the most meaningful example where the theories of influencers that have been developed in Chap. 1 can solve an important societal problem. We first use big data to build the network and then we check which algorithm from Chap. 1 is the most suitable to dismantle the GCC of disease transmission.

Mobile phones and smartphones provide digital sources of information on human mobility and therefore offer a promising way to automate outbreak location detection. Mobile datasets consist of an ID associated with each user (e.g., the Google Advertisement ID, GAID, used in advertisement), a timestamp of the user location, and its location provided as latitude/longitude, that track the user in space and time. In [15] the authors propose a method to identify outbreak locations of point-source outbreaks from geo-located GPS movement data of affected individuals as recorded from mobile phones. In [16] the authors investigate whether the observed discrepancies between mobile phone datasets affect the results of epidemic simulations. Ferretti et al. [17] showed that a contact tracing App can achieve epidemic control if used by enough people without resorting to mass quarantines. Other works combined cross-sectional survey and GPS data. For example, in [18] the authors define a contact tracing strategy that is likely to identify a sufficient proportion of infected individuals such that subsequent spread could be prevented. The solutions proposed often rely on using GPS data alone or combining GPS with self-reported infections (through a mobile app or questionnaire). Crawford et al. [19] determined the contact rates using geolocation data and proved that this can accurately predict the number of COVID-19 cases. Klein et al. [20] used mobility data to assess the consequences of lockdowns, and Chang et al. [21] showed that mobility networks models of COVID-19 can predict how the virus spreads in cities. Baker et al. [22] investigated contact tracing protocols where individuals can exchange messages of information showing that the risk of contracting COVID-19 could be better estimated than contact tracing protocols without exchanges.

Serafino et al. [11] obtained the proper approvals and consents to use two independent datasets to study COVID-19. The study was approved by the Institutional

Review Board (IRB) at City College of New York (Approval No: 2020-0423) and by the Comitê de Ética em Pesquisa at Universidade Federal do Ceará in accordance with Resolution CNS No 510. The patient data was used with the approval and consent from the Epidemiological Surveillance Department of the Fortaleza Health Secretariat (CEVEPI), and the Mayor of the Prefeitura de Fortaleza, Ceará, Brazil.

The first dataset included data from the '*Grandata-United Nations Development Programme partnership to combat COVID-19 with data*' [23]. It was composed of anonymized Global Positioning System (GPS) data acquired through data providers who collect this type of information from hundreds of mobile Apps across Latin America that allow to track the trajectories of users. The data identified each mobile phone device with a unique encrypted mobile ID and specifies its latitude and longitude location through time, encoded by geohash with 12 digits precision. Typically, this dataset generates $\sim$450 million data points of GPS location per day across Latin America in particular in the state of Ceará, Brazil, where the study was performed. The analysis is focused in the state of Ceará where the geolocation of over a quarter million unique users generating over half a billion of GPS datapoints are tracked during the three months period of the study.

The second dataset consisted of an anonymized list of COVID-19 patients with the date of first symptoms and a geocode of the residential address of the patient in the city of Fortaleza obtained from the Health Department authorities from the State of Ceará. It included the geohash of the residential address, the SARS-CoV-2 test detection date and first day of symptoms of COVID-19. This dataset is used with the consent of the local health authorities in Fortaleza, Ceará and constrained the possibility of retrieving the chain of transmission of the virus only to the state of Ceará.

Both datasets were then cross-checked to perform contract tracing on each patient. The location of the residential address of each patient is matched with the GPS geolocation from the mobile phone dataset, thus obtaining the encrypted mobile ID of the patients in the GPS dataset. The geolocalized trajectories of COVID-19 patients are then traced during a period $-14/+7$ days from the onset of symptoms to look for exposures from the infected person to secondary cases that define the transmission network using the model described below.

4.3 Contact Tracing Model

The contact tracing model starts with a model for a single exposure. A very short exposure time may not be enough to transmit the virus. The question is, what is the typical exposure time interval and closeness that is needed to transmit the virus? Social distance policies estimate that one should not get closer than two feet from each other. Data analysis on mobility can provide a more precise estimation by first developing a probabilistic model for what constitutes an exposure and then calculating the time of exposure and closeness by fitting the contact model to real data of reproduction number.

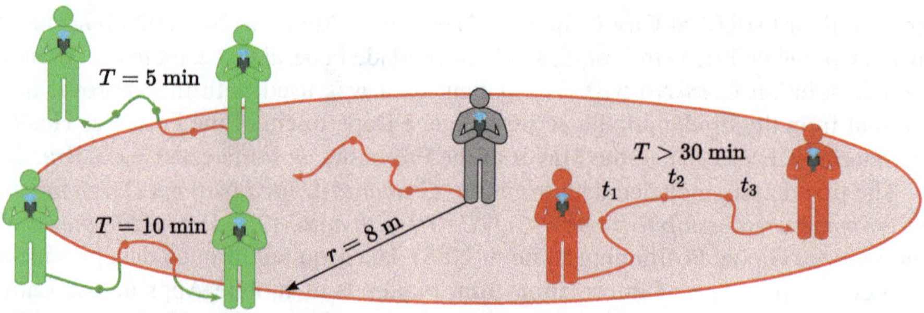

Fig. 4.2 Contact model for contact tracing algorithm. The grey person is at the first datapoint of the source of infection at t_0. All datapoints are collected for every user in a $T = 30$ min forward window ($t_1, t_2, t_3, \ldots, t_0 + T$) within an 8 m circle from the initial position. For each target (green and red), the position is tracked with the time spent inside the contact area (red part of the trajectory line). Figure reproduced from [11]. Copyright © 2022, Serafino et al.

Using the geolocation of the infected person several layers of contacts with other users are tracked. At each time stamp n the contact area is defined as a circle centered in the position of the source with a radius r, see Fig. 4.2. From each time stamp we gather all the targets that are within the circle in a T-minutes forward time window. Once all the targets are detected, we compute for them and the infected source the average position and the interval of time they have been within the contact area $\Delta t = t_f - t_0$, where t_0 and t_f are the first and last data points a user (target or source) was within the circle.

The contact probability of infection has a space and a time component. The space component probability is:

$$p_d[n] = 1 - \frac{d[n]}{2r}, \tag{4.1}$$

where $d[n]$ is the distance between the source and target average positions of the data points within the contact area in the next Δt from the time stamp n and r.

The time component probability is proportionally related to the amount of time target and source coexist within the contact area in the Δt time window:

$$\tau(\Delta t_s, \Delta t_t)[n] = \begin{cases} 0 \ \text{if}(\max(t_0^t, t_0^s) > \min(t_f^t, t_f^s)) \\ \max(t_0^t, t_0^s) - \min(t_f^t, t_f^s) \ \text{otherwise} \end{cases} \quad p_t[n] = \frac{\tau(\Delta t_s, \Delta t_t)[n]}{T}, \tag{4.2}$$

where $\tau(\Delta t_s, \Delta t_t)[n]$ is the overlap time in the timestamp n that depends on the Δt for the source (Δt_s) and the target (Δt_t), t_f^s and t_0^s are the final an initial time for Δt_s respectively, t_f^t and t_0^t are the final and initial time for Δt_t, respectively, and T is the time window size. The total probability of infection is $p_i[n] = p_d[n] \cdot p_t[n]$. Finally, we use a recursive expression to obtain a unique probability of infection for each contact between source and target:

$$P_i[n] = p_i[n](1 - P_i[n-1]) + P_i[n-1], \tag{4.3}$$

where $P_i[n]$ is the total probability of infection for a contact between a specific target and source until time stamp n. We then apply a threshold p_c to this probability to define a contact.

Calibration with real data determines the hyperparameters of the model: T, r, and p_c to match the reproduction number, R_0 in the particular location of study. In the case of the State of Ceará, $R_0 = 2.78$ in the month of March 2020, which gives $T = 30$ min, $r = 8$ m and $p_c = 0.9$. Thus, a contact is defined with probability one when exposure is at least 30 minutes within a distance < 8 m.

Once the contact exposure model is defined, then the contact tracing protocol is defined by using the epidemiological profile of Fig. 4.1a. Infectiousness starts 2 days before and lasts up to 5 days after the onset of symptoms. Ref. [11] added two days to each of these limits to conservatively capture most transmissions. Thus, in principle, to trace those people potentially infected by COVID-19 patients, Ref. [11] considered to track contacts at least 4 days before and 7 days after the reported date of first symptoms of the patient.

In addition, the tracing period was extended back in time to also consider exposures that could come from asymptomatic cases. Exposures start the incubation period of the infected person which can occur up to 12.5 days before onset of symptoms (5.2 days on average, 95% percentile 12.5 days). To conservatively trace these exposure events, ∼2 days were added to this incubation period to obtain the widely used 14 days period for contact tracing. Hence, to trace transmission and exposure cases from asymptomatics, Ref. [11] performed contact tracing over $-14/+7$ days from onset of symptoms.

With these parameters, Ref. [11] was able to define an accurate spatiotemporal transmission network of coronavirus by considering several layers of contacts, a typical network after the quarantine is displayed in Fig. 4.3. The mobility of people is measured by the mean square displacement (MSD) change across the quarantine. The mass quarantine in the state of Ceará was applied on March 19th, 2020 and it resulted in a massive drop of mobility to 50% of its value before the quarantine, as can be seen in the MSD curve in Fig. 4.4a. The same drop in mobility was observed across Latin America with countries like Argentina showing a 80% of mobility reduction after the implementation of the lockdown.

However, this reduction in mobility was not enough to flatten the curve. That is, to reduce R_0 below 1. Why? The GCC of disease transmission in Fig. 4.4a shows a drastic reduction of its size to 10% of its value before quarantine. However, the GCC was not completely dismantled to zero to stop completely the transmission chain. The remaining 10% of the GCC is responsible for the persistence of the pandemic beyond the drastic quarantines applied in this case.

This persistence can be clearly seen when we visualize the transmission network as a function of time before and after the quarantine in Fig. 4.4b and c, respectively. We clearly see a hairball structure in the network before the quarantine, which is a very connected network. Just seven days into the quarantine, this hairball was reduced to a tenuous network characterized by clusters connected by chains of weak links, and this very sparse network persists as long as we kept measuring during the quarantine.

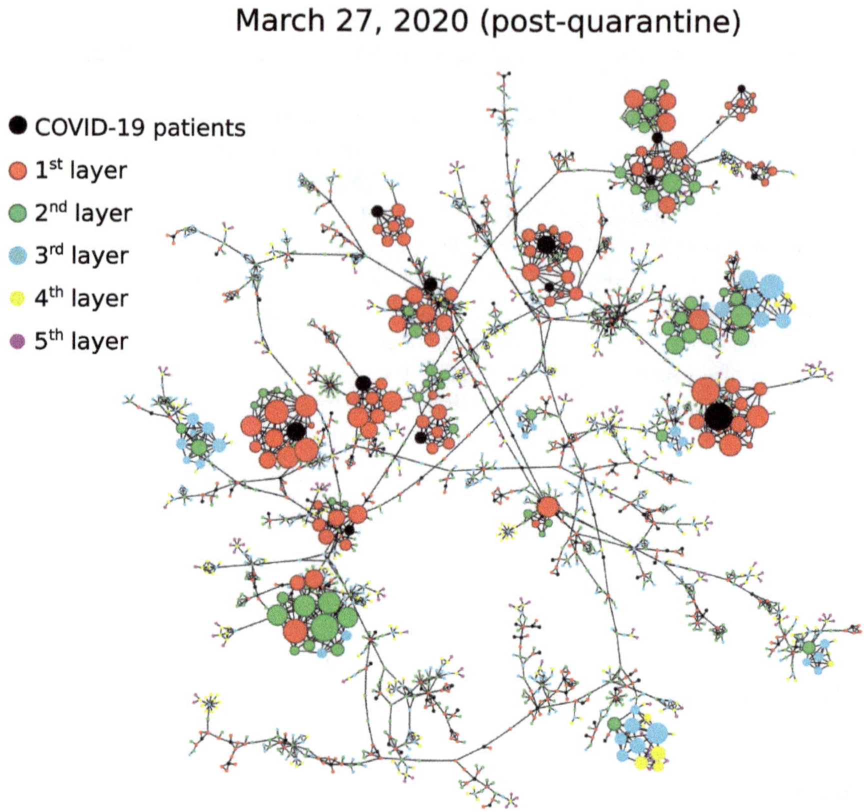

Fig. 4.3 **Typical contact network obtained with contact tracing.** The network corresponds to a time interval after the quarantine was applied in Fortaleza. The network construction starts at the patients (black nodes) and then performs a contact tracing within a time period to find the first layer of exposures, and then continues until the fifth layer to build the network of possible spreading events of COVID-19. The quarantine has reduced this network to a set of clusters related to the k-cores. This network can be effectively disintegrated by locating the weak-nodes along the tenuous links connecting the k-cores. Figure reproduced from [11]. Copyright © 2022, Serafino et al.

This cluster structure corresponds to k-cores, so the main question is how to dismantle these remaining k-cores. Figure 4.4a shows how these k-cores behave as a function of time after the application of the lockdown. The inner k-core (represented in the figure as the 0.55-core, see [11] for details) did not decay with the quarantine, to the opposite, it was reduced but then slowly increased. The peripheric shells of the contact network, on the contrary, where diminished almost to zero (0.55-shell in Fig. 4.4a).

This indicates that the effect of the quarantine was to reduce the outer k-shells of the network, but not the maximum k-core. This is the reason for the persistence of the pandemic after the massive lockdowns. The question is then how to dismantle this remaining persisting k-core. The reader familiar with the centralities of Chap. 1

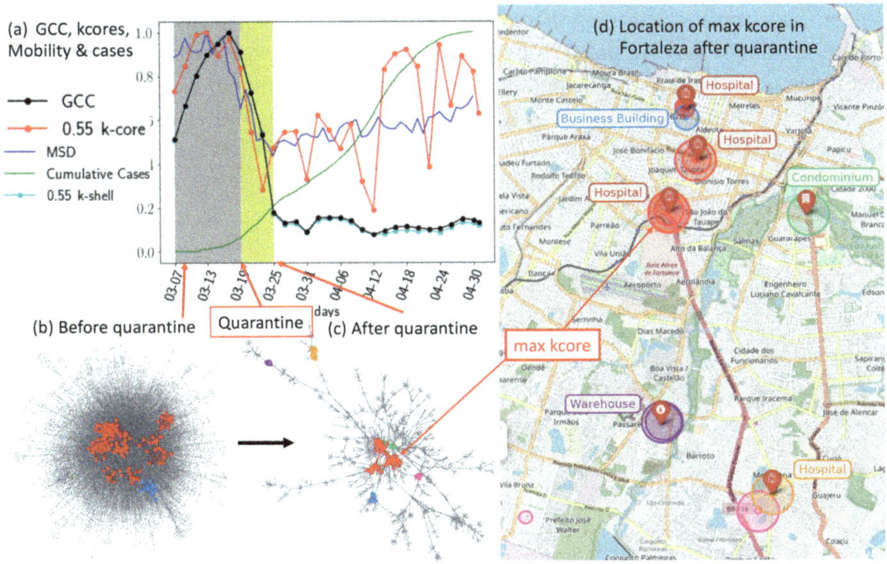

Fig. 4.4 Analysis performed in Fortaleza, Ceará. a Relative size of the GCC, max k-cores, and average user mobility measured by the MSD showing the massive drop when the quarantine was implemented on March 19, 2020. The mobility data is also showcased in the Grandata-United Nations Development Programme map shown in https://covid.grandata.com. The max k-core (0.55-core) keeps increasing, though, sustaining the pandemic. **b** Network of contacts before and **c** during the application of the quarantine. **d** The max k-core sustaining transmission are located in hospitals, large warehouses, and business buildings in Fortaleza

will recognize here the 'strength of weak ties', and how these ties are keeping the network alive. It should not be a surprise to anticipate that the best strategy to destroy the remaining components is to identify these weak ties with betweenness centrality. We explain this next.

4.4 Designing Intelligent Quarantines for COVID-19 with Minimal Isolations

The set of algorithms from Sect. 1.2 are then applied to dismantle the GCC with the minimal number of removals of people from the network. Here, a removal means that the chosen person (a node in the transmission network) goes into isolation/quarantine for a prudent period of time. The purpose is to break the remaining GCC with minimal isolation. It is clear that the algorithms of Chap. 1 are perfectly designed for this problem.

For this purpose, a set of influencer algorithms is used to search for the optimal dismantle of the network with minimal isolations by identifying the superspreaders. That is, we have a network and we want to find the minimal number of people to isolate. This implies removing them from the network, in such a way that the network collapses to zero and the pandemic stops. In this case, betweenness centrality is the best measure to identify the superspreaders in the most efficient way.

While the disease can persist in the maximum k-cores as shown in the simulation in Fig. 4.5a, see also Sect. 1.2.1.2, quarantining people by following a strategy based

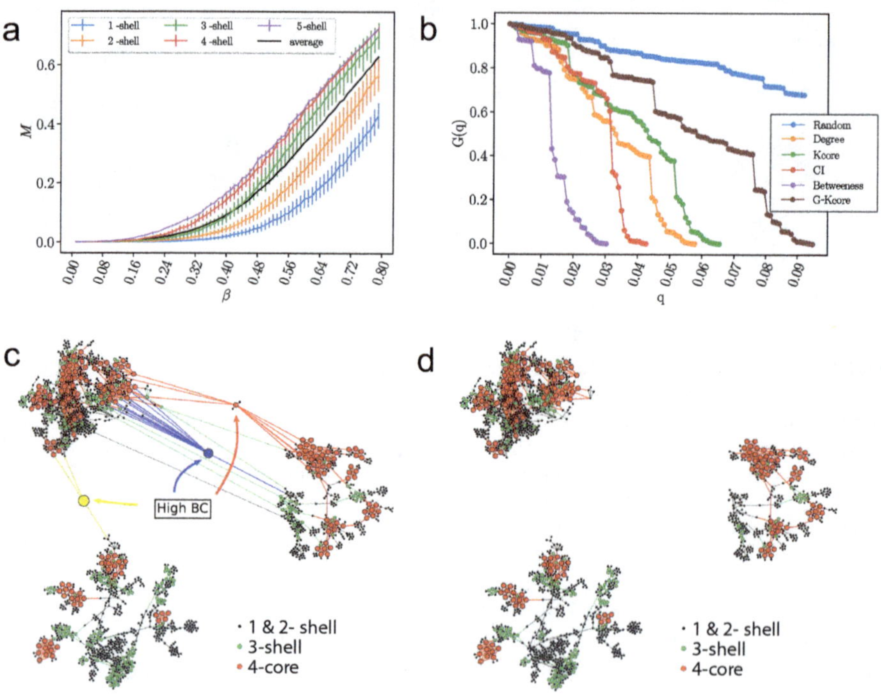

Fig. 4.5 Weak links and k-cores. a Average size of infected population, M, in an outbreak, averaged over all starting nodes in a k-shell as a function of the probability of infection β for a SIR model on the network during the lockdown. The black curve is the average value over all the network. The average divides the k-shell contribution to the spreading of the virus into two groups: above and below the average. The 0.5-cores have maximal spreading and the 0.5-shell has minimal spreading. Error bars correspond to a confidence interval of 95%. **b** Optimal percolation attack on a network in Fortaleza during the lockdown by following different attack strategies and their effect on the size of the largest connected component $G(q)$ versus the removal node fraction, q. Nodes are removed (in order of increasing efficiency): randomly (blue); by the highest k-shell followed by a high degree inside the k-shell; by the highest degree (orange); by collective influence (red); by the highest generalized k-core (brown); and by the highest value of betweenness centrality (green). After each removal, we re-compute all metrics. The most optimal strategy among those studied is removing the nodes by the highest value of betweenness centrality. **c, d** Effect of removing three high betweenness centrality nodes shown in Fig. 4.5b. (**c**) We show the 2-core component of the network after the removal of 12 high betweenness centrality nodes. The red node is the one with the highest betweenness centrality value (next node to remove, 13th) and the blue node is the 14th removal. Different k-cores and k-shells are in different colors. (**d**) Network k-cores are disintegrated after the removal of the high BC nodes. Figure reproduced from [11]. Copyright © 2022, Serafino et al.

on the k-core as in Sect. 1.2.1.2 is not optimal. The reason is that k-cores are populated by hyper-connected hubs that require many removals to break the GCC (around 7%, see Fig. 4.5b). For the same reason, removing directly the hubs is not the optimal strategy either, since the hubs are within the maximum k-core and not outside. The collective influence strategy from Sect. 1.6.1 improves a bit over hub-removal since it takes into account how hubs are spatially distributed, yet, it is far from optimal.

In Fig. 4.5c we clearly see the people with high betweenness centrality (BC) that represent the weak links. They are those that are strategically located connecting the different k-cores in the contact network (see Figs. 4.3 and 4.5d). Thus, the best possible targeted quarantines are achieved by removing from the contact network people with high BC. By doing so, we are effectively isolating the k-cores. Figure 4.5b shows that by isolating just 2% of the high BC people, the GCC is completely destroyed and the disease can be effectively stopped. In other words, this is the best possible and most effective quarantine protocol with the minimal disruptions. The high BC people in the network are the superspreaders of the disease whom should be quarantine in order to stop the spreading of the disease. They play the role of the influencers in the network.

While we have seen that the CI algorithm usually give a good performance, in this particular case, the CI strategy fails in comparison with BC. This fact shed light onto an important issue in the search of the best set of influencers: there is no predefined strategy, and different network structures will determine different features for the influencers. Thus, it is crucial to test a wide range of strategies for every situation and to choose the one accordingly. A scheme to find the best strategy is to compare them like in Fig. 4.5b and find the one that will dismantle the GCC of the problem with the minimal removals.

By mapping the network of contacts to the actual localization in the map we geolocalize the k-core contacts in the map of Fortaleza in Fig. 4.4d. We find that most of them are clustered in hospitals, warehouses and some condominiums. We find, for instance, that the three red clusters constituting the max k-core of the network of contacts are localized at the main hospitals in the city. This is corroborated by looking at the two trajectories of two people meeting at the main hospital in the map of Fig. 4.6. The people contributing to the maximum k-cores of the transmission network are essential workers going to hospitals, and cannot be quarantined.

We could not quarantine or isolate these people in the largest clusters of superspreaders because they are essential workers. The strategy is to quarantine people at the weak links as we can see in Fig. 4.5, who are the ones traveling between hospitals making a few contacts at each hospital.

This is an example of the power of the theory of influencers put to work to save humanity from a terrible calamity, COVID-19. It scientifically and rigorously proves that an intelligent quarantine based on a contact network obtained by geolocalization of individuals with GPS mobile phones and the subsequent application of the theories of Chap. 1 can effectively stop the transmission chain of the disease. The result shows that a global lockdown followed by contact tracing could have contained the pandemic at once. Whether these strategies can be implemented in society is

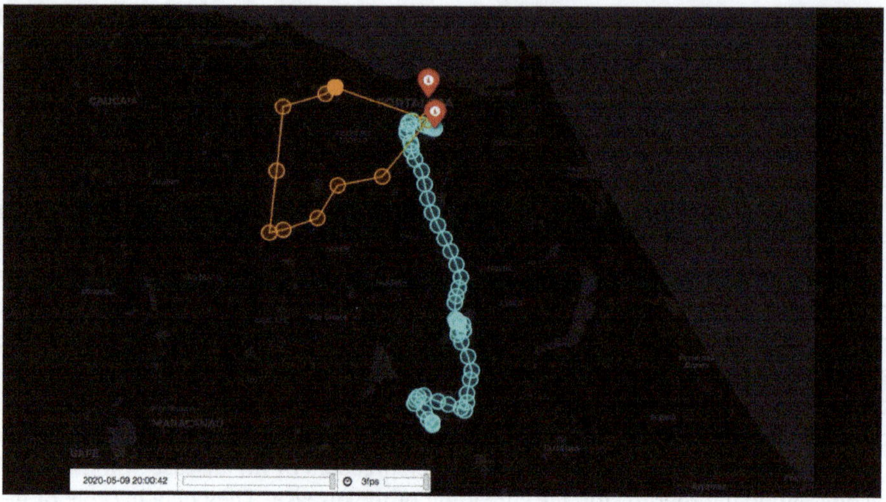

Fig. 4.6 Two trajectories contributing to a contact in the max k-core of the transmission chain. The contacts occur at the main hospital in Fortaleza as identified in Fig. 4.4d. They most probably are essential workers who cannot be quarantined to stop the spread of the disease. The network analysis then shows that the optimal directed quarantine strategy is to isolate people of high betweenness centrality connecting the k-core, and not the members of the k-cores

a question that we address next by elaborating on our own experience trying to implement these ideas in contact tracing apps around the world.

References

1. Zero-Covid, Wikipedia, Wikimedia Foundation, 31 Jan. 2022, https://en.wikipedia.org/wiki/Zero-COVID. Accessed Jan. 2022
2. Qin, A., Chien, A. C.: China Holds the Line on 'Zero Covid', but some wonder for how long, The New York Times, 21 Jan. 2022. https://www.nytimes.com/2022/01/21/world/asia/china-zero-covid-policy.html. Accessed Jan. 2022
3. Mallapaty, S.: China's zero-COVID strategy: what happens next? Nature **601**, 27 (2022). Jan
4. Pastor-Satorras, A., Vespignani, A.: Immunization of complex networks. Phys. Rev. E. **65**, 036104 (2002)
5. Pastor-Satorras, R., Castellano, C., Van Mieghem, P., Vespignani, A.: Epidemic processes in complex networks. Rev. Mod. Phys. **87**, 925–979 (2015)
6. Newman, M.E.J.: Spread of epidemic disease on networks. Phys. Rev. E. **66**, 016128 (2002)
7. He, X., Lau, E.H.Y., Wu, P., Deng, X., Wang, J., Hao, X., Lau, Y.C., Wong, J.Y., Guan, Y., Tan, X., Mo, X., Chen, Y., Liao, B., Chen, W., Hu, F., Zhang, Q., Zhong, M., Wu, Y., Zhao, L., Zhang, F., Cowling, B.J., Li, F., Leung, G.M.: Temporal dynamics in viral shedding and transmissibility of COVID-19. Nat. Med. **26**, 672–675 (2020)
8. Li, Q., et al.: Early transmission dynamics in Wuhan, China, of novel coronavirus-infected pneumonia. N. Engl. J. Med. **382**, 1199–1207 (2020)
9. Sanche, S., Lin, Y.T., Xu, C., Romero-Severson, E., Hengartner, N., Ke, R.: High contagiousness and rapid spread of severe acute respiratory syndrome coronavirus 2. Emerg. Infect. Dis. **26**, 1470–1477 (2020)

10. Anthes, E., Corum, J.: Charting an Omicron infection, The New York Times, 24 Jan. 2022. https://www.nytimes.com/interactive/2022/01/22/science/charting-omicron-infection.html. Accessed Apr. 2022
11. Serafino, M., Monteiro, H.S., Luo, S., Reis, S.D.S., Igual, C., Lima Neto, A.S., Travizano, M., Andrade, J.S., Jr., Makse, H.A.: Digital contact tracing and network theory to stop the spread of COVID-19 using big-data on human mobility geolocalization. PLoS Comput. Biol. **18**, e1009865 (2022)
12. O'Neil, P.H., Ryan-Mosley, T., Johnson, B.: A flood of coronavirus apps are tracking us. Now it's time to keep track of them. MIT Tech. Rev. (2020). https://bit.ly/2Y1NMet. Accessed Apr. 2022
13. Valentino-DeVries, J., Singer, N., Krolik, A.: A Scramble for Virus Apps That Do No Harm, The New York Times. https://nyti.ms/304uDLm. Accessed Apr. 2022. Singapore Government, TraceTogether, safer together. https://www.tracetogether.gov.sg/. Accessed Apr. 2022. Israel Ministry of Health, The Ministry of Health App for Fighting the Spread of Coronavirus. https://bit.ly/3gUMxGC. Accessed Apr. 2022. Helsenorge, Together we can fight coronavirus. https://helsenorge.no/coronavirus/smittestopp. Accessed Apr. 2022. Governor K. Noem (South Dakota Department of Health), Care 19 App. https://covid.sd.gov/care19app.aspx. Accessed Apr. 2022. O'Neil, P.H.: Apple and Google are building coronavirus tracking into iOS and Android, MIT Tech. Rev. https://bit.ly/2Y0gtZ0. Accessed Apr. 2022
14. Center for Disease Control and Prevention, COVID-19 Contact Tracing Communication Toolkit for Health Departments, 5 Nov. 2021. https://www.cdc.gov/coronavirus/2019-ncov/php/contact-tracing-comms.html
15. Schlosser, F., Brockmann, D.: Finding disease outbreak locations from human mobility data. EPJ Data Sci. **10**, 52 (2021)
16. Tizzoni, M., Bajardi, P., Decuyper, A., Kon Kam King, G., Schneider, C.M., Blondel, V., Smoreda, Z., González, M.C., Colizza , V.: On the use of human mobility proxies for modeling epidemics. PLoS Comput. Biol. **10**, e1003716 (2014)
17. Ferretti, L., Wymant, C., Kendall, M., Zhao, L., Nurtay, A., Abeler-Dörner, L., Parker, M., Bonsall, D., Fraser, C.: Quantifying SARS-CoV-2 transmission suggests epidemic control with digital contact tracing. Science **368**, 6491 (2020)
18. Keeling, M.J., Hollingsworth, T.D., Read, J.M.: Efficacy of contact tracing for the containment of the 2019 novel coronavirus (COVID-19). J. Epidemiol. Commun. Health **74**, 861–866 (2020)
19. Crawford, F.W. et al.: Impact of close interpersonal contact on COVID-19 incidence: Evidence from 1 year of mobile device data. Sci. Adv. **8**(1) (2022)
20. Klein, B., Rock, T.L., Cabe, S.M., Torres, L., Friedland, L., Privitera, F., Lake, B., Kraemer, M.U.G., Brownstein, J.S., Lazer, D., Eliassi-Rad, T., Scarpino, S.V., Vespignani, A., Chinazzi, M.: Reshaping a nation: mobility, commuting, and contact patterns during the COVID-19 outbreak (2020). https://www.mobs-lab.org/uploads/6/7/8/7/6787877/covid19mobility_report2.pdf. Accessed Jan. 2022
21. Chang, S., Pierson, E., Koh, P.W., Gerardin, J., Redbird, B., Grusky, D., Leskovec, J.: Mobility network models of COVID-19 explains inequities and inform reopening. Nature **589**, 82–87 (2021)
22. Baker, A., Biazzo, I., Braunstein, A., Catania, G., Dall'Asta, L., Ingrosso, A., Krzakala, F., Mazza, F., Mézard, M., Muntoni, A.P., Refinetti, M., Mannelli, S.S., Zdeborová, L.: Epidemic mitigation by statistical inference from contact tracing data. Proc. Nat. Acad. Sci. USA **118**, e2106548118 (2021)
23. Grandata-United Nations Development Programme partnership to combat COVID-19 with data. https://covid.grandata.com, https://www.latinamerica.undp.org/content/rblac/en/home/coronavirus/data-covid-region.html. Accessed Jan. 2022
24. O'Neill, P.H.: Apple and Google are building coronavirus tracking into iOS and android. MIT Tech. Rev. (2020). https://www.technologyreview.com/2020/04/10/999213/apple-and-google-are-building-coronavirus-tracking-into-ios-and-android. Accessed Apr. 2022
25. Apple and Google partner on COVID-19 contact tracing technology, Apple Newsroom, 10 Apr. 2020 https://www.apple.com/newsroom/2020/04/apple-and-google-partner-on-covid-19-contact-tracing-technology. Accessed Apr. 2022

26. Kermack, W.O., McKendrick, A.G.: A contribution to the mathematical theory of epidemics. Proc. Roy. Soc. Lond. Ser. A **115**, 700–721 (1927)
27. Foot, P.: Abortion and the Doctrine of Double Effect. Oxford Review, Number 5 (1967)
28. Trolley problem, Wikipedia, Wikimedia Foundation, 20 Apr. 2022. https://en.wikipedia.org/wiki/Trolley_problem. Accessed Apr. 2022
29. Code of Conduct, NAI. https://www.networkadvertising.org/code-enforcement/code. Accessed Feb. 2022
30. 2020 NAI Code of Conduct. https://www.networkadvertising.org/sites/default/files/nai_code2020.pdf. Accessed Feb. 2022
31. Guidance for NAI Members: Cross-Device Linking, May 2017. https://www.networkadvertising.org/pdfs/NAI_Cross_Device_Guidance.pdf. Accessed Feb. 2022

Chapter 5
Genetic Influencers in Gene Regulatory Networks

Abstract Genes encode the instructions to make proteins in their DNA sequences. When these sequences are mutated, the proteins do not work properly. But not all mutations are the same. Mutations to essential genes lead to loss of function that severely compromises the viability of the organism or result in genetic disorders like cancer. Identification of these essential genes, which play the central role as 'influencers' in the genetic network, is then fundamental. However, knowing the importance of individual genes is not enough to understand how to address the challenges of mutating genes. We need to understand how these genes interact to form building blocks and how these building blocks are further integrated into the full network. This chapter explains how to use the theory of influencers to finding those genes that are essential to the functioning of the cell, and how these genes organize themself into essential building blocks of the network through fibration symmetries that guarantee functional synchronization of gene expression.

5.1 Influencer and Essential Genes in Biological Networks

With the development of new technologies in biochemistry and molecular biology over the last decades, large numbers of detailed datasets on genes, mRNAs, proteins, metabolites and other small molecules have been accumulated. Instead of working in isolation, genes, proteins and metabolites achieve their functions through regulatory interactions and chemical reactions. As a consequence, these systems can be naturally described by networks [1–5]. Past research has shown the increasing relevance of understanding the properties of these networks to understand the function of biological systems. To understand and further control these biological networks, it is necessary to find key elements that allow to perturb the systems efficiently [6–8].

As the integration of disparate biological/biomedical data and analysis of networks gain importance for the description and regulation of the cell [9] and diseases [7, 8, 10, 11], molecular networks can serve as the basis for the determination of disease genes, essential genes, biomarkers [12, 13] and points of therapeutical intervention for drug development [14–16].

H. A. Makse and M. Zava, *The Science of Influencers and Superspreaders*, Understanding Complex Systems, https://doi.org/10.1007/978-3-031-78058-5_5

A frequently used model considers a disease as a perturbation of a cellular system [14, 17, 18]. Although a degree of robustness is needed to cope with a changing environment, a regulatory cellular system may be vulnerable toward specific perturbations that may trigger a cascade of failures. Such perturbations cause the gene regulatory network to malfunction and exhibit a disease phenotype [19, 20]. A few diseases, like sickle cell anemia, are monogenic in that mutation of a single gene triggers the disease, while most of diseases, like cancer and Alzheimer's disease, are multigenic in which many genes play a role [18], requiring a network approach for their understanding.

The hub-centric view considers hub genes as important since they are enriched with essential genes [6] and make a network more vulnerable upon their removal in a percolation-like process [21]. Back in 2000, Jeong et al. [6] showed that hubs in the protein interaction network of *Saccharomyces cerevisiae* contain more essential proteins than would be expected by chance. This 'centrality-lethality' rule has been confirmed by other studies [22–24], suggesting that a topological feature might be a good predictor for biological importance. Zotenko et al. [25] further demonstrated that the majority of hubs are essential due to their participation in Essential Complex Biological Modules. These are groups of densely connected proteins with shared biological function that are enriched with essential proteins, thus pointing to the importance of building blocks, rather than single genes, to capture the essentiality of these genes.

Hubs frequently occupy central positions in the biological network as they connect different network modules. Labelled as bottlenecks [26], such centrally placed nodes were shown to play a role in different cancers [27]. A percolation removal analysis of proteins in a human regulatory network identified key nodes that are important for network integrity, allowing the prediction of breast cancer outcomes [28]. This analysis emphasized the role of intra- and inter-modular hubs that mostly interact with proteins in the same clustered network neighborhood and connect different modules with each other. Inter-modular hubs that are more important for the integrity of the underlying network were more prone to be cancer related genes than intra-modular hubs.

Despite this evidence, inhibition of hubs often compromises network integrity, limiting their role as drug targets. That is, inactivating a hub is too detrimental for the organism, in that it affects not only isolated functions but also the integration of the network, and they are lethal to the organism [6]. Although topologically relevant, hubs frequently do not play a role in adult diseases, as such genes turn out to be embryonically lethal [7, 8, 10]. Therefore, the focus has shifted to the identification of weaker nodes that allow the control of the underlying network without its disintegration, pointing to the significance of less connected proteins [14, 29–32].

Notably, disease genes in more common diseases tend to be rather topologically peripheral, avoiding to be hubs [14] and are tissue specific [33]. We have already encountered these weak nodes in other settings all along this book. Essential genes that are low connected yet strongly relevant for the disease are the biological analogous of the social 'strength of weak ties' of Granovetter [34] and the 'connections are not everything' paradigm of social networks [35].

The theory of collective influencers from Sect. 1.6.1 is ideally suited to identify these essential genes irrespective to whether they are hubs or not [32]. The CI algorithm can provide the optimized set of nodes via optimal percolation [36], covering the influence of a node in relation to the others through their sphere of influence in the network of interactions. This approach emphasizes the role of weakly connected nodes with important impact on the dynamics in the underlying network.

As a consequence of these considerations the question as to which network feature can identify essential genes that could be good predictor of a drug target or the source of disease is a problem that can be tackled by influencer theory of Chap. 1 [37]. Assuming that a disease is a different robust state of a perturbed network, potential drugs need to target nodes that allow the re-transformation to the original non-disease state. Such a strategy, termed the 'network influence strategy' by Csermely et al. [14], aims at the determination of drug target candidates in therapeutics of polygenic, complex diseases.

Thus, network-based methods are relevant as the framework for identifying network influence strategy. Influential nodes located in vulnerable points of disease or tissue-specific networks and inter-modular low degree nodes that connect disease modules may be preferred targets for identifying essential nodes [37–40]. In signaling networks, target nodes may inhibit certain outputs of some signaling pathways while leaving others intact, redirecting signal flow [13, 14]. These requirements emphasize the collective influence of a node that can be taken into account by optimization frameworks that account for the interactions among all influential genes. The manipulation of gene network dynamics guided by a sound theoretical basis leads to more optimal therapeutic approaches.

The approach is based on the assessment of the stability and activity of the connected components of the networked system. Assuming that interactions between biological entities exist, mathematically guided drug-induced perturbations of genes can be designed to predict influencers that have the largest impact on the average activity of a given system. In the same way that we can plant an idea in social media by posting a tweet and monitor its transmission through the network, technology exists to modify the activity of a single biological entity, e.g., via optogenetics to change the activity of a neuron or a set of neurons, ablation of single neurons, gene knock-outs and CRISPR-Cas9 manipulations, that allows to manipulate the system in search of important nodes.

Optimization network approaches are transforming therapeutic and drug-targeting scenarios providing precise protocols (targets and perturbations) that are required to re-tune diseased molecular network dynamics in specific pathological conditions via the development of targeting scenarios for drugs. Therapeutic protocols can be designed to handle maladaptive network dynamics with interventions to core node's activity prescribed by optimization theory to reposition and reset diseased network dynamics in specific pathological conditions.

The framework to determine influential genes in biological networks includes two complementary optimization theories to identify two types of genetic influencers:

- **Structural influencers**. Identifies the minimum set of genes that upon deactivation (knock-out) produces the largest topological damage to the structure (e.g., the GCC) of the underlying genetic network. These are the influencer genes obtained by, for instance, the CI algorithm, the dismantling and decycling theories of Sect. 1.6.4, or any of the centralities studied in Sect. 1.2.
- **Dynamical influencers**. While most approaches to find influential genes consider only the underlying structural network and ignore knowledge of their dynamical mode of action, other approaches exist to assess the dynamics of perturbed systems [17], leading to dynamical influencers. These influencers are a minimal set of genes that upon perturbation produces the largest change to the dynamical activity (e.g., gene expression) in the genetic network, without necessarily dismantling the GCC. The predicted dynamical influential genes are responsible for abnormalities in gene regulation dynamics and disease states, indicating primary target for drug development.

Both types of influencers can provide complementary information. In particular, the structural influencer set is an 'upper bound' of the dynamical influencer set, as they produce the largest damage to the system. Dynamical influencers produce more subtle dynamical effects that can correlate with drug targets, while preserving the integrity of the giant connected component, and thus the life of the organism. Optimal percolation, decycling and network dismantling discussed in Chap. 1 are the major approaches relevant to finding structural genes. However, while percolation-based influencers carry important disease information, this approach misses two crucial components of gene regulatory systems, and thus can only be of partial utility in identifying disease genes:

1. While damage to the static connected components leads to the collapse of the system, control of the system dynamics is often more crucial. Even with a structurally intact giant connected component, the system may collapse by the deactivation or malfunction of a few nodes that control the dynamics. These dynamical influencers are more important than structural influencers in practical applications. While the optimization of influencers that affect the network structure has been widely studied [6, 21, 35, 36, 41–49], optimal disruption of the dynamics has been less investigated theoretically.
2. Interactions in biological networks are of positive and negative type, in that a weight indicates, e.g., a transcription factor activating (+), or repressing (−) a target gene. Such a feature is not captured by conventional percolation since it only admits positive interactions. Thus, applying percolation to biological networks simple ignores this fact, and therefore, it may not capture all the range of essential genes.

To go beyond percolation, approaches for dynamical influencers consider the dynamical system of equations defined on the network including inhibitory interactions. Influencers are then validated in genomic regulatory networks and drug induced gene perturbation data. Development of a set of disease bound genes and drug target genes uncover the disrupting processes that cause failures of gene regulation areas.

Such results could potentially lead to the discovery of new treatments to cure diseases in humans.

Gene essentiality is also closely associated with the notion of the minimal genome [50–53], which could be thought of as the "smallest autonomous self-replicating entity", or in biological terms, the "minimal genome content necessary and sufficient for a living cell" [50]. Systematic deletions of one or more genes in prokaryotes and simple eukaryotes have been performed to obtain rankings of genes for viability, fitness, robustness, redundancy and evolvability [54]. While gene essentiality is typically determined by assessing the viability of the corresponding mutant cells, this definition fails to account for the ability of cells to adaptively evolve after genetic perturbations. Then, Liu et al. [54] performed a more stringent screen to assess the degree to which *Saccharomyces cerevisiae* cells can survive the deletion of approximately 1,000 individual essential genes. They found 9% of these genetic perturbations could be overcome by adaptive evolution. Further progress in editing and silencing gene technologies extended the work to more complex organisms, like worms and mice, leading to the identification of essential genes that are shared across different species.

Supported by advances in genome sequencing technologies, Bartha et al. [53] review metrics of essentiality applied to large exome (the subset of the genome that is part of mature RNAs and translated into proteins) and genome sequence data sets to identify essential genes across species. They use three essential datasets:

1. Essential genes in knockout mice: enriched in recessive genes identified as essential by the International Mouse Phenotyping Consortium [55].
2. Essential genes in human populations: defined by organism needing the functions, not needed for simple cellular systems, and enriched in dominant disease genes. Contains the top 15% human essential genes as defined by RVIS, pLI, Phi, missense Z-score, LoFtool and s_{het} from human population exome sequence data [53].
3. Essential gene in human cell lines: defined by cell lines that are not representative of whole organisms, and enriched in recessive disease genes. Essential genes as identified by CRISPR-Cas9 screens on human cell lines [56–58].

The intersection of these genes leads to 188 genes scoring as essential in all three systems; it is depicted in Fig. 5.1. These scores are used to define gene essentiality with applications in clinical genetics and drug development. In the next section, we discuss algorithms for finding these disease-relevant genes and drug targets using structural and dynamic optimization theory and network symmetries.

5.2 Structural Influencers in Biological Networks

So far, we have discussed applications of structural influencers to opinion spreading in social media, epidemic spreading, and integrity of networks. Next, we will leverage this massive amount of work on influencers to develop a theory of influence in

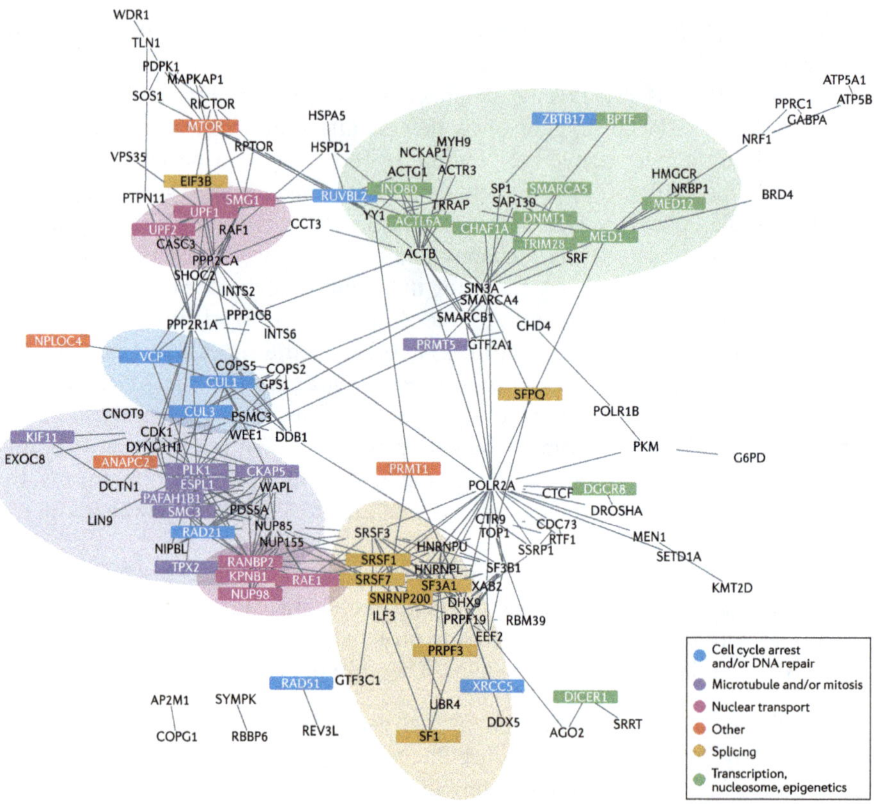

Fig. 5.1 Core network of essential genes in mice and humans. The network is obtained from human in vivo population, knockout mouse in vivo, and CRISPR-Cas9 human cell line data sets and contains 188 essential genes representing major cellular functions. Figure reproduced from [53].

biological networks to find influential nodes crucial for the integrity and functionality of gene regulatory networks.

Most biological networks are directed, and this will be treated in Sect. 5.2.2 and beyond. However, there are some cases where interactions in biology are undirected, a prominent case is protein-protein interactions which are bindings of proteins into complexes to regulate processes in the cell.

5.2.1 Structural Influencers in Protein Interaction Networks and Disease

A widely studied problem in biological networks is to identify important proteins in protein-protein interaction (PPI) networks. As a crucial element that exists in all living creatures on earth, proteins participate in the metabolism extensively as enzymes and also serve as transcription factors intermediaries of genetic regulations. Identifying crucial proteins in PPI networks not only helps to understand the mechanism of genetic regulation, but also contributes to locate the virulence genes and develop a target cure.

Inspired by the large number of PPI networks available, the heuristics metrics of Sect. 1.2 have been applied. One of the first studies was done by Jeong et al. in 2001 [6], for the degree-based lethality in PPIs. This study was inspired by scale-free theories that were being developed at the same time by Barabási and collaborators [21]. These theories pose that there are hubs in the network that are rare, yet crucial for integration. Thus, these theories naturally led to identifying those hubs as the essential nodes. Other centralities like betweenness centrality [59], closeness centrality [60], subgraph centrality [61], and eigenvector centrality [62] have been considered in later works to determine the essential nodes in biological networks. The intuitive motivation of these metrics was supported by the observation that highly connected proteins in PPI network are more likely to be essential, which is referred to as the centrality-lethality rule.

The collective influence algorithm from Sect. 1.6.1 was applied by Boltz et al. [32] to human protein interaction networks. They found that high-CI proteins are enriched with essential, regulatory, signaling and disease genes as well as drug targets, indicating their biological significance. Furthermore, CI proteins are evolutionarily conserved across different organisms suggesting that they play a relevant role in the species evolution.

Based on CI, Sun et al. [63] proposed another metric for signal spreading pathways. Given the adjacent matrix A_{ij} of the PPI network, the influence of protein i to j is defined as:

$$p_{i\to j} = \left[\frac{A_{ij} + |\theta(i, j)|}{k_j}\right] \times A_{ij}, \tag{5.1}$$

where $\theta(i, j)$ is the number of shared neighbors between i and j. Then the signal propagation network P is constructed by leaving edges larger than a threshold $p_{i\to j} > p_c$. The influence of each protein is quantified by its collective influence score in P. Considering five species in the IntAct database, including *Homo sapiens* (human), *Caenorhabditis elegans* (worm), *Drosophila melanogaster* (fly), *Saccharomyces cerevisiae* (yeast) and *Escherichia coli* (E. coli), it was found that the influential proteins tend to be located in the core region of PPI networks. In addition, by cross-validating this algorithm using Fisher's exact test on real data, it was demonstrated that the influential genes are enriched in essential yeast genes, human tumor suppressor genes, and drug target genes.

Although topological centralities were found effective in finding key proteins, their accuracy can be significantly limited by the incomplete and even erroneous information on interactions in PPI networks. Since this false information is almost inevitable in biology, the pure topological metrics have to be compensated from another perspective. Noticing that essential proteins tend to form densely connected clusters, and proteins in the same cluster are more likely to co-express, Li et al. [64]

proposed a new metric, PeC, that incorporates the topology PPI network with the gene expression data. The PeC centrality quantifies the importance of a protein v by

$$PeC(v) = \sum_{u \in \partial_v} ECC(u, v) \times PCC(u, v), \tag{5.2}$$

where $ECC(u, v)$ is the edge clustering coefficient and $PCC(u, v)$ is the Pearson correlation coefficient of the express level of u and v.

However, as reported by Han et al. [65], there exist two kinds of hubs in PPI networks: date hubs and party hubs, which are equally likely to be essential proteins. Since the PeC metric mainly emphasizes the co-clustering and co-expression properties, the importance of date hub proteins tends to be neglected because of their low degree. Fortunately, although date hubs and party hubs exhibit significant difference in the co-clustering property, the co-clustering properties of their neighbors were found to be similar. This leads to another centrality, CoEWC, that incorporates the local clustering coefficient with the Pearson correlation coefficient [66]. Compared with PeC, CoEWC has no bias in the selection of party hubs and date hubs. More importantly, it can discriminate those non-essential hubs whose neighbors are mainly disconnected. Comparisons made on *Saccharomyces cerevisiae* show that the mixed metric of PeC and CoEWC is superior than those pure topological centralities. Furthermore, CoEWC achieves higher prediction precision in the identification of essential proteins compared with PeC. The calculations of both PeC and CoEWC are efficient, hence they are scalable to large PPI networks.

Apart from the PPI networks, influencer identification has also been applied to the study of genetic regulations and cell cycles [67]. In the application of influence maximization on the regulatory network of *Saccharomyces cerevisiae*, Gibbs and Shmulevich [67] mapped gene regulation to the flow of information and framed the influence maximization problem in gene regulation as an information diffusion problem in networks. Defining the influence of a set of source nodes as the size of the network covered by a diffusion process starting from them, an influence ranking of genes were generated. It was found that influential genes act as critical sources of information flow, with the potential to affect the global network state.

Genetic regulations can be modeled by Boolean networks [68, 69]. Within this model, cycled orbits formed by the evolution of Boolean network dynamics, termed attractors, are believed to be relevant to a large number of biological phenomena, for instance, a cell cycle including growing, cell division and dying [68]. In applications, the stability of these attractors in response to external perturbations is crucial for organisms to adapt to a changing environment. Without such stability, organisms may suffer from serious illnesses such as cancer triggered by even minor changes of external environment [70, 71].

Wang et al. [72] studied the strategy to identify and control a minimal set of influential nodes that is capable of stabilizing an unstable Boolean network. A greedy algorithm based on collective influence theory was developed, aimed to minimizing the largest eigenvalue of a modified non-backtracking matrix that controls the stability of a Boolean network. Simulation results indicate that the proposed algorithm outperforms commonly used heuristic metrics by stabilizing the system with a smaller number of nodes.

Another variant of the collective influence algorithm was applied to excitable networks, a neural network model that mimics the formation of sensation in sensory systems [73–76]. The functionality of an excitable network is measured by its dynamic range, which quantifies the range of external stimulus intensities that can be robustly distinguished by the system [77–79]. In applications, maximizing the dynamic range under the minimal control of the network is crucial in designing excitable systems that achieve optimal performance. In previous works, it was found that the dynamic range of an excitable network is maximized at the critical state, characterized by the largest eigenvalue of the adjacency matrix [74].

Zhang and Pei [80] formulated the activation propagation in excitable networks as a message passing process and found that the critical state occurs when the largest eigenvalue of the weighted non-backtracking matrix is close to one. This result is in agreement with previous findings on directed networks but provides a better criterion on undirected networks. The authors generalized the collective influence theory and developed an algorithm to find the optimal set of links whose removal would drive the system to the critical regime, thus maximizing the dynamic range. Comparing with several competing heuristic strategies, the proposed method maximizes the dynamic range by removing the smallest number of links, and at the same time maintains the integrity of the network structure.

The relevance of influential genes can be tested in the human interactome, which is the set of all protein-protein interactions that occur in the human cell, potentially allowing to find disease relevant genes. Since interaction networks differ between different human tissues [81–83], tissue specific interactions considerably improve the prioritization of disease genes [11, 83]. Tissue specific network is dominated by a core-subnetwork that is common to all tissues. In turn, a small portion is entirely tissue specific and rather located at the periphery of networks [81–83]. Notably, the focus on interactions between proteins that were expressed in certain tissues considerably improved the prioritization of disease genes [11]. Furthermore, genes that cause hereditary diseases are involved in protein-protein interactions that occur exclusively in such tissues, highlighting disease mechanisms as well as disease etiology [83]. Thus, the presence of disease and drug targets is mostly a function of different tissues. Then regulatory interaction networks need to be obtained from integrating tissue-specific expression and genomic data.

Tissue specific human regulatory interaction networks [84] can be constructed by integrating genomic data. Specifically, transcription factor-target gene interactions in different tissues, where transcription factors bind corresponding binding sites in promoters and enhancers of target genes, can be considered. Influencer genes in such directed, unweighted regulatory networks are expected to show significant eQTL (expression quantitative trait loci) associations to their promoter regions and enhancers in tissue specific diseases. The first step is then to reconstruct human transcriptional networks from available data, and then apply influencer finding algorithms for directed networks, as discussed in Sect. 5.2.2.

Furthermore, we can assume that a small set of influencer genes exist that appear in many different tissue-specific regulatory networks and *vice versa*. Assuming that disease genes are tissue-specific [33], such genome-wide associations will significantly involve influential genes that are rather limited to similar tissues. In turn, influential genes that appear frequently in different tissues are enriched with essential genes, as they may be necessary for survival. While this procedure tests the hypothesis that structural influential genes are related to disease; it is always important to also test all other heuristics in the same way, like hubs, betweenness, coreness, etc., as well as composite indices obtained from the composition of all these rankings in search for the best possible network indicator for disease.

5.2.2 Optimal Percolation and Collective Influence in Directed Networks

Biological networks are composed of directed links. We have mentioned that a theory of influence should take into account the directional nature of the interactions. This includes for instance, transcriptional interactions turning genes on and off, or the directional nature of action potentials between two neurons, to the directional sense of a metabolic reaction. This does not mean that all biological interactions are directed, for instance an equilibrium chemical reaction is bidirectional, neural interactions mediated by gap-junctions are undirected too, as well as the protein-protein interactions of the previous section, since they correspond to physical bindings. However, all regulatory interactions follow a direction, then any biological theory should take into account this fact.

The theory of optimal percolation developed in Sect. 1.5 is for undirected networks. Below we extend this theory to identify structural influencers in directed networks. Concurrently, one can extend the heuristics in Sect. 1.2 to the case of directed networks, and then rank the hubs according to their in-degree or out-degree, or consider other centralities like PageRank, or a directional k-core. These rankings for directed networks could then be tested for their relevance to identify essential genes in biological networks and disease phenotypes.

Optimal percolation theory was generalized to directed networks by Furbish [85]. In undirected networks, the objective is to optimally destroy the giant connected component (GCC), but in directed networks there are multiple connected components of interest. The Weakly-Connected Component (WCC) is the largest connected subgraph that can be reached traversing edges in either direction. We can think of the WCC as the component that is obtained in a directed graph when we ignore the directionality of the edges. In this component, there is a connected path from a node to another, but this path does not necessarily comes back to the source node.

A stronger version of a connected component is the Strongly-Connected Component (SCC). It is the largest subgraph from which all nodes can be reached from all other nodes by traversing edges only in the direction they are oriented. In the SCC, there is a cycle that starts at every node. That is, we can always find at least one path that begins in a node of the SCC and comes back to the same node. Thus, the SCC is an important component in any directed network. Notice that there could be more than one SCC in the graph, and therefore, one is interested in the largest SCC. This is usually called LSCC or the giant GSCC if the network is taken in the thermodynamics limit of an infinite number of nodes as in a theoretical study.

The SCC has further components. The Outbound Component of the SCC is the union of the SCC and all vertices that can be reached from within the SCC by traversing edges along the direction they are oriented. The Inbound Component of the SCC is the union of the SCC and all vertices that can be reached from within the SCC by traversing edges against the direction they are oriented. The Outbound and Inbound components of the SCC may contain smaller strongly-connected subgraphs which are not present in the SCC. The Outbound and Inbound components of the SCC are a combination of tree-like portions (which can be destroyed trivially). By its definition, destroying the WCC is exactly the same task as destroying the GCC of the undirected network obtained by treating all edges as bidirectional. Optimal percolation in directed network attempts to derive an efficient method for destroying the largest of giant strongly-connected components, SCC, of the network.

The SCC is the intersection of its Outbound and Inbound components. The order parameter of the problem is the relative size of the SCC, which can be estimated by message passing algorithms based on the formulation for undirected networks [36] (Sect. 1.5), measuring the cavity probabilities of a node to be connected to the outbound, $r_{i\to j}$, and inbound, $s_{i\to j}$, SCC, respectively, in the absence of node j [85]:

$$r_{i\to j} = n_i \left[1 - \prod_{k\in\partial_{\text{in}} i\setminus j} (1 - r_{k\to i})\right], \tag{5.3}$$

$$s_{i\to j} = n_i \left[1 - \prod_{k\in\partial_{\text{out}} i\setminus j} (1 - s_{k\to i})\right], \tag{5.4}$$

where n_i are the entries of $\mathbf{n}$, the binary vector designating which vertices are present in the graph and which have been removed. Each message passing equation admits the solution $\{r_{i\to j} = 0\}$, which is stable if the $2M \times 2M$ stability matrix on $k \to l, i \to j$ given, again, by:

$$\mathcal{M}_{k\to l, i\to j} = \frac{\partial \nu_{i\to j}}{\partial \nu_{k\to l}}\Big|_{\nu_{i\to j}=0}, \tag{5.5}$$

where $\partial \nu_{i\to j}$ is $\partial_{\text{in}} r_{i\to j}$ for Eqs. (5.3) and $\partial_{\text{out}} s_{i\to j}$ for Eq. (5.4), is irreducible and has largest eigenvalue $\lambda_{\text{max}} < 1$. Here, we denote a change in the message $i \to j$ by $\partial_{\text{in}} r_{i\to j}$ using the superscript “in” (“out”) to designate that j is a predecessor (successor) of i. If the components defined by Eqs. (5.3) and (5.4) are locally tree-like, then:

$$\mathcal{M}_{k\to l, i\to j} = n_i \mathcal{B}_{k\to l, i\to j}, \tag{5.6}$$

where $\mathcal{B}$ is the non-backtracking matrix of the graph traversed only along edges for Eq. (5.3) or against edges for Eq. (5.4).

This allows to rephrase the task of destroying the SCC as setting $\lambda_{\max} < 1$ for both Eqs. (5.3) and (5.4) by removing an optimal set of vertices. To do this, the Power Method iteration is used, the established method to calculate the largest eigenvalue of a matrix as used in Sect. 1.6.1, to obtain $\lambda_{\max}$ in terms of $\mathcal{M}$ to minimize the number of vertices removed.

The operator $\mathcal{M}_{k\to l,i\to j}$ for each of the Outbound and Inbound equations can be combined into a single operator. Denoting the Outbound operator as $\mathcal{R}_{k\to l,i\to j}$ and the Inbound operator as $\mathcal{S}_{k\to l,i\to j}$, one can obtain the combined expression [85]:

$$\begin{bmatrix} \mathcal{R}_{k\to l,i\to j} & 0 \\ 0 & \mathcal{S}_{k'\to l',i'\to j'} \end{bmatrix} = \begin{bmatrix} \frac{\partial_{\text{in}} r_{i\to j}}{\partial_{\text{in}} r_{k\to l}} & 0 \\ 0 & \frac{\partial_{\text{out}} s_{i'\to j'}}{\partial_{\text{out}} s_{k'\to l'}} \end{bmatrix}, \tag{5.7}$$

whose eigenvalues are the union of the eigenvalues of the Inbound and Outbound operators. When the leading eigenvalue of Eq. (5.7) is less than 1, then both the Inbound and Outbound components have been independently destroyed, which is sufficient to demonstrate that the SCC is destroyed. The directed dCI index of each node i in the directed network was then proposed by Furbish [85] to be:

$$\text{dCI}(i) = k_i^{\text{in}} \left[\sum_{j\in\partial^{\text{out}} Ball(i,\ell)} k_j^{\text{out}} \right] + k_i^{\text{out}} \left[\sum_{j\in\partial^{\text{in}} Ball(i,\ell)} k_j^{\text{in}} \right]. \tag{5.8}$$

Efficient algorithms from undirected networks [41] can be used to calculate dCI. The algorithm outputs the contribution of each gene to this eigenvalue via the dCI index assigned to each node to rank genes according to their largest contribution to the destruction of the SCC. To define the set of structural influencers , one then considers the largest dCI index as a proxy of the largest influence, as in the undirected case. One can then test these dCI influencers for their importance in regulatory networks in health and disease against the set of heuristics like hubs and other centralities, as well as against the set of dynamical influencers which are explained next. A directed dCI study was performed on social networks. In Ref. [86] a restricted version of CI-directed, considering only the inbound component has been used to detect influencers of fake news during the 2016 US presidential election. The results are discussed in Sect. 2.4.

5.3 Dynamical Influencer Theory in Genetic Networks to Predict Disease Genes

Having developed a theory to find structural genes that may be relevant for disease by affecting the static SCC in directed networks, we now turn to the problem of dynamics. The idea is to find the set of nodes that maximize the damage to the overall activity of the network, which may not necessarily affect the underlying static giant connected component. Therefore, the set of dynamical influencers is expected to be smaller than the structural ones.

Biological networks are all about inhibition and directionality that provides the homeostatic balance between inhibitory and excitatory activity [87, 88]. Thus, a biological theory should admit any kind of directed interaction terms between two genes via the interaction strength w_{ij}, with $w_{ij} \neq w_{ji}$ and $w_{ij} > 0$ for activators and $w_{ij} < 0$ for inhibitors. These interactions are inferred from real data in Sect. 5.4. For the sake of exposition, we consider nonlinear dynamical equations for the expression of gene i, $x_i(t)$, in a gene regulatory network is given by the Michaelis-Menten equation or Hill function with cooperative exponent n [89]. A generic form of the dynamical system is:

$$\frac{dx_i}{dt} = f(x_i) + \sum_j w_{ij} G(x_j), \tag{5.9}$$

which, in the particular case of the Hill function, is:

$$\frac{dx_i}{dt} = -x_i + \sum_j w_{ij} \frac{x_j^n}{1 + x_j^n} . \tag{5.10}$$

The interaction strength w_{ij} is asymmetric admitting directed and weighted interactions. We are interested in the properties of the fixed points of the dynamical equations:

$$\frac{dx_i}{dt} = 0, \tag{5.11}$$

given by the solution of:

$$x_i = \sum_j w_{ij} \frac{x_j^n}{1 + x_j^n}. \tag{5.12}$$

While a fixed point with $x_i = 0$ for all i exists, there is also another fixed point with $x_i \neq 0$. We define the average gene activity as:

$$\langle x \rangle = \frac{1}{N} \sum_i x_i, \tag{5.13}$$

where x_i is computed at the fixed point. Assuming that we knock out a fraction q (much smaller than the total number) of genes, we then consider the change in the activity $\langle x \rangle$. We want to find the Nq genes whose removal decreases the most the activity $\langle x \rangle$. This set of genes is the most influential set to affect the dynamics of the genetic network.

Morone developed the following formulation of the problem in [90]. As in percolation, the problem is formulated with a binary variable $n_i = 0, 1$ which encodes the knockout of gene i: $n_i = 1$ for occupancy, $n_i = 0$ for knock out. The fixed point solution is transformed into:

$$x_i = n_i \sum_j w_{ij} \frac{n_j x_j^n}{1 + x_j^n}. \tag{5.14}$$

For any configuration of gene knock-outs $\mathbf{n} = (n_1, \cdots, n_N)$ of the network, the corresponding fixed point solution $\mathbf{x}(\mathbf{n})$ can be computed through the fixed point equation, Eq. 5.14. As we want to minimize the activity $\langle x \rangle$ with the minimal removal cost, we minimize the energy inspired by statistical physics. A natural energy cost-function for this problem can be defined as [91]:

$$E(\mathbf{x}, \mathbf{n}) = \sum_i x_i - \mu \sum_i n_i \tag{5.15}$$

where μ is analogous to the chemical potential that determines the number of genes to knock-out. Following a statistical mechanics approach, the problem is formulated at a finite "temperature" $\beta = 1/T$, and then taking the limit of zero temperature to minimize the energy. The partition function is used to minimize this problem, and it is given by:

$$Z = \sum_{\mathbf{n}} \sum_{\mathbf{x}} e^{-\beta E(\mathbf{x}, \mathbf{n})} \prod_i \delta \left(x_i - n_i \sum_j w_{ij} \frac{n_j x_j^n}{1 + x_j^n} \right), \tag{5.16}$$

where the sum in the partition function is constrained by the fixed point solution in the delta-function.

The partition function is written as:

$$Z = \sum_{\mathbf{n}} P(\mathbf{n}). \tag{5.17}$$

What we want to find are the marginals $P_i(n_i)$ that will tell us which genes are the influencers. This can be performed via the conventional cavity equations [91], as done in [90].

By marginalizing, one obtains the best ranking. The interaction function is defined as:

$$f(x_i) \equiv w_{ij} \frac{x_i^n}{1 + x_i^n} \equiv f_i, \tag{5.18}$$

a collective coordinate as $y_i = (x_i, n_i)$, and the energy as:

$$E(\mathbf{y}) = \sum_i \epsilon(y_i). \tag{5.19}$$

The partition function can be rewritten using this notation as:

$$Z = \sum_{\mathbf{y}} e^{-\beta E(\mathbf{y})} \prod_i \delta \left(x_i - n_i \sum_{j \in \partial i} n_j f(x_j) \right). \tag{5.20}$$

The marginal of the variable y_i is given by:

$$P_i(y_i) \propto \sum_{\{y_j\}_{j\in\partial i}} e^{-\beta\epsilon(y_i)}\delta\left(x_i - n_i\sum_{j\in\partial i} n_j f(x_j)\right)\prod_{j\in\partial i}\nu_{j\to i}(y_j, y_i)\,, \tag{5.21}$$

where the cavity marginals satisfies the self consistent equations:

$$\nu_{i\to j}(y_i, y_j) \propto \sum_{\{y_k\}_{k\in\partial i\setminus j}} e^{-\beta\epsilon(y_i)}\delta\left(x_i - n_i\sum_{k\in\partial i} n_k f(x_k)\right)\prod_{k\in\partial i\setminus j}\nu_{k\to i}(y_k, y_i)\,. \tag{5.22}$$

The marginal $\nu_{i\to j}(y_i, y_j)$ depends on both y_i and y_j due to the loops created by the factors in the graph. Using the pairs (y_i, y_j), instead of the single variable y_i only, allows to de-loop the graph. The cavity method of statistical mechanics [91] can then be applied to solve the message passing cavity equations (5.22) which are then used to calculate the marginals associated with each node Eq. (5.21).

Once the cavity marginals in Eq. (5.21) are calculated, the most important gene is identified by the largest marginal probability. The idea is to fix the chemical potential μ that fixes the number of influencers and then minimize the energy cost via Belief Propagation (BP) algorithms to calculate the marginal distributions [91]. Starting from a random distribution of marginals, BP assumes approximate global knowledge of all marginal probabilities in each iteration step to locally update individual marginals. The iterations terminate once the algorithm converges on a stable set of marginals.

This approach can be generalized to any kind of dynamical system with an asymptotically stable fixed point. We notice though, that for limiting cycles and chaotic attractors the analysis must be different, since the object to minimize is different, so these cases need to be treated separately.

This procedure produces a ranking of dynamical influential genes. The gene with the highest ranking is removed from the gene network and the cavity marginals are recalculated again. In this way a ranking of genes can be obtained that maximally affect the dynamics of the network, that is, the ranking of dynamical influencers. This set of nodes can be used then to test for disease genes in gene networks and can be evaluated against heuristic rankings as well as in comparison with the structural influencers from Sect. 5.2. The next section outlines applications of this ranking to predict drug targets.

5.4 Dynamical Influencer Theory in Genetic Networks to Predict Drug Targets

The theory to identify dynamical influencers in weighted, directed networks can be used as the backbone to determine drug targets that upon treatment with a corresponding drug could significantly change the gene expression profile of the underlying human tissue.

The first problem to solve in this application is how to reconstruct a human interaction network on which to apply the influencer theory [2]. To infer regulatory networks in disease cell lines, we can use gene expression profiles of different tissue-specific cell lines that have been treated with a variety of different drugs. The LINCS L1000 datasets [92] is a source of such perturbation data. The L1000 Connectivity Map is a Library of Integrated Network-based Cellular Signatures (LINCS) profiling gene expression changes following pharmacologic perturbation of cell lines in high-throughput screens. Approximately, more than 150 different cell lines (including normal and diseased) have been screened with 20,000 compounds, monitoring the expression of more than 12,000 human genes. Using such perturbation data, it is possible to determine weighted, directed regulatory networks in disease cell lines. Dynamical influencers are then prime drug targets; drugs aiming at these influencers will nudge the cell from the disease state with the goal to bring it back to a healthy state.

The next step is to claculate the response to perturbation, which is quantified in terms of relative changes in the measured levels of RNA expression. The regulatory activity between genes are derived using the same differential equations used in Sect. 5.3 to find dynamical influencers modified to the case where the system is perturbed.

The following theory has been developed by Molinelli et al. [17]. They model the system $\{x_i(t)\}$ in a set of perturbed conditions defined as an external force u by drugging a gene i with drug μ $\{u_i^\mu\}$ as a series of coupled non-linear differential equations. We can consider the general form of the dynamics of a regulatory network that was used in the theory of dynamical influencers $\frac{dx_i}{dt} = f(x_i) + \sum_j w_{ij} G(x_j)$ in Eq. (5.10) in Sect. 5.3.

Such a step allows *(i)* to infer the matrix of interactions w_{ij} that we need in Eq. (5.10), and *(ii)* find regulatory networks with the aid of perturbation data using the same consistent model. The following system of equations is considered:

$$\frac{dx_i^\mu}{dt} = -x_i^\mu + \sum_{j \neq i}^{N} w_{ij} f(x_j^\mu) + u_i^\mu, \tag{5.23}$$

where $f(z)$ could be the Hill function, like in Eq. (5.10), or a sigmoidal function $f(z) = \tanh(z)$ as used in [17]. The system variables x_i represent the expression values of corresponding genes in the presence of a perturbation μ, and the matrix $\mathbf{W} = \{w_{ij}\}$ quantifies the interactions between genes i and j. The matrix $\mathbf{W}$ is inferred from real data, and we use these interactions w_{ij} in Eq. (5.10), allowing us to find dynamical influencers with the methods of Sect. 5.3. As a consequence, perturbations to gene expression in the system that is introduced by a drug perturbing gene i propagate in time through the underlying network, producing expression trajectories of genes $x_i(t)$.

The goal of solving this dynamical system is to find the matrix $\mathbf{W} = \{w_{ij}\}$ to classify interactions as a gene activating (+), repressing (−) or behaving neutrally (0) toward the corresponding target genes. These interactions can be found by using a model configuration cost function as:

$$C(\mathbf{W}) = \beta \sum_{l}^{L} \sum_{i}^{N} \sum_{\mu}^{M} \left((x_i^{\mu}(t_l) - x_i^{\mu *}(t_l) \right)^2 + \lambda \sum_{i}^{N} \sum_{j \neq i}^{N} \delta(w_{ij}), \tag{5.24}$$

where $\delta(w_{ij}) = 1$ if $w_{ij} \neq 0$ and *vice versa.* Such a function minimizes: *(i)* the discrepancy between the predicted $x_i^{\mu}(t_l)$ and experimentally measured $x_i^{\mu *}(t_l)$ over L measurements and M drug targets, and *(ii)* the non-zero interactions in W at different time points (i.e. perturbations) $\{t_l\}$, via the regularization term controlled by λ.

Since the solution space of such a dynamical space is prohibitive large to be exhaustively investigated, methods from statistical physics can be used to find approximate solutions. Following [17], the probability of a network configuration $\mathbf{W}$ can be calculated using:

$$P(\mathbf{W}) = \frac{1}{Z} e^{-C(\mathbf{W})} = \frac{1}{Z} e^{-\beta \sum_i^N \sum_\mu^M \left(x_i^{\mu}(t_l) - x_i^{\mu *}(t_l)\right)^2} e^{-\lambda \sum_i^N \sum_{j \neq i}^N \delta(w_{ij})}, \tag{5.25}$$

where Z is the partition function that sums over all model configurations, allowing us to find the maximally probable network configuration. To solve such a system, we apply the methods to solve optimization problems in Chap. 1. As done in [17], the iterative optimization method of Belief-Propagation (BP) algorithm can be used to find the marginal probabilities of the different parameters in the underlying model.

Experimentally obtained abundances are discretized as discretized values allow better convergence characteristics. Furthermore, variables are evaluated at steady state, decoupling variables from each other. Therefore, predicted value x_i^{μ} only depends on the parameters in the ith row of $\mathbf{W}$ and on the set of experimentally measured abundances of genes $x_i^{\mu *}$ of perturbation μ, suggesting that $P(\mathbf{W}) = \prod_i^N P(W_i)$.

As a consequence, each row of $\mathbf{W}$ can be written as:

$$P(W_i) = \frac{e^{-\lambda \sum_{j \neq i}^N \delta(w_{ij})}}{Z_i} \left(\prod_{\mu} F^{\mu}(W_i) \right) = \frac{e^{-\lambda \sum_{j \neq i}^N \delta(w_{ij})}}{Z_i} \left(e^{\sum_\mu -\beta (x_i^{\mu} - x_i^{\mu *})^2} \right). \tag{5.26}$$

To apply the BP algorithm to this framework, local updates to local marginals need to be established. As the marginals of each entry in the $\mathbf{W}$ matrix can be written as:

$$P^{\mu}(w_{ij}) = \frac{e^{-\lambda \delta(w_{ij})}}{Z_{ij}} \prod_{\nu \neq \mu} \rho^{\nu}(w_{ij}), \tag{5.27}$$

where $\rho(w_{ij})$ will be updated as

$$\rho^{\mu}(w_{ij}) = \sum_{W_{i \setminus j}} F^{\mu}(W_{i \setminus j}) P^{\mu}(W_{i \setminus j}). \tag{5.28}$$

Specifically, $W_{i\setminus k}$ is a configuration of the ith row of $\mathbf{W}$ where the parameter w_{ik} is fixed and

$$F^{\mu}(W_{i\setminus j}) = e^{-\beta(x_i^{\mu}-x_i^{\mu*})^2}. \tag{5.29}$$

$\rho_{\mu\rightarrow k}(w_{ik})$ can be considered as passing a message when perturbation μ is passed on protein k. Furthermore, the probability $P^{\mu}(W_{i\setminus j})$ can be factorized into independent contributions as indicated before. Utilizing a mean field approach, $F^{\mu}(W_{i\setminus j})$ can be evaluated by averaging over the experimentally observed abundances of proteins that interact with the corresponding protein i in perturbation experiment μ. Furthermore, probabilities P^{μ} are estimated as gaussian distributed functions of average abundances [17].

As a deliverable of this first part of the method, the BP algorithm allows to find configurations of the underlying networks as represented by the matrix $\mathbf{W}$. The second step in the method is to use the inferred matrix of interactions w_{ij} in Eq. (5.10), to find the dynamical influencers as explained in Sect. 5.3.

In contrast to other approaches such as Monte-Carlo methods, the application of this BP message passing framework scales to large networks as shown by Braunstein et al. [93]. BP outperforms other inference algorithms since BP takes into account the collective influence from multiple inputs. Thus BP offers a tremendous speed advantage over MC at no observable loss of accuracy [93].

The formulation to determine a regulatory network in Eq. (5.23) is aimed at the inference of causal relationships between genes, ignoring any existing interactions, resulting in tremendous computational costs. The problem can be mitigated by accounting for existing molecular interactions, as the determination of matrix $\mathbf{W}$ can be limited to account for the strength between pairs of nodes w_{ij} if i and j are known to physically interact. The application of this needs, of course, network data. Regulatory interaction networks defined above augmented with tissue and disease specific protein-protein and kinase-substrate interactions [9, 82, 83] can be used. Instead of accounting for all node combinations, the BP algorithm now considers a relatively smaller number of interacting node pairs, significantly mitigating the computational costs.

Drug specific perturbation profiles of tumor specific cell lines of lung, kidney, breast, skin and brain tissues as LINCS also provide unperturbed, non-disease control profiles of these cell lines. For a given cell line, perturbation profiles can be used to infer directed, weighted regulatory networks that are subject to the determination of dynamical influencers.

The relevance of influential genes as drug target is reflected by their ability to provide gene signatures that resemble the non-disease state. To test the importance of the influential genes as disease targets, the Characteristic Direction method (CD) [94], a variant of a linear discriminant analysis (LDA), can be used allowing to filter a cell-specific and drug specific gene signature. Differences between such signatures are evaluated by calculating their cosine difference: *(i)* Measuring the difference of signatures of cell lines that were treated with drugs on the dynamical non-influential gene, significant differences are expected to non-influencers. The comparison can be performed against the heuristics for validation, such as hubs, betweenness, etc. *(ii)* An assumption underlying the fact that dynamical influencers may be potential drug targets is the notion that a disease is considered a perturbation of a cellular system [18]. Therefore, signatures of disease cell lines that have been treated with drugs that act on influential genes can be compared with the unperturbed signatures of non-disease cell lines. The influencer-specific signatures will resemble signatures of non-disease signatures compared to non-influential genes.

In summary, the pipeline for using influencer theory to identify potential drug targets are as follows:

- Infer a network of interactions **W** from a subset of perturbations (i.e., drugs) using the dynamical system Eq. (5.23)
- Predict dynamical influencers for drug targets using the same dynamical system Eq. (5.10) with the inferred **W** to determine influential nodes.
- Then validate this choice with a perturbation profile with a drug that targets an influential gene from the LINCS L1000 datasets [92].

Utilizing such inferred regulatory networks, previous work by Molinelli et al. [17] computationally investigated the efficacy of certain drug targets through the simulation of hypothetical targets. Their pipeline is shown in Fig. 5.2. While obtaining certain success, the choice of targets in these studies followed qualitative considerations. The approach explained above finds influential drug targets from the network, which allows potential targets that can be computationally assessed for their efficacy directly from optimization theory to be found. In particular, the dynamic changes that are triggered by targeting an influential gene can be investigated by modeling the underlying systems through the system of equations that describe a perturbed system. That is, the approach is to guide the discovery of drug targets with sound optimization theory rather than qualitative considerations.

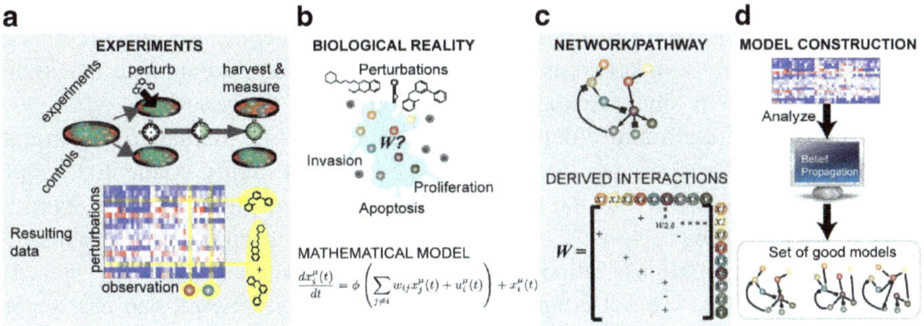

Fig. 5.2 Perturbation cell biology model of Molinelli et al. [17]. **a** A dynamic mathematical models of the cells' response to drugs is used to perturb cancer cells with targeted drugs to infer protein interactions. **b** The inference algorithm uses both quantitative parameters, and **c** qualitative network interpretation. **d** The predicted models are obtain with the Belief Propagation (BP) inference algorithm. Figure reproduced from [17]. Copyright © 2013, Molinelli et al.

5.5 Building Blocks of Genetic Networks

Having described biological networks at the level of influential nodes, we will now describe biological networks at a mesoscopic level of building blocks. Building blocks are small genetic circuits with well-defined functions that are the bricks of genetic machineries. Finding these building blocks is usually done by searching for overrepresented circuits [89], called "network motifs", and there is a large literature dedicated to identifying these motifs in, not only biological networks, but networks in general [89].

In all biological networks [95], some simple network motifs appear more often than they would by pure chance [89, 96, 97]. This statistical regularity has been interpreted as evidence that these network motifs are basic building blocks of biological machineries, a proposal that has had a major impact in systems biology [3, 89]. Searching for the designing principles underlying how these building blocks are arranged into an emergent biological networks is a basic goal of systems biology [3, 89, 95].

Network motifs introduced by Alon and collaborators [89, 96, 97] are the most used to find building blocks. They are overrepresented when they are compared to a null model usually defined by randomly rewiring the original network preserving the degree sequence. One of the most abundant motifs is the feed-forward loop (FFL) shown in Fig. 5.3a consisting of a regulator gene X regulating two genes, Y and Z, and with an additional regulation from gene Y to Z. Despite the 'loop' in the name, it is a fully feed-forward circuit, with no feedback among the genes.

However, mere statistical abundance by itself does not imply that the network motifs are funtional core bricks of biological systems. There is no experimental evidence that shows a direct link between network motifs to function [98], and how these motifs may have a functional role in biological computation remains unknown and controversial [98–101].

The same can be said about the metric-based centralities from Chap. 1 which are used to characterize the structure of biological networks like hubs, k-cores, collective influence or rich-clubs, for example. Despite of being essential for the integrity of the network as shown in previous sections, there is no direct evidence that these centralities can tell us anything about the functionality of the genes.

Modularity and community detection are very useful in understanding communities in social networks since they capture the idea that highly connected people who are weakly connected to the rest of the network form tightly-knight communities in social settings. However, this idea transported to biological networks to identify functional modules, called colloquially 'guilty by association' [3], presents problems. Genes that are not connected in the regulatory network can easily perform similar functions, and genes that are directly connected may perform different functions. Thus, using community detection algorithms to find functional modules from regulatory networks is problematic: such biological communities would only capture functional modules made of genes that are directly connected, while missing important modules of genes that are not.

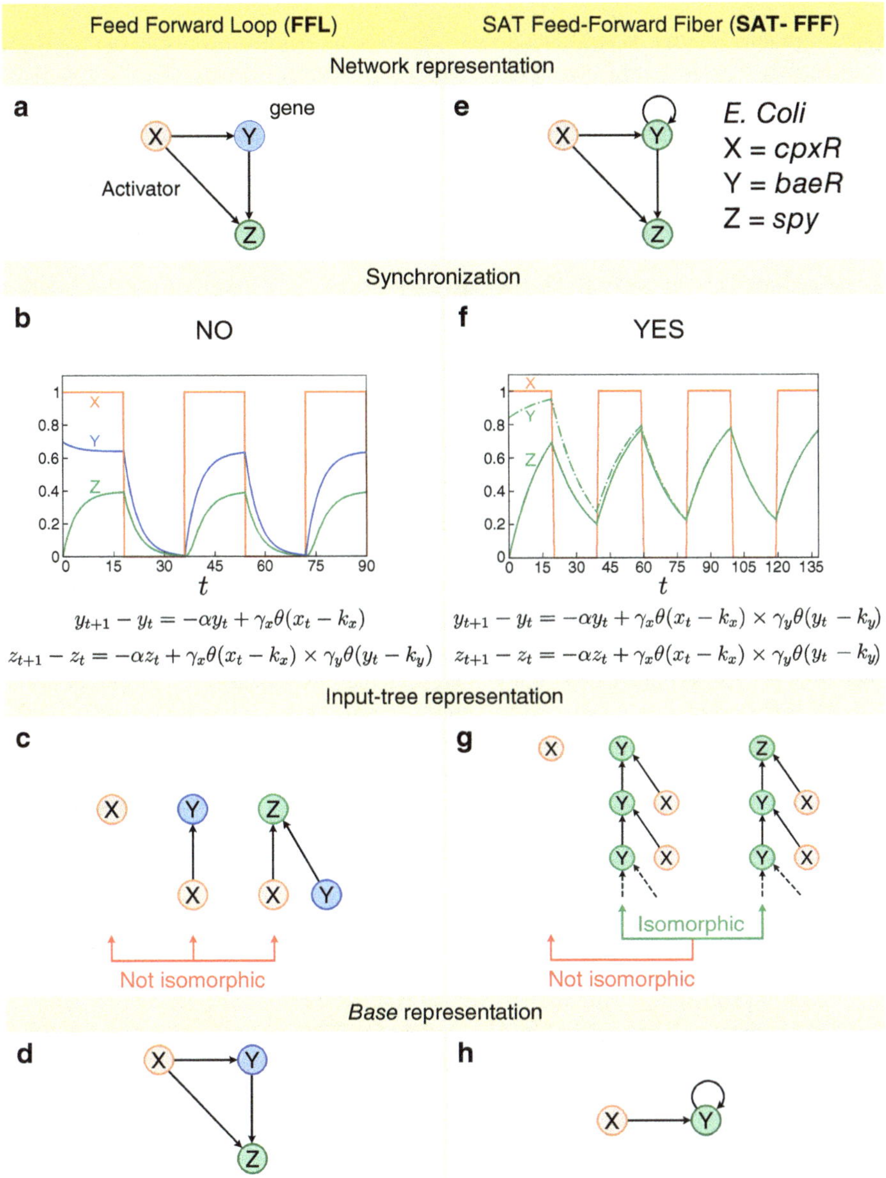

Fig. 5.3 (Continued)

◂**Fig. 5.3** (Continued) **Feed-Forward Loop (FFL) and Feed-Forward Fiber (FFF).** **a** FFL network representation. **b** Numerical solution of FFL dynamics. The expression levels of genes Y and Z do not synchronize. Dynamical equations in the figure with $\alpha = 2.0$, $\gamma_x = 0.12$, $\gamma_y = 0.7$, $k_x = 0.5$, $y_0 = 0.7$ and $z_0 = 0.0$. **c** Input tree representation of the FFL. The input trees of genes X, Y, and Z are not isomorphic; as a consequence, their expression levels do not synchronize, as shown in **(b)**. **d** The base of the FFL is the circuit itself since it has no symmetries. **e** FFF in E. coli made of genes *cpxR* = X, *baeR* = Y, and *spy* = Z. **f** The numerical solution of the equation shown in the figure for the FFF shows cluster synchronization of the expression levels of genes Y and Z. We use $\alpha = 0.06$, $\gamma_x = 0.775$, $\gamma_y = 0.775$, $k_x = 0.5$, $k_y = 1.0$, $y_0 = 0.85$ and $z_0 = 0.0$. **g** Genes Y and Z of the FFF have isomorphic input trees, form a fiber, and synchronize. The input tree of the external regulator *cpxR* is not isomorphic to Y nor Z, despite the fact that it directly regulates the fiber. **h** Since Y and Z form a fiber, they synchronize, and Z can be collapsed into Y, resulting in a reduced base. Figure reproduced from [123]. Copyright © 2020, Leifer et al.

Thus, network motifs, while providing a summary description of the structure of the networks, are too specific for understanding the emerging laws underlying the structure and function of biological networks. Network motifs, like the feed-forward loop [96, 97] are incapable of achieving the simplest form of coherent functionality, i.e., network motifs do not synchronize nor oscillate, and may not constitute functional building blocks. They cannot be traced back to fundamental principles on how the biological networks evolve over time, nor they have a functional interpretation since they are based purely on statistical significance. Application of graph theory to identify building blocks is very promising but requires more sophisticated methods of analysis than mere statistics to capture the complexity of the data.

We then ask what are the minimal canonical circuits that can achieve coherent function: an exceedingly simple topological principle, symmetry invariance [102], is necessary and sufficient to assure that a network configuration will result in coordinated function.

Network symmetries allow to find building blocks with functional significance. It has been shown in a series of papers starting with the work of Golubitsky and Stewart [103–105] that nodes in a network with symmetries show cluster synchronization of their activity [106, 107]. The mathematical principle of network symmetry can predict synchronization patterns in biological systems [108–110]. Symmetry principles are at the basis of important advances in abstract algebra [102, 111, 112] and physics [113–115], but their application to biological networks, or biology in general, is not well developed, except for seminal contributions by J. Monod [116] and D'Arcy Thompson [117].

Morone et al. [118, 119] showed that biological networks can be decomposed into building blocks identified by a particular symmetry of the graph called symmetry fibration. These symmetries can explain the regularities observed in the structure of biological networks made of genes, proteins, metabolites, and neurons. They are related to invariances in the input trees of the nodes specifying the dynamical state.

What information about the functionality of biological circuits can be gleaned from its decomposition into these building blocks? Symmetry fibrations predict coherent dynamics of cluster synchronization in their building blocks. Fibration building blocks, being uniquely defined by the symmetries of the network, determine the cluster synchronization of neural populations and gene co-expression profiles, and, thus, predict a common functionality of these blocks.

Interpreting genetic networks as composed of coherently synchronized genes, we can envision the biological system as an orchestra in which each instrument is a node in the network. When the instruments play coherently, with synchronized, rhythmical and structured temporal patterns, the network becomes functional. These synchronized patterns are observed in all biological systems [108–110]. For instance in circadian rhythms, neural oscillations and synchronization, cell-division cycle, sleep-wake cycle, respiration, locomotion, cardiac rhythm [3], and as gene co-expression patterns in gene regulatory networks [2, 3]. Next, we elaborate on how to use these symmetries to define essential building blocks in biology. A more comprehensive review of symmetries in biological systems appears in [120].

5.5.1 Fibration Symmetries to Identify Building Blocks

To set the grounds of our investigations, we first consider the example of the most abundant network motif in gene regulatory networks, the feed-forward loop (FFL) [89, 97]. A cFFL motif is shown in Fig. 5.3a and consists of three genes (X, Y and Z; c refers to coherent where all regulators are activators) where the transcription factor expressed by gene X positively regulates the transcription of Y and Z, and, in turn Y regulates Z. This network motif appears in the regulatory network of E. coli where X=*cpxR*, Y=*baeR*, and Z=*spy* genes. Numerical and analytic solutions for the expression levels of the genes in the cFFL shown in Fig. 5.3b show that the FFL does not produce coherent functions among the genes, since the gene expression do not reach synchronization (unless for ad-hoc setting of parameters) nor oscillations in expression levels. This is consistent with analysis by Mangan et al. [121] who have interpreted the functionality of the FFL as a sign-sensitive delay element in transcriptional networks.

Input tree In the FFL, gene Z receives input from X and Y, while gene Y, instead, only from X, and therefore the inputs are not symmetric (Fig. 5.3c). But, when looking for this motif in E. coli it is found that the gene Y in the FFL usually does not appear alone as in the FFL, but gene Y appears with an autoregulation (AR) loop. This minimal inclusion symmetrizes the circuit and this symmetry results in a circuit with a first and minimal form of function: synchronization in gene expression between Y and Z. This can be formalized by analyzing invariances in the '*input tree*', which represents all the paths that converge to a given gene.

In practice, the input tree of a gene is constructed as in Fig. 5.3e [119]. Consider the circuit in Fig. 5.3e. Gene X receives no input, so it has an empty input tree. The input tree of gene Y $=$ *baeR* (Fig. 5.3g) starts with Y at the root. Gene Y receives input from itself and X, so the first layer of its input tree is Y and X. Then, Y receives inputs also from Y and X and so on. This input tree has infinite number of layers, as shown in Fig. 5.3g. The input tree of Z starts at Z, with a first layer made of Y and X, and then repeated to infinity. Since there is loop in the circuit, the input tree has an infinite number of layers. We notice that the input trees of Y and Z have the same "shape".

The input tree formalism provides a framework to search for symmetries in information-processing networks [119]. These symmetries preserve the information flow in the network, in that it replaces the canonical notion of a single trajectory with the set of all possible 'histories' from an initial to a final state of the network, and this makes, in practice, reasonably straightforward to 'guess' a type of symmetry which is not apparent in the classical network framework. Simply put, if two input trees have the same 'shape', then the genes at the root of the input trees receive the same information through the totality of all paths ending at the root, they share the same dynamical state and they synchronize their activity, as shown in Refs. [119, 122, 123].

Isomorphic input trees This informal notion of equivalence is formalized by isomorphisms. An isomorphism between two input trees is a bijective map that preserves the topology of the input trees including their links (notice that this takes into account activators and repressors). Specifically, a map $\tau : T \to T'$ is an isomorphism iff for any pair of nodes a and b of T connected by a link, the pair of nodes $\tau(a)$ and $\tau(b)$ of T' is connected by the same type of link, too [119]. In practice, this means that isomorphic input trees are 'the same' except for the labeling of the nodes. Two input trees related by an isomorphism are isomorphic or equivalent and genes with isomorphic input trees are symmetric because they receive the same information and share their dynamical state.

It is important to note that genes Y and Z are symmetric and synchronized despite the fact that there is no automorphism in this circuit. An automorphism is a permutation symmetry of nodes in the network that leaves the adjacency matrix invariant. This is a much stringent condition than the one imposed by isomorphic input trees. Isomorphisms between input trees preserve the dynamics via the inputs of the nodes, while automorphisms preserve also the outputs, i.e., all the network connectivity. Biological networks prefer the less stringent symmetry fibrations to the stronger automorphisms forming symmetry groups that are prevalent in physics.

A principle of symmetry, called **symmetry fibration** [119, 122, 124] predicts this, since symmetries Y $\leftrightarrow$ Z that leave invariant the input trees (but not necessarily the outputs as imposed by automorphisms) are necessary and sufficient to achieve synchronization. When two genes have isomorphic input trees, their expression levels are synchronized [105, 119].

Symmetry Fibration and Fibers The set of all input tree isomorphisms defines the symmetries of the genetic network, which are extensions of '*Grothendieck fibration*'; [124], introduced by Grothendieck in the framework of algebraic geometry (see [122] for a historical account). The original Grothendieck definition of fibration is between categories [124]. The passage to a definition of fibrations between graphs requires to associate a category with a graph and rephrase Grothendieck's definition in elementary terms. Different categories may be associated with a graph, giving rise to different notion of fibration between graphs. The notion of fibration that we use henceforth has been introduced in computer sciences by Boldi and Vigna in [122] as a '*graph fibration*'.

A graph fibration of a network G is a homomorphism:

$$\psi : G \to B \tag{5.30}$$

that maps a graph G to a graph B, called the '*base*' of the fibration ψ. Two genes in G are mapped to the same node in B iff they have the same input tree.

Any two genes i and j with isomorphic input trees are mapped onto the same base which is called a '*fiber*' of B by a fibration ψ. Below we will always consider minimal fibrations: ψ is a minimal fibration, then ψ maps all the genes with isomorphic input trees to a common node in B. That is, a minimal fibration collapses a graph to its minimal base with the minimal number of fibers.

A minimal (and surjective) graph fibration is called a '*symmetry fibration*' in [119] to highlight the fact that this fibration captures the symmetries in the dynamical state (synchronization) of the nodes in the fiber. The minimal base of a network, where all the genes in a fiber have been collapsed into one node by the fibration, leads to a dimensional reduction of the network into its irreducible components or '*fiber building blocks*'. Then, the fibration is a dimensional reduction that preserves the information in the network, and thus the fibers are the synchronized building blocks of the network.

Feed-Forward Fiber (FFF) Gene *baeR* in Fig. 5.3e forms a symmetry with gene *spy* characterized by the isomorphism between their input trees (Fig. 5.3g). These genes belong to the same fiber, and their expression levels are synchronized (Fig. 5.3f). Genes in the same fiber are redundant, and are collapsed into the base shown in Fig. 5.3h by the symmetry fibration.

Instead, the FFL in Fig. 5.3a [89] cannot be reduced since it has no symmetry. Thus, the base of the FFL is the same FFL, see Fig. 5.3d. Numerical simulations and analytical solutions for the expression activity shown in Fig. 5.3f confirm that the addition of the autoregulation loop (AR) loop to the FFL at gene Y changes its functionality qualitatively, leading to synchronization of genes Y and Z into coherent co-expression. This synchronized fiber is called a Feed-Forward Fiber (FFF) in [119].

The FFF in Fig. 5.3e has all activator regulations and leads to a simple dynamics converging to a fixed point: all interactions are satisfied, meaning that the Heaviside step functions in Fig. 5.3f evaluate to 1. This circuit is called SAT-FFF for satisfied. Instead, when the autoregulation loop at Y is a repressor, the loop behaves as a logical NOT gate (see Fig. 5.4a, b). When expression is high, it inhibits itself shifting to a low state. Instead, if it is low, it promotes itself to change to a high state. Hence, the activity of this gene oscillates indefinitely [125, 126]. This is the simplest expression of *frustration* [126, 127], a core concept in complexity and spin glasses [91] which refers to a system which is always in tension and thus never reaches a stable fixed configuration.

Hierarchy of building blocks To understand the functional rationale of symmetric circuits made of repressors and feedback loops, they can be mapped to a hierarchy of

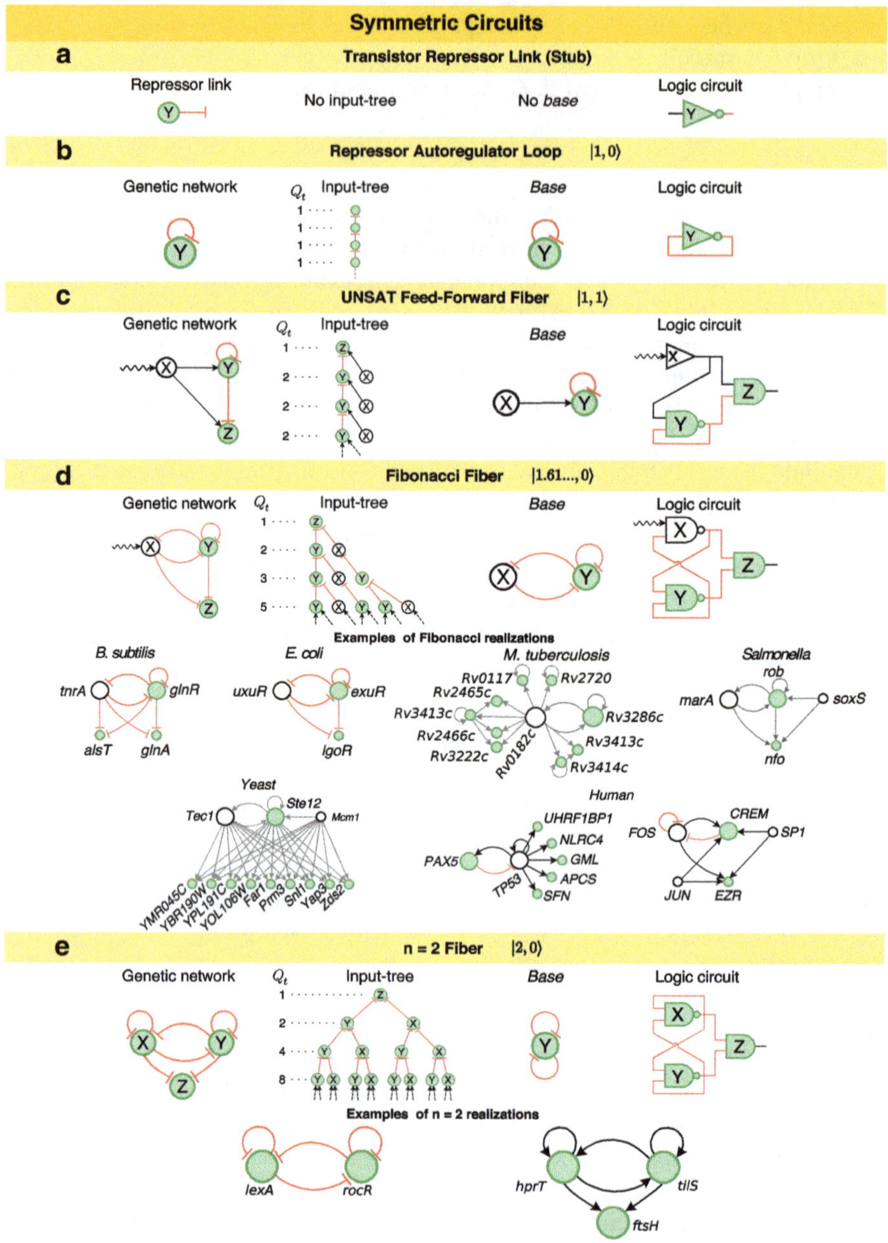

◄**Fig. 5.4 Hierarchy of symmetric building blocks of increasing complexity**. **a** A repressor link turns into a **b** repressor autoregulation loop on an input tree that feeds its own expression levels with $Q_t = 1$ and branching ratio $n = 1$ analogous to an electronic logic circuit ring oscillator made of a NOT gate. **c** The addition of an external regulator creates the UNSAT-FFF where genes Y and Z synchronize and oscillate verified in its corresponding input tree and logic circuit. **d** The (Continued)

◂**Fig. 5.4** (Continued) addition of a second feedback loop results in an input tree that follows the Fibonacci sequence $Q_t = 1, 2, 3, 5, 8, \ldots$. The branching structure of the input tree implies that the toggle switch at the base of Fibonacci Fiber is a memory storage device analogous to an electronic flip-flop. Examples shown from different species. **e** The second autoregulation at X results in a symmetric input tree with $Q_t = 2Q_{t-1}$ and $n = 2$ from the regulatory network of *B. subtilis*. Examples are shown with activator links (black), repressor (red), and interactions with unknown functionality (grey). Figure reproduced from [123]. Copyright © 2020, Leifer et al.

electronic analogues (see Fig. 5.4) [123]. The idea of using electronic circuitry and devices to mimic aspects of gene regulatory networks has been in circulation almost since the inception of regulatory genetics itself [116], and has been a driving force in synthetic biology; with several demonstrations showing that engineered biological circuits can perform computations such as a toggle switch [125, 128–130], logic [131], memory storage [128, 132, 133], pulse generators [134] and oscillators [125, 135–137].

The repressor AR genetic circuit of Fig. 5.4b becomes a one-stage ring oscillator logic circuit (Fig. 5.4b right), where the collector of the transistor in the NOT gate connects to the base forming the minimal signal feedback loop. This autoregulator loop (Fig. 5.4b) is very abundant in E. coli as shown in the work of Alon and others [89, 138].

An example of the AR is the tryptophan repressor *trpR* in the E. coli transcriptional network. The repressor AR loop can be extended to the FFF by symmetrizing it and adding gene Z so that it synchronizes with Y to express an enzyme that catalyzes a biochemical reaction. The circuit is completed with the external regulator X, which maintains the symmetry between Y and Z. The resulting circuit made of repressors is called UNSAT-FFF (since it is frustrated and some interactions are unsatisfied, see Fig. 5.4c). The UNSAT-FFF maps to the Widlar current-mirror electronic circuit, a building block of integrated electronic circuits used since the beginnings of the semiconductor industry (1967 US patent [139, 140]).

The UNSAT-FFF circuit serves two key functions: mirror synchronization of $y_t = z_t$ and oscillatory activity due to its frustration of interactions [123]. Thus, the repressor autoregulation converts the circuit into a clock, a primary building block in any logic computational device. Results in [119] indicate that these FFFs (including SAT-FFF and UNSAT-FFF) are abundant in gene regulatory networks across species spanning *A. thaliana*, *M. tuberculosis*, *B. subtilis*, E. coli, *salmonella*, *yeast*, mouse and humans.

This constructive procedure can be extended through an algebra of circuits that adds external regulators and loops to grow the base of symmetric circuits (Fig. 5.4). In this space, the AR (Fig. 5.4b) is the core loop unit referred to as $|n = 1, \ell = 0\rangle$ since it has $n = 1$ loop and $\ell = 0$ external regulator. The next step in the hierarchy is the FFF (Fig. 5.4c) which has also one loop $n = 1$ but also one regulators, $\ell = 1$, then: $|n = 1, \ell = 1\rangle$.

From this starting point, one can grow the number of regulator genes of the FFF, $|1, \ell\rangle$ with $\ell > 1$. This does not affect the complexity of the circuit, measured by n, which quantifies the divergence or branching ratio:

$$n = \left(\frac{Q_{t+1}}{Q_t}\right)_{t\to\infty}, \tag{5.31}$$

of the input tree, where Q_t is the number of source genes with paths of length $t - 1$ to the target gene, see Fig. 5.4c), because all the relevant dynamics remain constrained to the sole loop in the FFF circuit. This changes as soon as the fiber feeds information back to the external world. This is the case of the circuits in Fig. 5.4d, where the gene Y now regulates its own regulator gene X, creating a feedback loop. The inclusion of this feedback loop results in the coexistence of two time-scales in the network: one from the AR loop, another from the feedback loop. This, in turn, increases the diversity of trajectories and delays in the network; a dynamic complexity that is measured from the divergence of the input tree Q_t (Fig. 5.4d).

A quantitative analysis of this measure yields exactly the golden ratio:

$$\left(\frac{Q_{t+1}}{Q_t}\right)_{t\to\infty} = \varphi = \left(1 + \sqrt{5}\right)/2 = 1.6180... \tag{5.32}$$

for the circuits in Fig. 5.4d, revealing that the input tree is a Fibonacci sequence $Q_t = Q_{t-1} + Q_{t-2}$ updating the current state two steps backwards. This class of building blocks are called "*Fibonacci Fibers*", Fig. 5.4d.

For example, the repressor interactions between genes X = *uxuR*, Y = *exuR*, and Z = *lgoR* from the E. coli network function exactly as a Fibonacci Fiber (Fig. 5.4d, see also Fig. 5.5). Morone and Leifer et al. [119, 123] find examples of Fibonacci Fibers in the transcriptional networks across species from bacteria to yeast and humans shown in Fig. 5.4d. A list of FFF and Fibonacci fibers circuits appear in [141, 142] showing that the circuits are ubiquitous in E. coli. Figure 5.5 top shows the rich fibers that are found in the carbon processing component in E. coli regulated by the master regulator genes *crp* and *fis*, whose regulon includes hundreds of genes. The figure also shows how the fibration works to reduce this component to its base. A p-value statistical test indicates that these circuits are statistically significant [119, 123].

The complexity of the Fibonacci Fiber with feedback to the regulator is 1.6180..., which is lower than the number of loops in the circuit (two). The intuition of what this reveals is that, in this circuit the regulator X is still not part of the base (see Fig. 5.4d), since it does not receive input from itself and hence it is not within the symmetry of the fiber. This, in turn, indicates naturally that the next element in the hierarchy of fibers results from the inclusion of an AR loop in X. This creates a fully symmetric circuit (Fig. 5.4e) with a core $|2, \ell = 0\rangle$ that feeds the reporter/enzyme

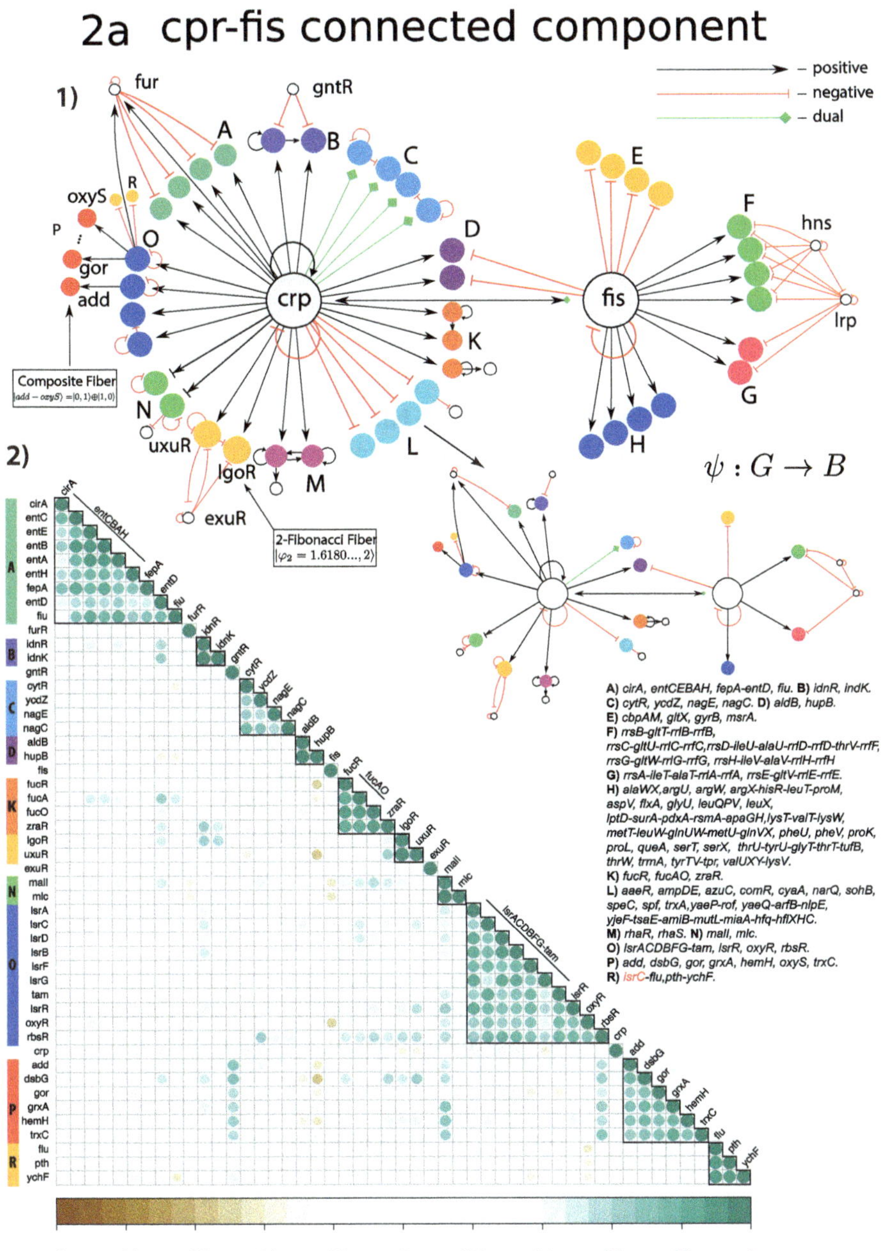

◂**Fig. 5.5 Gene synchronization of gene co-expression in the fibers regulated by the master regulators** *crp* **and** *fis* **in** E. coli **carbon catabolism regulatory network.** Top: Two-gene connected component of *crp-fis*. This component controls a rich set of fibers: FFFs, Fibonaccis (*uxuR-lgoR*), as well as the double-layer composite $|add - oxyS\rangle = |0, 1\rangle \oplus |1, 1\rangle$. **Bottom**: Gene co-expression (correlation) (Continued)

◂**Fig. 5.5** (Continued) matrix calculated from the Pearson coefficient between the expression levels of each pair of genes in the top. Synchronization of the genes in the respective fibers appears as the block structure of the matrix. The genes that form operons trivially synchronize since they are transcribed by the same polymerase, and are indicated in the figure by a line along the diagonal. Figure reproduced from [123]. Copyright © 2020, Leifer et al.

Z. In this case, the complexity of the circuit is exactly two (the number of autoregulators within the fiber). Examples in bacteria are shown in Fig. 5.4d. The important component of these circuits is the delay in the feedback loop through the regulator from Y → X and back to Y captured by the Q_{t-2} term in the Fibonacci sequence.

Oscillators, clocks and memory devices as building blocks All symmetric circuits shown in Fig. 5.4 work as clocks with varying levels of sophistication and robustness given by their time-scales or loops. Additionally, the more complex Fibonacci Fibers store RAM memory by integrating sequences of its two immediate past states, according to the input tree, Q_{n-1} to its previous one, Q_{n-2}, forming a dynamical memory counter $Q_n = Q_{n-1} + Q_{n-2}$ with dynamical two-step memory, which can compute temporal convolutions.

Leifer et al. [123] have shown that when the symmetry of the core base of the Fibonacci circuit and the $n = 1$ circuits is broken, the resulting circuit is isomorphic to a flip-flop electronic circuit, which is the universal storage device of ROM computer memories [140]. In fact, the feedback loop that appear between genes X and Y in the Fibonacci fiber (Fig. 5.4d) works exactly as a bistable toggle switch which is the simplest electronic device to store a bit of information in computer memory and has been reproduced in synthetic biological circuits in [125, 128–130] for memory storage [128, 132, 133]. Therefore, breaking the symmetry in a Fibonacci fiber can have the function to store memory, and thus perform a key logic computation in the genetic network, see [123].

Synchronization Lets consider a nonlinear dynamical equations for the expression of gene i, $x_i(t)$, in a gene regulatory network with Hill function interactions [89] with cooperative exponent n. The directed network is characterized by interaction strength $w^A_{j\to i}$ for activators and $w^I_{j\to i}$ for inhibitor interactions. A simple interaction form for the sake of argument assumes OR gates, but AND multiplicative gates can be used analogously without changing the argument, see [89]:

$$\frac{dx_i}{dt} = -\alpha x_i + \sum_j w^A_{j\to i} \frac{x_j^n}{1+x_j^n} + \sum_{j'} w^I_{j'\to i} \frac{1}{1+x_{j'}^n} \,. \tag{5.33}$$

When these equations are applied to the FFF in Fig. 5.3e, cluster synchronization is observed since for every solution $x(t), y(t), z(t)$ of this system exists a solution $x(t), y(t), y(t)$ of the system where $y(t) = z(t)$, thus predicting cluster synchronization of these two variables. This is so because the interaction term depends on the incoming links, and these nodes are isomorphic through the symmetry fibration that reduces the original set to its base.

Thus, symmetries in gene regulatory networks reveal the functions of cluster synchronization, clocks and memory through AR, FFFs, Fibonacci and $n = 2$ fibers which are electronic analogues of simple circuits like ring-oscillators and (with broken symmetries) flip-flops. They result in a set of building blocks with progressively more complex dynamics obtained by iterating a procedure of replication and symmetry. Beyond the circuits discussed here, the biological hierarchy can be extended to any number m of loops of length d and autoregulators in the fiber n, to form ever more sophisticated circuits whose complexity is expressed in generalized Fibonacci sequences $Q_t = nQ_{t-1} + mQ_{t-d}$. This complex biological circuitry can then be seen as a building process guided by the laws of symmetry that determine the biological computations.

The symmetry analysis allows to:

- systematically classify biological building blocks from first principles,
- propose principles of design for synthetic genetic circuits following biological symmetry, and
- systematically map the structure and function of biological networks, from genetic to metabolic, signaling and protein network to the connectome [118] following a principled theoretical approach.

Algorithms Studies of fibration symmetries [119] find that the transcriptional network of E. coli is organized into a hierarchy of fibers [141, 142]. The algorithm to obtain minimal fibers in networks is based on the balanced coloring algorithm developed by Kamei and Cock [143] and Cardon and Crochemore [144] which are provided in Appendix A. A fast version is proposed by Monteiro et al. in [145], and extensions to algorithms for network reconstructions via pseudo-balanced colorings and quasi-fibrations have been proposed in Leifer et al. [146] and Boldi et al. [147], respectively.

Using these algorithms a rich variety of topologies in the input trees is found. Despite this diversity, the input trees present common topological features that allow to classify all circuits into concise classes of fiber building blocks (Fig. 5.4). We define a fiber building block as the genes in the fiber plus all external regulators of the fiber and their regulators in turn that are needed to determine the synchronization of the genes in the fiber. Thus, the 'fiber building block' is composed by the genes in the fiber plus the minimum number of external regulator genes and loops needed to establish the local equivalence or isomorphism in the fiber.

Evidence of the existence of fibration symmetries in biological networks ranging from transcriptional regulatory networks, signaling networks, protein-protein interactions, metabolic networks, cellular processes networks and pathways, disease networks, and neural networks is presented in [119, 123]. These studies span a large variety of species ranging from *A. thaliana*, *M. tuberculosis*, *B. subtilis*, E. coli, *salmonella*, *yeast*, mouse and humans. These biological datasets (listed below) reveal the existence of fiber building blocks underlying and synchronization. This study was done on a approximately 400 biological networks across available species with data are gathered from publicly available sources [148–156].

Using these datasets, fibers are found to be ubiquitous and statistical significant in biological networks [123]. The obtained building blocks are compared with randomized samples to determine their statistical significance as done in similar analysis of motifs [89, 96]. This is done with p-value and Z-score statistics by comparing the occurrence of fibers with those in a random null model obtained by randomizing the links in the network preserving the degree sequence. This analysis classifies all the symmetries found in biological networks [119, 120]. Thus symmetry fibrations provide a classification of genetic circuits from first principles, and they are significant as compared with randomized null models obtained by randomizing the network structure preserving the degree sequence.

Fiber building blocks and gene co-expression patterns of synchronization The existence of isomorphic input trees within the fiber implies cluster synchronization. This result applies to all fibers, irrespective of the dynamical system law, as long as the set of dynamical equations imposed on the graph is admissible [120], i.e., it is compatible with the symmetries of the underlying graph.

Leifer et al. [157] used gene expression data profiles in E. coli compiled at Ecomics [158] to confirm the predicted genetic cluster synchronization in the fiber building blocks, see Fig. 5.5. Fibrations have direct implications for gene co-expression patterns in the regulatory network. They make sure that genes are turned on and off at the right amount to assure the synchronization of expression levels in the fiber needed to execute cellular functions. Since, correlated co-expression profiles are common to all species, fibrations and input trees topologies occur in biological networks of all species. The analysis is also extended to any type of information-processing networks like social networks and others in [123] providing a fibration landscape across complex networks. Fibration symmetries may help to systematically organize the building blocks of biological networks in general from the invariances in the information flow encoded by the topologies of the constitutive input trees that lead to the synchronization needed for cellular function.

5.5.2 Essential Building Blocks of Genetic Networks

The practical application of fibration symmetries is to classify the building blocks of genetic networks. The output of the theory provides a ranking of genes into fibers. The prediction is that perturbation of these genes will affect the synchronization and functionality of the cell. Concurrently, we can classify the genes with other network metrics for building blocks, like modularity and community detection [159–161] and network motif [89], hubs, and centralities. These rankings identify vulnerable network areas according to each theory and can be tested in tissue specific networks and related disease phenotypes.

Having these theories in hand to identify building blocks, the question arises as to which structures are most relevant for loss of function under targeted gene knockouts and manipulations. The theory developed in Sect. 5.4 can be used to identify the '*essential fibers*', which, under malfunction, most greatly affect the synchronization of the regulatory networks and may potentially lead to identifying genes that could be responsible for disease or potential drug targets. This optimization problem tries to find the most vulnerable genes fibers under knockouts in a biological network. The idea is to find the minimal set of fibers, not just nodes, as done in the influencer theory of Sect. 5.4, that maximize the damage to the synchronous activity of the network, and thus maximally affect function of the cell under malfunction.

The fibration symmetry reduces all synchronized redundant genes to its base. So, before applying the optimization theory one can apply the fibration and reduce the network to its minimal base. This pre-processing step performs a dimensional reduction that allows to find the essential fiber building blocks. The optimization theory finds those essential base genes controlling the essential fibers, that is, the essential base of the network whose knockout leads to the largest perturbation to the network. To test the relevance of these essential fibers in biological networks, extensive test can be performed in regulatory interaction networks, potentially allowing to find disease relevant building blocks using the methods explained in Sect. 5.2 or possible drug-target building blocks with the methods of Sect. 5.4. The full theoretical framework can be tested using genomic regulatory data and gene knockouts targeted to produce desynchronization in predicted essential fibers of genes. These essential fibers provides information to determine disease genes and specific drug-target pathways in the genetic network.

5.6 Outlook for Essentiality in Genetic Networks

In this chapter we have tackled a recurring theme in complexity science: how the function of the network emerges from the interactions of the building blocks of the complex system. This problem permeates math, physics and biology. For instance, in the physical science, the building blocks of matter are electrons and quarks, that can be assembled into atoms, which in turn combine to form molecules. In biology, these molecules are assembled to form genes and proteins that are the basic building blocks of all living organisms.

In the abstract realm of arithmetic, prime numbers are the building blocks of all natural numbers, since any natural number can be represented as a unique product of primes. Lastly, in the highly abstract area of group theory [111, 112], the building blocks of symmetry groups are represented by normal subgroups, which behave in much the same way as prime numbers, since all (finite) groups can be formed as a product of them in a unique way. In group theory, the symmetry group of an object is the group of all transformations under which the object remains invariant (e.g., the symmetry permutations of nodes that leaves the connectivity of the network invariant).

Interestingly, it is the latter, and more abstract, example that has found broad and profound implications for biological circuits. However, we have seen that symmetry groups that are ubiquitous to describe all physical phenomena, from classical, relativity and quantum mechanics to the Standard Model of particle physics and almost all phase transitions in nature, are not the right symmetry for biology. A more relaxed symmetry is found in the symmetry fibration which is ubiquitous across biological networks.

We have seen also that even percolation from classical physics is not enough to describe the influencer problem in biological networks in its full glory. To go beyond structural influencer from percolation approaches in biology, the dynamical system of equations for gene expression need to be taken into account with their proper activators and inhibitory interactions among genes.

When the system contains inhibition, the coupling between the network structure and the dynamics makes this problem extremely difficult to solve. However, there are certain circumstances where the dynamics can be decoupled from the network structure, leading to predictions of the dynamical state. This is the case of symmetric networks predicting synchronization in terms of only the network topology, i.e., ignoring the dynamics. Breaking apart the dynamical behavior from the structure is a feature that it is always thought after, but rarely happen in the study of dynamical systems, unless a very specific structure dominates the network: this is the case of symmetries.

But, we have also seen other examples of this decoupling. We have seen in Sect. 1.5.1 that when the dynamics of information spreading in a social network can be modeled with the LTM model with threshold $\theta_i = k_i - 1$, where k_i is the degree, then the problem to find the best spreaders is the same as optimal percolation. This is quite a nice result since it reduces a complicated dynamical problem to a structural problem that does not depend on the dynamics. This result implies that the dynamics is decoupled from the structure of the network, and this is a very nice result.

But this nicely behaved system was obtained at the expense of narrowing the dynamics to a particular model of information flow and with a specific threshold. For any other dynamics or threshold in the LTM or any other model of propagation, the mapping to percolation ceases to exist, and the problem needs to be solved at the dynamical level, which is always much more difficult than the percolation static level.

Another example of 'nicely behaved' system where the dynamics decouples from the static is the mutualistic system of species that will be analyzed in Chap. 7. In this case, we find again, that the influential species are determined purely by the static network, indeed by the k-core, and in principle there is no need to solve for the dynamics in this case. The most influential mutualistic species, the keystone species, are determined purely by a static metric: they are located in the maximum k-core of the network.

Again, this lucky decoupling is obtained with some approximations. The most important is that a mutualistic system represent only positive interactions among the species; mutualistic species can only benefit from each other, ergo, percolation works well for this system. Ecosystems, of course, are not only made of mutualistic species, but of all sort of different interactions. If we were to consider, for instance, predator-prey interactions which are of inhibitory type, then we would not be able to decouple the structure from dynamics, and the k-core solution would not be valid anymore.

Furthermore, mutualistic models of Chap. 7 assume reciprocal interactions, like an action-reaction principle in physics. That is, the benefit from species i to j is the same that from j to i. Reciprocity is again crucial to obtain the decoupling into the k-core.

Reciprocal interactions dominates physics, but not biology. They are built in classical mechanics in Newton's third law of action-reaction principle and are the basis of equilibrium in statistical mechanics via detailed balanced relations. Without reciprocity, equilibrium is not achieved, and the system will be in an out-of-equilibrium state. Biological systems rarely satisfy reciprocity in the interactions. Neurons and genes interact in a directional way, i interacts with j in a different way as j with i and most of the time, one of these interactions does not even exist. These non-reciprocal interactions show up also in social groups. They need to be treated with a different language than equilibrium statistical mechanics. Recent progress in physics [162–164], however, has shown that non-reciprocal systems show phase transitions, like those that are reciprocal, that can be characterized by exceptional points with bifurcations and synchronized behavior more proper of biological systems. Current investigations point to applications of these theories to the brain [165].

Overall, when the underlying graph over which the dynamical system of equations is defined contains symmetries, then the dynamics are determined by the topology of the network, i.e., the synchronized building blocks are determined by the topological symmetries. Symmetries predict the existence of cluster synchronization and this predicted synchrony state is independent of the type of dynamics as long as the dynamical equations are admissible with the graph structure. On the other hand, the stability of these synchronous solutions, called synchronizability, is determined by the type of dynamical system and needs to be treated separately for each dynamics in turn [107, 166].

References

1. Buchanan, M., Caldarelli, G., De Los Rios, P., Rao, F., Vendruscolo, M. (eds.): Networks in Cell Biology. Cambridge University Press, Cambridge (2010)
2. Palsson, B.Ø.: Systems Biology: Properties of Reconstructed Networks. Cambridge University Press, New York (2006)
3. Klipp, E., Liebermeister, W., Wierling, C., Kowald, A., Herwig, R.: Systems Biology: A Textbook. Wiley-VCH, Weinheim (2016)
4. Karlebach, G., Shamir, R.: Modelling and analysis of gene regulatory networks. Nat. Rev. Mol. Cell Biol. **9**, 770–780 (2008)
5. Lee, T.I., Rinaldi, N.J., Robert, F., Odom, D.T., Bar-Joseph, Z., Gerber, G.K., Hannett, N.M., Harbison, C.T., Thompson, C.M., Simon, I., Zeitlinger, J., Jennings, E.G., Murray, H.L., Gordon, D.B., Ren, B., Wyrick, J.J., Tagne, J.B.. Volkert, T.L., Fraenkel, E., Gifford, D.K.,

Young, R.A.: Transcriptional regulatory networks in Saccharomyces cerevisiae. Science **298**, 799–804 (2002)
6. Jeong, H., Mason, S.P., Barabási, A.-L., Oltvai, Z.: Lethality and centrality in protein networks. Nature **411**, 41–42 (2001)
7. Barabási, A.-L., Gulbahce, N., Loscalzo, L.: Network medicine: a network-based approach to human disease. Nat. Rev. Genet. **12**, 101–113 (2011)
8. Vidal, M., Cusick, M.E., Barabási, A.-L.: Interactome networks and human disease. Cell **144**, 986–998 (2011)
9. Yeger-Lotem, E., Sharan, R.: Human protein interaction networks across tissues and diseases. Front. Genet. **6**, 257 (2015)
10. Goh, K.I., Cusick, M.E., Valle, D., Childs, B., Vidal, M., Barabási, A.-L.: The human disease network. Proc. Natl. Acad. Sci. U.S.A. **104**, 8685–8690 (2007)
11. Magger, O., Waldman, Y.Y., Ruppin, E., Sharan, R.: Enhancing the prioritization of disease-causing genes through tissue specific protein interaction networks. PLoS Comput. Biol. **8**, e1002690 (2012)
12. Erler, J.T., Linding, R.: Network-based drugs and biomarkers. J. Pathol. **220**, 290–296 (2010)
13. Pawson, T., Linding, R.: Network medicine. FEBS Lett. **582**, 1266–1270 (2008)
14. Csermely, P., Korcsmaros, T., Kiss, H.J.M., London, G., Nussinov, R.: Structure and dynamics of molecular networks: a novel paradigm of drug discovery: a comprehensive review. Pharmacol. Ther. **138**, 333–408 (2013)
15. Hopkins, A.L.: Network Pharmacology: the next paradigm in drug discovery. Nat. Chem. Biol. **4**, 682–690 (2008)
16. Kotlyar, M., Fortney, K., Jurisica, I.: Network-based characterization of drug-related genes, drug targets, and toxicity. Methods **57**, 499–507 (2012)
17. Molinelli, E., Korkut, A., Wang, W., Miller, M.L., Gauthier, N.P., Jing, X., Kaushik, P., He, Q., Mills, G., Solit, D.B., Pratilas, C.A., Weigt, W., Braunstein, A., Pagnani, A., Zecchina, R., Sander, C.: Perturbation biology: Inferring signaling networks in cellular systems. PLoS Comput. Biol. **9**, e1003290 (2013)
18. del Sol, A., Balling, R., Hood, L., Galas, D.: Diseases as network perturbation. Curr. Opin. Biotechnol. **21**, 566–571 (2010)
19. Stelling, J., Sauer, U., Szallasi, Z., Doyle, F.J., 3rd., Doyle, J.: Robustness of cellular functions. Cell **118**, 675–685 (2004)
20. Zhu, X., Gerstein, M., Snyder, M.: Getting connected: analysis and principles of biological networks. Genes Dev. **21**, 1010–1024 (2007)
21. Albert, R., Jeong, H., Barabási, A.-L.: Error and attack tolerance of complex networks. Nature **406**, 378–382 (2000)
22. Batada, N.N., Hurst, L.D., Tyers, M.: Evolutionary and physiological importance of hub proteins. PLoS Comput. Biol. **2**, e88 (2006)
23. Hahn, M.W., Kern, A.D.: Comparative genomics of centrality and essentiality in three eukaryotic protein-interaction networks. Mol. Biol. Evol. **22**, 803–806 (2005)
24. Yu, H., Greenbaum, D., Lu, H.X., Zhu, X., Gerstein, M.: Genomic analysis of essentiality within protein networks. Trends Genet. **20**, 227–231 (2004)
25. Zotenko, E., Mestre, J., O'Leary, D.P., Przytycka, T.M.: Why do hubs in the yeast protein interaction network tend to be essential: reexamining the connection between the network topology and essentiality. PLoS Comput. Biol. **4**, e1000140 (2008)
26. Yu, H., Kim, P.M., Sprecher, E., Trifonov, V., Gerstein, M.: The importance of bottlenecks in protein networks: Correlation with gene essentiality and expression dynamics. PLoS Comput. Biol. **3**, e59 (2007)
27. Jonsson, P.F., Bates, P.A.: Global topological features of cancer proteins in the human interactome. Bioinformatics **22**, 2291–2297 (2006)
28. Taylor, I.W., Linding, R., Warde-Farley, D., Liu, Y., Pesquita, C., Faria, D., Bull, S., Pawson, T., Morris, Q., Wrana, J.L.: Dynamic modularity in protein interaction networks predicts breast cancer outcomes. Nat. Biotechnol. **27**, 199–203 (2009)

29. Wuchty, S.: Controllability of protein interaction networks. Proc. Natl. Acad. Sci. USA **111**, 7156–7160 (2014)
30. Khuri, S., Wuchty, S.: Essentiality and centrality in protein interaction networks revisited. BMC Bioinf. **16**, 109 (2015)
31. Wuchty, S., Boltz, T., Kuecuek-McGinty, H.: Links between critical proteins drive the controllability of protein interaction networks. Proteomics **17**, 1700056 (2017)
32. Boltz, T.A., Devkota, P., Wuchty, S.: Collective influencers in protein interaction networks. Sci. Rep. **9**, 3948 (2019)
33. Jiang, X., Liu, B., Jiang, J., Zhao, H., Fan, M., Zhang, J.: Modularity in the genetic disease-phenotype network. FEBS Lett. **582**, 2549–2554 (2008)
34. Granovetter, M.S.: The strength of weak ties. Am. J. Sociol. **78**, 1360–1380 (1973)
35. Kitsak, M., Gallos, L.K., Havlin, S., Liljeros, F., Muchnik, L., Stanley, H.E., Makse, H.A.: Identifying influential spreaders in complex networks. Nat. Phys. **6**, 888–893 (2010)
36. Morone, F., Makse, H.A.: Influence maximization in complex networks through optimal percolation. Nature **524**, 65–68 (2015)
37. Kitano, H.: Biological robustness. Nat. Rev. Genet. **5**, 826–835 (2004)
38. Kitano, H.: A robustness-approach to systems-oriented drug design. Nat. Rev. Drug Discov. **6**, 202–210 (2007)
39. Kitano, H.: Cancer as a robust system. Nat. Rev. Cancer **4**, 227–235 (2004)
40. Cho, D.Y., Kim, Y.-A., Przytycka, T.: Chapter 5: network biology approach to complex diseases. PLoS Comput. Biol. **8**, e1002708 (2012)
41. Morone, F., Min, B., Bo, L., Mari, R., Makse, H.A.: Collective influence algorithm to find influencers via optimal percolation in massively large social media. Sci. Rep. **6**, 30062 (2016)
42. Cohen, R., Erez, K., Ben-Avraham, D., Havlin, S.: Resilience of the Internet to random breakdown. Phys. Rev. Lett. **85**, 4626 (2000)
43. Buldyrev, S.V., Parshani, R., Paul, G., Stanley, H.E., Havlin, S.: Catastrophic cascade of failures in interdependent networks. Nature **464**, 1025–1028 (2010)
44. Cohen, R., Erez, K., Ben-Avraham, D., Havlin, S.: Breakdown of the Internet under intentional attack. Phys. Rev. Lett. **86**, 3682 (2001)
45. Callaway, D.S., Newman, M.E.J., Strogatz, S.H., Watts, D.J.: Network robustness and fragility: percolation on random graphs. Phys. Rev. Lett. **85**, 5468–5471 (2000)
46. Morone, F., Roth, K., Min, B., Stanley, H.E., Makse, H.A.: A model of brain activation predicts the neural collective influence map of the human brain. Proc. Natl. Acad. Sci. US **114**, 3849–3854 (2017)
47. Pei, S., Makse, H.A.: Spreading dynamics in complex networks. J. Stat. Mech. P12002 (2013)
48. Pei, S., Muchnik, L., Andrade, J.S., Jr., Zheng, Z., Makse, H.A.: Searching for superspreaders of information in real-world social media. Sci. Rep. **4**, 5547 (2014)
49. MIT Technology Review: Emerging Technology from the The Emerging Science of Superspreaders (And How to Tell If You're One Of Them), 13 May 2014. https://www.technologyreview.com/2014/05/13/13350/the-emerging-science-of-superspreaders-and-how-to-tell-if-youre-one-of-them/. Accessed Apr. 2022
50. Maniloff, J.: The minimal cell genome: 'on being the right size'. Proc. Natl. Acad. Sci. USA **93**, 10004–10006 (1996)
51. Hutchison, C.A., III., Peterson, S.N., Gill, S.R., Cline, R.T., White, O., Fraser, C.M., Smith, H.O., Venter, J.C.: Global transposon mutagenesis and a minimal Mycoplasma genome. Science **286**, 2165–2169 (1999)
52. Hutchison, C.A. III, et al.: Design and synthesis of a minimal bacterial genome. Science **351**, aad6253 (2016)
53. Bartha, I., di Iulio, J., Venter, J.C., Telenti, A.: Human gene essentiality. Nat. Rev. Genet. **19**, 51–62 (2018)
54. Liu, G., et al.: Gene essentiality is a quantitative property linked to cellular evolvability. Cell **163**, 1388–1399 (2015)
55. Dickinson, M.E., et al.: High-throughput discovery of novel developmental phenotypes. Nature **537**, 508–514 (2016)

56. Blomen, V.A., et al.: Gene essentiality and synthetic lethality in haploid human cells. Science **350**, 1092–1096 (2015)
57. Hart, T., et al.: High-resolution CRISPR screens reveal fitness genes and genotype-specific cancer liabilities. Cell **163**, 1515–1526 (2015)
58. Wang, T., Wei, J.J., Sabatini, D.M., Lander, E.S.: Genetic screens in human cells using the CRISPR-Cas9 system. Science **343**, 80–84 (2014)
59. Joy, M.P., Brock, A., Ingber, D.E., Huang, S.: High-betweenness proteins in the yeast protein interaction network. BioMed. Res. Int. **2005**, 96–103 (2005)
60. Wuchty, S., Stadler, P.F.: Centers of complex networks. J. Theor. Biol. **223**, 45–53 (2003)
61. Estrada, E., Rodríguez-Velázquez, J.A.: Subgraph centrality in complex networks. Phys. Rev. E **71**, 056103 (2005)
62. Bonacich, P.: Power and centrality: a family of measures. Am. J. Sociol. **92**, 1170–1182 (1987)
63. Sun, P.G., Quan, Y.N., Miao, Q.G., Chi, J.: Identifying influential genes in protein-protein interaction networks. Inf. Sci. **454**, 229–241 (2018)
64. Li, M., Zhang, H., Wang, J.-X., Pan, Y.: A new essential protein discovery method based on the integration of protein-protein interaction and gene expression data. BMC Syst. Biol. **6**, 15 (2012)
65. Han, J.-D.J., Bertin, N., Hao, T., Goldberg, D.S., Berriz, G.F., Zhang, L.V., Dupuy, D., Walhout, A.J.M., Cusick, M.E., Roth, F.P., Vidal, M.: Evidence for dynamically organized modularity in the yeast protein-protein interaction network. Nature **430**, 88 (2004)
66. Zhang, X., Xu, J., Xiao, W.-X.: A new method for the discovery of essential proteins. PLoS One **8**, e58763 (2013)
67. Gibbs, D.L., Shmulevich, I.: Solving the influence maximization problem reveals regulatory organization of the yeast cell cycle. PLoS Comput. Biol. **13**, e1005591 (2017)
68. Kauffman, S.A.: Metabolic stability and epigenesis in randomly constructed genetic nets. J. Theor. Biol. **22**, 437–467 (1969)
69. Kauffman, S., Peterson, C., Samuelsson, B., Troein, C.: Random Boolean network models and the yeast transcriptional network. Proc. Natl. Acad. Sci. USA **100**, 14796–14799 (2003)
70. González-García, I., Solé, R.V., Costa, J.: Metapopulation dynamics and spatial heterogeneity in cancer. Proc. Natl. Acad. Sci. USA **99**, 13085–13089 (2002)
71. Pomerance, A., Ott, E., Girvan, M., Losert, W.: The effect of network topology on the stability of discrete state models of genetic control. Proc. Natl. Acad. Sci. USA **106**, 8209–8214 (2009)
72. Wang, J., Pei, S., Wei, W., Feng, X., Zheng, Z.: Optimal stabilization of Boolean networks through collective influence. Phys. Rev. E **97**, 032305 (2018)
73. Kinouchi, O., Copelli, M.: Optimal dynamical range of excitable networks at criticality. Nat. Phys. **2**, 348 (2006)
74. Larremore, D.B., Shew, W.L., Restrepo, J.G.: Predicting criticality and dynamic range in complex networks: effects of topology. Phys. Rev. Lett. **106**, 058101 (2011)
75. Pei, S., Tang, S., Yan, S., Jiang, S., Zhang, X., Zheng, Z.: How to enhance the dynamic range of excitatory-inhibitory excitable networks. Phys. Rev. E **86**, 021909 (2012)
76. Larremore, D.B., Shew, W.L., Ott, E., Sorrentino, F., Restrepo, J.G.: Inhibition causes ceaseless dynamics in networks of excitable nodes. Phys. Rev. Lett. **112**, 138103 (2014)
77. Gautam, S.H., Hoang, T.T., McClanahan, K., Grady, S.K., Shew, L.W.: Maximizing sensory dynamic range by tuning the cortical state to criticality. PLoS Comput. Biol. **11**, e1004576 (2015)
78. Gollo, L.L., Kinouchi, O., Copelli, M.: Active dendrites enhance neuronal dynamic range. PLoS Comput. Biol. **5**, e1000402 (2009)
79. Pei, S., Tang, S., Zheng, Z.: Detecting the influence of spreading in social networks with excitable sensor networks. PLoS One **10**, e0124848 (2015)
80. Zhang, R., Pei, S.: Dynamic range maximization in excitable networks. Chaos **28**, 013103 (2018)
81. Souiai, O., Becker, E., Prieto, C., Bekhala, A., De las Rivas, J., Brun, C.: Functional integrative levels in the human interactome recapitulate organ organisation. PLoS One **6**, e22051 (2011)

82. Barshir, R., Basha, O., Eluk, A., Smoly, I., Lan, A., Yeger-Lotem, E.: The TissueNet database of human tissue protein-protein interactions. Nucl. Acids Res. **41**, D841-844 (2013)
83. Barshir, R., Shwartz, O., Smoly, I., Yeger-Lotem, E.: Comparative analysis of human tissue interactomes reveals factors leading to tissue-specific manifestations of hereditary diseases. PLoS Comput. Biol. **10**, e1003632 (2014)
84. Marbach, D., Lamparter, D., Quon, G., Kellis, M., Kutalik, Z., Bergmann, S.: Tissue-specific regulatory circuits reveal variable modular perturbations across complex diseases. Nat. Meth. **13**, 366–370 (2016)
85. Furbish, G.: Unpublished (2017)
86. Bovet, A., Makse, H.A.: Influence of fake news in Twitter during the 2016 US presidential election. Nat. Commun. **10**, 7 (2019)
87. van Vreeswijk, C., Sompolinsky, H.: Chaos in neuronal networks with balanced excitatory and inhibitory activity. Science **274**, 1724–1726 (1996)
88. van Vreeswijk, C., Sompolinsky, H.: Chaotic balanced state in a model of cortical circuits. Neural Comput. **10**, 1321–1371 (1998)
89. Alon, U.: An Introduction to Systems Biology: Design Principles of Biological Circuits (Chapman & Hall. Taylor & Francis, CRC (2006)
90. Morone, F., Makse, H.A.: Unpublished (2018)
91. Mézard, M., Montanari, A.: Information, Physics, and Computation. Oxford University Press, USA (2009)
92. Duan, Q., Flynn, C., Niepel, M., Hafner, M., Muehlich, J.L., Ferandez, N.F., Rouillard, A.D., Tan, C.M., Chen, E.Y., Golub, T.R., Sorfger, P.K., Subramanian, A., Ma'ayan, A.: LINCS Canvas Browser: interactive web app to query, browse and interrogate LINCS L1000 gene expression signatures. Nucl. Acids Res. **42**, W449-460 (2014)
93. Braunstein, A., Pagnani, A., Weigt, M., Zecchina, R.: Gene-network inference by message passing. J. Phys.: Conf. Ser. **95**, 012016 (2008)
94. Clark, N.R., Hu, K.S., Feldmann, A.S., Chen, E.Y., Duan, Q., Maayan, A.: The characteristic direction: a geometrical approach to identify differentially expressed genes. BMC Bioinf. **15**, 79 (2014)
95. Hartwell, L.H., Hopfield, J.J., Leibler, S., Murray, A.W.: From molecular to modular cell biology. Nature **402**, C47–C52 (1999)
96. Milo, R., Shen-Orr, S.S., Itzkovitz, S., Kashtan, N., Chklovskii, D., Alon, U.: Network motifs: simple building blocks of complex networks. Science **298**, 824–827 (2002)
97. Shen-Orr, S.S., Milo, R., Mangan, S., Alon, U.: Network motifs in the transcriptional regulation network of Escherichia coli. Nature Genet. **31**, 64–68 (2002)
98. Ingram, P.J., Stumpf, M.P., Stark, J.: Network motifs: Structure does not determine function. BMC Genomic. **7**, 108 (2006)
99. Payne, J.L., Wagner, A.: Function does not follow form in gene regulatory circuits. Sci. Rep. **5**, 13015 (2015)
100. Macía, J., Widder, S., Solé, R.: Specialized or flexible feed-forward loop motifs: a question of topology. BMC Syst. Biol. **3**, 84 (2009)
101. Ahnert, S.E., Fink, T.M.A.: Form and function in gene regulatory networks: the structure of network motifs determines fundamental properties of their dynamical state space. J. R. Soc. Interface **13**, 20160179 (2016)
102. Stewart, I.: Symmetry: A Very Short Introduction. Oxford University Press, Oxford (2013)
103. Golubitsky, M., Stewart, I.: The Symmetry Perspective: From Equilibrium to Chaos in Phase Space and Physical Space. Berkhäuser, Basel (2002)
104. Stewart, I., Golubitsky, M., Pivato, M.: Patterns of synchrony in coupled cell networks. SIAM J. Appl. Dynam. Sys. **2**, 609–646 (2003)
105. Golubitsky, M., Stewart, I.: Nonlinear dynamics of networks: the groupoid formalism. Bull. Am. Math. Soc. **43**, 305–364 (2006)
106. Pecora, L.M., Sorrentino, F., Hagerstrom, A.M., Murphy, T.E., Roy, R.: Cluster synchronization and isolated desynchronization in complex networks with symmetries. Nat. Commun. **5**, 4079 (2014)

107. Sorrentino, F., Pecora, L.M., Hagerstrom, A.M., Murphy, T.E., Roy, R.: Complete characterization of the stability of cluster synchronization in complex dynamical networks. Sci. Adv. **2**, e1501737 (2016)
108. Strogatz, S.: Nonlinear Dynamics and Chaos: with Applications to Physics, Biology, Chemistry, and Engineering. Westview Press, Boulder (2000)
109. Arenas, A., Díaz-Guilera, A., Kurths, J., Moreno, Y., Zhou, C.: Synchronization in complex networks. Phys. Rep. **469**, 93–153 (2008)
110. Abrams, D.M., Pecora, L.M., Motter, A.E.: Focus issue: patterns of network synchronization. Chaos **26**, 094601 (2016)
111. Hamermesh, M.: Group Theory and Its Application to Physical Problems. Dover, New York (1989)
112. Dixon, J.D., Mortimer, B.: Permutation Groups, Graduate Texts in Mathematics, vol. 163. Springer, New York (1996)
113. Landau, L.D., Lifshitz, E.M.: Quantum Mechanics: Non-Relativistic Theory, vol. 3, 3rd ed. Butterworth-Heinemann, Oxford (1977)
114. Weinberg, S.: The Quantum Theory of Fields. Cambridge University Press, Cambridge (2005)
115. Georgi, H.: Lie Algebras in particle physics: from isospin to unified theories. Frontier in Physics, vol. 54, 2nd ed. Westview Press (1999)
116. Monod, J.: Symmetry and function of biological systems at the macromolecular level. In: Engström, A., Strandberg, B. (eds.) Proceedings of the Eleventh Nobel Symposium held on 26–29 Aug. 1968. Södergarn, Sweden (1969)
117. D'Arcy Wentworth Thompson: On Growth and Form. Cambridge University Press, UK (1917)
118. Morone, F., Makse, H.A.: Symmetry group factorization reveals the structure-function relation in the neural connectome of Caenorhabditis elegans. Nat. Commun. **10**, 4961 (2019)
119. Morone, F., Leifer, I., Makse, H.A.: Fibration building blocks of biological networks. Proc. Nat. Acad. Sci. USA **117**, 8306–8314 (2020)
120. Makse, H.A., Boldi, P., Sorrentino, F., Stewart, I.: Symmetries of Living Systems: Symmetry Fibrations and Synchrony in Biological Networks. Cambridge University Press (2024)
121. Mangan, S., Zaslaver, A., Alon, U.: The coherent feedforward loop serves as a sign-sensitive delay element in transcription networks. J. Mol. Biol. **334**, 197–204 (2003)
122. Boldi, P., Vigna, S.: Fibrations of graphs. Disc. Math. **243**, 21–66 (2001)
123. Leifer, I., Morone, F., Reis, S.D.S., Andrade, J.S., Jr., Sigman, M., Makse, H.A.: Circuits with broken fibration symmetries perform core logic computations in biological networks. PLoS Comput. Bio. **16**, e1007776 (2020)
124. Grothendieck, A.: Technique de descente et théorémes d'existence en géométrie algébrique, I. Généralités. Descente par morphismes fidélement plats. Séminaire N. Bourbaki **5**, Talk no. 190, pp. 299–327 (1958–1960). http://www.numdam.org/article/SB_1958-1960__5__299_0.pdf. Accessed Apr. 2022
125. Atkinson, M.R., Savageau, M.A., Myers, J.T., Ninfa, A.J.: Development of genetic circuitry exhibiting toggle switch or oscillatory behavior in Escherichia coli. Cell **113**, 597–607 (2003)
126. Glass, L., Kauffman, S.A.: The logical analysis of continuous, non-linear biochemical control networks. J. Theor. Biol. **38**, 103–129 (1973)
127. Anderson, P.W.: The concept of frustration in spin glasses. J. Less-Common Met. **62**, 291–294 (1978)
128. Gardner, T.S., Cantor, C.R., Collins, J.J.: Construction of a genetic toggle switch in Escherichia coli. Nature **403**, 339–342 (2000)
129. Kramer, B.P., Fussenegger, M.: Hysteresis in a synthetic mammalian gene network. Proc. Natl. Acad. Sci. USA **102**, 9517–9522 (2005)
130. Kramer, B.P., Viretta, A.U., Baba, M.D.-E., Aube, D., Weber, W., Fussenegger, M.: An engineered epigenetic transgene switch in mammalian cells. Nat. Biotech. **22**, 867–870 (2004)
131. Guet, C.C., Elowitz, M.B., Hsing, W., Leibler, S.: Combinatorial synthesis of genetic networks. Science **296**, 1466–1470 (2002)
132. Ajo-Franklin, C.M., Drubin, D.A., Eskin, J.A., Gee, E.P., Landgraf, D., Phillips, I., Silver, P.A.: Rational design of memory in eukaryotic cells. Genes Dev. **21**, 2271–2276 (2007)

133. Ham, T.S., Lee, S.K., Keasling, J.D., Arkin, A.P.: Design and construction of a double inversion recombination switch for heritable sequential genetic memory. PLoS One **3**, e2815 (2008)
134. Basu, S., Mehreja, R., Thiberge, S., Chen, M.T., Weiss, R.: Spatiotemporal control of gene expression with pulse-generating networks. Proc. Natl. Acad. Sci. USA **101**, 6355–6360 (2004)
135. Elowitz, M.B., Leibler, S.: A synthetic oscillatory network of transcriptional regulators. Nature **403**, 335–338 (2000)
136. Stricker, J., Cookson, S., Bennett, M.R., Mather, W.H., Tsimring, L.S., Hasty, J.: A fast, robust and tunable synthetic gene oscillator. Nature **456**, 516–519 (2008)
137. Tigges, M., Marquez-Lago, T.T., Stelling, J., Fussenegger, M.: A tunable synthetic mammalian oscillator. Nature **457**, 309–312 (2009)
138. Madar, D., Dekel, E., Bren, A., Alon, U.: Negative auto-regulation increases the input dynamic-range of the arabinose system of Escherichia coli. BMC Sys. Biol. **5**, 111 (2011)
139. Widlar, R.J.: Some circuit design techniques for linear integrated circuits. IEEE Trans. Circuit Theory. **4**, 586–590 (1965)
140. Horowitz, P., Hill, W.: The Arts of Electronics, 3rd edn. Cambridge University Press, New York (2015)
141. Leifer, I., Makse, H.A.: List of fibers in E. coli transcriptional networks. https://bit.ly/2kONqrI. Accessed Feb. 2022
142. Leifer, I., Makse, H.A.: List of biological network datasets. https://docs.google.com/spreadsheets/d/1-RG5vR_EGNPqQcnJU8q3ky1OpWi3OjTh5Uo-Xa0PjOc/edit#gid=1973910256. Accessed Feb. 2022
143. Kamei, H., Cock, P.J.A.: Computation of balanced equivalence relations and their lattice for a coupled cell network. SIAM J. Appl. Dyn. Syst. **12**, 352–382 (2013)
144. Cardon, A., Crochemore, M.: Partitioning a graph in $O(|A| \log_2 |V|)$. Theor. Comput. Sci. **19**, 85–98 (1982)
145. Monteiro, H.S., Leifer, I., Reis, S.D.S., Andrade, J.S., Jr., Makse, H.A.: Fast algorithm to identify cluster synchrony through fibration symmetries in large directed networks. Chaos **32**, 033120 (2022)
146. Leifer, I., Makse, H.A., Phillips, D., Sorrentino, F.: Symmetry-driven network reconstruction through pseudobalanced coloring optimization. J. Stat. Mech. 073403 (2022)
147. Boldi, P., Leifer, I., Makse, H.A.: Quasifibrations of graphs to find symmetries and reconstruct biological networks. J. Stat. Mech. 113401 (2022)
148. Jin, J., He, K., Tang, X., Li, Z., Lv, L., Zhao, Y., Luo, J., Gao, G.: An Arabidopsis transcriptional regulatory map reveals distinct functional and evolutionary features of novel transcription factors. Mol. Biol. Evol. **32**, 1767–1773 (2015)
149. Balázsi, G., Heath, A.P., Shi, L., Gennaro, M.L.: The temporal response of the Mycobacterium tuberculosis gene regulatory network during growth arrest. Mol. Syst. Biol. **4**, 225 (2008)
150. Zhu, B., Stülke, J.: SubtiWiki in 2018: from genes and proteins to functional network annotation of the model organism Bacillus subtilis. Nucleic Acids Res. **46**, D743–D748 (2017)
151. Gama-Castro, S., et al.: RegulonDB version 9.0: high-level integration of gene regulation, coexpression, motif clustering and beyond. Nucl. Acids Res. **44**, D133–D143 (2015)
152. Métris, A., et al.: SalmoNet, an integrated network of ten Salmonella enterica strains, reveals common and distinct pathways to host adaptation. NPJ Syst. Biol. Appl. **3**, 31 (2017)
153. Cherry, J.M., et al.: Saccharomyces Genome Database: the genomics resource of budding yeast. Nucl. Acids Res. **40**, D700–D705 (2011)
154. Yang, T.-H., Wang, C.-C., Wang, Y.-C., Wu, W.-S.: YTRP: a repository for yeast transcriptional regulatory pathways. Database **2014**, bau014 (2014)
155. Han, H., et al.: TRRUST v2: an expanded reference database of human and mouse transcriptional regulatory interactions. Nucl. Acids Res. **46**, D380–D386 (2017)
156. Kanehisa, M., Sato, Y., Kawashima, M., Furumichi, M., Tanabe, M.: KEGG as a reference resource for gene and protein annotation. Nucl. Acids Res. **44**, D457–D462 (2015)

157. Leifer, I., Sánchez-Pérez, M., Ishida, C., Makse, H.A.: Predicting synchronized gene coexpression patterns from fibration symmetries in gene regulatory networks in bacteria. BMC Bioinf. **22**, 363 (2021)
158. Kim, M., Rai, N., Zorraquino, V., Tagkopoulos, I.: Multi-omics integration accurately predicts cellular state in unexplored conditions for Escherichia coli. Nat. Commun. **7**, 13090 (2016)
159. Girvan, M., Newman, M.E.J.: Community structure in social and biological networks. Proc. Natl. Acad. Sci. USA **99**, 7821–7826 (2002)
160. Blondel, V.D., Guillaume, J.-L., Lambiotte, R., Lefebvre, E.: Fast unfolding of communities in large networks. J. Stat. Mech. **10**, P10008 (2008)
161. Fortunato, S.: Community detection in graphs. Phys. Rep. **486**, 75–174 (2010)
162. Hanai, R., Edelman, A., Ohashi, Y., Littlewood, P.B.: Non-Hermitian phase transition from a polariton Bose-Einstein condensate to a photon laser. Phys. Rev. Lett. **122**, 185301 (2019)
163. Fruchart, M., Hanai, R., Littlewood, P.B., Vitelli, V.: Nature **592**, 363–369 (2021)
164. Reichhardt, C.J.O., Reichhardt, C.: An exceptional view of phase transitions in non-equilibrium systems. Nature **592**, 355–356 (2021)
165. Ornes, S.: A new theory for systems that Defy Newton's Third Law, Quanta Magazine, 11 Nov. 2021. https://www.quantamagazine.org/a-new-theory-for-systems-that-defy-newtons-third-law-20211111/. Accessed Feb. 2022
166. Nazerian, A., Panahi, S., Leifer, I., Phillips, D., Makse, H.A., Sorrentino, F.: Matryoshka and disjoint cluster synchronization of networks. Chaos **32**, 041101 (2022)

Chapter 6
Neural Influencers in the Brain

Abstract Brain network analysis is a crucial advance for the understanding of brain circuits and information processing in the brain. The techniques developed in the context of network theory offer a unique toolkit for the investigation of the brain. Of particular interest is the identification of essential brain areas for the performance of cognitive tasks, i.e., areas that if inactivated would cause a considerable impairment to performance. In this chapter we discuss theoretical frameworks to rigorously predict these areas in information-processing networks with specialized modules such as the brain. Brains are composed of tight local clusters exhibiting strongly correlated neural activity reflecting the functional specialization of different areas. How do these local clusters interact at the systems level to generate an integrated unity for global computation without loosing local independence? Here, we show that influencer theory is able to identify the crucial integrators of the modular brain network. We discuss how the identification of essential neural populations is of importance for the treatment of brain diseases and prevention of brain deficits. Evidence suggests that a number of brain dysfunctions and neurological disorders might be caused by alterations in the functional network connectivity across brain regions. This chapter aims to discuss advances in this direction. The ability to accurately predict the localization of core nodes for a given cognitive function and how these nodes reorganize under pathological conditions can improve our understanding of how the brain recovers function after an insult. We finally show how influencer theory is specifically designed to understand the consequence in the brain to insults such as tumor resection, and can guide brain tumor surgery in the clinical setting.

6.1 How the Brain Integrates Information

Brains consist of modular structures implementing highly distributed and efficient coding [1–3]. For example, in the visual modality, more than thirty visual areas analyze simultaneously distinct features of the visual scene: motion, color, orientation, space, form, luminance and contrast among others [4]. These features, as well as information from different sensory modalities, have to be integrated, as one of the main aspects of perception is its unitary nature [2, 5]. This leads to a basic conundrum

H. A. Makse and M. Zava, *The Science of Influencers and Superspreaders*, Understanding Complex Systems, https://doi.org/10.1007/978-3-031-78058-5_6

of brain networks: modular processors in the brain ought to be sufficiently isolated to achieve independent computations, but also globally connected to be integrated in coherent functions [2, 3, 6–8].

This problem, known as the 'binding problem' [5, 9] or 'segregation versus integration [6, 10–12], represents a fundamental open issue in systems neuroscience. It refers to the apparent tension between the highly modular topology of the brain necessary for parallel computation and the global integration of these modules necessary for unitary computation at the global level. Since the brain is composed of massively connected elements arranged in modules [1–3, 13–16] connected by weak correlations over long distances, forming a system of networks [7, 8, 17], the question arise as to how these local modules interact at the systems level to generate an integrated whole without losing local independence [1, 2, 5, 7, 9, 11]?

A widely held view is that small-world (SW) networks [18] provide a solution to this puzzle since they combine high local clustering (as measured by the clustering coefficient, C, quantifying the fraction of triangles in the network), and short path lengths (specified by the scaling of the average path length with number of nodes: $\ell \sim \ln N$) [15, 18, 19]. This viewpoint also includes the scale-free (SF) aspect of brain connectivity [20–23] that postulates a broad probability distribution of the degree, $P(k) \sim k^{-\gamma}$, which further results in a ultra small-world with $\ell \sim \ln \ln N$ [24].

This view has been fueled by the systematic finding of SW and SF topologies in a wide range of brain networks derived from structural [25], functional [21–23], and diffusion tensor MRI [26]. Small-world topology has been identified at the cellular scale in functional cortical neuronal circuits in mammals [27, 28], and in the connectome of the nematode *Caenorhabditis elegans* [18]. It has been also found that SW is disrupted by disease [29], normal aging and by pharmacological blockade of dopamine neurotransmission [23], suggesting that SW may be relevant for normal brain function. The SF topology has been found in a number of studies involving functional brain networks obtained from functional magnetic resonance (fMRI) under different conditions [14, 21–23].

Despite these systematic findings, small worlds cannot offer a final solution to the binding problem, because it cannot fully capture the coexistence of highly modular structures with broad global integration. The local clustering coefficient, C (measuring the fraction of triangles in the network) and the degree of modularity (measured by the Q-factor as the ratio of links inside/outside modules [30, 31]) are independent network features. The clustering coefficient, the typical measure of clustering in SW, is a purely local quantity counting triangular motifs, which can be assessed inspecting the immediate neighborhood of a node. On the contrary, modularity is a more global property of the network, determined by the existence of strongly connected groups of nodes that are loosely connected to the rest of the network [3, 30, 31].

In principle, modularity cannot be inferred from local clustering and vice versa. In fact, it is easy to construct modular and unclustered networks ($Q \approx 1, C \approx 0$) or, reciprocally, clustered networks without modules ($Q \approx 0, C \approx 1$). While a clustered network preserves its clustering when a small fraction of shortcuts are added, converting it into a small-world as shown in the Watts-Strogatz model [18] (see Fig. 6.1), the persistence of modules is not equally robust: shrinking of the network diameter

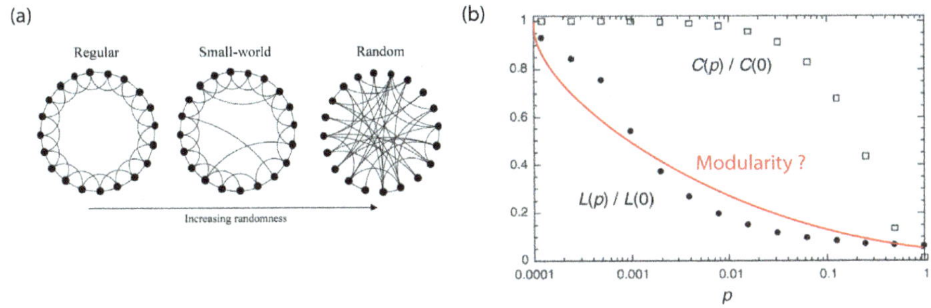

Fig. 6.1 The conundrum of modularity versus integration [8]. **a** The basic Watts-Strogatz model from [18]: a small fraction p of shortcuts in a lattice transforms it to a small-world network. **b** Strong modularity and small-worlds may not coexist at the same level of connectivity p. While the addition of a small number of shortcuts reduces the network diameter $\ell(p)$ (measured by the average minimum path over all nodes) but not the clustering coefficient $C(p)$ (measured by the fraction of triangles in the network), a modular network may rapidly loss its modularity (measured by the Girvan-Newman Q factor [30, 31]) as seen in the red curve (see Fig. 3b in [7, 8] for details). Figure adapted from [18]. Copyright © 1998, Macmillan Magazines Ltd

by the creation of a few short cuts may quickly destroy the modularity (see Fig. 6.1b and Fig. 3b in [7]).

Given this intrinsic tension between a modular and small-world architecture, other studies [7, 8, 17, 32, 33] have proposed hierarchical modularity as an emergent feature of brain organization that provides modularity together with integration, robustness, adaptability, and evolvability of network function. The crux of the matter is to understand the simultaneous emergence of two seemingly contradicting features: strong modularity and short path length. In other words, how the brain modules can be integrated closely without collapsing into a SW structure? Influencer theory applied to the brain network can be used to discover the essential nodes that integrate the segregated modules into a whole [14, 16, 34–41]. It is found that these influential nodes integrate the brain network through weak correlations [7, 8], and thus can be interpreted as the weak links in the sense of Granovetter [42].

Accumulating evidence suggests that alterations in core nodes in the functional network may lie behind a number of neurological and neuropsychiatric conditions [14, 34, 43–49, 51–54]. However, treating network dynamics is not addressed directly by conventional therapeutic approaches. Identifying these essential nodes may support the design of medical interventions to restore or anticipate dysfunction or disease patterns in the brain. But brain disorders are complex diseases that neuroscientists study using multiple methods. This includes magnetic resonance imaging (MRI), functional MRI (fMRI), electroencephalogram (EEG), diffusion tensor imaging (DTI), and others, and essential nodes in brains to complement an accurate prediction and classification of brain disorders and therefore better inform intervention protocols.

Thus, a therapeutic window may exist to handle maladaptive network dynamics with targeted interventions to core node's activity prescribed by network analysis.

Existing techniques with clinical presence in humans, such us deep brain stimulation, transcranial magnetic stimulation or direct current stimulation [55–57], are currently being applied without taking into account the alterations in network dynamics thought to be at the root of neurological maladies [14, 43–45, 58–60]. Advances in the network theory will allow researchers and treating clinicians to identify new targets for focal therapy and to move away from the "trial and error" approach often employed in biology [55, 56].

6.2 Neural Influencers in the Brain

For decades, essential nodes in the brain have been associated with hubs [14–16, 34–41]. These frameworks consider that network hubs provide the ability to integrate the different modules in the brain at multiple cognitive domains. The network neuroscience view of essential areas in the brain has been shaped by this hub-centric viewpoint for many years.

Network hubs also play an important role in disease states of the brain [34, 43, 44]. It has been hypothesized that disruption of hubs connectivity could be a hallmark of neurological and psychiatric disorders. For instance, memory disorders occur when certain parts of the brain are damaged, situation that prevents its ability to reason, store, retain, recall memories, communicate and make decisions; memory disorders include Alzheimer's disease, vascular dementia, dementia with Lewy bodies, frontotemporal dementia, mild cognitive impairment and mixed dementia. Disruption in hub nodes and their connections might be associated with Alzheimer's disease [45–50], schizophrenia [51–53], and disorders of consciousness [54].

The importance of central hubs and pathways emanating from such hubs has been studied for anatomical hubs (such as those obtained from direct structural connections, for instance, in diffusion imaging tractography and connectomes [32, 61]), and functional hubs as well (as obtained by functional techniques such as fMRI or EEG) and is consistent with structure-function relationships in the brain [62, 63].

This hub-centric picture of brain networks follows the scale-free models [20]. As discussed in Chap. 1, these models consider that the hubs mediate the information transfer in the network [16, 21], and, if disrupted, would cause the malfunction of the entire network [64]. Other heuristics have been also proposed as essential nodes in the brain [16].

Even though these heuristics are widely used, they lack a first principle formulation so that they do not optimize a global function of influence in the brain. Heuristics are based on ad-hoc assumptions about the importance of an individual node and, inevitably, they fail to optimize the influence since they do not take into account the dynamics of information flow in the brain network. As a consequence, there is no guarantee of its performance to find a minimal set of influencers in the brain. The construction of a theory of influence in brain networks demands theoretical models of how the brain activity propagates in the network. The problem is, in its most general formulation, intractably NP-hard as the IMP [65].

Since it is difficult to find an optimal solution to the problem of influence in general, and in particular in the brain, heuristic methods have gained a foothold in brain studies. They have provided valuable information on the structure of the brain and its relation to function [11, 14–16, 34–41, 43–45, 66].

In the next sections we review approaches to identify influencers in brain networks. The identification of neural influencers in the brain may transform therapeutic scenarios by providing protocols (targets and perturbations) required to control the brain dynamics and retuning diseased network dynamics in specific pathological conditions. Applications include: 1) rehabilitation programs based on increasing focal brain activity; 2) transcranial magnetic stimulation that can stimulate or deactivate focal brain activity, or even 3) implantable devices which have had success in Parkinson's disease and even motor weakness.

The approach may lead to schemes to reprogram central network dynamics by acting on influential nodes to induce controllable and on demand network transitions. The rational is that the activation/deactivation patterns applied to core nodes will propagate through the nervous system with the possibility to impact and modulate brain dynamics. The results discussed below provide the scaffold for this program.

6.3 Functional and Structural Brain Networks

In order to identify influential nodes in the brain, we need to first reconstruct the brain network. Generally speaking, it is important to distinguish between two networks measuring different aspects of the brain [14, 62, 67]:

- The functional brain network,
- The structural brain network, also called anatomical network or connectome.

The functional brain network refers to any statistical interaction between units in the brain, i.e., neurons, ensemble of neurons, region of interests (ROI), or large brain anatomical areas, like Broca or Wernicke in the human network. Functional interactions are obtained by statistical analysis of the dynamical activity of the units. For instance, by monitoring the activity of neuronal populations using calcium imaging techniques in either human cortical neurons or the 302 neurons in the worm *C. elegans* brain, one can test their functional relations with several techniques, like phase synchronization or Pearson correlations. At a more global level, fMRI and EEG offer the possibility to build functional brain networks at larger spatial scales. A functional relation does not imply that the neurons are physically connected. The physical connections between neuron through axons, or between neural areas and ROIs through white matter tracts are measured by the structural network.

While there is a certain level of correspondence between functional connectivity (derived, e.g., from resting state fMRI) and structural connectivity (derived, e.g., from diffusion tensor imaging, DTI, white matter tractography) in humans as shown in [14, 16, 62], both networks are not necessarily one-to-one. Relating these two networks is called the structure-function problem. It attempts to understand the underlying structural origin of the observed neural dynamics at the functional level. We will elaborate more on this in Sect. 6.7.

Next, we discuss techniques to build functional networks from existing data. There are many ways to measure a statistical association between two variables. The simplest (linear) one is the Pearson correlation coefficient between the activity of the units. More sophisticated methods focus on different measures of coherence (linear and non-linear) and synchronization. It has been counted that there exists more than 50 different ways to characterize these functional associations [68]. Researchers use concepts like coherent coefficients, mutual information, phase lag index, phase synchronization, phase locking value, phase slope index, Granger causality, so on and so forth. Different metrics capture different aspects of signal synchronization, from phase to amplitude, linear and non-linear, in real space or frequency domain, and have different applications and pitfalls. Detailed explanations and comparisons among these methods are provided in the reviews of Pereda et al. [69], and Bastos et al. [68]. Below, we focus on correlation based methods and their variants using maximum entropy methods and regularization schemes, which are the simplest, yet provide valuable information.

At the large scale of the whole human brain, functional MRI, including both task based and resting state, have provided a plethora of data for researchers to study. The different methods of analysis can be applied to obtain a functional brain network from fMRI to subsequently obtain core nodes. Standard functional network analysis of blood-oxygenation-level-dependent (BOLD) susceptibility contrast from fMRI time series defines a node as the voxel (the 3d analogous of a 2d pixel) where the BOLD signal is measured. When these voxels are aggregated they are called region of interest, ROI. Typically, in a 7 T high-magnetic field scanner, voxel resolution can be $0.25 \times 0.25 \times 1$ mm, while other scanners, 3T and 5T, provide lower resolution in the 1-4 mm range. The temporal resolution is limited by the hemodynamic BOLD response which peaks at around 6 s after stimulus and has a width of 3 s. Thus, a typical time resolution between two consecutive meaningful measurements is of the order of 10 s. This time scale is way much longer than the typical time scale for action potential propagation between neurons, which is of the order of ms and it is captured by EEG. While fMRI offers lower time resolution than EEG, it provides higher spatial resolution than EEG. Yet, each fMRI voxels may contain of the order of one million neurons.

A functional brain network can be obtained from the time series of the BOLD signal, $x_i(t)$, of a voxel or ROI i. Functional correlations between nodes are represented as undirected links from the correlation function or pair-wise Pearson correlation coefficient as:

$$C_{ij} = \frac{\langle x_i x_j \rangle - \langle x_i \rangle \langle x_j \rangle}{\sqrt{\langle x_i^2 \rangle - \langle x_i \rangle^2} \sqrt{\langle x_j^2 \rangle - \langle x_j \rangle^2}}, \tag{6.1}$$

which defines the cross-correlation between two BOLD signals x_i and x_j, in voxels i, j. Here, $\langle \cdot \rangle$ represents temporal averages taken over the time series of the BOLD signal: $\langle x_i \rangle = (1/T) \sum_{k=1}^{T} = x_i(t_k)$, if T measurements are taken, each one at times t_k. The correlation C_{ij} ranges from -1 to 1. $C_{ij} > 0$ corresponds to positive correlations, $C_{ij} < 0$ corresponds to negative correlations, and $C_{ij} = 0$ indicates the lack of correlation between a pair of voxels, i and j. There are many alternatives to measure the functional network, some of them will be discussed below, including mutual information, maximum entropy methods, Bayesian networks and different types of synchronization measures [14, 70, 71].

In structural networks, anatomical networks or connectomes [32, 72] the network information is specified by the adjacency matrix, $A_{i\to j} = A_{ij}$ measuring the presence of a direct physical connection from neuron i to neuron j via axons, or a direct white matter tract (a bunch of axons or fiber) from a ROI to another ROI in the brain. The adjacency matrix can be unweighted $A_{i\to j} = \{0, 1\}$, signaling the absence or presence of a direct link, or weighted $A_{i\to j} \in \mathbb{R}$ by the strength of the physical interaction, for instance, the number of fiber tracts or the number of chemical synapses between two neurons. Further information includes the type of synaptic connection between neurons, whether it is excitatory (positive) or inhibitory (negative).

Structural networks of the human brain can be obtained from non-invasive methods like diffusion tensor imaging (DTI) fiber-tract tracing [32, 72, 73]. Synapse-level connectomes from electron micrography can be also measured; the most celebrated connectome is the full map of the nematode *C. elegans* containing 302 neurons obtained in 1986 by White et al. [74]. The Allen Mouse Brain Connectivity Atlas provides a high-resolution 3D map of physical axonal projections throughout the mouse brain built from transgenic mice genetically engineered to target specific cell types labeled by viral tracers and providing a mesoscale connectome of the mouse brain [75]. More connectomes are discussed in [32].

6.3.1 Functional Brain Network by Thresholding the Correlation Function

The simplest way to obtain a functional brain network is by thresholding the correlation function. This creates an undirected link between voxels i and j if the correlation exceeds a given threshold value T: $C_{ij} > T$ [7, 8, 15–17, 21, 76]. This provides a measure of the synchronization between the nodes. Figure 6.2 shows the process of thresholding a correlation function in resting state data in the human brain from [17]. A large threshold value T reveals strongly connected independent networks via strong links. As we lower the value of T, these networks connect (percolate) to each other through weak links. By applying a percolation analysis to C_{ij}, a hierarchical structure emerges where critical values T_c can be identified with a concomitant modular structure as shown in Fig. 6.2, see [7, 17].

We can monitor the size of the largest connected component as a function of T as a percolation process [77]. The jump in the largest connected component at T_c in Fig. 6.2 signals the merging of two sub-networks or modules, A and B, formed by

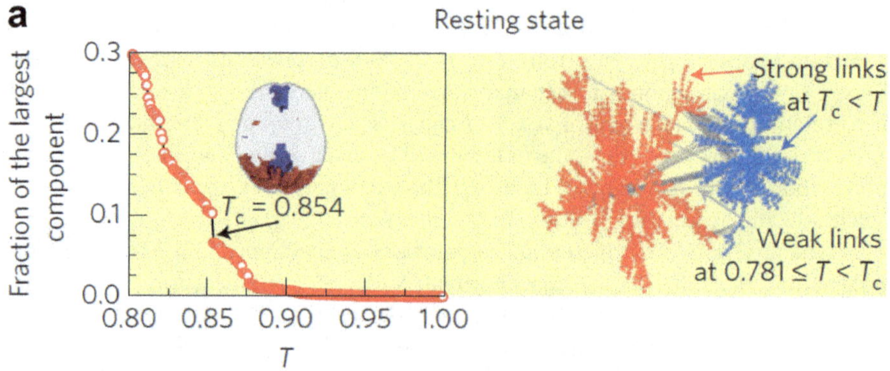

Fig. 6.2 Example of a modular functional network from [17]. Human functional connectivity obtained by thresholding the correlations between fMRI signals in resting-state. Clustering analysis to obtain the modular functional network for resting state from [7, 8, 17]. Left plot shows giant connected component versus T, a threshold imposed to the voxel correlations in fMRI. At $T_c = 0.854$, two largest networks (right panel and inset) merge in a percolation-like transition. Strong ingoing links, k^{in}, correspond to $T > T_c$. Interconnecting weak outgoing links, k^{out}, are defined for $0.781 \leq T < T_c$. Figure reproduced from [17]. Copyright © 2014, Nature Publishing Group

strong links for $T > T_c$ and interconnected by weak links with $T < T_c$. The modules A and B are composed of nodes with k^{in} links inside their clusters, defined by their strong links. Each node in A or B has a number of k^{out} weak links connecting to nodes in the other sub-networks. There is a difference between weak links and weak nodes: a weak node inside network A (with low in-degree k^{in}) may be connected to many other nodes in other modules via many weak links (large out-degree k^{out}).

Further modules can be obtained by lowering the threshold. Figure 6.3 shows the detailed behavior of a modular network obtained from fMRI BOLD signal in an individual performing a dual visual and auditive task obtained in [78] and analyzed in [7]. The hallmark of a hierarchical modular brain network structure is represented by the jumps in the percolation process seen in Fig. 6.3. The figure shows the size of the largest connected component during percolation for the correlation function as we lower a threshold called p in this study (rather than T as in Fig. 6.2). At larger values than $p = 0.98$, three large sub-networks are formed localized to the medial occipital cortex (red), the lateral occipital cortex (orange) and the anterior cingulate (green). At $p_c = 0.978$, the orange and red networks merge as revealed by the first jump in the percolation plot. As p continues to decrease this mechanism of network formation and absorption repeats, defining a hierarchical process as depicted in the tree in the top panel of Fig. 6.3.

This analysis further reveals the existence of 'stubborn' modules for this particular task. For instance, the anterior cingulate cluster (green, Fig. 6.3), known to be involved in cognitive control [79, 80] and which hence cannot commit to a specific functional module, remains detached from the other modules down to low p values. Even at the lower values of p, when a massive region of the cortex including motor,

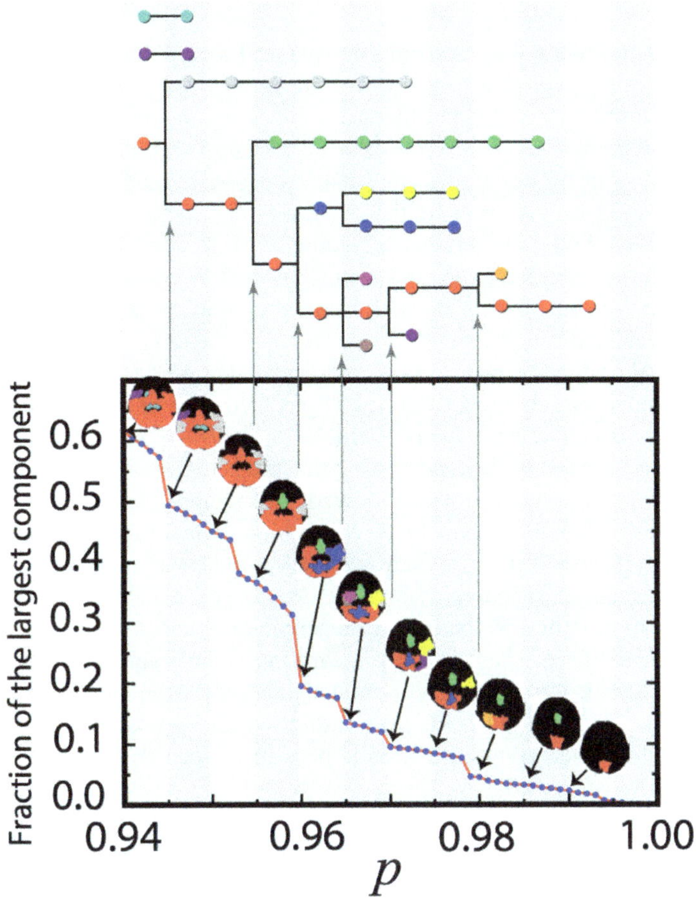

Fig. 6.3 Hierarchical formation of a modular brain network from human fMRI in a dual task experiment from [78] **as analyzed in** [7]. Size of the largest connected component of nodes (as measured by the fraction to the total system size) as a function of percolation threshold p. As we lower p the size of the largest component increases in jumps when new modules emerge, grow, and finally get absorbed by the largest component. We display the resulting networks for the indicated p. The jumps make clear the identification of sub-networks. The tree in the upper figure shows the hierarchical merging of sub-networks into a hierarchical network of networks. The topography of these networks reflects coherent patterns for a visual/auditory activity. In virtually all subjects we observe a network covering the anterior cingulate (AC) region, a cluster covering the medial part of the posterior parietal cortex (PPC) and a cluster covering the medial part of posterior occipital cortex (area V1/V2), along the calcarine fissure. This consistent topographic projection provides validation of the method (see [7, 8]). Figure reproduced from [7].

visual and auditory regions has formed a single incipient cluster (red, $p \approx 0.94$), two new clusters emerge; one involving subcortical structures including the thalamus and striatum (cyan) and the other involving the left frontal cortex (purple).

This mechanism reveals the iteration of a process by which modules form at a given p value and merged by comparably weaker links. This process is recursive. The weak links of a given transition become the strong links of the next transition, in a hierarchical fashion, resulting in a hierarchical percolating process where networks merge with each other and sharp transitions are observed in the size of the largest network (red in Fig. 6.3) as it absorbs smaller ones. The modules have consistent topographic projection in the map of the brain, see [7, 8]. The signature of a modular brain structure is, then, the hierarchical sequence of networks as a function of the threshold p as shown in the jumps in Fig. 6.3.

This clustering analysis can be performed also with other unsupervised methods like hierarchical clustering or k-means. Using the modular structure of the brain obtained from the percolation process, Russo et al. [81] studied the critical properties of neuronal avalanches showing that criticality is recovered only when inter-module connections are added to the network. Bardella *et al.* [76] applied a modified percolation method where rather than monitoring the jumps in the largest connected component, the number of components is followed. This method leads to better statistics. While the usual percolation analysis detects a hierarchical modular structure even in a random null model, the modified percolation analysis enables discrimination between the real and the random cases [76].

The hierarchical structure obtained from percolation represents a natural solution to the conundrum of segregation versus integration in the brain. It provides strongly correlated modules made of strongly correlated links, and at the same time, it provides the integration of these modules via percolating weakly correlated links [7, 8]. This two-level structure then allows the brain to be well segregated in modules, while at the same time be integrated.

We encounter again the idea inspired by the fundamental notion of Granovetter: 'the strength of weak ties' [42] by which the strong ties clump together forming brain modules, and the weak ties become the crucial bridge between the modules. To complete the model of segregated-integrated functional brain network, below we will use suitable modified influencer theory to identify the nodes that provide these weak links. They are the integrators of the brain.

6.3.2 Maximum Entropy Methods to Infer Functional Brain Networks

By thresholding the correlation matrix and building a functional brain network with the thresholded links, one obtains a sparse representation of the network. This sparse representation avoid spurious low-correlated links. There are more sophisticated methods to build a sparse network that are more elegant and systematic than just a simple thresholding. We denote the sparse representation of C_{ij} as J_{ij}. The problem of inferring pair-wise sparse interactions J_{ij} from functional data C_{ij} is treated by different methods as we show next.

The maximum entropy method is used to infer a representation of the couplings between units [82–87]. The method has been applied to neuronal populations by Schneidman *et al.* [82] and it is similar to methods to infer the weights of the paths connecting two brain areas used in computational neuroscience [88–91]. The weight of the links that are inferred are analogous to what is called direct effective connection matrix (deCM) in [90]: they embody the strength of each functional connection between points in a given brain state. However, they should not be confused with the actual structural connection.

These methods are based on the maximum entropy principle developed by Jaynes [92, 93]. It states that the probability distribution which best represents the data is the one with largest entropy (more random), when constrained by known priors. The prior in this case is the pair-wise correlation function, so maximum entropy provides the probability distribution of the data where all momenta of the distribution are randomized except for the pair-wise correlation function. The method is an Occam's razor since it provides the most parsimonious description (minimal assumptions) consistent with the known information about the system.

This method receives as an input the set of cross-correlations $\{C_{ij}\}$ of the fMRI signals between pair of voxels measured from the BOLD response in the brain, and outputs the effective functional weights $\{J_{ij}\}$ of the link between i and j, also called interaction strengths or couplings in statistical physics. The problem of learning pairwise interactions from the data can be formulated in a statistical physics language using the Hamiltonian

$$H = -\frac{1}{2}\sum_{ij} J_{ij} x_i x_j, \tag{6.2}$$

for the dynamical variables $x_i(t)$. A value $J_{ij} \neq 0$ means that there exists an effective interaction between the pair of voxels i and j and the weight of this link is given by the value of J_{ij}, while if $J_{ij} = 0$ (or smaller than a pre-defined small threshold) then there is no effective interaction between i and j.

Since we map the system to a Hamiltonian system, the coupling are symmetric, i.e., the resulting network is undirected $J_{ij} = J_{ji}$. This is an assumption of any equilibrium statistical mechanics approach to the problem, since statistical mechanics requires detailed balanced. Thus, this inference model ignores the directionality of direct synaptic connections in the brain, and therefore it provides an effective representation of the functional connectivity and not a representation of the directed structural connection.

In order to implement the maximum entropy method, we first calculate the cross-correlation C_{ij} between the BOLD response for each pair of voxels i and j as in Eq. (6.1). The method is based on the Maximum Entropy Principle. It infers the most general joint probability distribution $P(x_1, \ldots, x_N|\hat{J})$ of the signal variables assuming solely the knowledge of the correlations C_{ij}, contains only pairwise (i.e. two body) interactions (or equivalently weights) J_{ij}, and is explicitly given by the following expression:

$$P(x_1, \ldots, x_N|\hat{J}) = \frac{1}{Z(\hat{J})} \prod_{i<j} e^{J_{ij}[(x_i - x_j)^2]} . \tag{6.3}$$

The goal is to estimate the interactions $\{J_{ij}\}$ such that the correlations computed with the measure in Eq. (6.3) match the observed quantities C_{ij}, i.e.:

$$\langle (x_i - x_j)^2 \rangle \equiv \int d\boldsymbol{\phi}\, P(x_1, \ldots, x_N|\hat{J})(x_i - x_j)^2 = C_{ij} . \tag{6.4}$$

Inferring the interaction matrix $\hat{J}$ from the correlation matrix $\hat{C}$ is solved by maximizing the likelihood $\mathcal{L}(\hat{J}|\hat{C})$:

$$\mathcal{L}(\hat{J}|\hat{C}) = \sum_{i<j} J_{ij} C_{ij} - \log Z(\hat{J}) , \tag{6.5}$$

from which the inferred $\hat{J}^*$ is obtained as:

$$\hat{J}^* = \text{argmax}_{\hat{J}}\, \mathcal{L}(\hat{J}|\hat{C}) . \tag{6.6}$$

Indeed, by extremizing $\mathcal{L}(\hat{J}|\hat{C})$ with respect to J_{ij} we find:

$$0 = \frac{\partial}{\partial J_{ij}} \mathcal{L}(\hat{J}|\hat{C}) = C_{ij} - \langle (x_i - x_j)^2 \rangle \longrightarrow C_{ij} = \langle (x_i - x_j)^2 \rangle. \tag{6.7}$$

The main difficulty of this method is to compute the quantity $\log Z(\hat{J})$, the negative of which is the free energy in statistical physics. Unfortunately there is no known closed form for $\log Z(\hat{J})$, and, as a consequence, no closed form to estimate the interactions J_{ij} that maximize the likelihood Eq. (6.5), either. In principle, it would be wise to use a mean-field approximation to assess $\log Z(\hat{J})$, for the model Eq. (6.3) with continuous variables known as XY model in statistical physics. But this would be just an approximation.

In general, to solve the problem, a Montecarlo sampling method is used to compute the averages $\langle (x_i - x_j)^2 \rangle$, and then we use an approximate iterative gradient ascent algorithm to update the current estimate of the couplings J_{ij}. In practice, we start from an initial guess $\{J_{ij}^0\}$ at the initial time $t = 0$ of the learning algorithm, and then we update the J_{ij}'s using the following rule:

$$J_{ij}^{t+1} = J_{ij}^t - \eta \left[\langle (x_i - x_j)^2 \rangle^t - C_{ij} \right] + \alpha (J_{ij}^t - J_{ij}^{t-1}) , \tag{6.8}$$

where the quantities $\langle x_i x_j \rangle^t$ are the correlations computed via Montecarlo sampling using the current estimate of the couplings J_{ij}^t at time t; η is the learning rate, and α is a damping factor that we use to help the convergence. Typically, one can choose the initial $\{J_{ij}^0\}$ all equal to 0.1, the learning rate $\eta = 0.01$ and the damping factor $\alpha = 0.7$ [94].

After estimating the couplings J_{ij} the modular brain network can be built in two steps. First of all, we establish the intra-links between nodes (i.e. voxels) belonging to the same module, separately for each module, and then we connect the nodes in different modules through the inter-links. Ideally we would like to put a link between two nodes i and j if and only if the corresponding J_{ij} is different from zero. However, the inference of the couplings J_{ij} is affected by noise (both because of the uncertainties in the measurements of the C_{ij} and in the Montecarlo sampling), and thus we do not have a sharp classification of zero and non-zero couplings. Therefore, we define the connections by thresholding the J_{ij} with the following criterion. First we compute the standard scores Z_{ij} of the raw couplings J_{ij}, defined as $Z_{ij} = (J_{ij} - \langle J \rangle)/\sigma$, where $\langle J \rangle$ and σ are the mean and the standard deviation of the pool $\{J_{ij}\}$. Then, for each module separately, we consider a threshold T, and we create an intra-link between two nodes in the same module if $Z_{ij} > T$.

The question of what threshold value T precisely defines the network is resolved using the following procedure. First we add intra-links independently in each module by choosing T to be such that the average $\langle k^{\text{in}} \rangle$ of the degree of the intra-links is the same for each module, and equal to $\langle k^{\text{in}} \rangle = 5$. Once the intra-links have been established, we proceed to add inter-links between pair of voxels in different modules. Again, we consider a threshold T and we create an inter-link between two nodes i and j in two different modules if $Z_{ij} > T$. The threshold T is chosen to be such that the average $\langle k^{\text{out}} \rangle$ of the degree of the inter-links is $\langle k^{\text{out}} \rangle = 0.5$.

This procedure has been developed and applied by Morone et al. [94] to analyze fMRI data on the dual-task experiment in humans involving a visual and an auditory task from [78]. The method identifies predominant modules emerging in all 16 studied subjects involving the anterior cingulate (AC), posterior parietal cortex (PPC), and posterior occipital cortex (V1/V2) (see Fig. 6.4a, b) and compared well with the simple thresholding methods used in Fig. 6.3. The total sizes of the three main networks varies between 1500-2000 nodes out of the total number of voxels $\sim 60,000$. The average size of each module over 16 subject is respectively 641 ± 275, 692 ± 439, and 652 ± 361 for the AC, PPC, and V1/V2. The average in-degree is $\langle k^{\text{in}} \rangle = 5$ and out-degree $\langle k^{\text{out}} \rangle = 0.5$.

Alternatively, using a sparse approximation on the resulting network, the inference problem can be also solved by message passing to set up a faster algorithm that takes the functional correlations C_{ij} and outputs the couplings J_{ij}. The algorithm is described by:

$$
\begin{aligned}
u_{i\to j} &= (J_{ii} - \sum_{k\in\partial i\setminus j} J_{ik}^2 u_{k\to i})^{-1} \quad \forall\, i \to j\,, \\
J_{ij} &= \text{sign}(C_{ij})[-\frac{1}{2|C_{ij}|} + \sqrt{\frac{1}{4C_{ij}^2} + \frac{1}{u_{i\to j}u_{j\to i}}}] \quad i \neq j\,, \\
J_{ii} &= \frac{1}{C_{ii}} + \sum_{k\in\partial i} J_{ik}^2 u_{k\to i}\,,
\end{aligned}
\tag{6.9}
$$

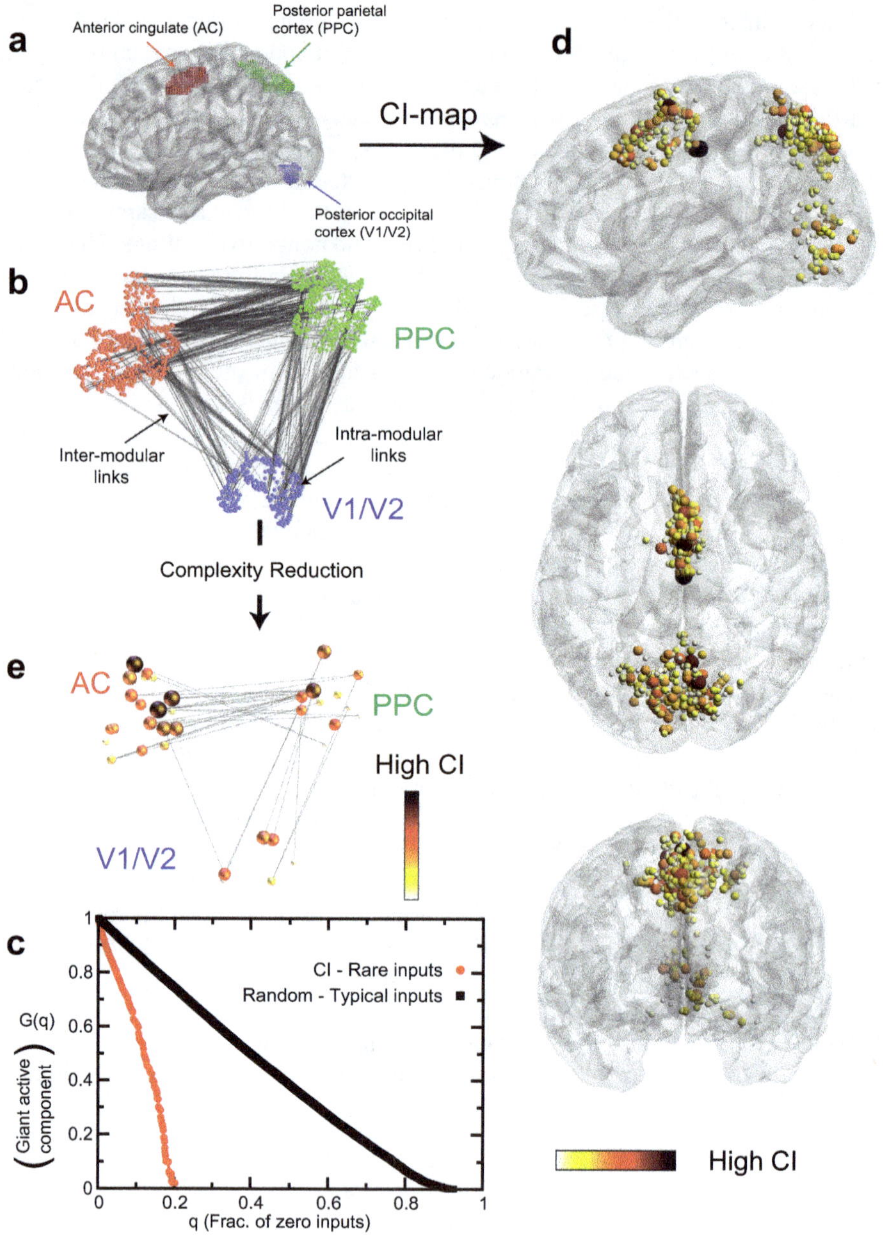
a
Anterior cingulate (AC)
Posterior parietal cortex (PPC)
Posterior occipital cortex (V1/V2)
CI-map
d
b
AC
PPC
Inter-modular links
Intra-modular links
V1/V2
Complexity Reduction
e
AC
PPC
V1/V2
High CI
c
CI - Rare inputs
Random - Typical inputs
G(q)
(Giant active component)
q (Frac. of zero inputs)
High CI

◀**Fig. 6.4 Neural influencers in a modular brain network. a** 3-module network in dual-task fMRI experiment from [94]. Spatial location of the 3 main modules for a typical subject showing the anterior cingulate (AC, red), posterior parietal cortex (PPC, green), and posterior occipital visual areas V1/V2 (blue). This 3-modular structure appears consistently for all 16 subjects studied in [7, 78, 94]. Nodes represent voxels in the fMRI BOLD signal of normalized size $2 \times 2 \times 2\,\text{mm}^3$. **b** Topology of the nodular network. Same as (**a**), but in the network representation with inter-links in gray. **c** Robustness and collective influence in the brain. Size of the largest active cluster $G(q)$ as a function of the untargeted ($n_i = 0$) nodes q following removal of brainCI nodes (red curve, $\ell = 3$) and following removal of random nodes (black, random percolation). **d** Influence-map of the human brain, averaged over 16 subjects. The color bar represents how many subjects a node ranks with a given brainCI. High brainCI influential regions are located mainly in the AC module for processing top-down influences, while the influential nodes are rarely located in the lower-level V1/V2 region. The PPC region contains a portion of influential nodes next to AC. **e** Complexity reduction by top brainCI nodes. Inter-links between different networks are mainly mediated by top influencers. Figure reproduced from [94].

which are solved by iteration starting from the initial condition $u_{i\to j} = J_{ii}^{-1}$, with $J_{ii} = C_{ii}^{-1}$ and $J_{ij} = 0$, otherwise.

The variables $u_{i\to j}$ can be interpreted as messages exchanged by the nodes. They are the cavity fields in statistical physics [83, 87]. The field $u_{i\to j}$ is the outgoing message from node i traveling along the link connecting i and j and incoming into node j. Outgoing messages are computed from the incoming ones in a self-consistent way. The message $u_{i\to j}$ quantifies the fluctuations of the signal x_i when node j is removed from the network (whence the name cavity).

6.3.3 *Graphical Lasso to Infer a Sparse Representation of Functional Brain Networks*

Beyond using simple thresholding of the correlation matrix or maximum entropy methods, to obtain a sparse and significant representation of the functional network, more common tools from machine learning can be used to reconstruct the network. To avoid assigning spurious links introduced by random covariates and to reduce the dimensionality of the problem, well-established regularizations, like the graphical lasso [95, 96], can be used to infer the sparse representation $\hat{J}$ of the correlation matrix C_{ij}. Graphical lasso assumes a multi-variate Gaussian distribution for the dynamical variables $x_i(t)$, and then infers the model's parameters from experimental data. The estimation of this matrix is made by minimizing the log-likelihood of a multivariate normal distribution:

$$\log \mathcal{L}(\hat{J}) = \log \det[\hat{J}] - \text{Tr}[\hat{C}\hat{J}] - \lambda|\hat{J}|, \tag{6.10}$$

where the penalization parameter λ controls the sparsity of the resulting network as estimated by the number of total connections.

To select the appropriate value of λ, two considerations, integration versus segregation, apply. Brain networks are integrated, i.e., signals are exchanged through the whole architecture within and between different brain modules. Therefore, the functional network must guarantee global connectivity. Second, the functional network is segregated into modules. Based on these considerations, the value of λ was fixed by Del Ferraro et al. [97] such that all nodes are connected through a path (to allow for brain integration) and, at the same time segregated into modules. The procedure follows a similar percolation process as thresholding the correlation matrix discussed in Sect. 6.3.1 and shown in Fig. 6.3 [98].

We decrease the value of λ to regularize the functional network less and less until all relevant modules appear and they are integrated in a global network. For this, we monitor the jumps in the size of the largest connected component to identify the weak and strong links defining the modular structure. This percolation process is analogous to the thresholding of the correlation matrix. It has been proven by Sojoudi [99] that the resulting functional networks obtained by thresholding C_{ij} or the glasso representation J_{ij} are the same.

Functional and structural networks can be incorporated in brain theory developed concurrently to test the implications of these correlated structures regarding the resilience to failure, optimality of information flow and existence of core nodes in the brain structure. Conclusions that can be drawn from this analysis include:

- Which are the integrator nodes of the network that serve to effectively propagate information?
- Which are the core nodes of the network, and are they connected locally or across networks?
- Which cascades of activity (or changes thereof) are anticipated if the strength of connections within the pathways is greatly suppressed or enhanced by a perturbation?

We investigate these issues next.

6.4 Optimal Percolation in the Brain

Chapter 1 shows that IMP [65], statistical physics, spin glasses [100–102] and optimal percolation [103] provide first principle formulations for finding the best spreaders in complex networks. Below we review extensions of these approaches to study robustness, vulnerability and navigability in networked neural systems.

Typically, the output of these theories provide a ranking of influential nodes whose inhibition could produce the largest disrupting effect extending to the largest portion of the brain network. These rankings identify particularly vulnerable areas in the brain according to each theory. Concurrently, the influential nodes are optimized as the minimal set of nodes for broadcasting information to the whole brain. These rankings can thus guide experimental studies to understand the impact obtained in the interventions to different nodes in the brain.

Influencer identification in the brain requires generalizations to a modular system interconnected as a network of networks [7, 8, 14, 17, 104, 105]. In contrast to interdependent man-made systems that can be extremely fragile to external shocks, for instance power grids as shown by Buldyrev et al. [104] and Gao et al. [105], a brain network is a robust interdependent modular system against failures of its components [17, 106]. In order to understand the mechanism of such robustness, a model of a brain as a system of modules connected in a network of networks is required [94].

Considering for simplicity a brain model with two different modules, each node i can be connected to nodes in the same module via intra-modular links, or nodes in another module via inter-modular links. Each node can be present or removed, and, if present, it can be activated or inactivated. We introduce the binary occupation variable $n_i = 1, 0$ to specify whether node i is present ($n_i = 1$) or removed ($n_i = 0$). By virtue of inter-module control/dependency connections, the functioning of a node in one module controls/depends on the functioning of nodes in other modules. In order to conceptualize this form of control, the activation state σ_i takes values $\sigma_i = 1$ if node i is activated and $\sigma_i = 0$ if not. A node i with one or more inter-module control links ($k_i^{\text{ctr}} \geq 1$) is activated ($\sigma_i = 1$) if and only if it is present ($n_i = 1$) and at least one of its inter-module neighbors j is also present ($n_j = 1$), otherwise it is not activated ($\sigma_i = 0$). In other words, a node with one or several inter-module dependencies is inactivated when the last of its inter-module neighbors is removed.

The rationale for this control rule is that the activation ($\sigma_i = \sigma_j = 1$) of two nodes connected by, for instance, one inter-link occurs only when both nodes are occupied ($n_i = n_j = 1$). If just one of them is unoccupied, let's say $n_j = 0$, then both nodes become inactive. Thus, $\sigma_i = 0$ even though $n_i = 1$, and we say that j exerts a control over i. This rule models the way neurons control the activation of other neurons in distant brain modules (via fibers through the white matter) in a process known as top-down influence in sensory processing [107]. Mathematically, σ_i is defined as [94]:

$$\sigma_i = n_i \left[1 - \prod_{j \in \mathcal{F}(i)} (1 - n_j) \right] \quad \text{for } k_i^{\text{ctr}} \geq 1 \,, \tag{6.11}$$

where $\mathcal{F}(i)$ denotes the set of nodes connected to i via a control link. Conceptually, the control links define a mapping from the configuration of occupation variables $\mathbf{n} \equiv (n_1, ..., n_N)$ to the configuration of activated states $\boldsymbol{\sigma} \equiv (\sigma_1, ..., \sigma_N)$, as given by Eq. (6.11). It should be noticed that the results obtained regarding robustness of these network model depend on the rule of transmission of information, Eq. (6.11). Other stricter rules considered in [104] lead to fragile models with catastrophic transitions.

Not all nodes participate in the control of other nodes via dependencies, however, i.e. a certain fraction of them does not establish inter-module control connections. If a node does not have inter-module links, it activates as long as it is present:

$$\sigma_i = n_i \quad \text{for } k_i^{\text{ctr}} = 0 \,. \tag{6.12}$$

This property also guarantees that we recover the single network case for vanishing inter-module connections ($\langle k_i^{\text{ctr}} \rangle = 0$), i.e. when considering the limiting case of one isolated module only.

The above control rule can alternatively be expressed by the McCulloch-Pitts model of neural activation [108]:

$$\sigma_i = 0 \qquad \text{direct inactivation} \tag{6.13}$$

$$\sigma_i = \Theta\left(\sum_{j \in \mathcal{F}(i)} \sigma_j\right) \qquad \text{indirect inactivation} \tag{6.14}$$

where a node can be inactivated directly (in which case we set $\sigma_i = 0$) or it can be inactivated indirectly as a result of lacking input from its inactivated neighbors [106].

The sum over $\mathcal{F}(i)$ in the second equation reflects the integration of incoming activity from nodes j in other modules connected to node i via control links. The threshold operation via the Heaviside step function Θ indicates that a minimum of incoming activity is needed for activation to propagate, similar to the perceptron.

The global communication efficiency of the brain modular network is quantified by the size of the largest mutually connected active component $\mathcal{G}$ consisting of active nodes $\sigma_i = 1$. The brain network is considered robust if $\mathcal{G}$ remains extensive for typically sampled configurations of the external inputs $\mathbf{n} = (n_1, \cdots, n_N)$. Induced by $\mathbf{n}$, the size of the largest component is a function of the fraction of zero inputs $q = 1 - \sum_i n_i/N$. As q increases continuously from 0 to 1, there exists a percolation threshold q_c^{ran} where $\mathcal{G}$ becomes subextensive in the thermodynamic limit, that is, it is disintegrated as in typical percolation. Marking the transition between a percolating phase and a non-percolating one, the threshold provides a measure for the system robustness under failure of a set of nodes chosen at random. An ideal robust network requires that $q_c^{\text{ran}} \approx 1$, which maintains global activity for almost all external inputs.

The behavior of the mutually connected component under random failure has been extensively studied by Buldyrev et al. [104, 105]. Although these works use a different rule of control than Eq. (6.11), which leads to fragile interdependent structures with catastrophic first-order transitions of component disintegration. Instead, Eq. (6.11) models activation dynamics that are more robust and are characterized by second-order transitions under failure.

When a fraction of nodes is removed, the percolated network breaks into isolated components of activated nodes. One then focuses on the largest (giant) mutually connected activated component $\mathcal{G}$, which encodes global properties of the system. The size of the largest component $\mathcal{G}$ can be obtained by a set of self-consistent message passing equations [103]. In order to distinguish the messages propagating along different links, $\rho_{i \to j}$ and $\varphi_{i \to j}$ are used to denote the information along an intra-modular link and an inter-modular link, respectively. Following the activation dynamics in the modular brain, the message passing equations are given by:

$$\rho_{i\to j} = \sigma_i \left[1 - \prod_{k\in\mathcal{S}(i)\setminus j} (1-\rho_{k\to i}) \prod_{\ell\in\mathcal{F}(i)} (1-\varphi_{\ell\to i}) \right], \tag{6.15}$$

$$\varphi_{i\to j} = \sigma_i \left[1 - \prod_{k\in\mathcal{S}(i)} (1-\rho_{k\to i}) \prod_{\ell\in\mathcal{F}(i)\setminus j} (1-\varphi_{\ell\to i}) \right], \tag{6.16}$$

where $S(i) \setminus j$ represents the set of neighbors of node i in the same module excluding node j. In an alternative model proposed in Ref. [17], a node can be activated iff it is connected to the giant component, thus the size of $\mathcal{G}$ can be calculated by the fraction of active nodes $G = \langle \sigma_i \rangle \equiv \sum_i \sigma_i / N$. In contrast, model Eq. (6.11) from Ref. [94] allows the activation of a node even if it is not connected to the largest component, which tends to prevent the catastrophic cascading effects observed in the model of Ref. [104] and enhance network robustness. Based on the solution of message passing equations, an additional quality $\rho_i \in \{0, 1\}$ is introduced encoding the probability that node i is connected to the giant component:

$$\rho_i = \sigma_i \left[1 - \prod_{k\in\mathcal{S}(i)} (1-\rho_{k\to i}) \prod_{\ell\in\mathcal{F}(i)} (1-\varphi_{\ell\to i}) \right]. \tag{6.17}$$

Following the collective influence theory in optimal percolation for a single network [103], the influencers in the modular brain can be identified by minimizing the largest eigenvalue of the stability matrix $\mathcal{M} = \{\partial\rho_{i\to j}/\partial\varphi_{k\to l}\}_{\rho=\varphi=0}$, which leads to an influencer identification algorithm through the collective influence index for the modular brain in terms of non-backtracking random walks, this time defined in the coupled network, see Fig. 6.5:

$$\text{brainCI}_\ell(i) = z_i \sum_{j\in\partial\text{Ball}(i,\ell)} z_j + \sum_{j\in\mathcal{F}(i):k_j^{out}=1} z_j \sum_{m\in\partial\text{Ball}(j,\ell)} z_m, \tag{6.18}$$

where $z_i \equiv k_i^{in} + k_i^{out} - 1$. By following the ranking of influencers given by brainCI, one disintegrates the mutually connected component employing the (close to) smallest possible fraction of removed nodes, q_c^{brain}. This set represents the brain influencers.

This influencer theory has been applied in Ref. [94] to the modular brain networks constructed by fMRI data obtained from individuals performing a dual visual and auditory task in [78], analyzed in Fig. 6.3 and Ref. [7]. Figure 6.4 illustrates how to identify influencers in this functional and modular brain network composed of 3 modules: AC-PPC-V1/V2 (Anterior cingulate-Posterior parietal cortex-Posterior occipital cortex).

Specifically, the giant component can be dismantled by blocking the external inputs to $q_c^{\text{brain}} \approx 0.2$ of the optimal neural influencers, compared with $q_c^{\text{ran}} \approx 0.9$ for random blocking (Fig. 6.4c). These neural influencers can be associated with

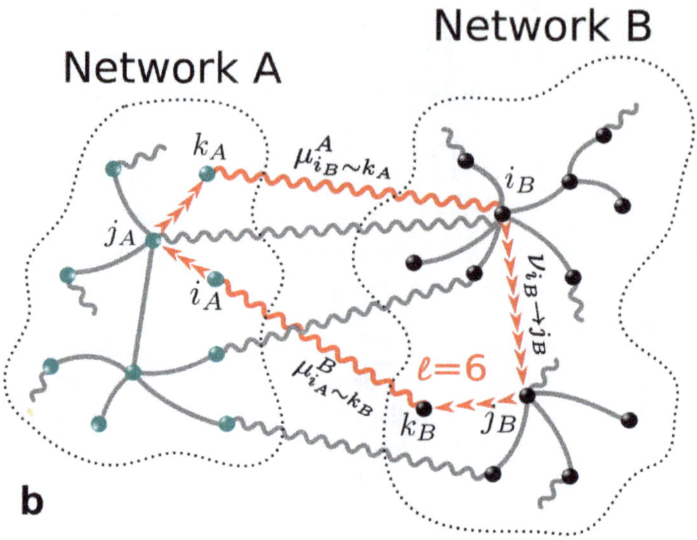

Fig. 6.5 Generalized non-backtracking (NB) random walks. They are random walks that cannot come immediately back. NB walks in a single network were discussed in Sect. 1.4.2 and Fig. 1.11b, and are parametrized by a single message $\nu_{i \to j}$. The generalization of the NB walks to coupled brain networks has two types of steps, in and out of the modules, and they are parametrized by four messages. The most influential nodes are the sources of NB random walks that cover the largest area of the network. This mapping allows to solve optimal percolation in a modular brain network as a diagrammatic expansion of increasingly larger NB walks of length ℓ (one NB walk is shown in red with $\ell = 6$)

pathological states of the brain and neurological disorders upon their failure. Averaging the neural influencer map over 16 different subjects, it is illustrated in Fig. 6.4d that the neural influencers are located mainly in the AC module, while they are also present in a portion of the PPC module, which are the central part of human brain control system. The neural influencer map suggests that brain control is deployed from the higher level module (AC) toward certain strategic locations in the lower ones (PPC-V1/V2), in agreement with previous research [78].

6.5 Neural Influencers in a Memory Network

Instrumental to the experimental finding of influencers in the brain is the capacity to exquisitely manipulate neuronal activity in specific nodes of the network while simultaneously recording the functional consequences of those manipulations. Specifically, optogenetic and pharmacogenetic manipulations [109–112] of genetically and anatomically-defined neuronal populations are tools with the potential to condition network dynamics. Furthermore, stimulation protocols to induce longlasting effects (for instance long-term potentiation, LTP [113–115]) on specific brain areas [116–119] can induce potentially durable changes in neuronal network dynamics that are ideal to investigate the theories of influencers in the brain.

Optogenetics is a powerful technology in which light-activated trans-membrane conductance regulators (proteins from microbial origin) are introduced into specifically targeted cell types to allow temporally precise, millisecond-scale activity modulation in-vivo [109, 110]. Optogenetic technology combines perfectly with fMRI since it is intrinsically nonmagnetic (optic fibers guide the appropriate wave-length light to the brain target from a laser source located far enough from the magnetic field) and therefore can be used to simultaneously perturb a brain network and inform on the dynamical response function. By combining optogenetic control with fMRI readout, neural activity arising from specific circuit elements defined by genetic identity, cell body location, and axonal projection targets can be monitored in-vivo across the whole brain. An advantage of optogenetics over micro-stimulation is that it cannot only activate but also suppress neuronal activity, in genetically specified cell populations in target brain regions.

Similarly, pharmacogenetic manipulations can be performed following a protocol designed by Carames et al. [120]. Designer Receptors Exclusively Activated by Designer Drugs (DREADDs) are used to inactivate neurons (subtype hM4D(Gi)) in the selected brain areas upon intraperitoneal administration of Clozapine-N-Oxide (CNO, 1mg/kg) [120, 121]. More specifically, with the aid of adenoviral vectors, the expression of a Gi-DREADD (hM4Di) protein is directed into the targeted area, which, under intra-peritoneal administration of the otherwise inert ligand clozapine-N-oxide (CNO), activates the receptor, inducing neuronal silencing and blocking the area output. The pharmacogenetic approach is more favorable than an optogenetic strategy because it avoids implanting bilateral cannula and optic fibers across cortical regions from which fMRI signals are collected.

Del Ferraro et al. [97] employed pharmacogenetic interventions *in vivo* to test the influential nodes identified by theory that are essential for global integration in a memory network in rodents. To define the rat brain network; they leveraged previous work on memory acquisition and consolidation by the Canals group [116–119] induced by long-term potentiation (LTP) in the hippocampus.

LTP represents the cellular model of memory acquisition based on associative inputs and coincidence detection [113–115]. The main finding of the Canals group [116–119] is that local modulation of synaptic strength induced by LTP of the perforant pathway using high-frequency electric stimulation leads to large scale changes of fMRI activity throughout the brain (Fig. 6.6a). This system of memory networks can be used to manipulate nodes to investigate the robustness and vulnerability under different influencer scenarios.

Changing synaptic strength by LTP protocols using electrical microstimulation locally in the dentate gyrus in the hippocampal circuit on sedated rats induces a long-lasting change in the functional connections that propagate long distances in the brain, enhancing hippocampal-neocortical interactions during more than 3 h. Thus, after LTP activation, hippocampal activity spreads to the prefrontal and perirhinal cortex [117–119] which are brain regions involved in memory consolidation [122, 123].

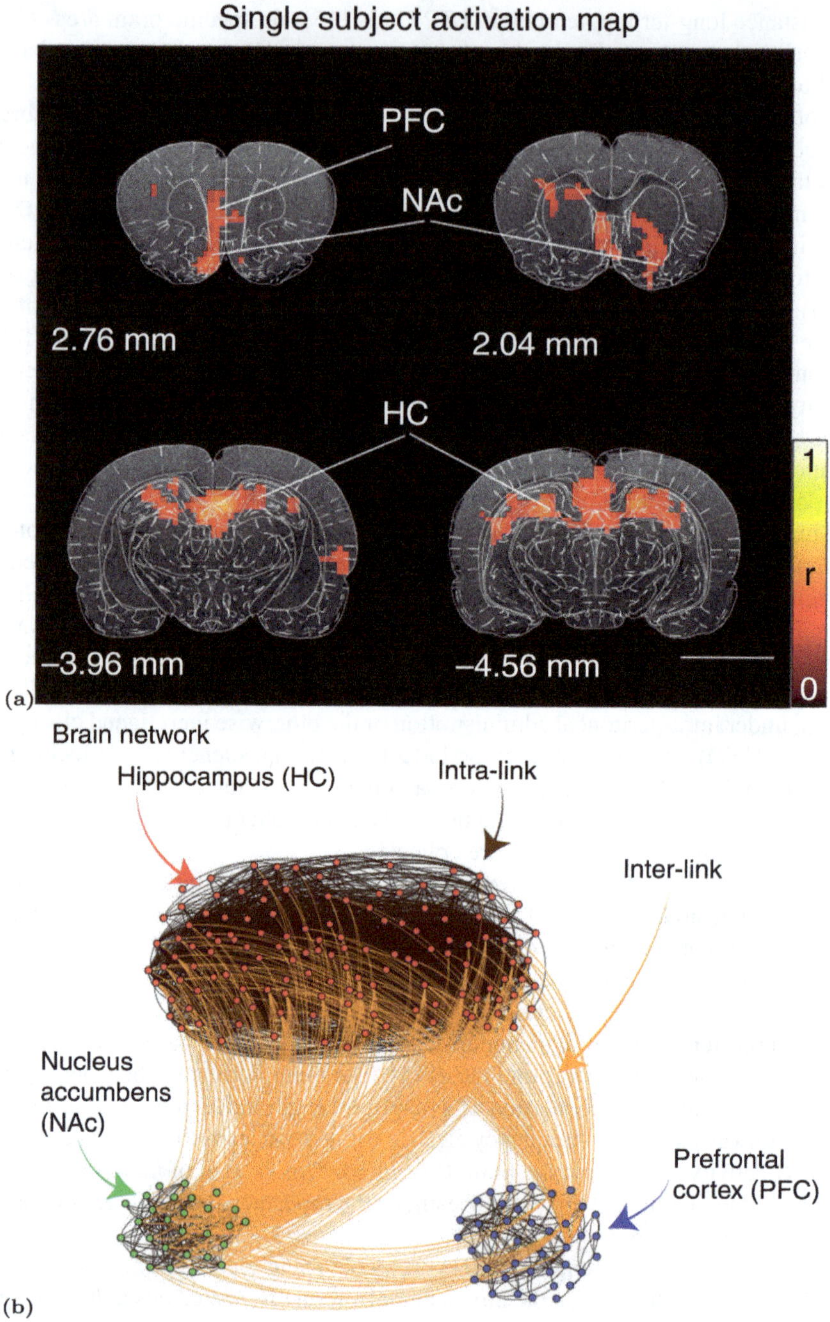

Fig. 6.6 Finding core nodes using influence theory in the rat brain [97]. **a** fMRI activation analysis results in **b** functional network used to determine which of the activated areas (red and yellow) are the core nodes for the integration of memory. Figure reproduced from [97]. Copyright © 2018, The Authors

Network analysis shows a modular structure (obtained from resting-state fMRI before and after LTP) appearing 3 h after the application of the LTP electrical stimulation connecting the hippocampus and long-range structures at the prefrontal cortex, nucleus accumbens and perirhinal cortex (see Fig. 6.6b). This result further confirms in the rat brain the typical structure of coupled interconnected networks, unveiling weak links that are (1) critical for hippocampal-neocortical interactions and (2) modulated by synaptic plasticity, providing a framework for investigating information storage (memory consolidation) and influencers in brain networks.

Analyzing the fMRI signal of the activated voxels, three dominant clusters of nodes were identified, which overlapped with the anatomical location of the hippocampus (HC), the prefrontal cortex (PFC), and the nucleus accumbens (NAc) (Fig. 6.6a, b). The fMRI correlation matrix is modified with a penalization parameter introduced to tune the sparsity of the network using the graphical lasso method of Eq. (6.10). The links descriptive of short-range interactions within each cluster are the intra-links, while links corresponding to long-range interactions between different modules are inter-links.

By characterizing the global integration of brain network as the formation of the largest mutually connected component (Fig. 6.7a), the essential integrators of the brain network can be identified according to different centralities, such as hubs and brainCI (Fig. 6.7b). It was found that the centralities based mostly on connections, including hubs, k-core and eigenvector, and betweenness centrality, favor the hubs in the HC region (Fig. 6.7g). As a comparison, brainCI from Eq. (6.18) predicts that the integration of the memory network is determined by a set of low-degree nodes in the nucleus accumbens (Fig. 6.7h). For $\ell > 1$, the collective influence manages to capture the effects emerging from the long-range structure, which tends to be neglected by other centralities.

As illustrated in Fig. 6.7c and f, attacking a node with a high brainCI value produces larger damage to brain integration than a hub. Figure 6.7c presents the relative size of giant mutually connected component as a function of the fraction of inactivated nodes, selected by two strategies: Hub inactivation and brainCI inactivation. Attacking only 7% high brainCI nodes in the nucleus accumbens prevents the formation of the memory network while inactivations of hubs in other brain areas leave the network intact (hubs are mainly located in the HC).

The specific predictions of the influencer theory for this specific pathway strengthened in the hippocampal-neocortical interactions are: (1) it is mediated by a core node in the NAc module, and (2) it has a crucial role in the network even prior to its potentiation. This concrete prediction is then tested with pharmacogenetic interventions. To experimentally test these predictions, each of the areas in the brain involved in the memory network were pharmacogenetically inactivated in vivo in order to test the theoretical predictions. Experimental results confirmed that the NAc has a dominant role in memory consolidation.

Adenoassociated viruses (AAV) were used to block particular targeted area of the NAc shell of several animals *in vivo*. Under such intervention, the formation of the long-range network involving HC, PFC and NAc are prevented completely. More specifically, the long-range Inter-network links from the HC to the PFC are missing,

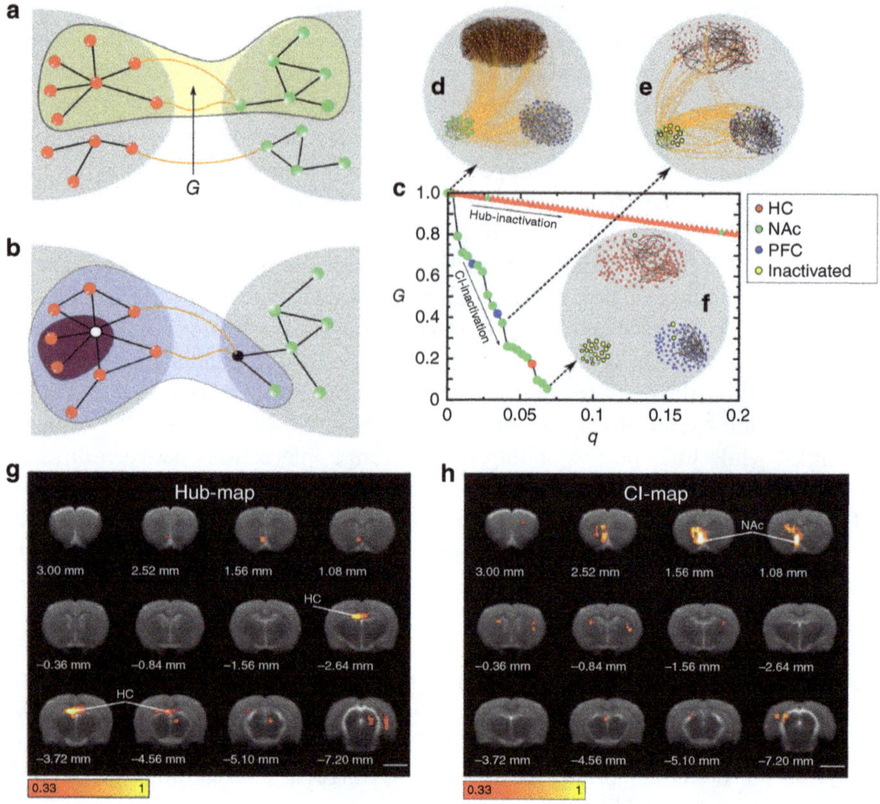

Fig. 6.7 Collective influence map of the rat brain. a The mutually connected largest connected component G in yellow captures the integration of two modules into a brain network. **b** The influence of a hub (white node) and a brainCI node (black) is shown: inactivation of the hub produces less damage to brain integration as measured by the size reduction of G, than the inactivation of the brainCI node. **c, f** Two strategies for choosing the essential nodes: hub inactivation (red triangles) and brainCI inactivation (Eq. (6.18), green circles). We plot the size of the GCC as nodes are removed one by one according to their degree (red curve) or brainCI score (green), respectively, from high to low. Most hubs (red symbols) are located in HC, yet they are not essential for integration: their removal makes minimal changes to the connectivity (vertical axis). On the contrary, by inactivating 7% of high brainCI nodes (horizontal axis), connectivity collapses to almost zero. Most brainCI nodes are in the NAc (green symbols), indicating that NAc is the essential structure in this process, even though it is a weak node compared to the HC. These theoretical predictions were confirmed by pharmacogenetic inactivation of the NAc. **g** Hub map of the rat brain indicating the location of the top hub nodes average over six animals. The yellow to white areas correspond to top essential nodes predicted by the degree. The top hubs are all located in the HC since this is the area of LTP induction. **h** brainCI map of the rat brain indicating the top CI nodes average over six animals. Most brainCI essential nodes result are located in the NAc and are generally not hubs. This prediction is then confirmed by pharmacogenetic inactivation of these essential areas. Figure reproduced from [97].

even though these sub-networks have not been directly deactivated. Thus, experimental results confirmed that the NAc has a dominant role in memory consolidation, validating predictions based on influencer theory in the rat brain. These predictions are more accurate (have larger experimental feedback) than predictions based on hubs and other centrality scores.

6.6 Neural Influencers to Guide Brain Tumor Surgery

Classification of essential brain areas based on their role in cognitive tasks are of interest for the understanding of brain diseases, their cure, and prevention [44]. By definition, perturbation of these core areas leads to devastating consequences: disturbances in the functional connectivity, and decreased global integration and neurological performance [44, 53]. In the case of human language, it is known that an infarct in the Broca's area leads to mutism, indicating that this is an essential area for language [124, 125].

Identifying these essential areas has practical consequences for neurosurgery. In the neurosurgical setting, fMRI is used to plan the neurosurgical approach and can guide the actual resection of a brain lesion during the operation [126–128], for instance, the neurosurgical resection of a tumor, which is a mainstay of the treatment of patients with brain tumors. One of the most important functions of the brain that the neurosurgeon uses fMRI to localize is language. Performing pre-operative fMRI presents distinct advantages.

- fMRI can influence the decision of whether to attempt a resection, a stereotactic biopsy or not to operate.
- fMRI may influence the approach that a surgeon will take in approaching the tumor and limiting the size of the craniotomy.
- fMRI can guide the neurosurgical resection since the fMRI data can be displayed in the operating room using a neurosurgical navigational system, which allows the neurosurgeon to view the relationship of the lesion to the adjacent eloquent cortex in real time.
- fMRI data can guide the cortical stimulation, which can shorten the operation [127].

However, in the pre-operative brain tumor setting, study of fMRI activation alone cannot differentiate between essential functional areas and non-essential functional areas (Fig. 6.8). Rather, fMRI identifies all activated areas associated with the function being tested. The distinction is crucial, as neurosurgical resection of essential areas results in severe language compromise, such as speed arrest, whereas resection of non-essential areas (that also activate on fMRI) will not.

Occasionally, areas of the brain that demonstrate fMRI activation, but do not exhibit speech arrest during cortical stimulation are resected to obtain a more complete control of the tumor. For example, in pre-operative fMRI for inferior, left, frontal lobe tumors, language fMRI often demonstrates activation of the middle frontal gyrus. This activation probably corresponds to 'verbal working memory' [129] and does not result in speech arrest when stimulated by the neurosurgeon intra-operatively [124].

The functional relevance and clinical sequelae of loss of secondary or tertiary language centers are difficult to predict [130, 131]. Secondary and tertiary language centers include the pre-SMA (language supplementary motor areas), middle frontal gyrus left anterior, and the middle frontal gyrus left posterior (pre-motor or Exner's area). Using the language of graph theory, what happens when we physically delete (or neurosurgically resect) a non-core node from the language circuit can be understood. A clinical scenario allows for a unique, ethically permissible opportunity to apply the theory of influencers to decipher the exact meaning of the function of secondary and tertiary language centers by correlating graph theoretical predictions from fMRI and cortical stimulation data to pre- vs post-resection neuropsychological tests.

Thus, a theory of brain influencers should be able to predict the essential core nodes and should be able to guide direct, intra-operative cortical stimulation of core areas during tumor resection. Additionally, understanding essential areas is important in the management of other pathologies: strokes, traumatic brain injury (TBI), spinal cord injuries, and neurodegenerative disorders (Alzheimer's, Parkinson's). The ability to identify focal nodes in the brain has potential applications to guide focal interventions, too. These may include:

- Rehabilitation programs based on increasing focal brain activity [132],
- Transcranial magnetic stimulation (TMS) that can stimulate or deactivate focal brain activity [133–135],
- Implantable devices which have had success in Parkinson's disease [136] and even motor weakness [137].

In the next section, we review current efforts to apply influencer theory to guide interventions in the human brain in the clinical setting. We will discuss the use of fMRI signals from patients performing a language task in order to construct functional brain networks associated with expressive language. The theory explained in previous sections then identifies the core nodes in these networks. The importance of the influential nodes is confronted with direct brain stimulation to guide tumor resection during awake surgery in the operating room.

6.6.1 Brain Surgery with Minimal Impairment to Language

We discuss how network theory can guide protocols to stimulate areas predicted to be core nodes for language function during awake brain surgeries to patients with brain tumors. Typically, neurosurgeons are left to guess as to which of the multiple areas of fMRI activation are essential for language function (and will correspond to regions of speech arrest on direct cortical stimulation) versus those that are not essential (will not lead to speech arrest on direct cortical stimulation).

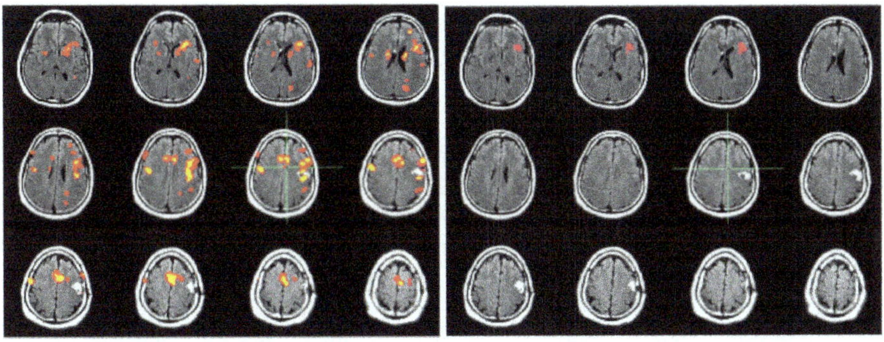

Fig. 6.8 To resect or not to resect. Use of fMRI and network theory to guide tumor resection in the operating room. **a** The figure shows a routinely performed pre-operative fMRI scan when the patient performed a language-related task that demonstrated multiple areas of activation (yellow and red). Some of these are near the tumor (white). Based on such data, the neurosurgeon does not know which areas are essential and whether or not the tumor can be resected safely. **b** However, the same data analyzed using influencer theory shows that the essential area for language in red is at some distance to the tumor so that the tumor can be resected with minimal impairment to the language function. Image data provided by Dr. A. I. Holodny, MSKCC

A clinical example is shown in Fig. 6.8a, obtained by Dr. Andrei I. Holodny and his team at the Department of Radiology of Memorial Sloan Kettering Cancer Center (MSKCC) in New York City. From the routinely performed pre-operative fMRI scan, it is impossible to tell which of the areas of fMRI activation represent essential language areas. Some of these are near the tumor. Based on such data, the neurosurgeon does not know whether the cancer can be resected safely. However, the same data analyzed for influential areas, Fig. 6.8b, shows that the essential area is at some distance to the cancer.

In the clinical setting, Dr. Holodny and his team of neuroradiologists present the neurosurgeons with fMRI activation maps, which depict the locations of eloquent cortices (such as language and motor areas) adjacent to tumors, which neurosurgeons use to plan and guide the resection of gliomas and other intracranial masses (Fig. 6.10). Notwithstanding its advantages, fMRI clearly has limitations. One of the main problems facing the pre-operative evaluation of brain tumors by fMRI is that this technology depicts activations of both essential and non-essential functional areas (Fig. 6.8. For a neurosurgeon, this distinction is of paramount importance. In this case, essential language areas are those whose direct, intraoperative stimulation by electrodes leads to speech arrest. Resection of such areas leads to severe language deficits. In contradistinction, non-essential language areas are defined as those whose direct, intraoperative stimulation by electrodes does not lead to speech arrest. Resection of such areas is thought to be safe.

While fMRI technology alone does not allow one to differentiate between core areas, influencer identification methods allow for this distinction. Li et al. [138] studied a functional brain network activated in a language task obtained from a patient at MSKCC with a brain tumor near the Broca area, as shown in Fig. 6.9. This right-handed patient shows a low-grade left frontotemporal insular glioma (in these images, the left hemisphere of the brain is on the right side of the fMRI image). In a right-handed subject with a healthy brain, the Broca area is located on the left side of the brain, which is the yellow cluster in Fig. 6.9b. The Broca area is translocated to the right side of the brain in the right-handed patient with a tumor shown in the yellow cluster in Fig. 6.9c.

Pre-operative fMRI activation map of Fig. 6.9a revealed robust activation in the left superior temporal gyrus (Wernicke's area, WA) and in the right inferior frontal gyrus (right anatomical homologue of Broca's Area, BA), which was confirmed by the neurosurgeon during intra-operative cortical stimulation in the operating room. While standard methods of fMRI activation data analysis demonstrated the translocation of BA to the right hemisphere, these methods are, in principle, not able to depict how the right-sided BA connects with WA (which remains on the left side). Hence, network methods are used to reconstruct the functional brain network to determine how these different brain areas connect, and which ones are essential and should be avoided in resection if function needs to be preserved.

Functional connectivity between the nodes is defined by starting with the pair-wise correlation matrix C_{ij} from Eq. (6.1) to obtain the sparse representation using graphical lasso Eq. (6.10) in order to eliminate spurious correlations. Li et al. [138] found that the right homologue of the Broca in this patient (Fig. 6.9c) functionally connected to the same areas as the left Broca area in a typical healthy control (Fig. 6.9b). As opposed to the functional connection of the left Broca in a healthy brain, the right Broca did not connect directly with the left Wernicke area, but connected indirectly, mediated by the Supplementary Motor Area and the Middle Frontal Gyrus.

Li et al. [139] established a functional healthy language core sub-network as a benchmark by analyzing healthy subjects, one example is shown in Fig. 6.9b. However, there are known functional differences in the language network of bilinguals compared to monolinguals, which may affect surgical management. Conducting a k-core analysis in the functional brain network Li et al. [140] revealed the most robust component of the functional language architecture in the healthy monolingual and bilingual subjects.

Important differences were found between the k-core decomposition in the bilingual and monolingual brains when performing the k-core decomposition between both networks. While Wernicke's area plays a crucial role in the activated language network, it was not known if this area plays a fundamentally different role in the bilingual brain as compared to the monolingual. In the monolingual subjects, Wernicke occupies both the innermost profound part of the brain network and the lower peripheral shells of the brain. However, in the bilingual subject, the Wernicke predominantly appears in the middle shells, which do not occupy the maximum k-core or the peripheral shells. This result is demonstrated in [140] to occur in all participants studied. Similar k-core analysis was performed by Arese Lucini et al. [141], which was able to discern the influential areas in the brain driving transitions between the state of consciousness and unconsciousness. The analysis then demonstrates the

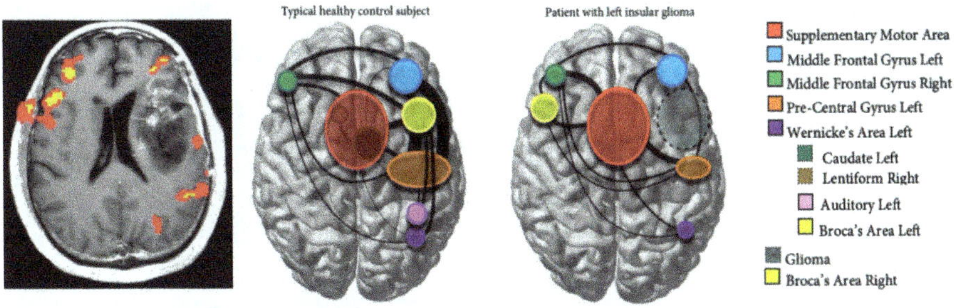

Fig. 6.9 Patient with a low-grade left frontotemporal insular glioma analyzed in [138]. **a** Pre-operative fMRI revealed reorganization of Broca's area (BA) to the right, confirmed by intra-operative cortical stimulation. Standard methods of fMRI activation alone are not able to depict how the right-sided Broca's area connects with Wernicke (which remains on the left side). Using graph theoretical methods demonstrated that (compared to a normal control) the right homologue of the BA did not connect directly with the left WA but connected indirectly through the Supplementary Motor Area and the Middle Frontal Gyrus. **b** Axial view of the brain with active fMRI areas and links obtained by statistical inference. Brain network topology inferred from a semantic fMRI tasks for a healthy control subject. The primary areas involved in a language task, BA and the Wernicke's area are identified in the left hemisphere and are functionally connected. **c** Shows the network for a patient with a tumor. The tumor (grey circular dashed-line area) infiltrates the left BA. The right BA homologue becomes statistically active and connects with several other active brain areas, resulting in a functional reorganization of the brain areas involved in a language task. This patient's right BA does not connect directly with the left WA. Functional communication between these two areas is conveyed through a common area to which they are both linked. The black lines represent functional connectivity, their thickness proportional to the strength of connectivity. Figure reproduced from [138]. Copyright © 2019, Li et al.

application of network theoretical methods to discern essential regions of the brain organization with clinical applications.

The results of the network analysis can be used by the neurosurgeon to guide the mapping of the brain areas important for language. Intra-operative mapping is often performed under 'conscious sedation' conditions, whereby the patient can be aroused during surgery and asked to participate in language tasks. The surgeon proceeds with the craniotomy and dural opening. Once the brain surface is exposed, the neurosurgical navigation system is used to identify the projected eloquent areas. This is facilitated by fusion of the pre-operative fMRI and anatomic MRI on the neurosurgical navigation system. This allows for immediate localization of the language area as predicted by network theory.

Stimulation is initiated at the site identified by theory pre-operatively and co-localized by the surgical navigator probe or within the gyrus just anterior to the site of phase reversal. The area is re-stimulated for confirmation and is then marked by a numbered paper tag (see bottom left Fig. 6.10). Typically, a minimum of twenty points are stimulated.

Mapping of the brain areas essential for language begins with cortical stimulation, with the patient counting aloud and reciting the days of the week and/or months of the year. The patient is monitored for speech arrest. Following simple arrest,

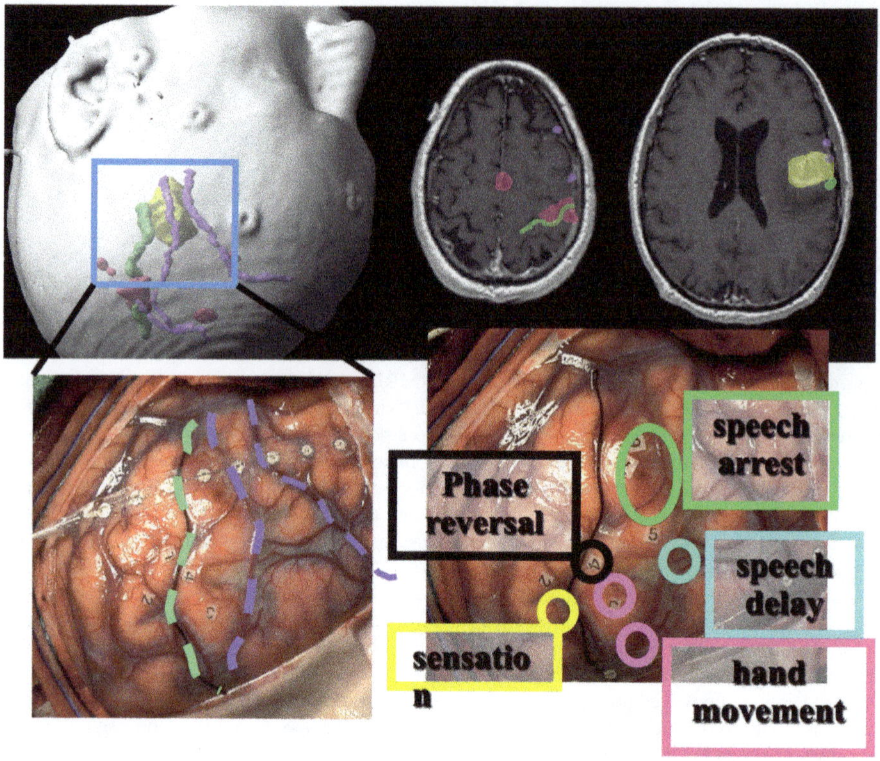

Fig. 6.10 Example of how fMRI data is presented to the neurosurgeon for glioma resection. A 3-D rendering (top row left) the projection the neurosurgeon will see during the craniotomy portrays the central sulcus (green), the hand motor homunculus (red), the tumor (yellow) and cortical veins (purple). The same information is seen on the 2-D images (top row right). The fMRI data correlated well with the intra-operative findings. The bottom row left image shows the exposed brain during the operation, with the green and purple lines corresponding to their counterparts on the 3-D MR image. The results of the direct cortical stimulation (bottom row, right image) again correspond to the locations predicted by fMRI. Image provided by Drs. A. I. Holodny and V. Tabar, MSKCC

the patient is asked to perform more cognitive language tasks, including phonemic fluency (generating words that start with a letter), semantic fluency (generating words that fit a category) as well as auditory responsive naming (e.g., What do you shave with? What color is grass?). Finally, the patient is asked to repeat a series of sentences. For all tasks, the current is escalated until an interruption of speech is measured or once the seizure threshold has been reached in order to rule out functional localization confidently.

Following identification of the appropriate eloquent cortex, the resection of the tumor proceeds guided by the theoretical analysis of the language network in Fig. 6.9. At the same time, the intraoperative stimulation data and the predictions of influencer theory from the functional network data can be compared in each of the ~ 20 points on the brain that are examined by direct cortical stimulation during the operation. The intra-operative stimulation is used as the ground truth and the response on each point on the brain that was stimulated can be compared to the influencer predictions. Obtaining such an ability is of great benefit to neurosurgical planning (where the neurosurgeon must balance the retention of neurological function versus obtaining a more complete resection of the tumor), as well as for the patient. By correlating the prediction of theoretical models to cortical stimulation and dedicated pre- and post-operative neuropsychological testing, one achieves this goal.

6.7 From Structure to Function in the Brain

The neural influencers obtained in Sect. 6.4 use the functional network and are not the same influencers obtained from a structural network. A more realistic way to find influencers is to search for those that could optimally transmit signals via real structural connections and not through functional ones. That is, we can use the structural network where the information is physically transmitted to identify the structural neural influencers. We shall then differentiate the functional influencers from the structural influencers according to which network is used to find them.

Finding structural influencers raises another problem: how to infer an structural network from a functional network. Every brain has one structural network (modulo its plasticity, which changes the brain connectivity on longer time scales than a performed task), but can sustain many spatiotemporal patterns, leading to different functional networks for each particular task. How does a single structural network supports many different spatiotemporal modes of propagation leading to different functional networks? This is a basic question in systems neuroscience, which is referred to as the structure-function problem, i.e., uncovering the underlying substrate in the brain for information propagation performing a function [12, 14, 62, 63, 67, 91, 142, 143].

Metaphorically speaking, one can think of the brain as a road system with the structural network representing the underlying system of highways where the signals can travel. Some highways are wider than others so that they can transmit more signals than others. Under a given task, let's say language, certain brain areas are activated, Broca, Wernicke, SMA, and others. These areas may transmit signals via certain pathways in the structural network and avoid others. Depending on the different inputs of the task, the same highway/anatomical substrate can support different traffic loads associated with different task-based functional networks by activating different pathways in the system. How the brain switches and modulates these multiple functional configurations is a topic of intense active research.

The different cognitive tasks require a coherent flow of information within the functionally interacting brain areas. It has been proposed that synchronized neuronal oscillations [144, 145] may facilitate a coherent flow of information between brain areas during cognitive tasks; a mechanism that is called 'dynamic coordination' [146–150]. The brain then dynamically coordinate the flow of information by changing the strength, pattern, or frequency of oscillatory synchrony between the brain areas, yet using a single underlying substrate.

We recall the existence of a link in a functional brain network between i and j obtained by any of the methods discussed in Sect. 6.3 involving correlation functions C_{ij} from fMRI or EEG, or sparse representation J_{ij}, or any methods measuring functional interactions, mutual information, or synchronization does not imply that there exists a direct physical (structural/anatomical) connection between i and j via a synaptic connection or white matter fiber track between the considered nodes.

It is clear then that a one-to-one relation between the structural network of the brain and the functional spatiotemporal network cannot exist because one single brain can sustain many functional networks. As shown by Honey et al. [62], structural weights are only weakly correlated with functional weights (Fig. 6.11) and many functional connections exist that have no corresponding structural connection and vice versa.

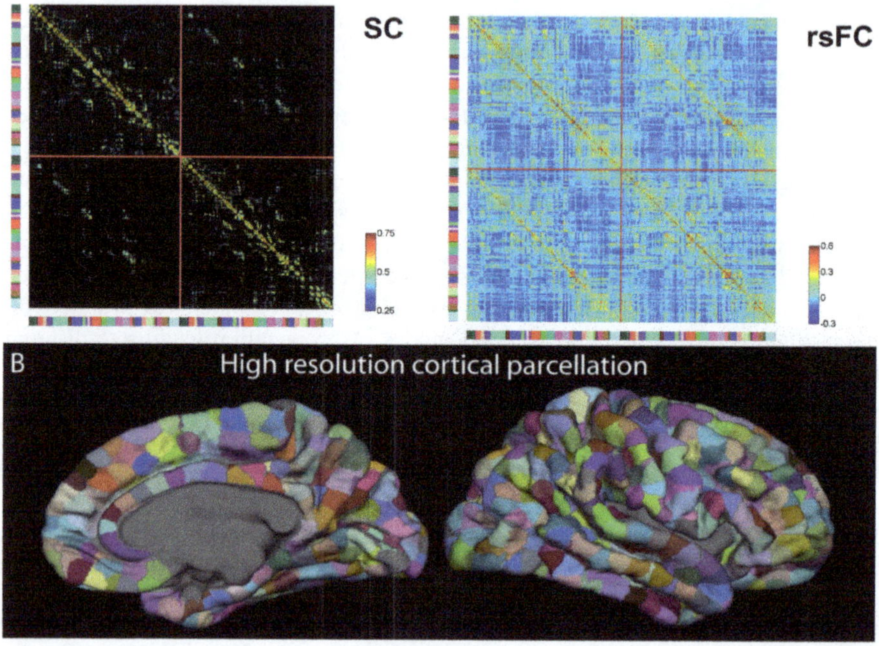

Fig. 6.11 Comparison of the structural network and the functional network in resting state from [62]. Matrices of the structural connectivity (SC) and empirical resting state functional connectivity (rsFC) in resting state. Plots represent averages across 5 participants. The rsFC was computed from the BOLD time series obtained for 998 anatomically defined ROIs at high resolution. The color bars on the axis indicate the ROI brain regions shown at the bottom (see Fig. S1 of [62]). While the SC resembles the rsFC, there is no relation between these two connectivity. This indicates that the SC cannot be predicted from the FC. The opposite might still be possible if a model of the structure of the structure is known, for instance, if there are symmetries in the SC. Bottom figure: High-resolution parcellation with 998 ROIs of the cerebral cortex used to measure the SC and the rsFC. Figure reproduced from [62]. Copyright © 2009, National Academy of Sciences U.S.A

Another example is the language circuits discussed in the previous section. The anatomical connectivity (white matter tracts) is reasonably well understood from diffusion tractography studies [151, 152]; however, resting-state fMRI studies have demonstrated that anatomical connectivity does not translate into functional connectivity [153, 154]. These results indicate that even when a functional connection exists between two ROIs, it is not guaranteed that a direct synaptic connection of fiber track will connect the two ROIs. Thus, in principle, it is impossible to infer the structural highways from the dynamical functional connectivity unless we know the dynamical model of how these signals propagate in the network.

Even though there is no direct relation between the structure and functional connections, neural elements that are anatomically connected display greater functional weights on average. Studies in visual and somatomotor networks [142, 155] support this view. Considering this weak correlation between structural and functional networks, functional influencers can then be deemed to be a first-order approximation to the structural influencers.

In Fig. 6.12, we see an example of how to combine dynamical fMRI data with DTI anatomical data to obtain the underlying structural networks that are activated in a given task, in this case, the language task investigated by Li et al. [139] and explained in Sect. 6.6.1. fMRI shows which areas are activated. Using these activated areas as ROIS, DTI white-matter probabilistic tractography then identifies the white matter pathways between those activated areas in the human brain. Figure 6.12 shows the language activation map for a healthy subject as analyzed in [139] with ROIs and

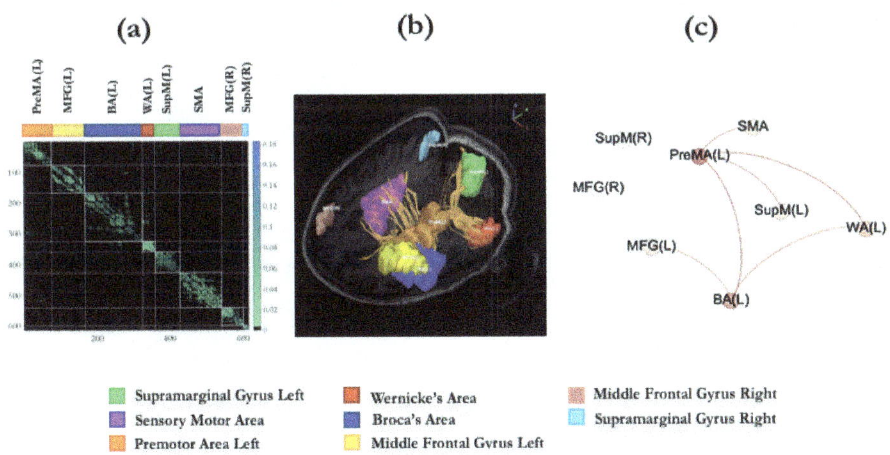

Fig. 6.12 Building a language structural network from combined fMRI-DTI data. **a** Functional connectivity matrix obtained from fMRI in [139]. **b** DTI fiber tracking between the active brain areas defined by fMRI. Each different color ROI corresponds to an activated ROI from fMRI, whereas fiber tracks depict structural connectivity obtained by probabilistic DTI tractography. **c** Combined fMRI-DTI network among the active brain areas during the fMRI tasks for this particular subject. Image provided by G. Del Ferraro, NYU

connectivity. The combined fMRI-DTI network is particular for each subject and forms the basis of a network analysis for influencers for each subject, where the rankings of core/non-core can be performed on the combined fMRI-DTI network. Analysis of the DTI data is facilitated as the same voxels that form the fMRI network are used to seed the probabilistic DTI tractography.

While the structural anatomical networks cannot be directly derived from the functional connections, if we know more information about the type of dynamics of signal propagation on the structural network, then we could use modeling to determine the structural pathways that support the given functional connectivity [91, 156, 157], see Fig. 6.13. We have encountered a similar problem in the study of gene regulatory networks in Sect. 5.5. Here, the role of the structural network is played by the physical binding of a transcription factor to the regulatory DNA of a target gene. The expression levels of mRNA play the role of the functional network. A strong functional connection, two genes highly synchronized, does not imply that they are directly regulated.

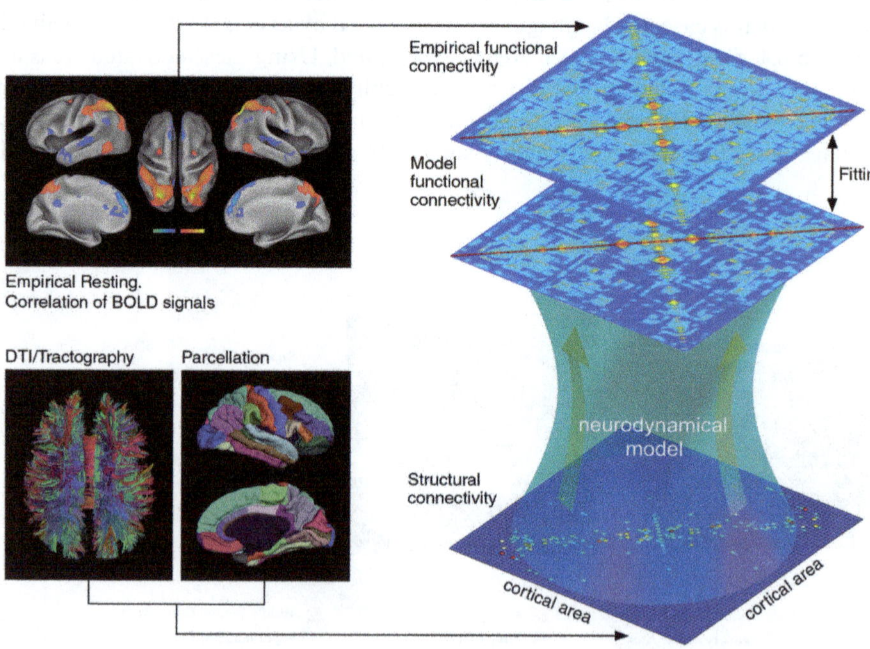

Fig. 6.13 Functional network from a structural network via modeling in the human brain. A framework to study the link between anatomical structure and resting-state dynamics as proposed in [156, 157]. DTI tractography and parcellation provide the neuroanatomical structural network averaged across subjects. Neurodynamical modelling is used to model BOLD signals in resting state and task-based cognitive states. The model is validated by comparing predicted spatiotemporal patterns with empirical functional connectivity data. A similar framework is used to test whether symmetries can be used to bridge the gap between structure and dynamical synchronization in section 6.8. Figure reproduced from [156]

However, we have seen that when the structural network contains symmetries, fibrations or automorphisms, then a direct relation between structure and dynamics exists. Symmetries in the structural network predict cluster synchronization, and this synchronization represents a functional relation between the nodes. When the network contains symmetries, the structure-function connection can be disentangled quite efficiently. As it turns out, biological networks have evolved to develop these symmetries that give rise to ubiquitous coherent synchrony [158–160]. In the next section we elaborate on how the symmetries of the connectome determine the function of the neural system.

6.8 Synchronized Building Blocks from Brain Symmetries

Fibration symmetries explained in Sect. 5.5 for genetic networks come into play to study the structure-function relation in the brain and to uncover its building blocks. These building blocks can then be used to find the essential circuits in the brain for synchronization. This section discusses a neural theory that takes into account the structure-function relation via the essential building blocks for synchronization in the structural network.

Extending the result from Sect. 5.5 to neural connectomes has been done by Morone and Makse [158]. A detailed account of symmetries and synchronization in biological networks appears in [160]. Finding symmetries in the connectome requires the knowledge of the complete wire map. If the wire diagram is not complete, then pseudosymmetries can also be studied, which are almost symmetries but important for the synchronization of the neural units.

The wiring diagram of the worm *C. elegans* has been fully mapped in the seminal work of White et al. [74], providing the synaptic connections and gap junctions between its 302 neurons [161]. The subgraph of the neural connectome containing neurons active during forward and backward locomotion circuits has been investigated in [158] where it demonstrated the symmetries of these circuits and its relation to function via neural synchronization. Results of the symmetry study in the locomotion circuits are shown in Figs. 6.14a for the forward gap-junction circuit and Figs. 6.14b for the chemical synapses forward circuit, and Fig. 6.15 for the backward chemical synapses circuit.

In these cases, most of the symmetries are automorphisms of the network (rather than symmetry fibrations as in genes discussed in Sect. 5.5; however, a fibration still can be observed in Fig. 6.15). Automorphisms are permutations of nodes that conserve not only the inputs, like fibrations, but also the outputs, i.e., they preserve the full adjacency matrix. Automorphisms are then more strict than symmetry fibrations found in genetic networks. The automorphisms appearing in the *C. elegans* brain might be indicative of its more primitive structure in comparison with more evolved genetic networks. These automorphisms form symmetry groups [162], while fibrations form symmetry groupoids, Golubitsky and Stewart [163]. In principle, automorphisms can be obtained using the Nauty program [164]. However, due to the

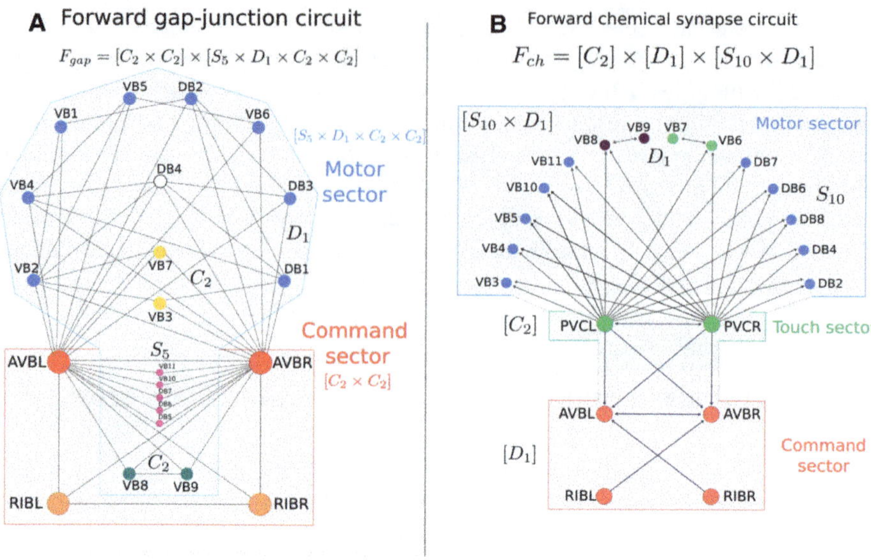

Fig. 6.14 **Symmetry group of** ***C. elegans*** **locomotion.** **a** Circuit of forward locomotion made of gap-junctions and **b** chemical synapses. Both circuits show automorphisms and the factorization of the connectome into normal subgroups, leading to symmetry building blocks. Figure reproduced from [158]. Copyright © 2019, The Authors

natural variability of the connectomes, symmetries are not perfect, and a reconstruction of the network needs to be performed to reveal the ideal symmetries behind them as done in [158, 165, 166].

The symmetries induce a partition into normal subgroups of the circuit into uniquely defined sectors or building blocks of neurons, one for each normal subgroup, such that no two building blocks have neurons in common. The building blocks defined by the symmetries show their biological significance in that they match specific anatomical classes of neurons. Specifically, the command interneuron sector in the forward gap-junction circuit shown in Fig. 6.14a is identified by the normal subgroup $(C_2 \times C_2)_{\text{inter}}$ which comprises interneurons that drive the forward locomotion: AVBL, AVBR, RIBL and RIBR. The normal subgroup identifies the motor sector $(S_5 \times D_1 \times C_2 \times C_2)_{\text{motor}}$ and captures all the motor neurons, which innervate the muscle cells responsible for the motion of *C. elegans* [161]. Each specific normal subgroup, like S_5, defines a particular building block of this connectome (the same results are found for chemical synapses, Fig. 6.14b and Fig. 6.15).

The existence of symmetries impose cluster synchrony in the orbits of the symmetry group [163]. Thus, symmetries lead directly to the ensemble of neurons that synchronize and are functionally related, providing a link from structure to function.

Experiments performed by Kato et al. [167] have shown synchronization of the neural activity during forward and backward locomotion in the worm. Calcium imaging techniques on transgenic worms expressing a nuclear-localized Ca++—sensor

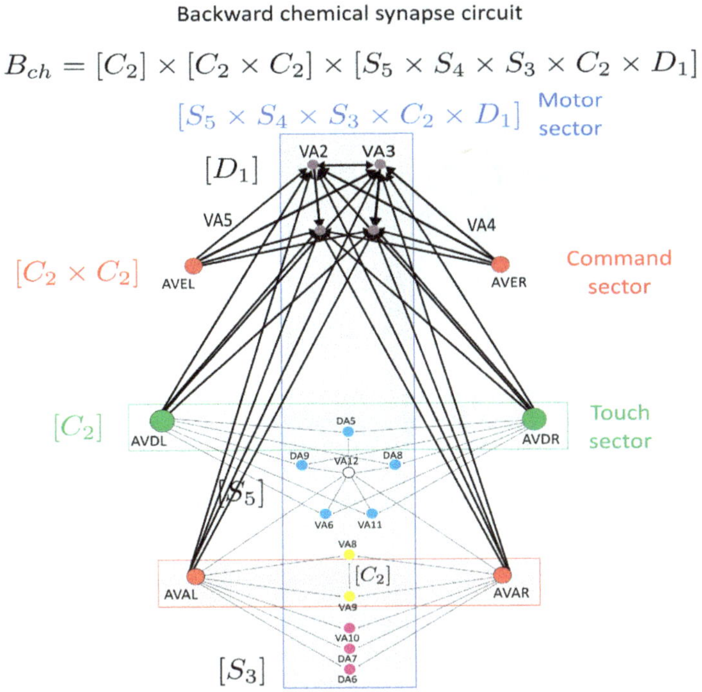

Fig. 6.15 Symmetry group factorization of the chemical synapse backward circuit. We highlight the ventral circuit VA2, VA3, VA4, VA5 with an interesting symmetry fibration unique to this circuit, see Sect. 5.5. Figure reproduced from [158]. Copyright © 2019, The Authors

(GCaMP) in a pan-neuronal fashion are used in immobilized worms in microfluidic devices. This enables volumetric fluorescence imaging at sufficient speed and optical resolution to capture the activity of most neurons in the worm with single-cell resolution in the animals' head ganglia. Results showing the neural synchronization are shown in Fig. 6.16 obtained from [167]. These synchronous neuronal ensembles are functionally related. Since these dynamics and synchronies are not acutely triggered or bound by fluctuating external inputs, they must emerge as a property of the network.

We notice that in this case, both networks, structural and functional are known, so it is the ideal case for testing the structure-function relation. Analysis of these data shows that the strength of functional interactions (measured as pairwise absolute correlations or covariances from Fig. 6.16b) are only poorly correlated with connection strength (number of direct synapses), showing that the relation of function from structure cannot be obtained directly from the connectome. Heuristics like hub theory or centralities are not able to capture this functionality either. On the other hand, links that are symmetric and capture the observed synchronization. These data show that direct connection strength is a poor predictor of functional interaction and

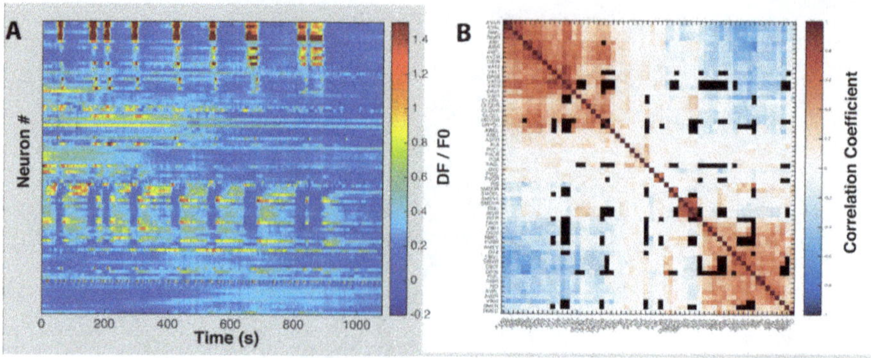

Fig. 6.16 Neural synchronization during *C. elegans* **locomotion. a** Spontaneous network activity in the *C. elegans* brain is dominated by synchronies as shown in the experiments by Kato et al. [167]. Heat map of an exemplary whole brain Ca2+ imaging dataset. **b** Matrix of pairwise correlation coefficients, averaged across $n = 4$ such datasets. Evidently, there are two large synchronous neuronal ensembles of backward and forward neurons. Figure reproduced from [167]. Copyright © 2015, Elsevier Inc

symmetry-induced synchronization, a high-level network features, perform better in predicting the structure-function relation.

The symmetry framework predicts synchronized neurons functionally related or genes with synchronized expression patterns. Such building blocks can be manipulated by breaking their symmetries in a controlled manner by testing the functionality of these synchronous circuits, providing the targets to identify functions by manipulating the building blocks and their symmetries. Controlled perturbation experiments (ablations, optogenetics perturbations, gene knockouts) on the identified building blocks can be used to manipulate network dynamics by acting directly on the building blocks. The activation/deactivation patterns applied to symmetric blocks can produce asymmetries and de-synchronizations that propagate through the neural to impact brain network dynamics. To this end, these theoretical results are the scaffold to guide neural interventions in experiments in the search for influential building blocks for neural synchronization.

References

1. Dehaene, S.: Consciousness and the Brain. Penguin Group, New York (2014)
2. Dehaene, S., Naccache, L.: Towards a cognitive neuroscience of consciousness: basic evidence and a workspace framework. Cognition **79**, 1–37 (2001)
3. Meunier, D., Lambiotte, R., Bullmore, E.T.: Modular and hierarchically modular organization of brain networks. Front. Neurosci. **4**, 200 (2010)
4. Felleman, D.J., van Essen, D.C.: Distributed hierarchical processing in the primate cerebral cortex. Cereb. Cortex **1**, 1–47 (1991)
5. Treisman, A.: The binding problem. Curr. Opin. Neurobiol. **6**, 171–178 (1996)

6. Tononi, G., Sporns, O., Edelman, G.M.: A measure for brain complexity: relating functional segregation and integration in the nervous system. Proc. Nat. Acad. Sci. USA **91**, 5033–5037 (1994)
7. Gallos, L.K., Makse, H.A., Sigman, M.: A small world of weak ties provides optimal global integration of self-similar modules in functional brain networks. Proc. Natl. Acad. Sci. USA **109**, 2825–2830 (2012)
8. Gallos, L.K., Sigman, M., Makse, H.A.: The conundrum of functional brain networks: small-world efficiency or fractal modularity. Front. Physiol. **3**, 123 (2012)
9. Tononi, G., Edelman, G.M.: Consciousness and complexity. Science **282**, 1846–1851 (1998)
10. Baars, B.J.: In the Theater of Consciousness: The Workspace of the Mind. Oxford University Press, USA (1997)
11. Sporns, O.: Network attributes for segregation and integration in the human brain. Curr. Opin. Neurobiol. **23**, 162–71 (2013)
12. Deco, G., Tononi, G., Boly, M., Kringelbach, M.L.: Rethinking segregation and integration in the human brain: Contributions of whole-brain computational modelling. Nat. Rev. Neurosci. **16**, 430–439 (2015)
13. Dosenbach, N.U.F., et al.: Distinct brain networks for adaptive and stable task control in humans. Proc. Natl. Acad. Sci. USA **104**, 11073–11078 (2007)
14. Bullmore, E., Sporns, O.: Complex brain networks: graph theoretical analysis of structural and functional systems. Nature Rev. Neurosci. **10**, 186–198 (2009)
15. Sporns, O., Chialvo, D.R., Kaiser, M., Hilgetag, C.C.: Organization, development and function of complex brain networks. Trends Cognit. Sci. **8**, 418–425 (2004)
16. Sporns, O.: Contributions and challenges for network models in cognitive neuroscience. Nat. Neurosci. **17**, 652–660 (2014)
17. Reis, S.D.S., Hu, Y., Babino, A., Andrade, J.S., Jr., Canals, S., Sigman, M., Makse, H.A.: Avoiding catastrophic failure in correlated networks of networks. Nat. Phys. **10**, 762–767 (2014)
18. Watts, D.J., Strogatz, S.: Collective dynamics of small-world networks. Nature **393**, 440–442 (1998)
19. Bassett, D.S., Bullmore, E.T.: Small-world brain networks. Neuroscientist **12**, 512–523 (2006)
20. Barabási, A.-L., Albert, R.: Emergence of scaling in random networks. Science **286**, 509 (1999)
21. Eguiluz, V.M., Chialvo, D.R., Cecchi, G.A., Baliki, M., Apkarian, A.V.: Scale-free brain functional networks. Phys. Rev. Lett. **94**, 018102 (2005)
22. Achard, S., Salvador, R., Whitcher, B., Suckling, J., Bullmore, E.T.: A resilient, low-frequency, small-world human brain functional network with highly connected association cortical hubs. J. Neurosci. **26**, 63–72 (2006)
23. Achard, S., Bullmore, E.T.: Efficiency and cost of economical brain functional networks. PLoS Comput. Biol. **3**, e17 (2007)
24. Cohen, R., Havlin, S.: Scale-free networks are ultrasmall. Phys. Rev. Lett. **90**, 058701 (2003)
25. He, Y., Chen, Z.J., Evans, A.C.: Small-world anatomical networks in the human brain revealed by cortical thickness from MRI. Cereb. Cortex **17**, 2407–2419 (2007)
26. Hagmann, P., Kurant, M., Gigandet, X., Thiran, P., Wedeen, V.J., Meuli, R., Thiran, J.P.: Mapping human whole-brain structural networks with diffusion MRI. PLoS One **2**, e597 (2007)
27. Song, S., Sjostrom, P.J., Reigl, M., Nelson, S., Chklovskii, D.B.: Highly nonrandom features of synaptic connectivity in local cortical circuits. PLoS Biol. **3**, e68 (2005)
28. Yu, S., Huang, D., Singer, W., Nikolic, D.: A small world of neuronal synchrony. Cereb. Cortex **18**, 2891–2901 (2008)
29. Stam, C.J., Jones, B.F., Nolte, G., Breakspear, M., Scheltens, P.: Small-world networks and functional connectivity in Alzheimer's disease. Cereb. Cortex **17**, 92–99 (2007)
30. Girvan, M., Newman, M.E.J.: Community structure in social and biological networks. Proc. Natl. Acad. Sci. USA **99**, 7821–7826 (2002)
31. Fortunato, S.: Community detection in graphs. Phys. Rep. **486**, 75–174 (2010)

32. Sporns, O.: Discovering the Human Connectome. MIT Press, Cambridge (2012)
33. Bullmore, E., Sporns, O.: The economy of brain network organization. Nat. Rev. Neurosci. **13**, 336–349 (2012)
34. van den Heuvel, M.P., Sporns, O.: Network hubs in the human brain. Trends Cogn. Sci. **17**, 683–696 (2013)
35. Sporns, O., Honey, C.J., Kotter, R.: Identification and classification of hubs in brain networks. PLoS One **2**, e1049 (2007)
36. Gong, G., He, Y., Concha, L., Lebel, C., Gross, D.W., Evans, A.C., Beaulieu, C.: Mapping anatomical connectivity patterns of human cerebral cortex using in vivo diffusion tensor imaging tractography. Cereb. Cortex **19**, 524–536 (2009)
37. Zalesky, A., Fornito, A., Harding, I.H., Cocchi, L., Yucel, M., Pantelis, C., Bullmore, E.T.: Whole-brain anatomical networks: does the choice of nodes matter? Neuroimage **50**, 970–983 (2010)
38. Bassett, D.S., Brown, J.A., Deshpande, V., Carlson, J.M., Grafton, S.T.: Conserved and variable architecture of human white matter connectivity. Neuroimage **54**, 1262–1279 (2011)
39. Hagmann, P., Cammoun, L., Gigandet, X., Meuli, R., Honey, C.J., Wedeen, V.J., Sporns, O.: Mapping the structural core of human cerebral cortex. PLoS Biol. **6**, e159 (2008)
40. Wang, J., Zuo, X., He, Y.: Graph-based network analysis of resting-state functional MRI. Front. Syst. Neurosci. **4**, 16 (2010)
41. Tomasi, D., Volkow, N.D.: Functional connectivity hubs in the human brain. Neuroimage **57**, 908–917 (2011)
42. Granovetter, M.S.: The strength of weak ties. Am. J. Sociol. **78**, 1360–1380 (1973)
43. Misic, B., Betzel, R.F., Nematzadeh, A., Goñi, J., Griffa, A., Hagmann, P., Flammini, A., Ahn, Y.-Y., Sporns, O.: Cooperative and Competitive Spreading Dynamics on the Human Connectome. Neuron **86**, 1518–1529 (2015)
44. Stam, C.J.: Modern network science of neurological disorders. Nat. Rev. Neurosci. **15**, 683–695 (2014)
45. Buckner, R.L., Sepulcre, J., Talukdar, T., Krienen, F.M., Liu, H., Hedden, T., Andrews-Hanna, J.R., Sperling, R.A., Johnson, K.A.: Cortical hubs revealed by intrinsic functional connectivity: mapping, assessment of stability, and relation to Alzheimer's disease. J. Neurosci. **29**, 1860–1873 (2009)
46. Fischer, F.U., Wolf, D., Scheurich, A., Fellgiebel, A.: Alzheimer's Disease Neuroimaging Initiative: Altered whole-brain white matter networks in preclinical Alzheimer's disease. Neuroimage Clin. **8**, 660–6 (2015)
47. Kim, H., Yoo, K., Na, D.L., Seo, S.W., Jeong, J., Jeong, Y.: Non-monotonic reorganization of brain networks with Alzheimer's disease progression. Front. Aging Neurosci. **7**, 111 (2015)
48. Toussaint, P.J., Maiz, S., Coynel, D., Doyon, J., Messe, A., de Souza, L.C., Sarazin, M., Perlbarg, V., Habert, M.O., Benali, H.: Characteristics of the default mode functional connectivity in normal ageing and Alzheimer's disease using resting-state fMRI with a combined approach of entropy-based and graph theoretical measurements. Neuroimage **101**, 778–86 (2014)
49. Sanz-Arigita, E.J., Schoonheim, M.M., Damoiseaux, J.S., Rombouts, S.A., Maris, E., Barkhof, F., Scheltens, P., Stam, C.J.: Loss of 'small-world' networks in Alzheimer's disease: graph analysis of FMRI resting-state functional connectivity. PLoS One **5**, (11), e13788 (2014)
50. Bozzali, M., Parker, G.J.M., Serra, L., Embleton, K., Gili, T., Perri, R., Caltagirone, C., Cercignani, M.: Anatomical connectivity mapping: a new tool to assess brain disconnection in Alzheimer's disease. NeuroImage **54**, 2045–2051 (2011)
51. Fornito, A., Zalesky, A., Pantelis, C., Bullmore, E.T.: Schizophrenia, neuroimaging and connectomics. Neuroimage **62**, 2296 (2012)
52. Lo, C.Y., Su, T.W., Huang, C.C., Hung, C.C., Chen, M.L., Lan, T.H., Lin, C.P., Bullmore, E.T.: Randomization and resilience of brain functional networks as systems-level endophenotypes of schizophrenia. Proc. Natl. Acad. Sci. USA **112**, 9123–9128 (2015)
53. Tomasi, D., Volkow, N.D.: Mapping small-world properties through development in the human brain: disruption in schizophrenia. PLoS One **9**, e96176 (2014)

54. Achard, S., Delon-Martin, C., Vertes, P.E., Renard, F., Schenck, M., Schneider, F., Heinrich, C., Kremer, S., Bullmore, E.T.: Hubs of brain functional networks are radically reorganized in comatose patients. Proc. Natl. Acad. Sci. USA **109**, 20608–20613 (2013)
55. Martin, P.I., Naeser, M.A., Ho, M., Doron, K.W., Kurland, J., Kaplan, J., Wang, Y., Nicholas, M., Baker, E.H., Alonso, M., Fregni, F., Pascual-Leone, A.: Overt naming fMRI pre- and post-TMS: two nonfluent aphasia patients, with and without improved naming post-TMS. Brain Lang. **111**, 20–35 (2009)
56. Garcia, G., Norise, C., Faseyitan, O., Naeser, M.A., Hamilton, R.H.: Utilizing repetitive transcranial magnetic stimulation to improve language function in stroke patients with chronic non-fluent aphasia. J. Vis. Exp. **77**, e50228 (2013)
57. Naeser, M.A., Martin, P.I., Theoret, H., Kobayashi, M., Fregni, F., Nicholas, M., Tormos, J.M., Steven, M.S., Baker, E.H., Pascual-Leone, A.: TMS suppression of right pars triangularis, but not pars opercularis, improves naming in aphasia. Brain Lang. **119**, 206–213 (2011)
58. Seeley, W.W., Crawford, R.K., Zhou, J., Miller, B.L., Greicius, M.D.: Neurodegenerative diseases target large-scale human brain networks. Neuron **62**, 42–52 (2009)
59. Bassett, D.S., Bullmore, E.T.: Human brain networks in health and disease. Curr. Opin. Neurol. **22**, 340 (2009)
60. van den Heuvel, M.P., Mandl, R.C., Stam, C.J., Kahn, R.S., Hulshoff, H.E.: Aberrant frontal and temporal complex network structure in schizophrenia: a graph theoretical analysis. J. Neurosci. **30**, 47 (2010)
61. Sporns, O., Tononi, G., Kotter, R.: The human connectome: a structural description of the human brain. PLoS Comput. Biol. **1**, e42 (2005)
62. Honey, C.J., Sporns, O., Cammoun, L., Gigandet, X., Thiran, J.P., Meuli, R., Hagmann, P.: Predicting human resting-state functional connectivity from structural connectivity. Proc. Nat. Acad. Sci. USA **106**, 2035–2040 (2009)
63. Honey, C.J., Thivierge, J.-P., Sporns, O.: Can structure predict function in the human brain? Neuroimage **52**, 766–776 (2010)
64. Albert, R., Jeong, H., Barabási, A.-L.: Error and attack tolerance in complex networks. Nature **406**, 378–382 (2000)
65. Kempe, D., Kleinberg, J., Tardos, E.: Maximizing the spread of influence through a social network. In: Proceedings of the 9th ACM SIGKDD International Conferences on Knowledge Discovery and Data Mining, pp. 137–143 (2003)
66. Zuo, X.N., Ehmke, R., Mennes, M., Imperati, D., Castellanos, F.X., Sporns, O., Milham, M.P.: Network centrality in the human functional Connectome. Cereb. Cortex **22**, 1862–1875 (2012)
67. Park, H.J., Friston, K.: Structural and functional brain networks: from connections to cognition. Science **342**, 6158 (2013)
68. Bastos, A.M., Schoffelen, J.-M.: A tutorial review of functional connectivity analysis methods and their interpretational pitfalls. Front. Syst. Neurosci. **9**, 175 (2016)
69. Pereda, E., Quian Quiroga, R., Bhattacharya, J.: Nonlinear multivariate analysis of neurophysiological signals. Prog. Neurobiol. **77**, 1–37 (2005)
70. Zalesky, A., Fornito, A., Bullmore, E.: On the use of correlation as a measure of network connectivity. Neuroimage **60**, 2096–2106 (2012)
71. Smith, S.M., Miller, K.L., Salimi-Khorshidi, G., Webster, M., Beckmann, C.F., Nichols, T.E., Ramsey, J.D., Woolrich, M.W.: Network modeling methods for fMRI. Neuroimage **54**, 875–891 (2011)
72. Van Essen, D.C., Smith, S.M., Barch, D.M., Behrens, T.E.J., Yacoub, E., Ugurbil, K.: The WU-Minn human connectome project: an overview. Neuroimage **80**, 62–79 (2013)
73. Gyengesi, E., Calabrese, E., Sherrier, M.C., Johnson, G.A., Paxinos, G., Watson, C.: Semi-automated 3D segmentation of major tracts in the rat brain: comparing DTI with standard histological methods. Brain Struct. Funct. **219**, 539–550 (2014)
74. White, J.G., Southgate, E., Thomson, J.N., Brenner, S.: The structure of the nervous system of the nematode Caenorhabditis elegans. Philos. Trans. R. Soc. Lond. B **314**, 1–340 (1986)
75. Oh, S.W., et al.: A mesoscale connectome of the mouse brain. Nature **10**, 207–214 (2014)

76. Bardella, G., Bifone, A., Gabrielli, A., Gozzi, A., Squartini, T.: Hierarchical organization of functional connectivity in the mouse brain: a complex network approach. Sci. Rep. **6**, 32060 (2016)
77. Bunde, A., Havlin, S.: Fractals and Disordered Systems, 2nd edn. Springer-Verlag, Heidelberg (1996)
78. Sigman, M., Dehaene, S.: Brain mechanisms of serial and parallel processing during dual-task performance. J. Neurosci. **28**, 7585–7598 (2008)
79. Zylberberg, A., Dehaene, S., Roelfsema, P.R., Sigman, M.: The human Turing machine: a neural framework for mental programs. Trends Cogn. Sci. **15**, 293–300 (2011)
80. Duncan, J.: The multiple-demand (MD) system of the primate brain: mental programs for intelligent behaviour. Trends Cogn. Sci. **14**, 172–179 (2010)
81. Russo, R., Herrmann, H.J., de Arcangelis, L.: Brain modularity controls the critical behavior of spontaneous activity. Sci. Rep. **4**, 4312 (2014)
82. Schneidman, E., Berry, M.J., Segev, R., Bialek, W.: Weak pairwise correlations imply strongly correlated network states in a neural population. Nature **440**, 1007–1012 (2006)
83. Mezard, M., Mora, T.: Constraint satisfaction problems and neural networks: A statistical physics perspective. J. Physiol. Paris **103**, 107–113 (2009)
84. Mora, T., Bialek, W.: Are biological systems poised at criticality? J. Stat. Phys. **144**, 268–302 (2011)
85. Roudi, Y., Aurell, E., Hertz, J.: Statistical physics of pairwise probability models. Front. Comput. Neurosci. **3**, 22 (2009)
86. Roudi, Y., Hertz, J.: Mean field theory For non-equilibrium network reconstruction. Phys. Rev. Lett. **106**, 048702 (2011)
87. Mezard, M., Montanari, A.: Information, Physics, and Computation. Oxford University Press, Oxford (2009)
88. Sarkar, S., Chawla, S., Xu, D.: On inferring structural connectivity from brain functional-MRI data. Comput. Intell. Neurosci. 5573740 (2021)
89. Robinson, P.A.: Interrelating anatomical, effective, and functional brain connectivity, using propagators and neural field theory. Phys. Rev. E **85**, 011912 (2012)
90. Robinson, P.A., Sarkar, S., Pandejee, G.M., Henderson, J.: Determination of effective brain connectivity from functional connectivity with application to resting state connectivities. Phys. Rev. E **90**, 012707 (2014)
91. Deco, G., McIntosh, A.R., Shen, K., Hutchison, R.M., Menon, R.S., Everling, S., Hagmann, P., Jirsa, V.K.: Identification of optimal structural connectivity using functional connectivity and neural modeling. J. Neurosci. **34**, 7910–7916 (2014)
92. Jaynes, E.T.: Information theory and statistical mechanics. Phys. Rev. **106**, 620–630 (1957)
93. Jaynes, E.T.: Information theory and statistical mechanics II. Phys. Rev. **108**, 171–190 (1957)
94. Morone, F., Roth, K., Min, B., Stanley, H.E., Makse, H.A.: A model of brain activation predicts the neural collective influence map of the human brain. Proc. Natl. Acad. Sci. USA **114**, 3849–3854 (2017)
95. Friedman, J., Hastie, T., Tibshirani, R.: Sparse inverse covariance estimation with the graphical lasso. Biostatistics **9**, 432–441 (2008)
96. Hsieh, C.-J., et al.: BIG and QUIC: Sparse inverse covariance estimation for a million variables. In: Advances in Neural Information Processing Systems (2013)
97. Del Ferraro, G., Moreno, A., Min, B., Morone, F., Pérez-Ramírez, Ú., Pérez-Cervera, L., Parra, L., Holodny, A., Canals, S., Makse, H.A.: Finding influential nodes for integration in brain networks using optimal percolation theory. Nat. Commun. **9**, 2274 (2018)
98. Mazumder, R., Hastie, T.: Exact covariance thresholding into connected components for large-scale graphical Lasso. J. Mach. Learn. Res. **13**, 781–794 (2012)
99. Sojoudi, S.: Equivalence of graphical lasso and thresholding for sparse graphs. J. Mach. Learn. Res. **17**, 1–21 (2016)
100. Altarelli, F., Braunstein, A., Dall'Asta, L., Zecchina, R.: Optimizing spread dynamics on graphs by message passing. J. Stat. Mech. P09011 (2013)

101. Altarelli, F., Braunstein, A., Dall'Asta, L., Wakeling, J.R., Zecchina, R.: Containing epidemic outbreaks by message passing techniques. Phys. Rev. X **4**, 021024 (2014)
102. Guggiola, A., Semerjian, G.: Minimal contagious sets in random regular graphs. J. Stat. Phys. **158**, 300–358 (2015)
103. Morone, F., Makse, H.A.: Influence maximization in complex networks through optimal percolation. Nature **524**, 65–68 (2015)
104. Buldyrev, S.V., Parshani, R., Paul, G., Stanley, H.E., Havlin, S.: Catastrophic cascade of failures in interdependent networks. Nature **464**, 1025–128 (2010)
105. Gao, J., Buldyrev, S.V., Stanley, H.E., Havlin, S.: Networks formed from interdependent networks. Nat. Phys. **8**, 40–48 (2012)
106. Roth, K., Morone, F., Min, B., Makse, H.A.: Emergence of robustness in networks of networks. Phys. Rev. E **95**, 062308 (2017)
107. Gilbert, C.D., Sigman, M.: Brain states: top-down influences in sensory processing. Neuron **54**, 677–699 (2007)
108. McCulloch, W., Pitts, A.: A logical calculus of the ideas immanent in nervous activity. Bull. Math. Biophys. **5**, 115–133 (1943)
109. Boyden, E.S., Zhang, F., Bamberg, E., Nagel, G., Deisseroth, K.: Millisecond-timescale, genetically targeted optical control of neural activity. Nat. Neurosci. **8**, 1263–1268 (2005)
110. Fenno, L., Yizhar, O., Deisseroth, K.: The development and application of optogenetics. Annu. Rev. Neurosci. **34**, 389–412 (2011)
111. Lee, J.H., Durand, R., Gradinaru, V., Zhang, F., Goshen, I., Kim, D.-S., Fenno, L.E., Ramakrishnan, C., Deisseroth, K.: Global and local fMRI signals driven by neurons defined optogenetically by type and wiring. Nature **465**, 788 (2010)
112. Weitz, A.J., Fang, Z., Lee, H.J., Fisher, R.S., Smith, W.C., Choy, M., Liu, J., Lin, P., Rosenberg, M., Lee, J.H.: Optogenetic fMRI reveals distinct, frequency-dependent networks recruited by dorsal and intermediate hippocampus stimulations. Neuroimage **107**, 229 (2015)
113. Bliss, T.V.P., Lomo, T.J.: Long-lasting potentiation of synaptic transmission in the dentate area of the anaesthetized rabbit following stimulation of the perforant path. Physiol. (Lond) **232**, 331–356 (1973)
114. Bliss, T. V. P., Collingridge, G.L., Morris, R.: Synaptic plasticity in the hippocampus. In: Andersen, P., Morris, R., Amaral, D., Bliss, T., O'eefe, J. eds., The Hippocampus Book. Oxford University Press, Oxford (2007)
115. Hebb, D.O.: The organization of Behavior. Wiley, New York (1947)
116. Canals, S., Beyerlein, M., Murayama, Y., Logothetis, N.K.: Electric stimulation fMRI of the perforant pathway to the rat hippocampus. Magn. Reson. Imaging **26**, 978–986 (2008)
117. Canals, S., Beyerlein, M., Merkle, H., Logothetis, N.K.: Functional MRI evidence for LTP-induced neural network reorganization. Curr. Biol. **19**, 398–403 (2009)
118. Alvarez-Salvado, E., Pallarés, V., Moreno, A., Canals, S.L Functional MRI of long-term potentiation: imaging network plasticity. Philos. Trans. R. Soc. Lond. B Biol. Sci. **369**, 20130152 (2014)
119. Moreno, A., Morris, R.G., Canals, S.: Frequency-dependent gating of hippocampal-neocortical interactions. Cereb. Cortex **26**, 2105–2114 (2016)
120. Carames, J.M., Pérez-Montoyo, E., García-Hernández, R., Canals, S.: Hippocampal dentate gyrus coordinates brain-wide communication and memory updating through an inhibitory gating (2020). https://doi.org/10.1101/2020.07.14.202218 BioRxiv
121. Alexander G.M., Rogan, A., Abbas, I., Armbruster, B.N., Pei, Y., Allen, J.A., Nonneman, R.J., Hartmann, J., Moy, S.S., Nicolelis, M. A., McNamara, J.O., Roth, B.L.: Remote control of neuronal activity in transgenic mice expressing evolved G protein-coupled receptors. Neuron **63**, 27–39 (2009)
122. Morris, R.G., Moser, E.I., Riedel, G., Martin, S.J., Sandin, J., Day, M., O'Carroll, C.: Elements of a neurobiological theory of the hippocampus: the role of activity-dependent synaptic plasticity in memory, Philos. Trans. R. Soc. Lond. B Biol. Sci. **358**, 773–786 (2003)
123. Squire, L.R., Stark, C.E.L., Clark, R.E.: The medial temporal lobe. Annu. Rev. Neurosci. **27**, 279–306 (2004)

124. Tate, M.C., Herbet, G., Moritz-Gasser, S., Tate, J.E., Duffau, H.: Probabilistic map of critical functional regions of the human cerebral cortex: Broca's area revisited. Brain **137**, 2773–2782 (2014)
125. Holodny, A.I., Schulder, M., Ybasco, A., Liu, W.C.: Translocation of Broca's area to the contralateral hemisphere as the result of the growth of a left inferior frontal glioma. J. Comput. Assist. Tomogr. **26**, 941–943 (2002)
126. Holodny, A.I., Schulder, M., Liu, W.C., Wolko, J., Maldjian, J.A., Kalnin, A.J.: The effect of brain tumors on BOLD functional MR imaging activation in the adjacent motor cortex: implications for image-guided neurosurgery. AJNR Am. J. Neurorad. **21**, 1415–1422 (2000)
127. Petrella, J.R., Shah, L.M., Harris, K.M., Friedman, A.H., George, T.M., Sampson, J.H., Pekala, J.S.: Preoperative functional MR imaging localization of language and motor areas: effect on therapeutic decision making in patients with potentially resectable brain tumors. Radiology **240**, 793–802 (2006)
128. Peck, K.K., Bradbury, M., Petrovich, N., Hou, B.L., Ishill, N., Brennan, C., Tabar, V., Holodny, A.I.: Presurgical evaluation of language using functional magnetic resonance imaging in brain tumor patients with previous surgery. Neurosurgery **64**, 644–652 (2009)
129. Hugdahl, K., Lundervold, A., Ersland, L., Smievoll, A.I., Sundberg, H., Barndon, R., Roscher, B.E.: Left frontal activation during a semantic categorization task: an fMRI-study. Int. J. Neurosci. **99**, 49–58 (1999)
130. Oda, K., Yamaguchi, F., Enomoto, H., Higuchi, T., Morita, A.: Prediction of recovery from supplementary motor area syndrome after brain tumor surgery: preoperative diffusion tensor tractography analysis and postoperative neurological clinical course. Neurosurg. Focus **44**, E3 (2018)
131. Chivukula, S., Pikul, B.K., Black, K.L., Pouratian, N., Bookheimer, S.Y.: Contralateral functional reorganization of the speech supplementary motor area following neurosurgical tumor resection. Brain Lang. **183**, 41–46 (2018)
132. Blesneag, A.V., Popa, L., Stan, A.D.: Non-invasive brain stimulation in early rehabilitation after stroke. J. Med. Life. **8**, 52–56 (2015)
133. Martin, P.I., Naeser, M.A., Ho, M., Doron, K.W., Kurland, J., Kaplan, J., Wang, Y., Nicholas, M., Baker, E.H., Alonso, M., Fregni, F., Pascual-Leone, A.: Overt naming fMRI pre- and post-TMS: Two nonfluent aphasia patients, with and without improved naming post-TMS. Brain Lang. **111**, 20–35 (2009)
134. Naeser, M.A., Martin, P.I., Theoret, H., Kobayashi, M., Fregni, F., Nicholas, M., Tormos, J.M., Steven, M.S., Baker, E.H., Pascual-Leone, A.: TMS suppression of right pars triangularis, but not pars opercularis, improves naming in aphasia. Brain Lang. **119**, 206–213 (2011)
135. Garcia, G., Norise, C., Faseyitan, O., Naeser, M.A., Hamilton, R.H.: Utilizing repetitive transcranial magnetic stimulation to improve language function in stroke patients with chronic non-fluent aphasia. J. Vis. Exp. **77**, e50228 (2013)
136. DeLong, M.R., Wichmann, T.: Basal Ganglia Circuits as Targets for Neuromodulation in Parkinson Disease. JAMA Neurol. **26**, 1–7 (2015)
137. Tekriwal, A., Baltuch, G.: Deep brain stimulation: expanding applications. Rev. Neurol. Med. Chir. (Tokyo) **55**, 861–877 (2015)
138. Li, Q., Dong, J.W., Del Ferraro, G., Petrovich Brennan, N., Peck, K.K., Tabar, V., Makse, H.A., Holodny, A.I.: Functional translocation of Broca's area in a low-grade left frontal glioma: graph theory reveals the novel, adaptive network connectivity. Front. Neurol. Appl. Neuroimaging **10**, 702 (2019)
139. Li, Q., Del Ferraro, G., Pasquini, L., Peck, K.K., Makse, H.A., Holodny, A.I.: Core language brain network for fMRI language task used in clinical applications. Netw. Neurosci. **4**, 134–154 (2020)
140. Li, Q., Pasquini, L., Del Ferraro, G., Gene, M., Peck, K.K., Makse, H.A., Holodny, A.I.: Monolingual and bilingual language networks in healthy subjects using functional MRI and graph theory. Sci. Rep. **11**, 10568 (2021)
141. Arese Lucini, F., Del Ferraro, G., Sigman, M., Makse, H.A.: How the brain transitions from conscious to subliminal perception. Neuroscience **411**, 280–290 (2019)

142. Van Den Heuvel, M.P., Mandl, R.C.W., Kahn, R.S., Pol, H.E.H.: Functionally linked resting-state networks reflect the underlying structural connectivity architecture of the human brain. Hum. Brain Mapp. **30**, 3127–3141 (2009)
143. Hens, C., Harush, U., Haber, S., Cohen, R., Barzel, B.: Spatiotemporal signal propagation in complex networks. Nat. Phys. **15**, 403–412 (2019)
144. Buzsáki, G., Wang, X.-J.: Mechanisms of gamma oscillations. Annu. Rev. Neurosci. **35**, 203–225 (2012)
145. Womelsdorf, T., Schoffelen, J.-M., Oostenveld, R., Singer, W., Desimone, R., Engel, A.K., Fries, P.: Modulation of neuronal interactions through neuronal synchronization. Science **316**, 1609–1612 (2007)
146. Singer, W.: Neuronal synchrony: a versatile code for the definition of relations? Neuron **24**, 49–65 (1999)
147. Varela, F., Lachaux, J.P., Rodriguez, E., Martinerie, J.: The brainweb: phase synchronization and large-scale integration. Nat. Rev. Neurosci. **2**, 229–239 (2001)
148. Fries, P.: A mechanism for cognitive dynamics: neuronal communication through neuronal coherence. Trends Cogn. Sci. **9**, 474–480 (2005)
149. Fries, P.: Rhythms for cognition: communication through coherence. Neuron **88**, 220–235 (2015)
150. Siegel, M., Donner, T.H., Engel, A.K.: Spectral fingerprints of large-scale neuronal interactions. Nat. Rev. Neurosci. **13**, 121–134 (2012)
151. Chang, E.F., Raygor, K.P., Berger, M.S.: Contemporary model of language organization: an overview for neurosurgeons. J. Neurosurg. **122**, 250–261 (2015)
152. Smits, M., Jiskoot, L.C., Papma, J.M.: White matter tracts of speech and language. Semin Ultrasound CT MR. **35**, 504–516 (2014)
153. Warbrick, T., Rosenberg, J., Shah, N.J.: The relationship between BOLD fMRI response and the underlying white matter as measured by fractional anisotropy (FA): A systematic review. Neuroimage **153**, 369–381 (2017)
154. Hu, M.L., Zong, X.F., Mann, J.J., Zheng, J.J., Liao, Y.H., Li, Z.C., He, Y., Chen, X.G., Tang, J.S.: A Review of the functional and anatomical default mode network in schizophrenia. Neurosci. Bull. **33**, 73–84 (2017)
155. Alves, P.N., Foulon, C., Karolis, V., Bzdok, D., Margulies, D.S., Volle, E., de Schotten, M.T.: An improved neuroanatomical model of the default-mode network reconciles previous neuroimaging and neuropathological findings. Commun. Biol. **2**, 1–14 (2019)
156. Deco, G., Ponce-Alvarez, A., Mantini, D., Romani, G.L., Hagmann, P., Corbetta, M.: Resting-state functional connectivity emerges from structurally and dynamically shaped slow linear fluctuations. J. Neurosci. **33**, 11239–11252 (2013)
157. Wein, S., Deco, G., Tomé, A.M., Goldhacker, M., Malloni, W.M., Greenlee, M.W., Lang, E.W.: Brain connectivity studies on structure-function relationships: a short survey with an emphasis on machine learning. Comput. Intell. Neurosci. 5573740 (2021)
158. Morone, F., Makse, H.A.: Symmetry group factorization reveals the structure-function relation in the neural connectome of Caenorhabditis elegans. Nat. Commun. **10**, 4961 (2019)
159. Morone, F., Leifer, I., Makse, H.A.: Fibration building blocks of biological networks. Proc. Nat. Acad. Sci. USA **117**, 8306–8314 (2020)
160. Makse, H.A., Boldi, P., Sorrentino, F., Stewart, I.: Symmetries of Living Systems: Symmetry Fibrations and Synchrony in Biological Networks. Cambridge University Press (2024)
161. Varshney, L.R., Chen, B.L., Paniagua, E., Hall, D.H., Chklovskii, D.B.: Structural properties of the Caenorhabditis elegans neuronal network. PLoS Comput. Biol. **7**, e1001066 (2011)
162. Dixon, J.D., Mortimer, B.: Permutation Groups, Graduate Texts in Mathematics, 163. Springer-Verlag, New York (1996)
163. Golubitsky, M., Stewart, I.: Nonlinear dynamics of networks: the groupoid formalism. Bull. Amer. Math. Soc. **43**, 305–364 (2006)
164. McKay, B.D.: Nauty user's guide (version 1.5), Technical report TR-CS-90-02, Australian National University (1990)

165. Leifer, I., Makse, H.A., Phillips, D., Sorrentino, F.: Symmetry-driven network reconstruction through pseudobalanced coloring optimization. J. Stat. Mech. 073403 (2022)
166. Boldi, P., Leifer, I., Makse, H.A.: Quasifibrations of graphs to find symmetries and reconstruct biological networks. J. Stat. Mech. 113401 (2022)
167. Kato, S., Kaplan, H.S., Schrödel, T., Skora, S., Lindsay, T.H., Yemini, E., Lockery, S., Zimmer, M.: Global brain dynamics embed the motor command sequence of Caenorhabditis elegans. Cell **163**, 656–669 (2015)

Chapter 7
Keystone Species are Influencers in Ecosystems

Abstract In the sixties, ecologist Robert Paine introduced the concept of *keystone species*, referring to certain species that exert a crucial influence on ecosystems. Their extinction can have far-reaching consequences for the survival of the entire system. Influencers can be compared to keystone species, as they play a pivotal role in maintaining the system's integrity. These influencers are the ones that hold the network together and ensure its sustainability. It is thus natural to map the problem of identifying these species in the ecological network to the problem of percolation. Given the importance of mutualistic species interactions in community dynamics due to their stabilizing and positive effect, the strategy that emerges to identify keystone species in mutualistic ecosystems is the k-core. Monitoring the k-core of the network anticipates the tipping point of ecosystem collapse. In this chapter, we review theoretical frameworks to identify keystone species within theories of influencers and assess their importance for stability, resilience, and vulnerability, as well as early warning systems to anticipate catastrophic collapses of ecosystems.

7.1 Ecosystems

Identifying keystone species [1–5] requires a mathematical framework able to predict which ones are the crucial ones. All that matters is where the keystone species are located in the network. To identify keystone species, we need to understand first the interplay between the network in conjunction with the dynamics of interactions among the related species, a key feature that requires more attention from ecologists or biologists.

The ecological literature has focused on identifying keystone species in ecosystems on a more qualitative basis, but fewer studies have analyzed the interactions of a network using quantitative methods. Libralato et al. [6], for instance, quantified the impact of a species by examining the relative change of biomasses in the food web resulted from a perturbation on the biomass of a given species using dynamical simulations. Jordán et al. [7] explored the performance of different centrality measures in finding keystone species in food webs, and later compared two approaches

H. A. Makse and M. Zava, *The Science of Influencers and Superspreaders*, Understanding Complex Systems, https://doi.org/10.1007/978-3-031-78058-5_7

using centrality metrics that quantify the pure structural importance and dynamical simulations that also, incorporate functional importance [8]. Berry and Widder used co-occurrence networks to detect keystone species among microbial interactions [9].

Morone et al. [10] developed a theoretical framework to link network structural properties with the stability of mutualistic ecosystems that allows the classification of species from keystone one to species of lesser importance in the ecosystem. The theory requires going beyond what has been discussed in Chap. 1, in that it is not only the network that is important, but the dynamics between the species. These interactions are now described by a set of dynamical equations defined on top of the network structure. The dynamic solution reflects both effects, and it is usually difficult to separate the effect of the network from the dynamics. This is related to the long-standing problem that we have already treated in the brain and genes: how the structure determines the function. In this case, how the structure of the network determines the function of the ecosystem as given by its dynamics of evolution towards a healthy state or its collapse.

Mutualism is defined as an interaction between two or more species for each other's benefit. Examples of mutual relationships include plant-pollinators such as bees and humming-birds. In the particular case of mutualistic networks, Morone et al. [10] was able to obtain analytical solutions which decouple the contribution of the network from the dynamics. By doing so one is able to classify the species that are important to the dynamical evolution of the network, in particular, its stability and collapse, solely using network properties and some parameters describing the dynamics. We will see in the next section that since the dynamics are characterized by a threshold-like model (Hill, sigmoidal or Theta-functions), analogous to the cascading models of Sect. 1.3.3 and bootstrap percolation [11–13], then the feature of the network that deciphers the connection between structure and dynamics is, naturally, the k-core; a result that was already anticipated in the work of García-Algarra et al. [14].

7.2 Robustness and Resilience of Ecosystems

Since the seminal work of Robert May [15] in the 70's, scientists have been trying to understand what makes a complex system stable, and what causes its ultimate collapse. Early warning signals have been designed [16–19], and heuristic metrics—including compartmentalization, nestedness, diversity, etc.—have been invoked to predict the tipping point of collapse and the conditions of resilience of the system [20–22]. Controversies have arisen as a consequence of contradicting results, sometimes producing conjectures from simulations with mixed results. These controversies can be partially strengthened by theoretical results, as shown by the seminal work of May [15]. However, these theoretical results are based on approximations and predict more controversial results, e.g., May's prediction that large ecosystems are unstable and that diversity is detrimental for ecosystems [15], creating even more controversies.

One way to describe the robustness of the ecosystem is by using its nestedness [22], which comes into play when considering both ecological and financial networks [19]. This feature can be measured graphically by determining how likely it is that those nodes with the fewest connections are linked to those nodes with the most connections, such that an adjacency matrix representing the network can be arranged to fill the upper triangle of the grid [22, 23]. There are several ways of quantifying nestedness, including the 'temperature' [24], in which a lower temperature denotes greater nestedness. This method draws an analogy between the ecological network and a physical system by describing the 'disorder' of the network. Since greater nestedness implies greater order in the interactions between species in the network, more nested systems have lower temperatures. Those systems that are maximally-nested (i.e., temperature $T = 0$) will tolerate no reordering of interactions, whereas less-nested systems can maintain the same degree of system order, even if there is some shuffling of interactions.

Another method of calculating nestedness, which does not demand that the matrix structure be rearranged to be as close as possible to upper-triangular, is that of Johnson et al. [25]. Here, the expected nestedness η of a bipartite matrix is calculated as:

$$\eta = \frac{n_1 \langle k^2 \rangle_2 + n_2 \langle k^2 \rangle_1}{\langle k \rangle_1 \langle k \rangle_2 (n_1 + n_2)^2}, \tag{7.1}$$

where n_1 and n_2 are the numbers of nodes in each set of the bipartite matrix (for the case of the ecological networks, these sets would be "plants" and "animals"); k is the number of edges; and $\langle \cdot \rangle_i$ is an average over set i. η takes values which increase with increasing order (nestedness), with the minimum value of η being 1 for a random network.

The nestedness of a network has been correlated with how mutualistic are the interactions between its members. The work of Bascompte et al. [22] has shown that nestedness increases with mutualism in ecological networks, such that highly-mutualistic networks are also highly nested (as opposed to having a compartmentalized structure). This, in practice, means that the most generalist of the species— those with the greatest number of interactions and, therefore, with numerous roles or a more general, fundamental role in the ecosystem— interact with the most specialist of the species— those that interact with the fewest other species and therefore have a more specialized function in the ecosystem. The system is then organized around a core of interactions, displaying a small number of species interacting with many others, and many species which interact only with a few others.

Similarly, May et al. [26] noted that, in financial networks [27, 28], smaller and more specialized banks and businesses tend to have few connections, which are mostly with larger, more generalist banks and businesses, thus drawing a comparison between ecology and economics. Due to this central core of interactions that are found in mutualistic networks, the question arises as to what structure of the network is related to its robustness and resilience. To study this problem, we examine both ecological (and financial networks) through a model of species interactions.

7.2.1 How to Identify Keystone Species

Finding and ranking the keystone species in the network of interactions allow us to understand the stability and robustness of ecosystems [29–35]. As put it by Bascompte and Jordano [23], the network is the 'architecture of biodiversity', and analyzing its properties in conjunction with the dynamics of species interactions allows us to predict which species are important for the stability of the ecosystem in response to ecosystem disturbance [36–38]. Using these results, ecologists and scientists can now combine both quantitative with qualitative analysis to fully understand exactly what cascading effects will trigger ecosystem collapses.

Dunne et al. [39] has ranked the species by degree to measure their importance to the resilience of food webs and Memmott et al. [40] studied the resilience of mutualistic networks by removing species in a percolation process. Domínguez-García and Muñoz [35] exploited the nested structure of mutualistic networks [22] to derive a ranking of species of importance in the network, as the nested structure provides robustness and promotes biodiversity [40, 41]. Analogous results for food webs were obtained by Allesina and Pascual [42] who applied Google's PageRank to rank species according to their importance for coextinctions in these ecosystems. The modularity of the network has also been used to understand the different roles of species in mutualistic networks [43].

In 2017, García-Algarra et al. [14] found that the k-core decomposition, together with other k-magnitudes like k-radius, k-degree, and k-risk, are effective ways to measure the vulnerability of the ecosystem to the loss of particular species. The k-radius describes the network compactness, the k-degree describes how a species is connected to the innermost shell, and the k-risk is the set of species whose removal from the network poses a greater risk to the entire network. By analyzing the Web of Life collection [44] of 89 mutualistic networks, they found that k-risk was the ranking method with the lowest average species removal percentage to destroy the giant connected component in the majority of the studied networks. This quantity provides a reliable ranking to identify keystone species by identifying species linked to a high number of nodes of the maximum k-core of the network. They conclude that these keystone species provide vulnerabilities to the network because their extinction may produce cascading effects in the lower k-shells they are linked to since they lack redundant paths to the maximal k-core.

The results of García-Algarra et al. [14] were confirmed by Morone et al. [10] by using a theory for catastrophic collapse of mutualistic ecosystems, encompassing an analytical determination of the tipping point. Drawing upon the keystone species classification, a mathematical analysis starts by classifying the type of interactions between species [36, 45–47]:

- Predator-prey $(-, +)$. The predator gets the benefit $(+)$, the prey is eaten and therefore does not receive a benefit $(-)$. Examples are multiple, including the bear salmon, wolf-elk, lion-gazelle.
- Mutualism $(+, +)$. Both species benefit from each other's interaction. E.g., bees, hummingbirds and plants.

- Commensalist (+, 0). One species benefit (+), the other is unaffected (0). It is not negative, as the unaffected species do not die; they just do not benefit from the interaction. E.g., a monkey eating from a tree.
- Amensalism (−, 0). One species is harmed (−), the other is unaffected (0). E.g., an elephant walking/destroying (0) on top of the grass (−).
- Competition (−, −). When two species compete against each other. For example, brown bears compete with packs of wolves to control the food supply.

Once the interactions between species in an ecosystem are defined, a network can be built. In this network, we could have all sorts of directed interactions, positive, negative, zero. After this, a dynamical system can be modeled using a first-order differential equation, taking into account a natural exponential decay of the species and the benefits (positive interactions) and detriments (negative interactions) encapsulated by the interaction terms in the dynamical system.

There are several ways to model an ecological network, one of which is the weighted adjacency matrix A_{ij} which includes all the interactions. For each interaction, the dynamics are defined typically by either linear terms or nonlinear terms with some form of sigmoidal function like a Theta-function or a Hill function. An important feature of these models is the use of thresholds and nonlinearities in the interactions and that species (or individuals) receive a benefit and process it through these interactions. Other examples include the gene regulatory networks studied in Chap. 5, and the brain neural networks (and artificial neural networks) studied in Chap. 6.

A theoretical framework to find the keystone species needs to link network structural properties with the stability of ecosystems. An analytical solution of such a model has been achieved so far only for mutualistic ecosystems [10], owing to the fact that all interactions are positive. We treat this case next, and then we discuss if it is possible to generalize this model to more complex interactions. It proceeds as follows. In the model, the dynamics of species densities, $x_i(t)$, is described by a set of nonlinear differential equations discussed in a series of works [36, 45–47]:

$$\dot{x}_i(t) = -dx_i - sx_i^2 + \sum_{j=1}^{N} A_{ij}\gamma_{ij}\frac{x_i x_j}{\alpha + \sum_{k=1}^{N} A_{ik}x_k}, \quad i \in \{1, \cdots, N\}, \tag{7.2}$$

where A_{ij} represents the interacting network. In principle, this network is directed, but to achieve an analytical solution of the dynamical system, the network is considered to be undirected, $A_{ij} = A_{ji}$. This represents a major approximation to a real ecosystem. Likewise, the interaction strength $\gamma_{ij} > 0$ is taken to be positive, representing the mutualistic interactions. The death rate $d > 0$, the self-limitation parameter $s > 0$, and the half-saturation constant α are taken all constant independent of the species as another approximation.

Morone et al. [10] obtained the fixed point solution of this model by employing an extra approximation: $\gamma_{ij} = \gamma$, i.e., all species have the same interaction strength. This approximation ignores the asymmetric and heterogeneous nature of the species interactions and allows us to obtain the kcore solution, as explained below. A realistic system will have all interactions different and have no reciprocity. In this case, the exact solution obtained below ceases to be valid. Yet, a system with reciprocal interactions can inform us of the exact solution that fully characterizes the system. Understanding realistic ecosystems needs numerical solutions. When the asymmetries and heterogeneities in interactions do not depart much from the uniform approximation, the solution is expected to have a vestige of the k-core as investigated in [10]. The simulations in [17] corroborate this point.

The fixed point of the solution x^* is obtained by the reduced density y^*:

$$y_i^* = \sum_{j=1}^{N} A_{ij} H_1\Big(y_j^* - \frac{\alpha d s}{(\gamma - d)^2}, \frac{\alpha \gamma s}{(\gamma - d)^2}\Big), \tag{7.3}$$

where

$$y_i^* = \frac{s}{\gamma - d} \sum_{j=1}^{N} A_{ij} x_j^*, \tag{7.4}$$

and

$$H_n(x, T) = \frac{x^n}{x^n + T^n} \tag{7.5}$$

is the Hill function characterizing the interaction among different species. Using the Boolean approximation ($n \to \infty$) in which the Hill function $H_n(x, T)$ is replaced by the Heaviside function $\Theta(x - T)$, the fixed point equations can be further recast in the following form:

$$y_i^* = \sum_{j=1}^{N} A_{ij} \Theta(y_j^* - K_\gamma), \tag{7.6}$$

$$K_\gamma = \frac{\alpha s(\gamma + d)}{(\gamma - d)^2}, \tag{7.7}$$

where K_γ is the threshold of the mutualistic benefit; the subscript emphasizes its main dependence on γ, $K_\gamma \propto 1/\gamma$ (see Fig. 7.1a) since we can consider $d \ll \gamma$.

Following the above equations, the reduced density y_i^* takes only integer values in the set $y_i^* \in \{1, 2, \ldots, k_i\}$ where k_i is the degree of species i. Those species with $k_i < K_\gamma$ can be removed from the system iteratively. The pruning process stops when the degree of each node is larger than K_γ. Thus, the nontrivial fixed point solution of the ecosystem is:

$$y_i^* = \mathcal{N}_i(K_\gamma), \tag{7.8}$$

where $\mathcal{N}_i(K_\gamma)$ is the degree of species i in the K_γ-core. As the mutual benefit γ decreases, the species located at the outer shell of the ecosystem tend to vanish first, followed by the species in the core region. The critical interaction strength for collapse, γ_c, is determined by the maximum k-core $k_{\rm core}^{\rm max}$:

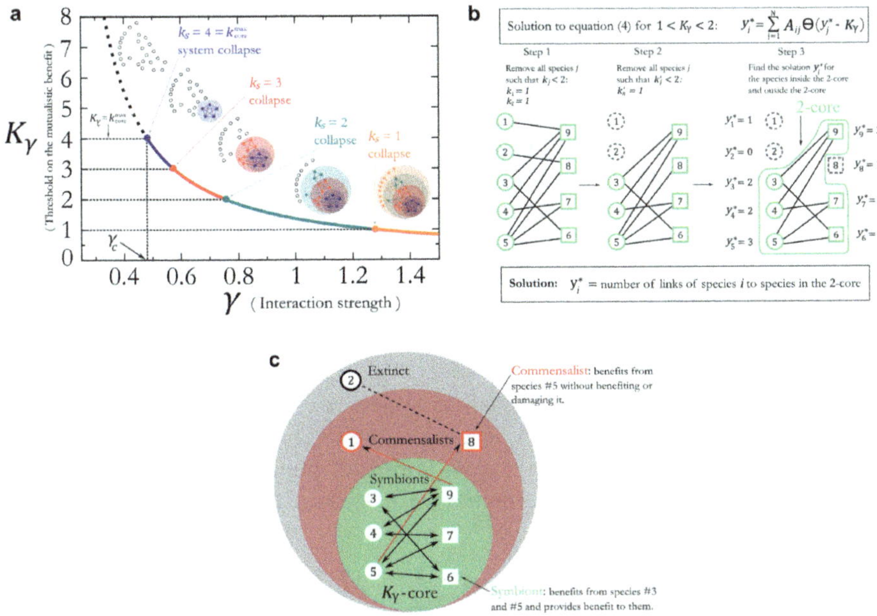

Fig. 7.1 **Solution scheme for the fixed point Eq.** (7.6). **a** Threshold K_γ as a function of a sample network's interaction strength γ. For large γ such that $K_\gamma < 1$, the network is most stable, and all species provide benefits to their interacting species. When γ is reduced such that K_γ is above 1, all the species with $k_s = 1$ cannot confer/receive their benefits to the others and go extinct, while the species in the 2-core keeps providing its benefits. When γ is further reduced so that K_γ becomes slightly larger than 2, the species with $k_s = 2$ go extinct. Further reducing γ inhibits the mutualistic benefit from species in the 3-shells and so one, so the k-shells go extinct one by one, and, eventually, causes the threshold K_γ to surpass the value of the max k-core of the network $K_\gamma > k_{\text{core}}^{\text{max}} = 4$. At this point, all the system collapses, since no species can provide the mutualistic benefit anymore. This series of collapses results in the staircase shape of species density shown in Fig. 7.2b. **b** A simple ecosystem network with a 2-core and species with interaction strength K_γ anywhere between $1 < K_\gamma < 2$ showing the pruning process of the solution of Eq. (7.6). **c** Hierarchy of species according to their k-shell. In ecological terms, species 1 and 8 are commensalists as opposed to the true symbionts (the keystone) living in the max 2-core, because they receive a benefit from the species in the 2-core but provide no benefit back, while symbionts receive and provide benefits as well. Species 2 has $y_2^* = 0$, since this species has no links to species in the 2-core, and they go extinct. The symbionts in the K_γ-core are the keystone species for this particular ecosystem with a given γ. Figure reproduced from [10]. Copyright © 2019, The Authors, under exclusive licence to Springer Nature Limited

$$K_{\gamma_c} \equiv \alpha s \frac{(\gamma_c + d)}{(\gamma_c - d)^2} = k_{\text{core}}^{\text{max}}. \tag{7.9}$$

The procedure is graphically explained in Fig. 7.1b for a sample ecosystem with a maximum 2-core and interaction strength that could be anywhere between $1 < K_\gamma < 2$. After these first removals are done (Step 1 in Fig. 7.1b), the species left in the network has a smaller degree k_j', and we perform a new wave of removals of species j' if $k_j' < K_\gamma$ (Step 2 in Fig. 7.1b). This pruning process stops when the degree of each remaining species is larger than K_γ. The process we just described is precisely the algorithm for extracting the K_γ-core of the network as defined in Sect. 1.2.1.2: iteratively removing all species from the network with degree $k < \lceil K_\gamma \rceil$, where $\lceil \cdot\cdot \rceil$ denotes the ceiling function. Thus, the nodes remaining at the end of this pruning process, if any, form a K_γ-core by construction, as shown in Step 3 of Fig. 7.1b.

How can we interpret this solution? We can think of an ecosystem that is being affected by a global phenomenon that make the species interact less and less. This could be a global effect of a temperature increase. For instance, we can think that climate change produces a global phenomenon of increasing temperatures that decrease the strength of interaction of mutual benefit between the species. The model indicates that when the beneficial mutual interactions γ decreases, then the system approaches a tipping point given by γ_c (and the corresponding K_{γ_c} in Eq. (7.9) past which the system irreversible collapses.

The collapse goes k-shell by k-shell as γ decreases. For $\gamma < \gamma_c$, even species in the maximum k-core are extinct, leading to the total collapse of the ecosystem. An example of such a collapse in a plant-pollinator mutualistic network is shown in Fig. 7.2. The collapse of the system as γ decreases (or K_γ increases) is done

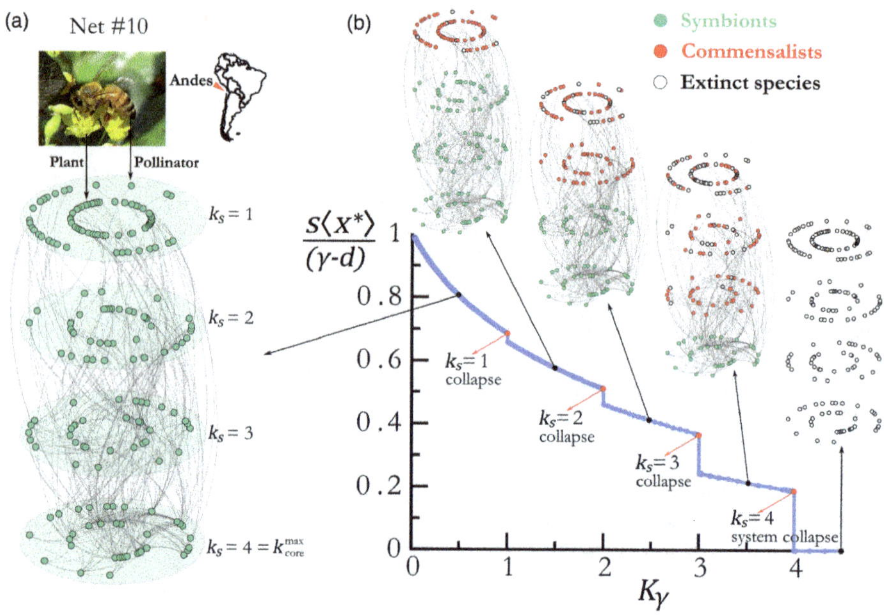

Fig. 7.2 Collapse of a plant-pollinator mutualistic network. **a**. A bipartite plant-pollinator ecosystem located in the Chilean Andes [49], formed by four rings containing species within the k-shell from $k_s = 1$ to $k_s = 4$. In each k-shell, plants and pollinators are located in the inner and outer rings, respectively. **b** Rescaled fixed point average density $\langle x^* \rangle$ as a function of the threshold K_γ. Green circles are symbionts that benefit from each other. Red circles are commensalists that receive benefits from other species but no longer provide benefits. Open circles are extinct species. As K_γ increases, more extant species turn into commensalism or go extinct whenever K_γ exceeds integer values of the successive k-shell. For $K_\gamma > 4$, the whole ecosystem suddenly collapses. Figure reproduced from [10]. Copyright © 2019, The Authors, under exclusive license to Springer Nature Limited

k-shell by k-shell. The actual variable that controls the transition is K_γ, which is the threshold of interaction in Eq. (7.6). Increasing this parameter can be achieved by a decrease of γ when the mutualistic benefits that the species receive get scarcer, Fig. 7.1a.

The ecosystem is more stable when $K_\gamma \ll 1$. As K_γ becomes larger, by a decrease of benefits, then the species located in the first outer shell with $k_s = 1$ are the first to go extinct. These are the commensalists that benefit mainly from the k-core but do not provide much benefit to the system. This is depicted in the hierarchy of species in Fig. 7.1c, which shows the commensalists that benefit from the species located in the K_γ-core without benefiting nor damaging them. This 1-shell species with a single link to the max k-core do not influence the ecosystem, meaning that their disappearance does not change the density of any other species. At the same time, these commensalists are the most exposed to any variation in the strength of benefit interactions. On the contrary, the symbionts located in the K_γ-core in Fig. 7.1c are the keystone species and the last species to go extinct by the decrease of the benefit interactions. The fact that keystone species are located in the innermost k-core has implications in other nonlinear dynamical systems, such as financial networks.

These considerations lead to the phase diagram of feasible and stable mutualistic ecosystems depicted in Fig. 7.3a in the space $(K_\gamma, k_{\text{core}}^{\text{max}})$. The phase diagram

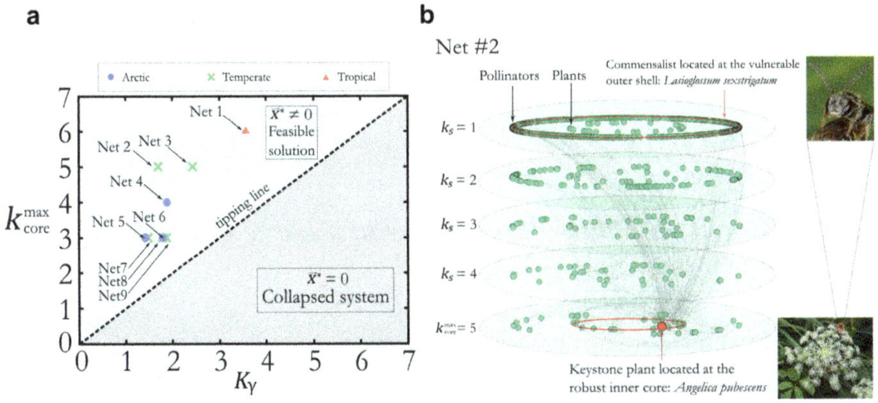

Fig. 7.3 Phase diagram of ecosystem stability. a, Predicted phase diagram for k-core number $k_{\text{core}}^{\text{max}}$ versus K_γ for nine empirical mutualistic networks (the the list appears in supplementary materials in [10]) corresponding to ecosystems at different latitudes: Arctic (blue point), Temperate (green cross), and Tropical (red triangle). All the networks lie in the stable feasible region predicted by Eq. (7.10), $k_{\text{core}}^{\text{max}} > K_\gamma$, i.e., above the tipping line defined by $k_{\text{core}}^{\text{max}} = K_{\gamma_c}$. **(b)** Shown is the network structure for the ecosystem Net #2 of plant-pollinator in Japan from Ref. [50]. Species are ordered by increasing k-shell number k_s from top to bottom, with plants in the inner circles and pollinators in the outer circles. The arrangement of species in k-shells shows the keystone species at the inner k-core, in this case $k_{\text{core}}^{\text{max}} = 5$, one example is the *Angelica pubescens*, which is pollinated by a commensalist located at the more vulnerable outer shell *Lasioglossum sexstrigatum*; suggestive names for the function of these two species in the ecosystem, which are corroborated by their position in the k-shells of the network. Figure reproduced from [10].

features the predicted 'tipping line' of instability defined by the condition (7.9), which separates the feasible-stable phase:

$$K_\gamma < k_{\text{core}}^{\max} \quad \text{(condition of existence of the feasible-stable state)}, \tag{7.10}$$

from the collapsed phase:

$$K_\gamma > k_{\text{core}}^{\max} \quad \text{(condition of collapsed state)}. \tag{7.11}$$

Morone et al. [10] tested this phase diagram by plotting the values of $(K_\gamma, k_{\text{core}}^{\max})$ obtained from real mutualistic networks of plant-pollinator and plant-seed dispersers where these parameters have been measured in Refs. [36, 48] (see Fig. 7.3a). All real mutualistic ecosystems lie in the feasible-stable region situated above the tipping line, in agreement with the theory.

From the graphical representation of a sample real ecological network in Fig. 7.3b, an interesting structure emerges. Many pollinators in the outer shell $k_s = 1$ interact with a single keystone plant species, the *Angelica pubescens*, located in the innermost core of the network and, therefore, is quite stable to external changes, since the inner core is the most stable part of the ecosystem. On the contrary, there are quite fewer plant species in the outer shell ($k_s = 1$) interacting with just a single pollinator in the inner core $k_{\text{core}}^{\max}$. Plants tend to populate the more robust inner k-shells, while pollinators concentrate more in the low k-shells (i.e., the upper levels). This result hints that plants are more needful for the survival of many pollinators than vice versa, a conclusion stemming directly from the k-core organization of the ecological network.

7.2.2 Diversity-Stability Paradox: "*Will a Large Complex System be Stable?*" Yes, It Will

Equation (7.9) provides a hierarchical classification of species in terms of their location in different k-cores of the network and the prediction of the tipping point of collapse. A mutualistic ecosystem with strong interactions between the species, that is, where the species receive a nice amount of nutrients and benefits from each other, is quite stable. This is because the parameter in Eq. (7.7) satisfies $K_\gamma \to \alpha s/\gamma \to 0$ as the interaction strength increases. This means that the parameter K_γ will never achieve the threshold of the tipping point, which is given by the max k-core of the network, Eq. (7.9) when K_γ is equal to $k_{\text{core}}^{\max}$. This stability is greater when the system size is larger. And since the system size N counts the number of different species, increasing diversity increases the stability of the ecosystem. This is because when we increase the number of species in the ecosystem, the max k-core increases, and then the tipping point of collapse is even more difficult to achieve. That is, this model of mutualistic ecosystem predicts that large systems are more stable: diversity of species is better for the stability of the ecosystem.

The reader familiar with the work of Robert May on the matter of ecosystem stability would recognize that this conclusion contradicts a cornerstone of theoretical ecology. In his seminal 1972 paper, *Will a large complex system be stable?*, which started the field of theoretical ecology (and even theoretical biology) [15], May concluded theoretically that the answer to this questions is *no*. By performing linear stability analysis on a system of equations like Eq. (7.2), May showed that increasing the diversity of the ecosystem makes the ecosystem more unstable, an statement known as the 'diversity-stability' paradox. This conclusion has created large headaches in the ecological community, as it is difficult to reconcile it with the fact that diversity is always good for any system, as we can anecdotally see in the beautiful diversity of nature and human society.

The diversity-stability paradox is resolved Arese et al. [51] by considering the actual fixed point solution in the stability analysis and the interaction term's saturation effect. The May paradox is a by-product of the linear approximation employed in the model of May that ignores the stability matrix's dependence on the system's fixed point solution. On the contrary, a nonlinear model Eq. (7.2), with its exact solution, allows the study of the stability matrix devoided of approximations to show the reverse of the May paradox: diversity increases the stability of the ecosystem.

Thus, the mapping of the dynamical system collapse to k-core allows the calculation of the tipping point. In this manner, one may, for example, determine the magnitude of harm that a given species' extinction or even endangerment will do to an ecosystem. Moreover, by dividing the ecosystem into shells, one may determine how dire the situation of the ecosystem, i.e., if many species in the shell closest to the tipping point are disappearing or in danger of doing so, it would be necessary to act quickly to save the ecosystem from irreparable harm or even collapse.

The structural and topological properties revealed by the k-core solution brings a different look to the problem of quantifying the resilience of complex natural systems. The notion of k-core provides an understanding of this kind of collapse phenomena in that it provides a structural warning signal to monitor the proximity to collapse. It also deciphers the connection between structure and dynamics. This analytic result could add itself to the list of exactly solvable models for complex systems, which, by revealing theoretical predictions, could unleash further scientific production.

Talking about further production, the problem of resilience, stability, and earlier warning signals for collapse covers a much larger swath of complex systems, from ecology biology to finance. It would be nice if the k-core solution would be valid for these systems to allow one to predict the collapse of systems as diverse as ecosystems, microbiomes, genetic networks, and markets. Thus, an important question is how general this result is.

Unfortunately, the answer to whether the k-core can be generalized to these systems is No. Why? Because this framework is only valid when all the interactions are positive. Moreover, in the framework, all interactions are undirected or reciprocal. These assumptions, while suitable for mutualistic systems, do not hold for the majority of systems of interest in biology, mostly all, like the brain and genetic networks. It does not hold for more complex ecosystems beyond pure mutualistic ones, e.g., predator-prey interactions. These systems are characterized by inhibitory or negative interactions, and all of them are directed and asymmetric in the sense that i interacts with j differently than j with i. This rules out the application of the k-core to any biological system beyond the mutualistic systems studied in this chapter.

We have discussed this issue already in Sect. 5.6. In the language of physics, these systems are non-reciprocal and do not satisfy a cornerstone of physics, which is the action-reaction principle. In physical systems, this is a critical issue, since non-reciprocity of interactions further voids the use of equilibrium statistical mechanics (no detailed balance), and the theoretical tools to treat these systems are reduced to a minimum.

The treatment of systems with inhibitory and directed non-reciprocal interactions require more serious thinking. When we tried to find influencers for these systems, we soon realized that k-core percolation or any percolation variant was not the answer. This has led us to investigate the conditions under which the dynamics can be decoupled from the network structure in inhibitory and directed networks until we concluded that for this to occur, symmetries must be realized in the network. This line of thinking was inspired by the work on groupoid symmetries by Golubitsky and Stewart [52]. And when we started searching for these symmetries in biological networks, we found they were ubiquitous. At this point, our search for influencers via k-cores, percolation, and collective influences ended up turning into a search for symmetrical biological building blocks. These ideas led to the elaborations in Chap. 5, Sect. 5.5 for genetic networks, and Chap. 6, Sect. 6.8 for the brain, and are reviewed in more detail in a forthcoming book [53].

References

1. Paine, R.T.: A note on trophic complexity and community stability. Am. Nat. **103**, 91–93 (1969)
2. Paine, R.T.: A conversation on refining the concept of keystone species. Conserv. Biol. **9**, 962–964 (1995)
3. Wagner, S.C.: Keystone species. Nat. Educ. Knowl. **3**, 51 (2010)
4. MacDonald, J.: The 'Keystone Species' Concept That Transformed Ecology, JSTOR Daily (30 Jun 2016). https://daily.jstor.org/how-the-keystone-species-concept-transformed-ecology. Accessed Apr 2022
5. Denchak, M.: Keystone Species 101. NRDC, 9 Sept. 2019. https://www.nrdc.org/stories/keystone-species-101. Accessed Apr 2022
6. Libralato, S., Christensen, V., Pauly, D.: A method for identifying keystone species in food web models. Ecol. Modell. **195**, 153–171 (2006)
7. Jordán, F., Benedek, Z., Podanic, J.: Quantifying positional importance in food webs: a comparison of centrality indices. Ecol. Modell. **205**, 270–275 (2007)
8. Jordń, F., Okey, T.A., Bauer, B., Libralato, S.: Identifying important species: linking structure and function in ecological networks. Ecol. Modell. **216**, 75–80 (2008)
9. Berry, D., Widder, S.: Deciphering microbial interactions and detecting keystone species with co-occurrence networks. Front. Microbiol. **5**, 219 (2014)
10. Morone, F., Del Ferraro, G., Makse, H.A.: The k-core as a predictor of structural collapse in mutualistic ecosystems. Nat. Phys. **15**, 95–102 (2019)
11. Chalupa, J., Leath, P.L., Reich, G.R.: Bootstrap percolation on a Bethe lattice. J. Phys. C **12**, L31 (1979)

12. Baxter, G.J., Dorogovtsev, S.N., Goltsev, A.V., Mendes, J.F.F.: Bootstrap percolation on complex networks. Phys. Rev. E **82**, 011103 (2010)
13. Baxter, G.J., Dorogovtsev, S.N., Goltsev, A.V., Mendes, J.F.F.: Heterogeneous k-core versus bootstrap percolation on complex networks. Phys. Rev. E **83**, 051134 (2011)
14. García-Algarra, J., Pastor, J.M., Iriondo, J.M., Galeano, J.: Ranking of critical species to preserve the functionality of mutualistic networks using the k-core decomposition. PeerJ **5**, e3321 (2017)
15. May, R.M.: Will a large complex system be stable? Nature **238**, 413–414 (1972)
16. Scheffer, M.: Anticipating critical transitions. Science **338**, 344–348 (2012)
17. Gao, J., Barzel, B., Barabási, A.-L.: Universal resilience patterns in complex networks. Nature **530**, 307–312 (2016)
18. Scheffer, M., Bascompte, J., Brock, W.A., Brovkin, V., Carpenter, S.R., Dakos, V., Held, H., van Nes, E.H., Rietkerk, M., Sugihara, G.: Early-warning signals for critical transitions. Nature **461**, 53–59 (2009)
19. Haldane, A.G., May, R.M.: Systemic risk in banking ecosystems. Nature **469**, 351–355 (2011)
20. Rohr, R.P., Saavedra, S., Bascompte, J.: On the structural stability of mutualistic systems. Science **345**, 1253497 (2014)
21. Staniczenko, P.P.A., Kopp, J.C., Allesina, S.: The ghost of nestedness in ecological networks. Nat. Commun. **4**, 1931 (2013)
22. Bascompte, J., Jordano, P., Melián, C.J., Olesen, J.M.: The nested assembly of plant-animal mutualistic networks. Proc. Natl. Acad. Sci. U.S.A. **100**, 9383–9387 (2003)
23. Bascompte, J., Jordano, P.: Plant-animal mutualistic networks: the architecture of biodiversity. Annu. Rev. Ecol. Evol. Syst. **38**, 567–593 (2007)
24. Atmar, W., Patterson, B.D.: The measure of order and disorder in the distribution of species in fragmented habitat. Oecologia **96**, 373–382 (1993)
25. Johnson, S., Domínguez-García, V., Muñoz, M.A.: Factors determining nestedness in complex networks. PLoS ONE **8**, 1–7 (2013)
26. May, R.M., Levin, S.A., Sugihara, G.: Ecology for bankers. Nature **451**, 893–895 (2008)
27. Mantegna, R.N.: Hierarchical structure in financial markets. Eur. Phys. J. B **11**, 193–197 (1999)
28. Bury, T.: Collective behaviours in the stock market–a maximum entropy approach, Ph. D Thesis (2014). arXiv:1403.5179v2
29. Mills, L.S., Soulé, M.E., Doak, D.F.: The keystone-species concept in ecology and conservation. Bioscience **43**, 219–224 (1993)
30. Dunne, J., Williams, R., Martinez, N.: Network structure and robustness of marine food webs. Mar. Ecol. Prog. Ser. **273**, 291–302 (2004)
31. Srinivasan, U., Dunne, J., Harte, J., Martinez, N.: Response of complex food webs to realistic extinction sequences. Ecology **88**, 671–682 (2007)
32. Solé, R., Bascompte, J.: Self-Organization in Complex Ecosystems. Princeton University Press, New Jersey (2006)
33. Montoya, J., Pimm, S., Solé, R.: Ecological networks and their fragility. Nature **442**, 259–264 (2006)
34. Estrada, E.: Food webs robustness to biodiversity loss: the roles of connectance, expansibility and degree distribution. J. Theor. Biol. **244**, 296–307 (2007)
35. Domínguez-García, V., Muñoz, M.A.: Ranking species in mutualistic networks. Sci. Rep. **5**, 8182 (2015)
36. Thebault, E., Fontaine, C.: Stability of ecological communities and the architecture of mutualistic and trophic networks. Science **329**, 853–856 (2010)
37. Solé, R., Montoya, J.: Complexity and fragility in ecological networks. Proc. Roy. Soc. Lond. B **268**, 2039–2045 (2001)
38. Dakos, V., Carpenter, S.R., van Nes, E.H., Scheffer, M.: Resilience indicators: prospects and limitations for early warnings of regime shifts. Phil. Trans. Roy. Soc. Lond. B: Biol. Sci. **370**, 1–10 (2014)
39. Dunne, J.A., Williams, R.J., Martinez, N.D.: Food web structure and network theory: the role of connectance and size. Proc. Natl. Acad. Sci. U.S.A. **99**, 12917–12922 (2002)

40. Memmott, J., Waser, N., Price, M.: Tolerance of pollination networks to species extinctions. Proc. Roy. Soc. Ser. B **271**, 2605–2611 (2004)
41. Kaiser-Bunbury, C.N., Muff, S., Memmot, J., Muller, C.B., Caflisch, A.: The robustness of pollination networks to the loss of species and interactions: a quantitative approach incorporating pollinator behaviour. Ecol. Lett. **13**, 442–452 (2010)
42. Allesina, S., Pascual, M.: Googling food webs: Can an eigenvector measure species' importance for coextinctions? PLoS Comp. Biol. **5**, e1000494 (2009)
43. Olesen, J.M., Bascompte, J., Dupont, Y.L., Jordano, P.: The modularity of pollination networks. Proc. Nat. Acad. Sci. U.S.A. **104**, 19891–19896 (2007)
44. Fortuna, M.A., Ortega, R., Bascompte, J.: The Web of Life (2014). arXiv:1403.2575, https://www.web-of-life.es
45. Bastolla, U., Fortuna, M.A., Pascual-García, A., Ferrera, A., Luque, B., Bascompte, J.: The architecture of mutualistic networks minimizes competition and increases biodiversity. Nature **458**, 1018–1020 (2009)
46. Holland, J.N., DeAngelis, D.L., Bronstein, J.L.: Population dynamics and mutualism: functional responses of benefits and costs. Am. Nat. **159**, 231–244 (2002)
47. Okuyama, T., Holland, J.N.: Network structural properties mediate the stability of mutualistic communities. Ecol. Lett. **11**, 208–216 (2008)
48. Bascompte, J., Jordano, P., Olesen, J.M.: Asymmetric coevolutionary networks facilitate biodiversity maintenance. Science **312**, 431–433 (2006)
49. Arroyo, M.T.K., Primack, R.B., Armesto, J.J.: Community studies in pollination ecology in the high temperate Andes of Central Chile. I. Pollination mechanisms and altitudinal variation. Am. J. Bot. **69**, 82–97 (1982)
50. Kato, M., Makutani, T., Inoue, T., Itino, T.: Insect-flower relationship in the primary beech forest of Ashu, Kyoto: an overview of the flowering phenology and seasonal pattern of insect visits. Contr. Biol. Lab. Kyoto Univ. **27**, 309–375 (1990)
51. Arese Lucini, F., Morone, F., Tomassone, M.S., Makse, H.A.: Diversity increases the stability of ecosystems. PLoS One **15**, e0228692 (2020)
52. Golubitsky, M., Stewart, I.: Nonlinear dynamics of networks: the groupoid formalism. Bull. Am. Math. Soc. **43**, 305–364 (2006)
53. Makse, H.A., Boldi, P., Sorrentino, F., Stewart, I.: Symmetries of Living Systems: Symmetry Fibrations and Synchrony in Biological Networks. Cambridge University Press (2024)

Chapter 8
Networks and Artificial Intelligence in Finance

Abstract This chapter explores the dynamic intersection of network science and artificial intelligence in modern finance. We trace the evolution of venture capital networks from Silicon Valley's inception to recent phenomena like the PayPal Mafia, revealing patterns of innovation and influence. The study then examines AI's transformative impact on mergers and acquisitions, highlighting enhanced efficiency in due diligence and decision-making. Our analysis of institutional investors' networks in the Italian initial public offers uncovers intriguing patterns of capital concentration. The research extends to novel applications of natural language processing in stock market analysis, leveraging unconventional data sources such as Google Trends and YouTube content. Through an interdisciplinary lens, we demonstrate how the integration of network science and AI is not only enhancing analytical capabilities but also reshaping the very nature of financial markets. This exploration offers fresh perspectives on market behavior and investment strategies, contributing to both theoretical understanding and practical applications in finance. The findings challenge conventional wisdom and open new avenues for research in this rapidly evolving field.

8.1 Introduction

> Cash moves around each day, like ocean currents circulate. Like ocean currents control the weather, circulating money controls the economy. Money must keep moving to fulfill its obligation.
>
> Uncle Scrooge, *Ducktails*.

Financial markets are complex systems characterized by intricate relationships between various actors, instruments, and information flows. This chapter delves into the evolving landscape of finance, exploring how network science and artificial intelligence are reshaping our understanding and approach to key aspects of the financial world.

We begin by examining the pivotal role of venture capital networks in fostering innovation and entrepreneurship. From the birth of Silicon Valley to the rise of influential groups such as the PayPal Mafia, we trace how interconnected networks

H. A. Makse and M. Zava, *The Science of Influencers and Superspreaders*, Understanding Complex Systems, https://doi.org/10.1007/978-3-031-78058-5_8

of investors and entrepreneurs have driven technological advancement and economic growth. We focus on how investors can catalyze funding in the companies they back and how they become influential.

The chapter then shifts focus to the transformative impact of artificial intelligence on mergers and acquisitions (M&A). We explore how AI is revolutionizing due diligence, valuation, and decision-making processes in M&A transactions, using the success story of the M&A Research Institute as a prime example. We then outline the Accelerated M&A Process.

Next, we examine the dynamics of initial public offerings (IPOs), with a particular emphasis on the network structure of institutional investors in the Italian market. This analysis provides insights into the concentration of capital and influence in public equity markets.

Finally, we investigate the application of natural language processing and sentiment analysis to stock market dynamics. By exploring novel data sources such as Google Trends and YouTube content, we demonstrate how AI-driven analyses of public sentiment can offer new perspectives on market behavior and investor decision-making.

Throughout this chapter, we highlight how the integration of network science and artificial intelligence is not only enhancing our analytical capabilities but also reshaping the very nature of financial markets and investment strategies. As we navigate this new frontier, we aim to provide readers with a comprehensive understanding of the cutting-edge tools and methodologies that are defining the future of finance.

8.2 Venture Capital Networks

Not everyone knows that the true essence of Silicon Valley, and most of the innovations you use in your daily life, can be traced back to one company: Fairchild Semiconductor. We will explore their story and then the (related) more recent development of the PayPal network. After the anecdotal introduction, we discuss the role of venture capital (VC) networks as analyzed by the academic literature. Further, the chapter examines recent publications on the role of influence in the venture capital industry, answering three questions: how does the investors' influence help companies enjoy success, how do investors become influential, and how investors might be missing out on opportunities by excessively focusing on their own network. Lastly, we discuss how artificial intelligence is becoming handy in newborn companies' valuation.

8.2.1 Silicon Valley Birth

In 1957, eight engineers and scientists defected from Shockley Semiconductor Laboratory, a company founded by Nobel Laureate William Shockley. Disillusioned by

Shockley's management style, these individuals sought to create an environment where innovation could thrive. They were later known as the "Traitorous Eight," and their departure marked the beginning of a revolution in the semiconductor industry. With financial backing from Sherman Fairchild, an influential inventor and entrepreneur, the group founded Fairchild Semiconductor. The company was located in Mountain View, California, an area that would eventually become the heart of Silicon Valley.

Fairchild Semiconductor quickly made its mark with groundbreaking innovations. Among the most significant was the development of the planar process, a method of manufacturing silicon transistors and integrated circuits. Invented by Jean Hoerni, one of the Traitorous Eight, the process dramatically improved the reliability and scalability of semiconductors. Robert Noyce, another member of the group, further advanced the technology by co-inventing the integrated circuit. This innovation allowed multiple transistors to be placed on a single chip, revolutionizing electronics and paving the way for the modern computer industry.

Fairchild Semiconductor's success created a culture of entrepreneurship and technological innovation. The company's environment encouraged employees to explore new ideas and take risks, leading to the birth of numerous influential companies.

The first-generation spin-offs include:

1. Intel: Founded by Robert Noyce and Gordon Moore in 1968, Intel became the world leader in microprocessors, essential components of modern computers.
2. National Semiconductor: Formed by former Fairchild employees, National Semiconductor played a crucial role in the development of analog and mixed-signal integrated circuits.
3. Advanced Micro Devices (AMD): Established in 1969 by Jerry Sanders and seven colleagues from Fairchild, AMD emerged as a significant competitor to Intel in the microprocessor market.
4. Intersil: Founded in 1967 by Jean Hoerni, Intersil focused on high-performance analog semiconductors.
5. Micron Technology: Founded by former Fairchild employee Ward Parkinson, Micron became a leading producer of memory and storage solutions.
6. SanDisk: Co-founded by former Fairchild engineer Eli Harari, SanDisk revolutionized data storage with its flash memory products.
7. Kleiner Perkins and Sequoia Capital: These venture capital firms, established by Fairchild alumni, became instrumental in funding and nurturing future tech giants.

Second and subsequent generations included companies such as:

1. LSI Logic: Founded by Wilfred Corrigan, a former Fairchild executive, LSI Logic specialized in custom semiconductors.
2. Seagate Technology: Established by Alan Shugart, Seagate became a global leader in data storage solutions.
3. NVIDIA: Founded by Jensen Huang, an alumnus of LSI Logic, NVIDIA transformed the graphics processing unit (GPU) market.

4. Google: Though not a direct spin-off, Google's founders, Larry Page, and Sergey Brin, benefited from the venture capital ecosystem established by Fairchild alumni.
5. Apple: Steve Jobs and Steve Wozniak were deeply influenced by the Silicon Valley culture, which Fairchild helped to establish.
6. YouTube, PayPal, Tesla, LinkedIn: These companies were either founded or significantly funded by alumni of Fairchild's network.

Fairchild Semiconductor's influence extends far beyond its direct contributions to technology. The company's culture of innovation, risk-taking, and collaboration laid the foundation for Silicon Valley's unique ecosystem. This environment fostered an unprecedented wave of technological advancement and entrepreneurship, making Silicon Valley synonymous with innovation. Fairchild's legacy is evident in the vast network of companies (displayed in Fig. 8.1) that trace their origins back to it, directly or indirectly. These ventures have collectively shaped the modern world, driving advancements in computing, data storage, communication, and countless other fields.

Fairchild Semiconductor is rightly hailed as the "Mother of Silicon Valley." Its pioneering work in semiconductor technology and its role in fostering a culture of innovation have left an indelible mark on the tech industry. The ripple effect of its influence can be seen in the success of its many spin-offs and the thriving ecosystem of Silicon Valley, which continues to be the global epicenter of technological innovation, intrinsically connected before its birth. Nowadays, the companies who are enjoying venture capital funding from the most influential individuals in the world are, in fact, connected at birth too [1], as we will see later in this chapter. Another influential group of entrepreneurs that has significantly shaped the technology industry in more recent times is the well-known...

8.2.2 PayPal Mafia

PayPal, a mobile payment company used by most of us, was founded in December 1998 as Confinity by Max Levchin, Peter Thiel, and Luke Nosek. Initially focused on developing security software for handheld devices, the company shifted its focus to digital payments. In March 2000, Confinity merged with X.com, an online banking company founded by Elon Musk. The combined entity adopted the PayPal name in 2001. PayPal revolutionized online payments by allowing users to send and receive money securely and easily over the Internet. Its growth was rapid, and in 2002, PayPal went public. Later that year, it was acquired by eBay for $1.5 billion in stock, marking a significant milestone in its history.

The term "PayPal Mafia" refers to the group of former PayPal employees and founders who went on to become influential figures in the tech industry. This group is notable for its extraordinary success and the creation of numerous successful companies after its tenure at PayPal. Table 8.1collects its historical members, outlining

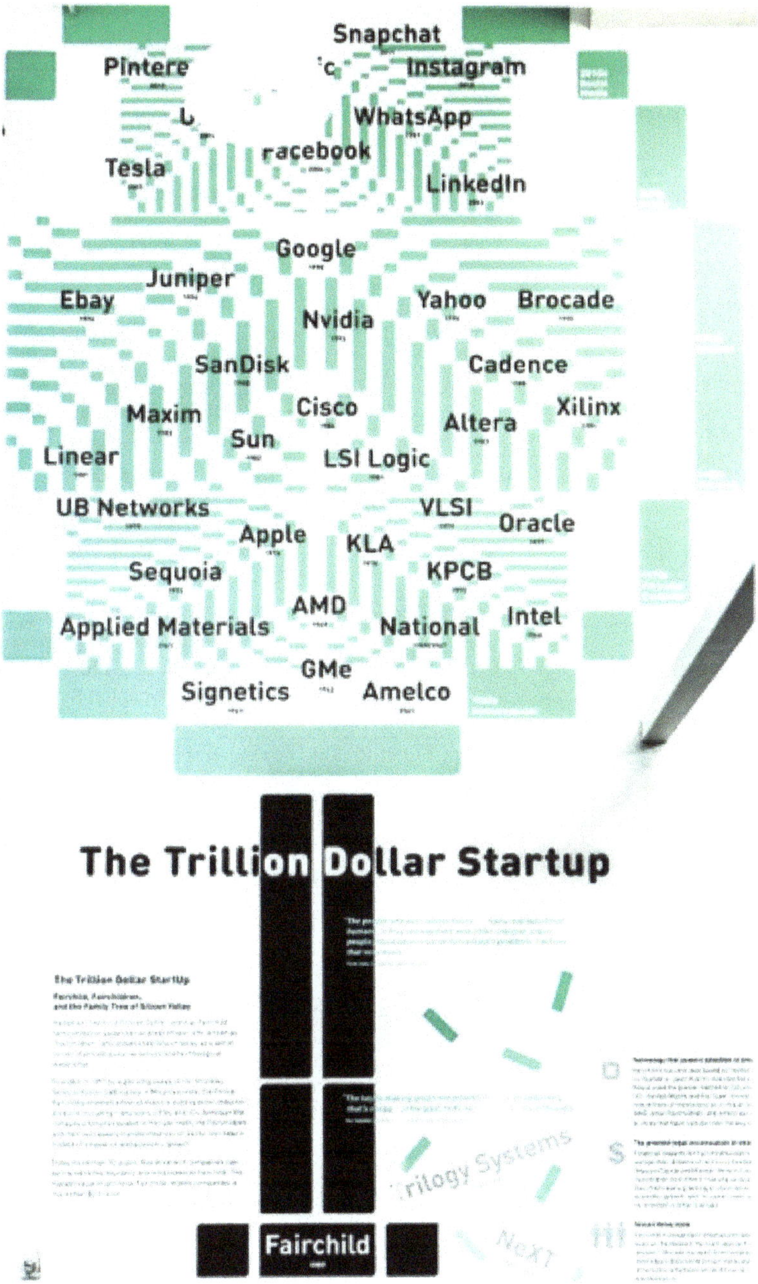

Fig. 8.1 Fairchild, Fairchildren. Exhibit display in CHM lobby, representing examples of the hundreds of Fairchildren established over six generations of technology ventures. Figure reproduced from Douglas Fairbairn Photography

Table 8.1 PayPal Mafia members and their contributions

Name	Companies founded/role	Contributions
Elon Musk	Tesla, SpaceX, Neuralink, The Boring Company, OpenAI (co-founder)	One of the most influential figures in tech and space exploration. Tesla revolutionized the electric vehicle market, and SpaceX made significant advancements in rocket technology and space travel
Peter Thiel	Palantir Technologies, Founders Fund (venture capital firm)	Prominent venture capitalist and entrepreneur. Palantir Technologies specializes in big data analytics, and the Founders Fund has invested in several successful tech startups
Max Levchin	Affirm, Slide	Affirm provides financial services and installment loans for consumers. Slide was a social media app acquired by Google
Reid Hoffman	LinkedIn	Co-founded LinkedIn, the world's largest professional networking site. Partner at venture capital firm Greylock Partners
Yishan Wong	Former CEO of Reddit	Served as the CEO of Reddit, one of the largest online communities and social news aggregation sites
Jeremy Stoppelman	Yelp	Co-founded Yelp, a platform connecting people with local businesses through user-generated reviews and ratings
Roelof Botha	Partner at Sequoia Capital	Prominent venture capitalist who has funded several successful companies, including YouTube and Instagram
Steve Chen and Chad Hurley	YouTube	Co-founded YouTube, the world's largest video-sharing platform, acquired by Google in 2006
Russel Simmons	Yelp	Co-founder of Yelp, alongside Jeremy Stoppelman
David Sacks	Yammer, Geni	Founded Yammer, a social network for businesses acquired by Microsoft. Also founded Geni, a genealogy website

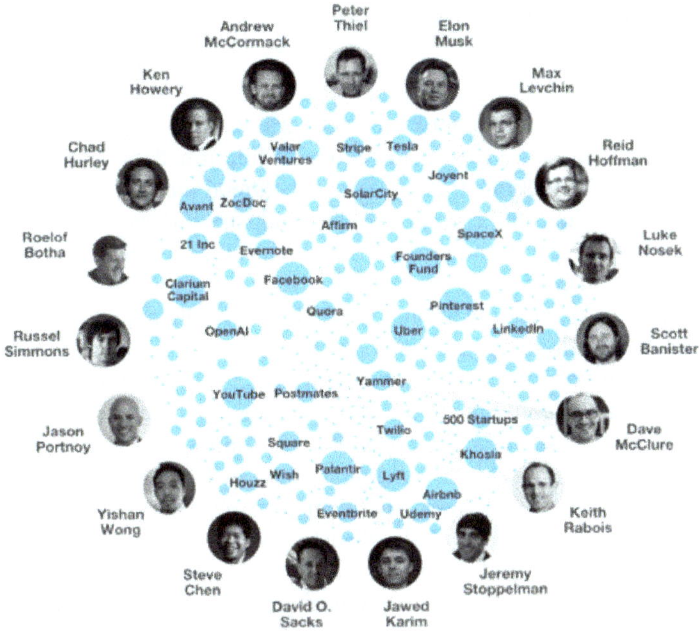

Fig. 8.2 **Paypal Mafia**. The tech entrepreneurs have continued to work and invest together since the company's 2002 sale to eBay. Figure reproduced from public domain

their contribution to the tech space and the most famous companies they founded afterward.

As the reader can have inferred by Table 8.1 and Fig. 8.2, the PayPal Mafia has also been credited with founding and leading a wide array of successful disruptive companies across various sectors. Their ventures have collectively reshaped the technology landscape, impacting industries ranging from space travel to social networking, finance, and artificial intelligence. Tesla has become the leader in electric vehicles and renewable energy, SpaceX focuses on reducing space transportation costs to enable the colonization of Mars. LinkedIn is the largest professional networking site. Palantir Technologies provides data analysis software for governments and businesses, while Yelp is a popular platform for business reviews. Most of us use YouTube daily, as it is the leading video-sharing platform, and a few enjoyed a purchase thanks to Affirm, which offers installment loans to consumers at the point of sale. The list goes on, but we will stop by saying that the very same OpenAI, one of the leading companies in the current artificial intelligence revolution, was founded by ex-Paypals. The PayPal Mafia's legacy continues to grow as it supports and inspires new generations of entrepreneurs and innovators.

8.2.3 Networks in Venture Capital

We have seen that networks are the elephant in the room in venture capital. Indeed, the startup ecosystem is a tightly interwoven network where entrepreneurs and venture capitalists (VCs) constantly seek connections. While founders dedicate significant time to securing introductions to VCs, investors attend numerous events hoping to discover the next billion-dollar company. This industry thrives on seemingly minor connections: shared event attendance, participation in the same accelerator programs, or co-investments in promising or struggling ventures. Personal relationships also play a crucial role. Co-founding a company or sharing an alma mater can forge strong bonds between individuals. Additionally, startups create interconnections by becoming customers of one another's products and services. At the core of this complex network are investors and entrepreneurs whose interactions drive company growth and innovation. This ecosystem facilitates the flow of funding, information, and expertise through a diverse array of connections formed in various ways. This overview examines the current state of research in this field, highlighting the challenges faced by scholars and suggesting potential avenues for future study. The curious reader could refer to [2–4] for further reference.

Evidence has shown that the very same genesis of companies is intricately tied to human connections. Becker et al. [5] delves into founders' motivations, uncovering a shared passion for innovation and funding. While the entrepreneurial journey can be solitary, the allure of funding is contagious among co-founders. Surveying a cohort of founders, 34% of observed similarity is explained by initial homophily selection, while social contagion accounts for 57%. This selection process begins early, as evidenced by Ebbers and Wijnberg [6], who found that children valuing commercial success over expert review tend to form networks that evolve into entrepreneurial networks. On the same line, relational cohesion theory, quantifying the impact of connections, highlights the pivotal role of equal power and mutual dependence in fostering positive exchanges and commitment in entrepreneurial team formation [7]. Furthermore, Ref. [8] demonstrates how gender can play a role in the impact of connections, proving the existence of gender-based differences in the degree to which increased exposure to investors can address networking frictions in venture capital.

Once a company is looking for capital, it takes part in a two-sided matching game. On one side, the company wants to take on board the investor who fits the most, on the other side, the investor who is searching for the next venture with the greatest potential. Of course, the entrepreneur is also willing to maximize the resources it collects (financial and not), while the venture capitalist aims at getting the most equity in the most interesting company at the lowest cost. Given all the constraints and the business of a market filled with lemons, the role of networks increases its relevance, as networks help build trust and share risk [9].

Venture capital investment is deeply rooted in complex networks that shape every stage of the process, from deal sourcing to evaluation and syndication (Fig. 8.3, [10]). Deal sourcing and screening mark the initial phase where investors seek out

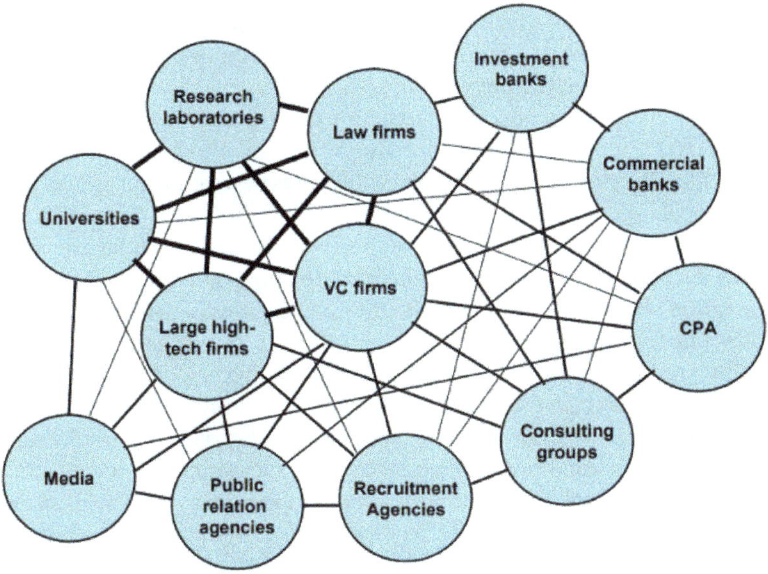

Fig. 8.3 The complex network of innovation of Silicon Valley. This figure illustrates the interconnected ecosystem of Silicon Valley, showcasing the diverse entities–from universities and research labs to VC firms and banks–that collaborate and contribute to the region's innovation dynamics. Figure reproduced from [10]. Copyright © 2009

potential opportunities, leveraging connections to access promising ventures and gather information [11, 12]. Entrepreneurs predominantly rely on informal contacts for finding resources, with cold contacts being the second most utilized source [9].

The evaluation and due diligence phases involve in-depth analysis of business plans, financial projections, and market potential. Investors tap into their networks to gather insights, validate assumptions, and assess risks [13]. As opportunities progress, lead venture capital firms may reach out to potential co-investors, initiating the syndication process [14].

Syndication dynamics play a crucial role in venture capital investments. Investors engage in syndication for refined selection processes and augmented value derived from diversified skills [15]. Research by Gompers et al. [16] revealed that venture capitalists with shared backgrounds tend to syndicate. However, high-affinity syndicates may demonstrate diminished decision-making efficacy in early-stage investments.

The evolution of syndication networks is influenced by both structural and relational embeddedness. Gu et al. [17] found that VC networks evolve based more on structural rather than relational embeddedness, emphasizing the importance of common interests over shared expectations and trust. Trust and identification become particularly crucial in the face of institutional uncertainty [18].

Syndication has been shown to enhance success rates and fund performance. Brander et al. [19] attributed this to complementary management skills, while Das et al. [20] confirmed positive effects on internal rate of return, successful exit, and time taken to exit. The role of structural holes within syndication networks introduces an intriguing competitive perspective, with Burt [21] emphasizing that their relative advantage depends on market uncertainty.

The consequences of syndication breakups can be significant and long-lasting. Zhelyazkov and Gulati [22] revealed enduring negative impacts on the abandoning party's likelihood of entering future co-investment relationships, affecting both global and local reputations.

Throughout the venture capital ecosystem, networks play a pivotal role in shaping investment decisions, facilitating information exchange, and influencing outcomes. From the initial stages of deal sourcing to the complexities of syndication dynamics, these interconnected relationships form the backbone of successful venture capital investments and entrepreneurial growth. Understanding and navigating these networks is crucial for both investors and entrepreneurs seeking to thrive in this dynamic landscape. However, the networks' role extends well beyond the initial investment, as does the venture capital involvement, encompassing strategic guidance, network leveraging, and active participation in company operations. Post-investment, VCs play a crucial role in shaping the trajectory of portfolio companies through their networks and expertise [23, 24].

Strategic alliances are a key aspect of this involvement, with VCs facilitating connections and fostering trust. Lindsey [25] found that companies sharing the same VC are more likely to form alliances. These connections can signal resource quality and facilitate future partnerships [26]. Furthermore, innovation exchanges within VC portfolios are another critical area, with ventures sharing assets, processes, and knowledge [27]. However, this exchange can be complex, with potential risks of competitive information leakage [28].

Also, follow-on investments are heavily influenced by VC networks, impacting due diligence, risk assessment, and future success forecasting. The decision to participate in later funding rounds is shaped by connections to other VCs, industry experts, and potential partners. Similarly, exit strategies, whether through IPO or acquisition, rely heavily on VC networks. Ties to investment banks, market makers, and potential corporate buyers can significantly impact the exit process and valuation [29].

The global perspective of VC networks reveals variations in network density across countries [30], with networks playing a crucial role in emerging economies with weak formal institutions [31]. Throughout these stages, the interplay between VCs' networks and portfolio companies' operations underscores the importance of understanding and leveraging these connections for sustained entrepreneurial success.

Advancing methodological frontiers in venture capital research is crucial to capturing the intricate dynamics of venture capital ecosystems, which is essential for creating a vibrant entrepreneurial connection in every country. The authors identify three key areas for advancement: multilayer network analyses, high-order interactions, tailored clustering techniques, and generative network models.

Multilayer network analyses explore interdependencies between different types of ties, such as social connections and co-investment relationships. This approach can reveal patterns of network alignment and misalignment across layers, offering a more comprehensive understanding of the ecosystem's dynamics [32].

High-order interactions move beyond traditional dyadic analyses by incorporating relationships involving three or more entities. This approach can uncover nuanced patterns in collaborative or competitive relationships, particularly in syndicate formation and strategic alliances. Our group developed a package to study higher-order dynamics, available at https://github.com/GiorgioMB/auto_dotmotif/tree/main.

Tailored clustering techniques capture the specific characteristics of the VC ecosystem by incorporating node attributes and handling bipartite or multipartite networks. These methods can reveal meaningful subgroups and patterns within the broader network structure.

Dynamic and generative network models offer valuable tools for understanding the temporal evolution of venture capital networks. These models can help identify critical events, turning points, and underlying mechanisms that shape network formation and evolution [33]. We refer to Zava and Caselli [4] for further discussion on the topic.

Cross-disciplinary collaboration between venture capital research and network science is emphasized as a means to drive theoretical and methodological innovations. Integrating diverse perspectives can inform policy decisions, industry practices, and entrepreneurial strategies. These methodological advancements provide deeper insights into the venture capital ecosystem, enabling researchers and practitioners to understand better and navigate its complexities. By leveraging these approaches, stakeholders can gain a more nuanced understanding of the drivers of success, innovation, and growth within the entrepreneurial ecosystem.

8.2.4 Do VCs Determine Companies' Access to Funding and Success?

In the paper titled "Who brings investments? Venture Capital through reputation, networks, and influence", Zava and Caselli [4] examined how reputation and networks of influence among venture capitalists (VCs) impact startups' access to funding. They coined the Funding Attraction Index (FAI), a novel metric designed to measure an investor's ability to attract subsequent investments to their portfolio companies. The research leverages the fact that quality in startups is challenging to quantify objectively, making the role of investor reputation and networks crucial in determining access to resources [34].

The study concludes that early investors play a crucial role in a startup's journey, with their reputation and networks significantly influencing access to follow-on funding. The FAI metric provides a more dynamic and accurate measure of an investor's influence, emphasizing the importance of continuous adaptation in the fast-evolving venture capital landscape.

The main findings are three. First, early influence relevance. It emerged that the influence of investors is most significant in the earliest funding stages (seed and series A). Further, early-stage investors' reputation plays a critical role when the startup's future is uncertain, and the company has yet to prove its viability. Second, long-term impact. While the influence of early investors fades in later funding stages (series B and C), the latent effects persist, often resulting in larger amounts of capital being raised. This indicates that the initial backing by reputable investors can have a long-lasting impact on a startup's ability to attract substantial follow-on funding. Lastly, comparison with traditional metrics. The FAI outperforms traditional reputation measures (such as experience and success metrics) in predicting access to follow-on funding and the amount of capital raised. Traditional metrics tend to accumulate success over time without accounting for the dynamic nature of the startup ecosystem.

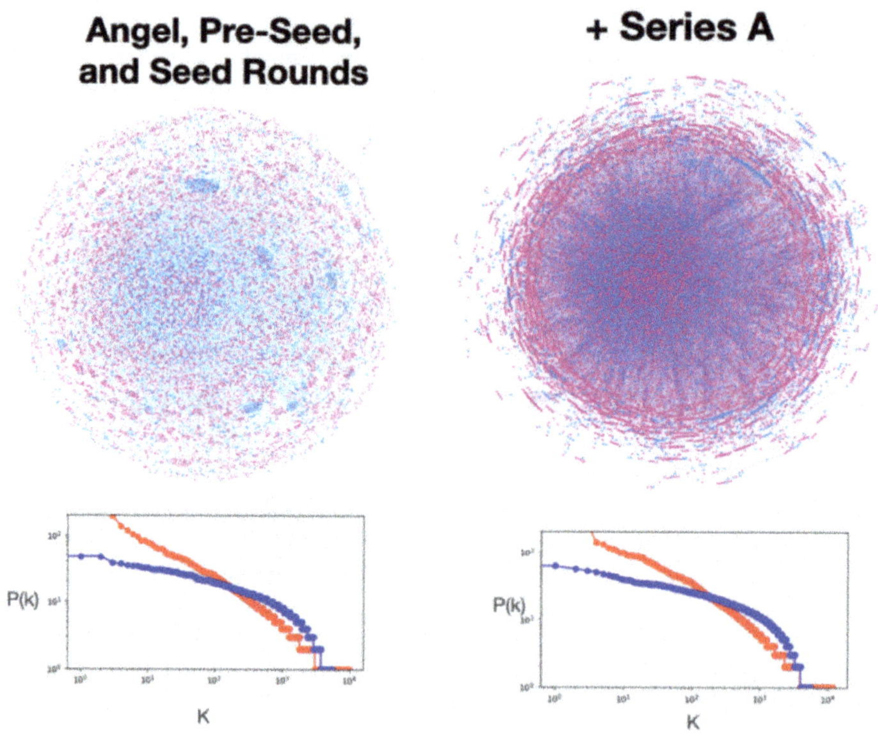

Fig. 8.4 Network representations of Seeding Funding (Angel, Pre-Seed, and Seed) and Startup Funding (Series A). Nodes are investors (in red) and companies (in blue) connected by the investment ties and the respective degree distribution

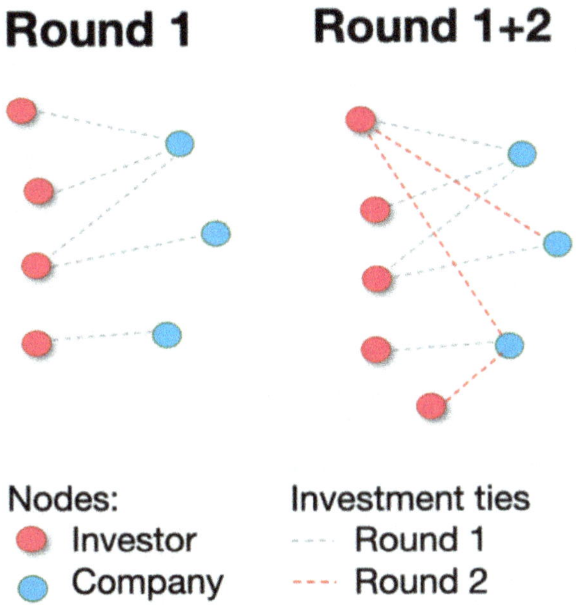

Fig. 8.5 Dynamic Bipartite Network. Theoretical Construction. The figure displays the evolution of investment ties between investors and companies across two funding rounds, demonstrating how the network structure changes as new investments are made. Zava and Caselli [4]

The authors use a robust dataset, primarily from Crunchbase, covering startups and investors located in the USA (networks depicted in Fig. 8.4). Moreover, the analysis includes company fixed effects and pre-money valuation variables to increase explanatory power. The robustness of the analysis is guaranteed as the results are consistent across different geographical regions and account for potential biases such as survivorship and nondisclosure of private deals.

Nevertheless, the study acknowledges that some investors possess superior skills in nurturing early-stage companies, which can contribute to their high FAI scores. These skills encompass mentorship, strategic planning, operational expertise, and insights into market trends. Further potential limitations include unobserved investor preferences and the difficulty in separating investors' impact from preexisting technological advancements.

From a methodological perspective, the paper describes a novel methodology for analyzing the dynamics of venture capital investments using network science. The authors introduce a dynamic bipartite network model (Fig. 8.5) to represent the relationships between early investors and companies over time. This model captures complex temporal patterns such as burstiness, memory, and non-stationarity in funding activities.

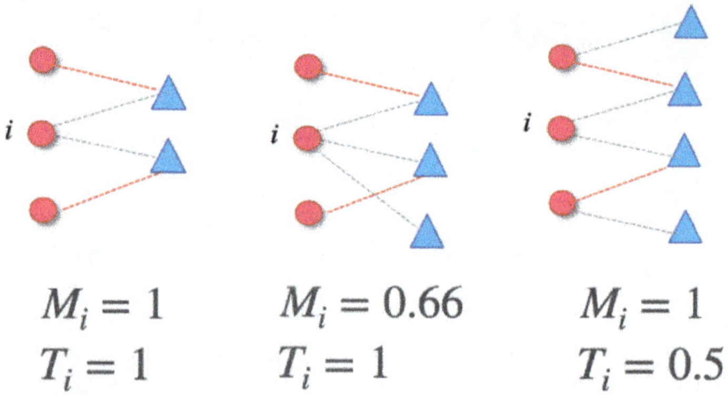

Fig. 8.6 Comparison between the Funding Attraction Index and the Adapted Topological Score. The grey ties indicate a connection at time t, and the red ones at $t + 1$. Investors are represented by the red circles and companies by the blue triangles [4]

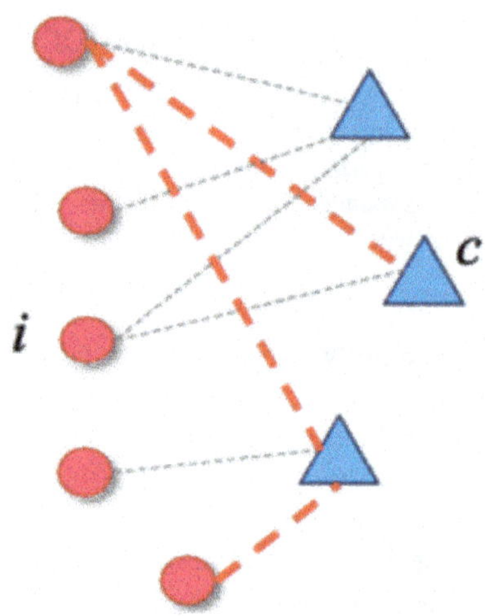

Fig. 8.7 Transitive Fitness Model. In the Transitive Fitness Model, at $t + 1$, node c takes the value of node i at time t. Nodes c and i connected at time t. Zava and Caselli [4]

The key innovation is the development of the FAI to quantify the investor's ability to catalyze future investments in their portfolio companies. This index is based on a modified version of the topological overlap matrix [35], extended to incorporate bipartite, temporal, and community dimensions. The FAI measures how often an investor's early-stage investments are followed by other investors in subsequent rounds. A comparison between the two scores is displayed in Fig. 8.6.

The authors also introduce a "transitive fitness measure" (Zc) for companies, which represents their appeal to investors based on the characteristics of their previous investors. This measure is inspired by fitness models in complex network theory [36] but innovates by determining a node's fitness through the properties of its connected nodes at an earlier time point, as displayed in Fig. 8.7.

The methodology allows for a month-by-month analysis of investment patterns from 2010 to 2021, capturing the evolution of investor influence and company attractiveness over time. This approach offers advantages over static network analyses and traditional reputation measures in venture capital research. This methodology provides a nuanced framework for understanding the complex dynamics of early-stage funding, offering insights into how investor reputation and network position influence future investment patterns. Capturing the temporal and relational aspects of venture capital networks has the potential to enhance our understanding of investment decisions and outcomes in entrepreneurial finance.

To validate their novel measures, the authors compare them to classical reputation metrics used in venture capital literature, such as investor experience [37, 38] and success rates [39]. They also incorporate industry notoriety by considering investors' presence on the Forbes Midas list.

Having assessed that some investors have indeed the power to catalyze investments in the startups they back, do they also have the power to make them succeed? Considering exit, intended as getting listed in the stock exchange or getting acquired by another company, as success, the answer is yes. A study conducted in 2011 by Lee et al. [40] shows that yes, they do. Specifically, they suggest that early involvement in an IPO firm's development significantly enhances the positive relationship between a VC's reputation and both initial market reactions and post-IPO operating performance. Nevertheless, rather than considering VC networks, their study focused on reputation as an index combining the total number of portfolio companies a VC invested in; the total funds invested in portfolio firms; the total dollar amount of funds raised; the number (count) of individual funds raised; the number of portfolio firms taken public; and VC age in the focal year (calculated as the IPO year minus the year a VC firm raised its first fund). Thus, the ex-post performance of startups backed by high-attraction investors, including outcomes like going public, being acquired, or achieving unicorn status, can still be explored.

8.2.5 How do Investors Become Influential?

Zava and Caselli [4] digs into the dynamics of venture capital (VC) networks and how investors can move from peripheral positions to more central, influential roles. The study focuses on the concept of k-shell decomposition (Sect. 1.2.1.2) in syndicated investment networks, which provides a measure of an investor's position within the network structure.

The research begins by establishing a correlation between an investor's k-shell position and their success rate in terms of portfolio company exits (either through public listing or acquisition). This finding builds upon previous work that linked degree centrality to investor success, suggesting that network position is indeed a crucial factor in VC performance.

The central question of the study is how investors in peripheral k-shells can move to core k-shells, effectively becoming more influential within the VC ecosystem. The author constructs a temporal syndication network of US VC institutions, analyzing it month by month to track changes in investors' positions.

The paper outlines three potential paths for a VC in a lower k-shell (periphery) to ascend to a higher, more central position:

1. Co-invest with an influential VC: This involves a peripheral investor participating in a syndicated investment alongside a more established core investor. While this can provide valuable connections and potential access to deal flow, it appears to be the least impactful method for gaining influence.
2. Back an investment of an influential VC: In this scenario, a peripheral investor provides follow-on funding to a company already backed by a core VC. This approach demonstrates the peripheral VC's ability to identify promising investments and align with established players, but its impact on influence is moderate.
3. Have one of its investments backed by an influential VC: This scenario, where a company in the peripheral VC's portfolio receives later-stage funding from a core VC, emerges as the most effective path to gaining influence. Interestingly, the success of the underlying connecting company is less critical than the fact that a core VC has chosen to invest in it.

The study finds that scenario 3 has the most significant impact on a peripheral VC's rise to influence, regardless of the eventual success of the connecting company. This suggests that the act of having a core VC validate a peripheral VC's investment choice carries more weight in the network than the ultimate financial performance of that investment.

This finding has profound implications for the VC industry. It highlights the niche-like nature of venture capital, where reputation and network effects can sometimes outweigh pure financial performance. For new or less established VCs, it suggests that careful portfolio curation and the ability to attract follow-on investments from core VCs may be more crucial for gaining influence than simply achieving a high success rate in exits.

The paper concludes by emphasizing the hope this offers to novel investors aiming to become successful. By understanding these network dynamics, new entrants to the VC world can strategically position themselves for growth and increased influence. The author anticipates that this work will serve as a foundation for developing new investment strategies that quantitatively incorporate both financial and social components.

This research not only provides insights into the inner workings of the VC industry but also opens up new avenues for studying and optimizing investment strategies. By demonstrating the importance of network position and the specific mechanisms through which influence is gained, the paper offers a roadmap for aspiring VCs and a fresh perspective on what drives success in the venture capital ecosystem.

8.2.6 Dear Investors, are You Missing Out?

Zava [41] examines the investment patterns of Midas investors, the top 100 venture capitalists nominated yearly by Forbes. Using a novel temporal network model, the study analyzes interactions among VCs, Midas investors, companies, and founders in the US venture capital ecosystem. The research reveals that Midas investors are responsible for about one-third of VC money invested in the US, and companies they back tend to have higher success rates and raise more funding in subsequent rounds. However, the study's most intriguing finding is that Midas investors exhibit a strong preference for companies within their "niche." This niche is primarily determined by pre-existing relationships: founders who previously received Midas funding, co-founders who worked with Midas-backed entrepreneurs, companies backed by investors who have syndicated with Midas before, and investors who have previously attracted Midas to their portfolio companies.

The title "Dear investors, you are missing out" aptly captures the paper's central insight. By focusing so heavily on their existing networks and relationships, Midas investors are potentially overlooking a vast array of promising startups outside their immediate circle. The authors argue that it's difficult to believe these investors are truly selecting the best companies in the ecosystem, given how limited their pool of consideration is.

This behavior suggests that a company's destiny can be set before its birth, based more on pre-existing connections than on merit alone. The paper concludes with a stark message to investors: they are missing out. By limiting themselves to a small, interconnected network of familiar entrepreneurs and companies, top investors may be passing up valuable opportunities in the broader startup ecosystem. In essence, the title encapsulates the paper's critique of the current investment practices among top VCs. It serves as a wake-up call, suggesting that by broadening their horizons and looking beyond their established networks, investors could potentially discover untapped sources of innovation and returns.

8.2.7 Valuation

Berre [42] addresses how the challenges and methodologies of valuing startups, particularly in the context of venture capital investments in the EU and EEA regions, can be addressed with the use of machine learning. It acknowledges the complexity and opacity inherent in startup valuations, driven by their short histories and intangible assets, which traditional economic theories often fail to account for accurately. The study aims to identify the most critical factors influencing startup valuations through the use of tree-based hierarchical models and supervised machine learning techniques.

Startups often exhibit significant valuation differences across various geographic locations and industries, disparities not fully explained by conventional metrics like revenues, growth rates, and risk-adjusted discount rates. The scarcity of comprehensive data exacerbates these valuation challenges, necessitating innovative empirical research approaches. Traditional econometric techniques, such as OLS, face limitations due to hidden-variable bias, particularly when accounting for diverse categories, regions, and clusters. This study proposes a novel methodology combining firm-performance indicators, market conditions, and predictive segmentation of categorical variables to enhance valuation accuracy.

The research is grounded in an extensive dataset comprising 1,089 venture capital investments involving 1,042 unique startups within the EU and EEA. The study employs fixed-effects and joint-fixed-effects regressions alongside non-parametric approaches that merge econometric and machine learning techniques. This dual approach enables a detailed exploration of microeconomic, sectoral, and local-level impacts on startup valuation.

Machine learning role in finance and entrepreneurial finance has been increasingly validated, with techniques such as LASSO, LDA, CART, XGBoost, and Bayesian approaches like NUTS and MCMC demonstrating significant predictive power. These methods are particularly valuable in handling high-dimensional data, providing accurate predictions even in complex macrofinancial and macroeconomic contexts. Policy-level applications, such as those by the ECB and in competition policy, further underscore the efficacy of decision-tree-based machine learning techniques.

The study's methodology section details the construction of the dataset, which integrates proprietary venture-capital deal data from Early Metrics with supplementary data from EIKON and Crunchbase. The dataset encompasses firm-performance metrics, industry-level data, and macroeconomic indicators, allowing for a comprehensive analysis of valuation determinants. The use of a European dataset offers the advantage of institutional and macroeconomic diversity, which is essential for examining the impact of legal origins and contextual factors on startup valuations.

The research employs pre-money valuation as the primary dependent variable, with valuations expressed in euros. The dataset spans various significant economic events, including the dotcom bubble's aftermath, the Eurozone crisis, and the onset of the COVID-19 pandemic, providing a robust temporal and sectoral distribution. The study's regression models accommodate multiple observations per startup, reflecting the diverse investor base and enhancing the analysis's robustness.

In summary, this paper contributes to the understanding of startup valuations by leveraging advanced machine learning techniques to dissect the factors that matter most. It addresses the limitations of traditional valuation methods and highlights the potential of hierarchical ranking and predictive segmentation in improving valuation accuracy. The integration of categorical variables and contextual factors provides a nuanced perspective on startup valuations, offering valuable insights for both academic researchers and industry practitioners in the venture capital space.

8.3 Artificial Intelligence in M&A

Mergers and Acquisitions (M&A) play a vital role in finance by facilitating corporate growth, market consolidation, and value creation. Through M&A transactions, companies can rapidly expand their market presence, achieve economies of scale, and access new technologies or talent pools. These strategic moves often result in increased operational efficiency, enhanced competitive positioning, and the potential for significant synergies. Moreover, M&A activities can drive shareholder value by enabling firms to diversify their revenue streams, mitigate risks, and capitalize on emerging market opportunities. As such, M&A serves as a crucial mechanism for corporate restructuring and economic dynamism in the global financial landscape. The integration of artificial intelligence technologies in the Mergers and Acquisitions process offers numerous opportunities for enhancing efficiency, accuracy, and strategic decision-making. From streamlining due diligence and improving valuation accuracy to facilitating post-merger integration and ensuring regulatory compliance, AI has the potential to transform every stage of the M&A lifecycle. Nowadays, the greatest success story when considering M&A and artificial intelligence is the one of Shunsaku Sagami and his M&A Research Institute.

The chapter develops by narrating the M&A Research Institute story and explaining its use of AI. Then, it continues by depicting the many business applications that AI has in the field of M&A. It then concludes by illustrating the Accelerated M&A Process developed by academics Parikh et al. [43].

8.3.1 The M&A Research Institute

Shunsaku Sagami, a prominent Japanese businessman, founded the M&A Research Institute in 2006. The institute has since become a cornerstone in the M&A landscape in Japan, known for its unique and strategic approach to mergers and acquisitions. Sagami's vision was to create a comprehensive advisory service that could leverage data-driven insights to facilitate successful M&A activities. Sagami established the M&A Research Institute with a clear mission: to support companies in their growth strategies through mergers and acquisitions and to help the aging entrepreneurial generation sell their companies at a fair price. The institute offers a wide range of services, including market research, financial analysis, strategic planning, and post-merger integration support. By focusing on a data-driven approach, the institute ensures that every transaction is backed by thorough research and analysis, minimizing risks and maximizing value for all parties involved. Shunsaku Sagami is now one of the youngest self-made billionaires.

Under Sagami's leadership, the M&A Research Institute has facilitated numerous high-profile mergers and acquisitions across various industries. For instance, the institute has helped several Japanese companies expand their market presence by acquiring foreign firms, enabling them to enter new markets and diversify their product offerings. Further, through strategic acquisitions, the institute has assisted technology companies in enhancing their capabilities, driving innovation, and staying ahead of the competition. Also, the institute has played a crucial role in industry consolidation efforts, helping companies achieve economies of scale and improve their competitive positioning through strategic mergers.

Through the M&A Research Institute operations, we can identify several interesting artificial intelligence applications to the M&A activity–such operations work beyond the realm of academic curiosity, having real-world business applications. Initially, the institute conducts comprehensive market research to identify potential acquisition targets that align with the client's strategic goals. This involves analyzing industry trends, competitive landscapes, and market opportunities to ensure that the target company fits well within the client's growth strategy. Consequently, a detailed financial analysis is performed to assess the financial health and valuation of potential targets. This includes examining financial statements, revenue streams, cost structures, and growth potential. The institute uses advanced financial modeling techniques to project future performance and assess the feasibility of the transaction.

Once a potential target is identified, the institute helps clients develop a strategic plan for the acquisition. This includes defining the objectives of the acquisition, identifying synergies, and outlining a roadmap for integration. The goal is to ensure that the acquisition aligns with the client's long-term vision and adds significant value to their business. As a last step, thorough due diligence is conducted to identify any potential risks or liabilities associated with the target company. This involves reviewing legal, financial, and operational aspects to ensure that there are no hidden issues that could impact the success of the transaction. Nevertheless, the job is not finished: the key aspects of post-merger integration are still missing. The institute provides support during the post-merger integration phase to ensure a smooth transition. This includes aligning corporate cultures, integrating operations, and realizing synergies. The focus is on creating a unified entity that can operate efficiently and achieve the desired outcomes.

Shunsaku Sagami's M&A Research Institute has made a significant impact on the M&A landscape in Japan. By combining rigorous research, financial expertise, and strategic planning, the institute has successfully guided numerous companies through complex transactions, helping them achieve their growth objectives. Sagami's visionary leadership and commitment to excellence continue to drive the institute's success, solidifying its reputation as a leader in the field of mergers and acquisitions and showing many business applications of AI.

8.3.2 *Business Applications of AI in M&A*

As we have seen in the previous section, in recent years, the field of M&A has witnessed a significant transformation due to the integration of artificial intelligence technologies. This section explores the various ways in which AI can be applied to improve the M&A process, focusing on key areas such as due diligence, market analysis, valuation, deal sourcing, post-merger integration, legal and compliance, decision support, and cybersecurity.

One of the most time-consuming and critical aspects of the M&A process is due diligence. AI has the potential to revolutionize this phase by automating document analysis and enhancing risk assessment. AI-powered tools can rapidly analyze vast amounts of documents, including contracts, financial statements, and legal records, providing a more comprehensive and efficient due diligence process. The application of Natural Language Processing (NLP) techniques allows AI systems to extract and understand information from both structured and unstructured data sources. This capability is particularly valuable when dealing with complex legal documents or financial reports. By automating the analysis of these documents, AI can significantly reduce the time and resources required for due diligence while minimizing the risk of human error.

Furthermore, AI algorithms can be employed for anomaly detection and risk assessment. By analyzing historical data and identifying patterns, AI systems can highlight potential red flags and assess risks more accurately. This enhanced risk assessment capability enables decision-makers to make more informed choices throughout the M&A process.

Moving forward to market analysis, AI's role is twofold: it enhances predictive analytics capabilities and streamlines competitor analysis. In terms of predictive analytics, AI algorithms can analyze market trends, customer behavior, and industry data to provide valuable insights into potential target companies and market conditions. This predictive capability allows organizations to make more informed decisions about potential acquisitions or mergers.

AI can also automate the process of gathering and analyzing data on competitors, helping companies better understand the competitive landscape. By monitoring and analyzing competitors' activities, including their marketing efforts and product developments, AI systems can provide real-time insights that inform strategic decision-making.

Another crucial aspect of market analysis in M&A is the assessment of intellectual property (IP) and patent data. AI technologies can efficiently process and analyze large volumes of patent information, helping organizations evaluate the IP portfolios of potential acquisition targets. This analysis can uncover valuable assets, identify potential synergies, and assess the overall innovation potential of target companies.

Furthermore, the valuation process in M&A can be significantly improved through the application of AI technologies too. AI algorithms can assist in predicting future financial performance, enabling more accurate company valuations. By leveraging machine learning techniques, these algorithms can analyze historical financial data, market trends, and other relevant factors to forecast future performance with greater precision.

Moreover, AI systems can process and analyze real-time data, ensuring that valuations are based on the most up-to-date information available. This real-time analysis capability is particularly valuable in dynamic market conditions, where traditional valuation methods may struggle to capture rapid changes in company performance or market sentiment.

AI technologies can also revolutionize the deal-sourcing process by leveraging NLP and algorithmic matching techniques. NLP-driven tools can analyze vast amounts of textual data, including news articles, social media posts, and industry reports, to identify potential acquisition targets or strategic partnership opportunities. This capability allows organizations to cast a wider net in their search for suitable targets and uncover opportunities that might otherwise be overlooked.

Furthermore, AI algorithms can be employed to match buyer criteria with potential targets, streamlining the deal-sourcing process. By automating the initial screening and matching process, AI can help organizations focus their resources on the most promising opportunities, improving the efficiency of the overall M&A process.

On a different note, the post-merger integration phase is critical for the success of any M&A transaction. AI can play a significant role in facilitating a smoother integration process by automating routine tasks and reducing operational inefficiencies. Workflow automation powered by AI can streamline various aspects of the integration process, from data migration to process harmonization.

AI-powered tools can also assist in managing and integrating the workforce during the post-merger phase. By analyzing employee data, skills, and performance metrics, AI systems can help predict potential challenges in workforce integration and provide solutions for a seamless transition. This capability is particularly valuable in large-scale mergers where cultural and operational differences can pose significant challenges.

Further, in the complex regulatory landscape of M&A transactions, AI can provide valuable support in ensuring legal compliance and mitigating risks. AI-powered contract review tools can analyze large volumes of legal documents, identifying potential issues and ensuring compliance with regulatory requirements. These tools can significantly reduce the time and resources required for comprehensive contract review while minimizing the risk of overlooking critical legal aspects.

Moreover, AI systems can help organizations track changes in regulations and ensure that the M&A process complies with evolving legal requirements. By continuously monitoring regulatory updates and analyzing their potential impact on the transaction, AI can help organizations stay ahead of compliance challenges and adapt their strategies accordingly.

Nevertheless, one of AI's most significant contributions to the M&A process is its ability to provide decision-makers with actionable insights derived from vast datasets. By processing and analyzing large volumes of structured and unstructured data, AI systems can uncover patterns, trends, and correlations that may not be apparent through traditional analysis methods.

These data-driven insights can inform various aspects of the M&A process, from target selection and valuation to post-merger integration planning. By providing decision-makers with a more comprehensive and nuanced understanding of the factors influencing the transaction, AI enables more informed and strategic decision-making throughout the M&A lifecycle.

In an era of increasing cyber threats, ensuring the security of sensitive information during the M&A process is paramount. AI can play a crucial role in assessing and enhancing the cybersecurity posture of both acquiring and target companies. AI-powered security tools can analyze network traffic, detect anomalies, and identify potential vulnerabilities in real-time, providing a more robust defense against cyber threats.

Furthermore, AI can assist in conducting thorough cybersecurity due diligence, helping organizations evaluate the security practices and potential risks associated with target companies. This assessment is critical for understanding the true value and potential liabilities of an acquisition target, particularly in industries where data security is a significant concern.

As AI technologies continue to evolve, their impact on the M&A process is likely to grow even further. Organizations that embrace these technologies and effectively integrate them into their M&A strategies will be better positioned to navigate the complexities of modern transactions and achieve successful outcomes.

However, it is important to note that while AI offers significant advantages, it should be viewed as a complement to human expertise rather than a replacement. The successful application of AI in M&A requires a thoughtful approach that combines technological capabilities with human judgment and industry knowledge. As the field continues to evolve, further research and practical applications will be necessary to fully realize the potential of AI in revolutionizing the M&A landscape.

Table 8.2 provides a useful summary of the ways in which AI can be applied to improve the M&A process and offers the names of key companies operating in the field as an example.

8.3.3 Accelerated M&A Process

Parikh et al. [43] worked on the Accelerated M&A Process (AMP), a protocol leveraging machine learning and data mining to automate key aspects of Mergers and Acquisitions (M&A), including legal, valuation, company identification, financials, and due diligence. The research explores case studies of AI in M&A transactions, particularly in due diligence, and includes interviews with industry experts discussing challenges and the future of traditional acquisition processes. The paper begins by outlining the history of M&A, detailing seven historical waves, each characterized by specific trends and economic conditions. It then proposes an eighth wave: The AI-Driven and Machine Learning Era, where AI technologies are not just part of the deal but are driving the deal-making process itself. The authors highlight several challenges in traditional deal-making, including time-consuming document examination, difficulty in gauging market perception and sentiment, and limitations in effectively managing and organizing knowledge. They underscore the challenges of deriving insights from unstructured data in M&A processes.

The paper then delves into leveraging NLP for streamlining M&A processes. NLP emerges as a powerful tool for extracting meaningful insights from vast amounts of textual data in M&A transactions. As we mentioned, it can automate tasks such as document classification, information extraction, sentiment analysis, and entity recognition.

Table 8.2 AI applications in M&A processes

Area/AI application	Examples
Due diligence	
Automated Document Analysis: AI-powered tools can quickly analyze vast amounts of documents, contracts, financial statements, and legal records, providing a more comprehensive due diligence process	• Kira Systems: AI-powered contract analysis, extracting relevant information from large sets of legal documents • Leverton: Uses machine learning to analyze and extract data from real estate and corporate documents • Luminance: Employs natural language processing to enhance due diligence by automating the review of legal documents
Risk Assessment: AI algorithms can assess potential risks by analyzing historical data, identifying patterns, and highlighting red flags, allowing for a more informed risk assessment	• IBM Watson: AI solutions provide risk assessment by analyzing historical data and identifying potential risks in M&A transactions • Deloitte's Risk Intelligent: Uses AI to assess financial, operational, and compliance risks in M&A deals • Ayasdi: Leverages machine learning for anomaly detection and risk assessment in the due diligence process
Market analysis	
Predictive Analytics: AI can analyze market trends, customer behavior, and industry data to provide insights into potential target companies and market conditions	• Kensho Technologies: Applies AI and analytics to predict market trends and assess the impact of events on financial markets • AlphaSense: Uses AI to analyze financial and market data, providing predictive insights for investors • Predata: Employs AI to predict geopolitical and economic events that may affect M&A activities
Competitor Analysis: AI can automate the process of gathering and analyzing data on competitors, helping companies understand the competitive landscape better	• Crayon: Offers AI-driven competitive intelligence, monitoring and analyzing competitors' activities across digital platforms • Kompyte: Uses AI to track and analyze competitors' marketing strategies and product changes • Cipher.ai: Leverages AI to provide competitive intelligence, analyzing patent and intellectual property data
Valuation	
Predictive Modeling: AI algorithms can assist in predicting future financial performance, aiding in more accurate company valuations	• Quantemplate: Uses machine learning for predictive analytics in insurance, helping with risk modeling and valuation • AlgoValue: Applies AI algorithms for business valuation, considering financial and non-financial factors • Drooms: Uses AI in financial modeling and valuation during M&A transactions

(continued)

Table 8.2 (continued)

Area/AI application	Examples
Real-time Data Analysis: AI can process real-time data, ensuring that valuations are based on the latest information available	• Kensho: Real-time event recognition platform uses AI to analyze and respond to market-moving events • Sentieo: Provides AI-driven financial research, enabling real-time analysis of market data • Kenshoo: Leverages AI for real-time analysis and optimization of digital marketing campaigns
Deal sourcing	
Natural Language Processing (NLP): AI-driven tools can analyze vast amounts of textual data, news articles, and social media to identify potential acquisition targets or strategic partnerships	• Mergermarket: Utilizes NLP to extract actionable insights from news and data for M&A professionals • Squirro: Applies AI and NLP for data discovery, helping businesses identify relevant information for deal sourcing • Ayfie: Offers text analytics and NLP solutions to extract insights from unstructured data sources
Algorithmic Matching: AI algorithms can match buyer criteria with potential targets, streamlining the deal-sourcing process	• Axial: AI-driven platform uses algorithms to match buyers and sellers based on their criteria and preferences • DealCloud: Employs AI to enhance deal origination by matching investors with potential opportunities • CapLinked: Utilizes AI algorithms to facilitate secure deal rooms and improve deal matching
Post-merger integration	
Workflow Automation: AI can automate routine tasks, facilitating a smoother integration process and reducing operational inefficiencies	• UiPath: Provides robotic process automation (RPA) solutions to automate repetitive tasks in post-merger integration processes • WorkFusion: Offers AI-driven automation for streamlining business processes, including post-merger integration • Celonis: Uses process mining and AI to optimize workflows during the integration phase
Employee Integration: AI-powered tools can assist in managing and integrating the workforce, predicting potential challenges, and providing solutions for a seamless transition	• IBM Watson Talent Insights: Applies AI to analyze workforce data, facilitating smooth employee integration post-merger • Glint: Uses AI for employee engagement and integration insights • Culture Amp: Employs AI to assess and manage cultural alignment during post-merger integration

(continued)

Table 8.2 (continued)

Area/AI application	Examples
Legal and compliance	
Contract Review: AI tools can review and analyze contracts, ensuring compliance and identifying potential legal issues	• Seal Software: Uses AI-powered contract analytics to review and extract key information from contracts, ensuring compliance • LawGeex: Employs AI to automate the review and approval of contracts, ensuring legal compliance • eBrevia: Utilizes natural language processing to analyze and extract relevant information from legal documents
Regulatory Compliance: AI can help track changes in regulations and ensure that the M&A process complies with legal requirements	• ComplyAdvantage: Applies AI for real-time monitoring of regulatory changes and ensuring compliance in financial transactions • Ayasdi: Regulatory compliance solutions use machine learning to analyze and adapt to changing regulatory requirements • KPMG: AI-driven compliance platform assists organizations in staying compliant with evolving regulations
Decision support	
Data-driven Insights: AI can provide decision-makers with actionable insights by processing and analyzing vast datasets, helping them make more informed strategic decisions	• Tableau: Uses AI to provide interactive and visual analytics, aiding decision-makers in gaining insights from complex datasets • ThoughtSpot: Employs AI-driven search and analytics to provide instant insights from data, supporting decision-making • IBM Cognos Analytics: Utilizes AI for data exploration and visualization, enhancing decision support capabilities
Cybersecurity	
Security Assessment: AI can be used to assess and enhance the cybersecurity posture of both acquiring and target companies, ensuring the protection of sensitive data during the M&A process	• Darktrace: Uses AI for autonomous threat detection and response, enhancing cybersecurity during M&A processes • Palo Alto Networks' Cortex XDR: Employs AI to detect and respond to cyber threats across the network • CrowdStrike: Utilizes AI and machine learning for endpoint security, ensuring the protection of sensitive data during M&A activities

The technical underpinnings of AMP are explained in detail. AMP utilizes advanced NLP models like BERT and GPT-3, implements domain-specific fine-tuning for understanding legal documents in M&A transactions, uses Named Entity Recognition (NER) to identify key entities in legal documents, employs information extraction and summarization techniques, and incorporates continuous learning and improvement through feedback loops.

The authors also address several challenges and considerations in implementing AI and NLP in M&A processes. These include data privacy and security concerns, language nuances and multilingual support, the training and expertise required for NLP models, integration and adoption challenges, the need for continuous improvement and model maintenance, and ethical considerations and bias mitigation.

The paper concludes that the application of NLP and AI in streamlining M&A processes presents a transformative opportunity for the industry. AMP and similar AI-based solutions have the potential to revolutionize the M&A landscape by enhancing efficiency, reducing costs, and improving transaction outcomes. The authors emphasize the importance of embracing these innovations to harness the full potential of NLP and AI in M&A legal processes.

In summary, this comprehensive paper provides an in-depth overview of the current state and future potential of AI and NLP in M&A processes, backed by expert interviews and case studies. It presents AMP as a promising solution to address the challenges in traditional M&A processes, particularly in due diligence, by leveraging advanced AI and NLP technologies. The research underscores the transformative potential of these technologies in reshaping the M&A landscape while also acknowledging the challenges and considerations that need to be addressed for successful implementation.

8.4 Initial Public Offerings (IPO)

To provide a key insight into how initial public offerings are often ruled by a few institutional investors, we offer a deep dive into the investors in Italian IPO deals by focusing on the deals that took place between 2020 and 2021 [44], acknowledging the extraordinary circumstances of the period considered.

As displayed in Fig. 8.8, we focused on the bipartite network created among investors and companies. In the period under examination, a total of 95 different investors played a role in 64 IPOs, often by means of distinct funds. The most connected investors are Azimut, Natam, and Lemanik: they invested in the highest number of different companies. Moreover, investors often invest in the same companies through various entities. On the other side, Ariston and Intercos are the companies that raised capital from the largest investor bases.

It is important to distinguish between the number of investments and the value of the investments. Indeed, the investors that devoted the largest amount of capital are Norges Bank (€129 million), if we consider overall invested capital, and Capital Research Global Investor (€93 million) if we consider the single investment. As a comparison, Azimut Holdings invested an overall €46 million with 26 funds, resulting in a total number of investments (and therefore connections fund-company) equal to 144 and an average investment size of €0.46 million. The fewer the investments, the higher the amounts. When it comes to companies, the situation turns out to be

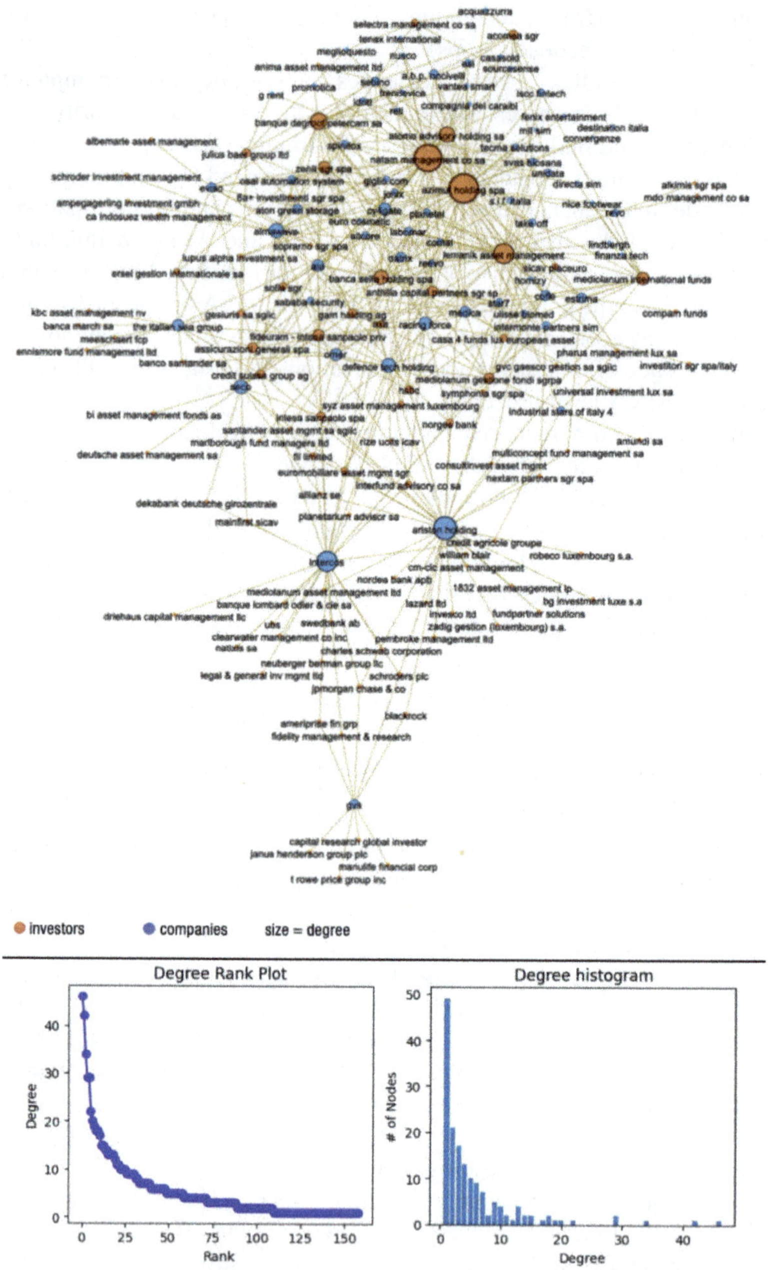

Fig. 8.8 Bipartite network of investors and companies. Data correspond to Italian IPOs (2020–2021). 95 different investors played a role in 64 IPOs, often by means of distinct funds. The most connected investors are Azimut, Natam, and Lemanik: they invested in the highest number of different companies. On the business side, Ariston and Intercos are the companies that raised capital from the largest investor bases. Zava [44]

the opposite: often, the IPOs that are able to gather a higher number of investors are also the ones raising higher amounts of capital. Indeed, Ariston raised €802 million and is the biggest IPO by size in our sample, while Intercos raised €317 million and resulted in the third IPO by size in the sample. GVS, the second IPO by size in our sample (€570 million), raised capital by nine different investors, through 66 funds, confirming our thesis.

The observed scale-free degree distribution among investors indicates a highly skewed participation pattern in IPO deals. This phenomenon manifests as a power-law distribution, where a small number of investors engage in a disproportionately large number of IPO deals while the vast majority participate in only a limited number of offerings. This finding aligns with the broader concept of preferential attachment in complex networks, suggesting that experienced or well-connected investors may have advantages in accessing and participating in a wider array of IPO opportunities.

A similar scale-free distribution is evident when examining the degree of companies in relation to their investors. This parallel structure implies that a select few companies attract a large number of investors while most companies secure investments from a relatively small pool of participants. This asymmetry in investor attraction may be attributed to factors such as company reputation, market sector, or perceived growth potential.

Interestingly, as mentioned, an inverse relationship emerges between the number of investments and the amount invested by individual investors. Those who participate in fewer IPO deals tend to commit larger amounts of capital per deal. This pattern could be indicative of different investment strategies, with some investors opting for a more concentrated approach, carefully selecting a limited number of high-stakes opportunities. In contrast, others pursue a more diversified strategy across numerous smaller investments.

Conversely, when examining the companies' perspective, a positive correlation is observed between the number of investors and the total capital raised. IPOs that successfully attract a larger number of investors often coincide with those raising higher amounts of capital. This relationship suggests that broad investor interest may serve as both a cause and effect of substantial capital accumulation, potentially reflecting market confidence in the offering or the company's ability to generate widespread appeal.

These network-based insights into IPO dynamics offer valuable implications for both academic research and practical applications in finance. They highlight the importance of considering relational structures and network effects in understanding market behaviors, investor decision-making processes, and the success factors of public offerings. Furthermore, these findings may inform strategies for companies planning IPOs, as well as for investors seeking to optimize their participation in such events.

Future research in this area could explore the temporal evolution of these network structures, investigate the role of investor characteristics in shaping network positions, and examine how these patterns may vary across different market sectors or geographical regions. Such studies would further enhance our understanding of the complex ecosystem surrounding IPOs and contribute to more nuanced financial theories and practices.

8.5 Stock Market

While artificial intelligence and network science applications to the stock market are infinite, for the sake of this book, the authors decided to focus mainly on Natural Language Processing (NLP) applications to the stock market, public sentiment extraction, and investment concentration. After accurately depicting the multiplicity of NLP techniques, the sections analyze how public sentiment can be extracted from unusual sources such as Google Trends and YouTube. Lastly, we discuss the stock market's network concentration.

8.5.1 Natural Language Processing (NLP)

As mentioned several times in this chapter, Natural Language Processing (NLP) has become a crucial tool in finance, enabling the analysis of vast amounts of unstructured text data such as news articles, financial reports, social media posts, and more.

The key NLP Models used in the financial field are the following:

1. **FinBERT**
   Description: FinBERT is a domain-specific variant of BERT tailored for financial text. Compared to general NLP models, it provides more accurate sentiment analysis and classification for financial documents.
   Applications: Sentiment analysis, financial text classification, event detection.
2. **BloombergGPT**
   Description: BloombergGPT is an advanced language model developed by Bloomberg for financial NLP tasks. It is trained on a vast corpus of financial data to understand the nuances of financial language.
   Applications: Sentiment scoring, information extraction, document summarization.
3. **GPT-3 and GPT-4**
   Description: General-purpose language models developed by OpenAI that are used for various NLP tasks. They can be fine-tuned for financial applications to generate reports, perform sentiment analysis, and answer financial questions.
   Applications: Text generation, summarization, Q&A systems, sentiment analysis.
4. **XLNet**
   Description: An autoregressive language model that improves upon BERT by capturing bidirectional context. It is used in finance for tasks requiring an understanding of context from both directions in a text.
   Applications: Text classification, sentiment analysis, document summarization.

5. **RoBERTa**
   Description: A robustly optimized version of BERT, which improves performance on a variety of NLP tasks by training with more data and computation. In finance, it is used for high-accuracy text analysis.
   Applications: Sentiment analysis, named entity recognition, text classification.

   Furthermore, other relevant NLP techniques are the ones mentioned below:

1. **Named Entity Recognition (NER)**
   NER Models: Used to identify and classify entities such as company names, stock tickers, financial instruments, and key figures from text data. This helps in structuring unstructured data for further analysis.
2. **Topic Modeling**
   Latent Dirichlet Allocation (LDA): A common technique for discovering the underlying topics in a collection of documents. It is used to analyze earnings calls, financial reports, and news articles to identify prevalent themes and trends.
3. **Text Classification**
   Support Vector Machines (SVM): Used for classifying documents into categories such as buy/sell recommendations, credit ratings, or financial reports.
   Neural Networks: Deep learning models, such as convolutional neural networks (CNNs) and recurrent neural networks (RNNs), are employed for more complex classification tasks.
4. **Information Extraction**
   Relation Extraction: Identifying relationships between entities in text, such as mergers and acquisitions, executive changes, or financial events.
   Event Detection: Detecting and classifying events from text data that could impact financial markets, such as earnings announcements, product launches, or regulatory changes.
5. **Machine Translation**
   Translation Models: Used for translating financial documents and news articles from one language to another. This is particularly important for global financial markets, where information needs to be accessible in multiple languages.
6. **Document Summarization**
   Extractive Summarization: Techniques that identify the most important sentences or paragraphs from a document to create a concise summary.
   Abstractive Summarization: More advanced methods that generate summaries in their own words, understanding the context and meaning of the text.
7. **Question Answering Systems**
   BERT-based Models: Leveraging models like BERT to build systems that can answer financial queries by understanding context and retrieving relevant information from large datasets.
8. **Textual Similarity**
   Embedding Models: Techniques such as word2vec or GloVe to represent words or documents as vectors, which can be used to measure similarity between texts. This is useful for finding comparable companies, documents, or news articles.

NLP techniques have become indispensable in the financial industry, providing insights and automation that enhance decision-making and operational efficiency. From sentiment analysis with FinBERT and BloombergGPT to the versatile applications of GPT-3 and GPT-4, these technologies enable financial institutions to extract value from vast amounts of unstructured text data. As these models and techniques continue to evolve, their impact on finance will undoubtedly grow, driving further innovation and transformation in the industry.

8.5.2 Sentiment in the Stock Market

In this section, we focus on using artificial intelligence to capture and interpret public opinion and sentiment about specific stocks, industries, or innovations. Public sentiment plays a crucial role in shaping stock market dynamics, influencing investor behavior and, consequently, stock prices. This phenomenon is rooted in behavioral finance theory, which posits that psychological factors significantly impact financial decision-making.

Firstly, public sentiment acts as a key driver of market momentum. When sentiment is positive, investors are more likely to buy stocks, driving prices upward. Conversely, negative sentiment can lead to selling pressure and price declines. This self-reinforcing cycle can create market trends that persist beyond what fundamental analysis might suggest is warranted. Moreover, sentiment influences risk perception and tolerance. During periods of high optimism, investors may underestimate risks and overvalue assets, potentially leading to market bubbles. Conversely, excessive pessimism can result in undervaluation and market crashes, as seen in historical events like the 1929 Wall Street Crash or the 2008 Financial Crisis.

Public sentiment also affects market liquidity. High confidence typically correlates with increased trading volume, enhancing market efficiency. In contrast, low confidence can lead to reduced liquidity, potentially exacerbating price volatility and widening bid-ask spreads. Furthermore, sentiment impacts corporate decision-making. Companies may time their initial public offerings (IPOs) or secondary offerings to coincide with periods of positive market sentiment, potentially affecting their ability to raise capital efficiently.

Indeed, public sentiment serves as a critical factor in stock market behavior, influencing prices, volatility, and overall market efficiency. Understanding and measuring sentiment through various indicators (e.g., consumer confidence indices, social media sentiment analysis) can provide valuable insights for investors, policymakers, and researchers seeking to comprehend and navigate financial markets effectively. It is important to note that public sentiment doesn't always align with economic fundamentals. For instance, Jacoby and Liao [45] investigated the influence of sentimental noise traders on the security price adjustment. This disconnect can create opportunities for value investors who capitalize on market inefficiencies caused by sentiment-driven mispricing.

Thus, to tackle the issue, professionals try to extrapolate and understand sentiment in a variety of ways, many of them powered by the introduction of artificial intelligence. The applications are infinite. For the purpose of this book, we will now analyze two interesting and (until a few years ago) unusual sources of financial information: Google Trends and YouTube.

8.5.2.1 Google Trends

The use of Google Trends data in financial market research has garnered significant attention in academia, particularly when forecasting is considered. A substantial body of literature has emerged exploring the relationship between search volume data and various financial market phenomena. This research spans multiple countries and markets, yielding diverse insights into Google Trends' potential and limitations as a predictive tool.

Several studies have demonstrated correlations between search volume and market behavior. Fan et al. [46] identified a relationship between the popularity of company names in the Taiwan 50 Index and fluctuations in the TAIEX. More recently, Huang et al. [47] conducted a comprehensive examination of Google Trends' validity as a measure of investor attention. Expanding on this concept, Petropoulos et al. [48] developed an innovative approach by extracting keywords from central bank speeches to construct a financial dictionary. This dictionary was then employed to gauge public interest in financial news, with the ultimate goal of predicting market turbulence.

However, the predictive power of Google search volumes appears to vary across different markets and time horizons. Bank et al. [49] conducted a study on stocks traded in the German market, revealing that search volume primarily captures the interest of non-institutional investors. Their findings suggest that while there is a positive impact on future stock returns, this effect is temporary and dissipates over longer holding periods. This conclusion aligns with the work of Takeda and Wakao [50], who analyzed a sample of Nikkei 225 stocks. Their research uncovered a strong positive relationship between higher Google search volumes and increases in trading volume but only a weak positive effect on stock prices.

Further scrutiny of Google Trends data has raised important methodological considerations. Challet and Bel Hadj Ayed [51] posit that Google Trends data are comparable to past price returns in terms of exploitable information, characterizing both as noisy aggregated signals generated by numerous individuals. Their work also highlights several biases inherent in Google Trends data, including reliability issues prior to August 6, 2008, due to infrequent updates. Additionally, they note significant changes in Google's data presentation methodology. In 2012, Google Trends transitioned from providing real numbers with a non-transparent normalization technique to displaying a scaled integer between 0 and 100, representing relative term popularity within a given time series. This change, along with increased data granularity triggered by new maximum values, has potentially compromised the reliability of the data for certain analytical purposes.

The application of Google Trends data extends beyond direct market prediction to broader economic indicators. Castelnuovo and Tran [52] developed a Google Trends Uncertainty index based on keywords frequently mentioned in the Federal Reserve's Beige Book. Their research demonstrated positive correlations (R^2 between 0.28 and 0.63) with other monthly measures of economic uncertainty. Employing vector autoregression models (VARs), they illustrated that variations in this index could predict movements in US unemployment rates, under scoring the potential of search data in macroeconomic forecasting.

In the realm of volatility modeling, Xu et al. [53] made significant strides by developing a multi-factor GARCH-MIDAS model capable of incorporating factors sampled at different frequencies. Their research focused on investigating the efficacy of Google Trends in predicting stock market volatility. The authors posit that the daily volatility of the Dow Jones Industrial Average (DJIA) is influenced by two primary components: a short-run component comprised of unpredictable market events and a long-run component explicable through event impacts and macroeconomic fundamentals. To model these components, they utilized macroeconomic variables to explain the fundamentals and employed the search trend of the keyword "stock" as an indicator of public concern to capture event impacts. Their findings suggest that "stock" search data contains valuable information for stock market volatility forecasting, and its predictive power is enhanced when combined with wholesale inflation and gross domestic product data.

This comprehensive body of research underscores the complex relationship between Google Trends data and financial markets. While search volume data offers intriguing insights into investor behavior and market dynamics, its predictive power varies across different contexts and timeframes. As such, future research in this field may benefit from integrating Google Trends data with other predictive models and considering the methodological challenges inherent in working with this type of data.

In order to investigate the efficacy of Google Trends data in financial markets, Timpone and Zava [54] studied three Exchange-Traded Funds (ETFs) traded on NYSE Arca, a subsidiary of the NYSE Group Inc. The selected ETFs are XLK, XLF, and XLV, which track indices of the Technology, Financials, and Healthcare sectors, respectively. These ETFs were chosen based on their high market capitalization and net assets among US sectorial ETFs, as well as their distinct characteristics, which could potentially provide more nuanced insights into the efficacy of Google Trends as a predictive tool.

To comprehensively assess the utility of Google Trends data, four distinct indices were devised:

1. GT: This index was constructed following the methodology of [53], serving as a measure of general public attention to the market.
2. GT_XLK: A sector-specific index for the Technology ETF (XLK).
3. GT_XLF: A sector-specific index for the Financials ETF (XLF).
4. GT_XLV: A sector-specific index for the Healthcare ETF (XLV).

Each of the sector-specific indices (GT_XLK, GT_XLF, GT_XLV) was constructed using eight equally weighted keywords pertinent to their respective sectors.

The selection of these keywords was a critical process aimed at capturing the essence of each sector while minimizing potential biases. The researchers endeavored to choose terms that were as generic and unbiased as possible, consciously avoiding the introduction of future-based bias during the selection process.

The keywords used were the following:

1. GT_XLK: technology, microchip, data, electricity price, computer, silicon, lithium, product launch;
2. GT_XLF: financials, financial regulations, merger, bankruptcy, crisis, inflation, earnings, economic policy;
3. GT_XLV: healthcare, flu, epidemic, medical discovery, healthcare regulations, diabetes, cancer, new drug.

The use of equally weighted keywords for each sector-specific index ensures a balanced representation of search trends within each sector. This approach mitigates the risk of overemphasis on any single aspect of a sector, potentially providing a more accurate reflection of overall sector-related search activity. Furthermore, the methodological approach allows for a multi-faceted analysis of Google Trends data in relation to specific market sectors, as well as broader market trends. By constructing both a general market attention index (GT) and sector-specific indices, the study aims to provide a nuanced understanding of how search volume data correlates with market behavior across different segments of the economy.

By employing this carefully constructed set of indices, the study seeks to elucidate the potential of Google Trends data as a predictive tool in financial markets while also exploring any sector-specific variations in the efficacy of this approach. This methodology sets the stage for a comprehensive analysis of the relationship between online search behavior and market dynamics across different economic sectors.

In addition to the aforementioned data sources, this study incorporates the Chicago Board Options Exchange's CBOE Volatility Index (VIX) as a benchmark for comparison with our constructed indices. The VIX, often referred to as the "fear index," provides a market estimate of expected volatility based on S&P 500 index options and serves as a widely recognized measure of market sentiment and risk perception [55].

The selection of macroeconomic variables for this study was informed by previous research in the field. Castelnuovo and Tran [52] considered unemployment, prices, and the shadow rate constructed by [56] in their analysis. Similarly, Xu et al. [53] explored various models and achieved optimal results by incorporating the producer price index and gross domestic product. Building upon these foundations, our study employs a carefully curated set of macroeconomic indicators.

As our primary economic indicator, we have opted for industrial production (IP) over gross domestic product (GDP) for several compelling reasons. First, the frequency of data release: IP data is published monthly, aligning more closely with our analytical framework compared to the quarterly release of GDP figures. Second, the specificity. Industrial production focuses explicitly on industrial activity and production levels, rendering it more sensitive to economic fluctuations that may directly or indirectly impact the sectors covered by our selected ETFs. Lastly, the relevance:

while GDP encompasses IP data, it also includes a broader spectrum of economic activities, such as services and government spending, which may dilute the relevance of our sector-specific analysis.

To provide a comprehensive view of the economic environment affecting the technology, financial, and healthcare sectors, we have also included the unemployment rate (U) and the Producer Price Index (PPI) in our analysis. The unemployment rate serves as a key indicator of economic health and can significantly influence market sentiment and consumer behavior, which in turn affects sector performance [57]. The PPI is incorporated as it reflects changes in selling prices received by domestic producers for their output and is widely recognized as a leading indicator of consumer price inflation. This measure can have substantial impacts on financial markets, as demonstrated in [58] and more recently in [59].

Data for this study were sourced from reputable and widely used databases: Yahoo Finance, which provided historical ETF prices and the VIX index; Google Trends, which supplied keyword search volume data; the U.S. Bureau of Labor Statistics was the source for the Producer Price Index (PPI); and the Federal Reserve Bank of St. Louis provided data on the unemployment rate (U) and Industrial Production index (IP).

This comprehensive dataset, combining financial market data, search volume information, and macroeconomic indicators, provides a robust foundation for investigating the relationships between Google Trends data, market volatility, and broader economic conditions across the technology, financial, and healthcare sectors.

The dataset spans from January 1, 2004, to April 30, 2024, with monthly frequency for most variables. An exception is made for ETF prices and the VIX index, which are collected daily to compute logarithmic returns. The choice of logarithmic returns is motivated by their desirable properties, including the ability to straightforwardly aggregate returns over multiple periods and handle proportionate gains and losses symmetrically [55]. Trends of time series used included in the model are displayed in Fig. 8.9.

The logarithmic returns are subsequently used to calculate realized monthly volatility for XLK, XLF, XLV, and VIX (denoted as XLK_rv, XLF_rv, XLV_rv, and VIX_rv, respectively). The computation of realized volatility adheres to the standard definition proposed by Corsi [60]. To enhance interpretability, the resulting values are multiplied by 100.

The formula for realized volatility over a month can be expressed as:

$$r_{d_i} = \log\left(\frac{p_{d_i}}{p_{d_{i-1}}}\right) \tag{8.1}$$

$$rv_{m_j} = 100 \cdot \sqrt{\frac{1}{\#m_j} \sum_{d_i \in m_j} r_{d_i}^2} \tag{8.2}$$

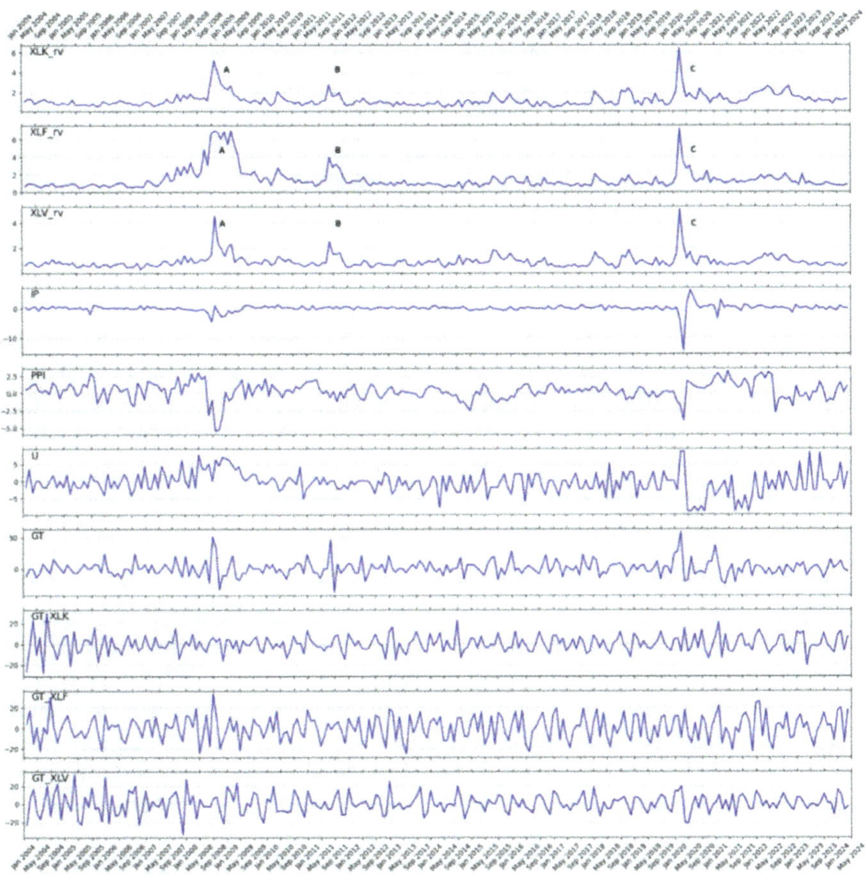

Fig. 8.9 **Trends of time series** used included in the model (Sample 2004M1–2024M4). Major spikes in volatility: A: 2007–2008 financial crisis; between A and B: European debt crisis; B: U.S. debt-ceiling crisis; C: COVID-19 pandemic

Where d_i indicates the ith trading day, m_j the month j, and $\#m_j$ the number of trading days for that month.

Every other variable is considered–congruently with current literature–in log-levels as follows:

$$Y_t = 100 \cdot \log\left(\frac{X_t}{X_{t-1}}\right) \tag{8.3}$$

The panel data satisfies the criteria of stationarity for all variables tested using the Augmented Dickey-Fuller Test and cointegration tests for all relevant pairs of variables [61, 62].

Moving forward, we conducted tests for autocorrelation in the residuals of the models with dependent variables XLK_rv$_t$, XLF_rv$_t$, and XLV_rv$_t$. The Durbin-

Watson statistics obtained were 2.00, 2.33, and 2.05, respectively. These results indicate the absence of first-order autocorrelation in the residuals [63], thereby supporting our decision to exclude the second lag in our models.

Following the autocorrelation assessment, we proceeded with the analysis using the standard reduced-form variance-covariance matrix. To isolate shocks of individual variables, we employed Cholesky decomposition. This technique involves finding a lower triangular matrix B such that:

$$\Omega = BB^\top \tag{8.4}$$

We then define $\tilde{\varepsilon} = B^{-1}\varepsilon$, which represents a linear transformation of our error ε_t. By construction, this new error is orthogonal, as its variance is diagonal [64].

Having orthogonalized the variance-covariance matrix, we examined the effects of orthogonal shocks of the Google Trends Indices (GTIs) on the ETFs' volatilities. The Impulse Response Functions (IRFs) from the three Vector Autoregression (VAR) models, displayed in Figs. 8.10 and 8.11, revealed several noteworthy findings:

1. GT_XLK: The only statistically significant effect of GT_XLK is a negative impact on unemployment, which aligns with existing literature. Castelnuovo and Tran [52] similarly found that a shock in their Google Uncertainty Index was associated with a negative response in unemployment. While modest, this effect is notable. Given the keywords associated with this index, GT_XLK may serve as a forecasting indicator for emerging technologies and sectoral growth. Interestingly, the volatility of the technology sector appears unaffected, suggesting that the observed effects might be spurious or influenced by confounding factors such as economic stimuli or incentives promoting technological advancements and job creation.
2. GT_XLF: Among the GTIs, GT_XLF uniquely demonstrates a positive, albeit not strong, effect on sectoral volatility. This effect persists for several months, though at a confidence interval below 95% after the initial month. This finding corroborates previous observations suggesting that GT_XLF may signal market fragility and periods of economic instability.
3. GT_XLV: This index appears to be the least informative for our models, confirming our earlier analyses that suggested limited explanatory power of GT_XLV in relation to the variables studied.
4. GT: Searches associated with "stock" are linked to negative responses in both the Producer Price Index (PPI) and Industrial Production (IP), with effects lasting approximately one month. These results are consistent with the coefficients and p-values reported in Tables 4–6. While these relationships are unlikely to be causal, they suggest potential indirect effects. Considering GT as a proxy for investor attention, an increase in "stock" popularity might serve as a leading indicator of broader economic concerns or optimism, such as a period of lower inflation. These effects could be indirect or due to confounding factors; however, the notion that such an effect would directly impact inflation and industrial production within a single month seems implausible. Current economic models

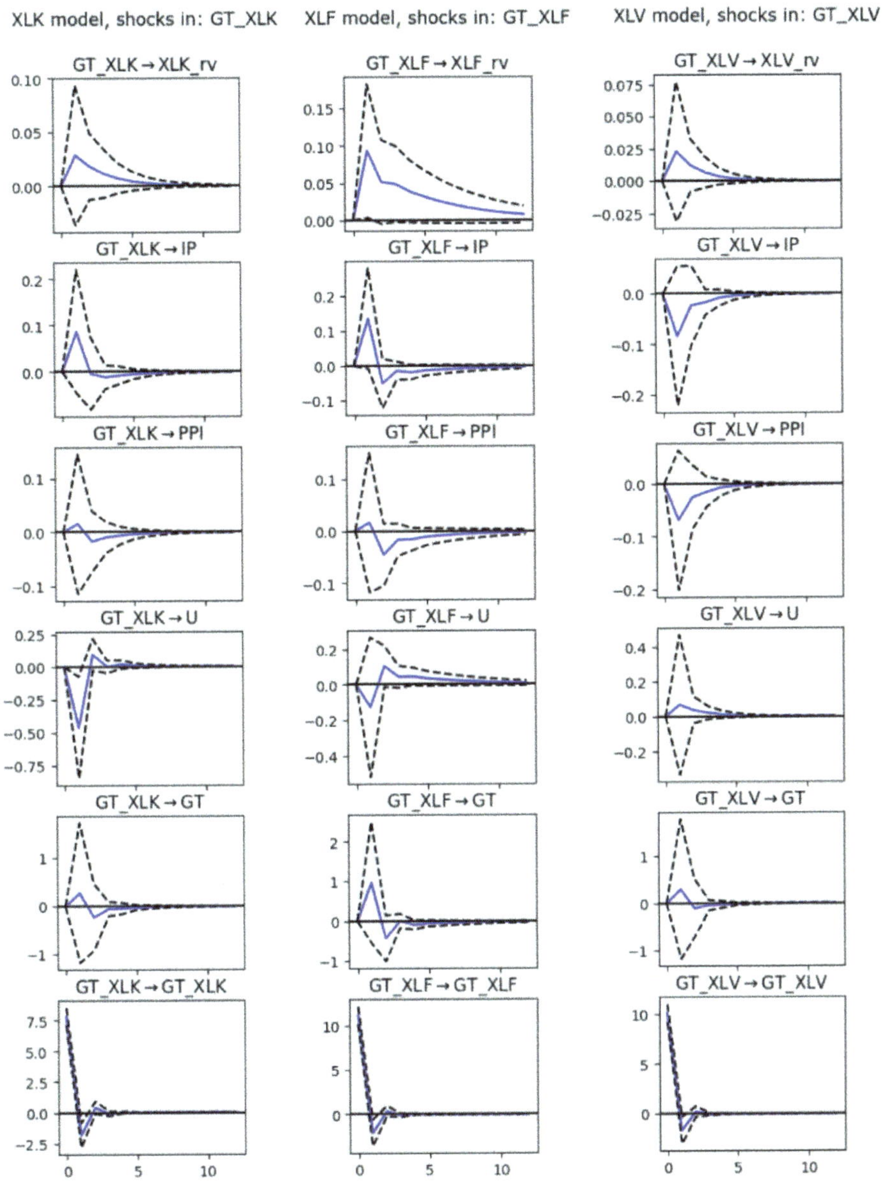

Fig. 8.10 Impulse response functions (IRFs). Impulse response functions to a one positive standard deviation shock in GTIs for the three VAR models; each subplot represents the response of a different variable over a 12-month horizon. Dashed black lines indicate the 95% confidence interval for the response

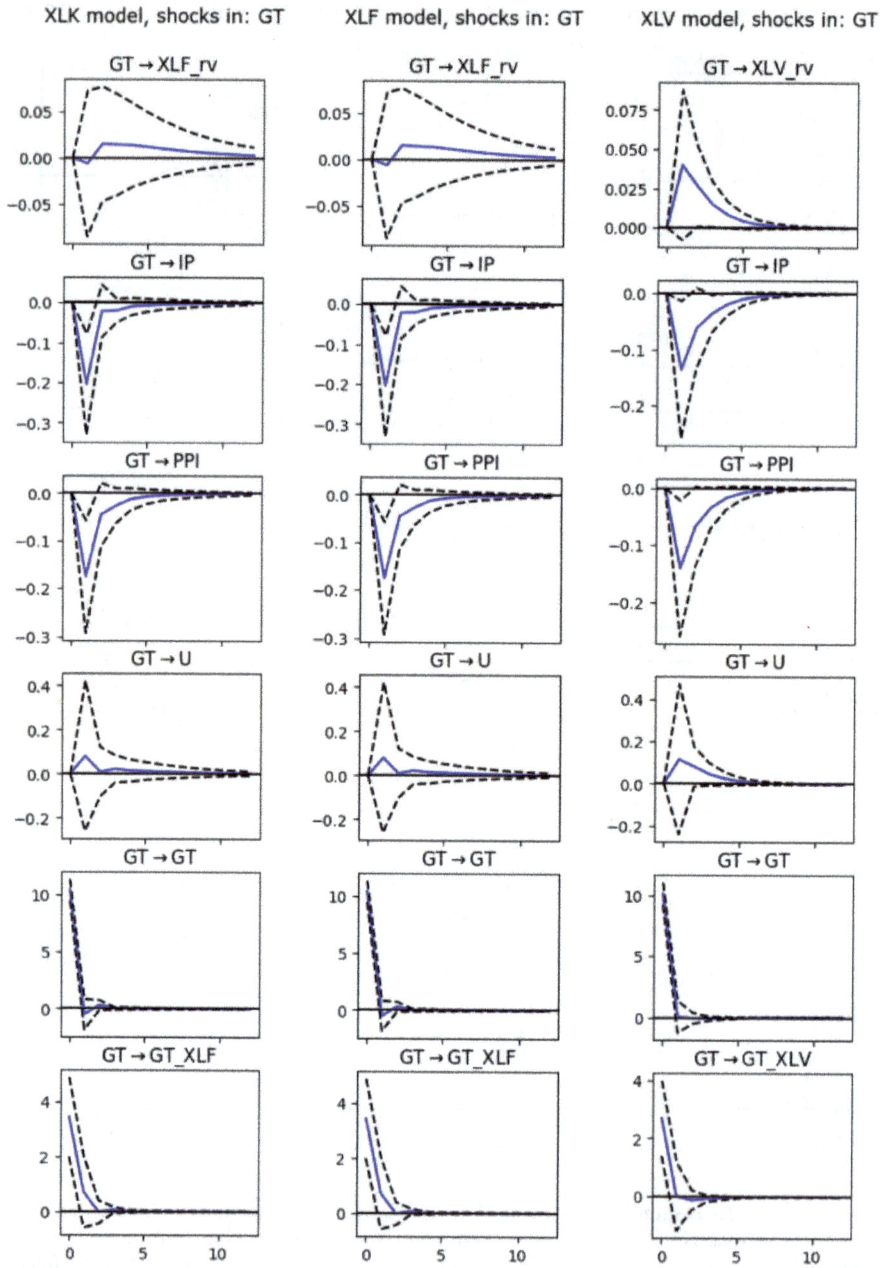

Fig. 8.11 Impulse response functions (IRFs). Impulse response functions to a one positive standard deviation shock in GTIs for the three VAR models; each subplot represents the response of a different variable over a 12-month horizon. Dashed black lines indicate the 95% confidence interval for the response

generally posit that conventional monetary policy affects inflation after 12–18 months (Federal Reserve Bank of San Francisco, 2024).

These findings contribute to our understanding of the complex relationships between online search behavior, market volatility, and macroeconomic indicators. However, further research is necessary to elucidate the causal mechanisms underlying these observed correlations.

This study sought to elucidate the complex relationship between Google Trends data and market volatility across different sectors, with the aim of enhancing our understanding of market dynamics and improving volatility forecasting models. The findings of this research contribute to the growing body of literature on the application of search volume data in financial markets while highlighting important limitations and considerations for future research.

A primary challenge encountered in this study, and indeed in the broader field of market volatility research, is the multifaceted nature of volatility itself. Market volatility comprises numerous unobservable components, rendering it difficult to capture and quantify through a limited set of measures. This inherent complexity often leads to significant challenges in distinguishing between noise and genuine volatility signals, a difficulty that was evident in our analysis of Google Trends data.

This research's key focus was the efficacy of sector-specific Google Trends Indices (GTIs) constructed from aggregated search signals. Our findings in this regard were mixed and nuanced. While one of the three constructed GTIs demonstrated some relevance to market volatility, the effects, even when statistically significant, were minimal in magnitude. This outcome suggests that while Google Trends data can offer certain insights into market dynamics, its utility as a stand alone predictor of volatility is limited.

These results underscore the importance of cautious interpretation and application of search volume data in financial modeling. The minimal effects observed, even when statistically significant, indicate that Google Trends data should not be overly relied upon for volatility forecasting, particularly in the absence of more robust data sources and advanced statistical techniques. This finding aligns with previous research suggesting that the predictive power of search volume data may be context-dependent and variable across different markets and time periods [49, 50].

Furthermore, our study reinforces the notion that market volatility is influenced by a complex interplay of factors that extend beyond public interest as measured by search volumes. The limited explanatory power of our GTIs suggests that other economic indicators, market-specific factors, and perhaps more sophisticated measures of public sentiment may be necessary to develop more accurate volatility forecasting models.

It is important to note that while our findings suggest limited utility for Google Trends data in direct volatility prediction, they do not negate the potential value of such data in other aspects of financial research. For instance, search volume data may still offer valuable insights into investor attention and market sentiment or as a complementary indicator when used in conjunction with other economic and financial metrics.

In conclusion, this study contributes to the ongoing dialogue about the role of alternative data sources in financial modeling and market analysis. While Google Trends data shows some promise, its application in volatility forecasting appears to be limited. Future research in this area might benefit from exploring more sophisticated methods of aggregating and analyzing search volume data, investigating potential non-linear relationships between search trends and market volatility, or combining Google Trends data with other alternative data sources to create more robust predictive models.

As financial markets continue to evolve and new data sources emerge, the quest for improved volatility forecasting models remains an important area of research. This study underscores the need for a balanced approach that leverages innovative data sources while maintaining a critical perspective on their limitations and potential biases.

8.5.2.2 YouTube

Financial markets are increasingly influenced by public sentiment and discourse, particularly in the age of social media and online content. By analyzing YouTube content, which serves as a significant platform for financial discussion and analysis, Pitta and Zava [65] provide a novel approach to gauging market sentiment. The granularity of the analysis, focusing on sentence-level sentiment rather than broad video-level assessments, allows for a nuanced understanding of how financial topics are discussed and perceived over time.

The study employs a sophisticated methodology for data collection and analysis in the realm of YouTube content, with potential applications in understanding sentiment dynamics in financial markets. The research process begins with a comprehensive data collection and preprocessing phase. Information pertaining to video attributes, including hyperlinks, durations, and publication dates, is methodically gathered from the selected YouTube channels. A critical aspect of this phase is the implementation of a data balancing procedure. This procedure involves a systematic adjustment of the dataset to ensure an equitable quantity of content is analyzed from each channel, thereby mitigating potential biases that could arise from asymmetrical data volumes.

The subsequent phase of the study focuses on transcript retrieval and analysis, which forms the core of the investigative process. For each video in the curated dataset, the study employs a multi-step analytical approach. Initially, the raw transcript is acquired using the YouTubeTranscriptApi. This is followed by a rigorous cleaning process to eliminate non-verbal elements, ensuring that the analysis is concentrated solely on spoken content. The study then employs an innovative technique to reconstruct sentence structure by inferring punctuation based on speech timing patterns. This reconstruction is crucial for enhancing the accuracy of subsequent sentiment analysis, as sentence structure significantly influences sentiment interpretation.

The cleaned and punctuated transcript is then subjected to a filtering process, isolating sentences that contain the predetermined keywords of interest, including algorithmically generated synonyms. This filtration process focuses the analysis on content directly pertinent to the chosen topic. Finally, the study conducts sentiment analysis on these relevant sentences utilizing the TextBlob library. This analysis yields quantitative measures of polarity and subjectivity for each sentence, culminating in an overall sentiment score derived from the product of these two metrics. The results for the blockchain and e-commerce sectors are displayed in Figs. 8.12 and 8.13. A direct correlation with the future valuation at IPO of startups founded in that period is found.

The importance of this methodological approach in the context of financial markets cannot be overstated. Firstly, it offers a method to track sentiment around specific financial instruments, companies, or economic indicators, potentially providing early signals of market trends or shifts in investor sentiment. Secondly, comparing content from different channels allows for the identification of divergences in sentiment across various segments of the financial community, which could indicate potential market inefficiencies or upcoming corrections. Lastly, the historical nature of the data collected enables the study of how sentiment evolves in response to various economic events or policy changes, providing valuable insights for both academic research and practical market analysis.

Moreover, the study's methodology in balancing data across channels and rigorously cleaning and structuring the text data addresses common challenges in sentiment analysis, such as bias and noise in unstructured data. This rigorous approach enhances the reliability and applicability of the findings to real-world financial analysis and decision-making processes.

In conclusion, this study presents a sophisticated and nuanced approach to sentiment analysis in YouTube content, with significant potential applications in understanding and predicting financial market dynamics. By leveraging advanced natural language processing techniques and a carefully structured analytical process, it offers a valuable tool for researchers and practitioners seeking to gain deeper insights into the complex interplay between public discourse and financial market behavior.

8.5.3 Concentration in the Stock Market–FTSE MIB

The FTSE MIB (Milano Indice di Borsa) is the primary benchmark stock market index for the Italian equity market. It comprises the 40 most liquid and capitalized stocks listed on the Borsa Italiana (Italian stock exchange). It is a capitalization-weighted index, meaning the larger companies carry more weight in the index. First calculated in 1994, it is derived from the previous MIB 30 index. The index aims to capture around 80% of the domestic market capitalization, and it is reviewed quarterly to ensure it remains representative of the Italian market. The performance of the FTSE MIB is considered a bellwether for the overall health and investor sentiment in the Italian economy: it is the primary equity benchmark that investors, fund managers, and analysts look to for tracking the performance of Italy's largest and most liquid publicly traded companies across key industries.

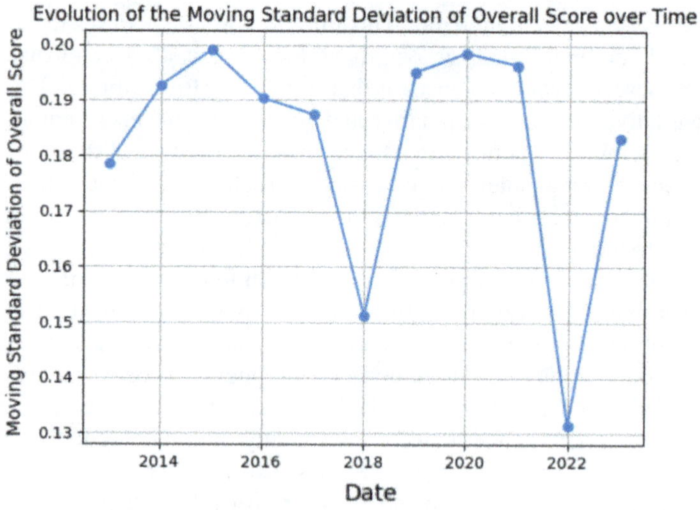

Fig. 8.12 **Blockchain sentiment evolution**. Pitta and Zava [65]

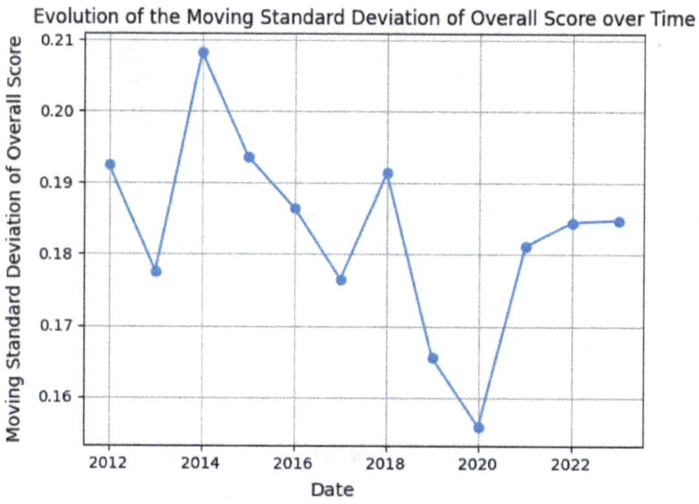

Fig. 8.13 **E-commerce sentiment evolution**. Pitta and Zava [65]

Figure 8.14 presents a fascinating network visualization of the institutional investors in FTSE MIB companies as of 2023 [66]. This network graph offers valuable insights into the structure and interconnectedness of Italy's primary stock market from a network science perspective. The visualization depicts a bipartite network where nodes represent two distinct entities: institutional investors (shown in blue) and FTSE MIB-listed stocks (shown in red). The edges connecting these nodes represent

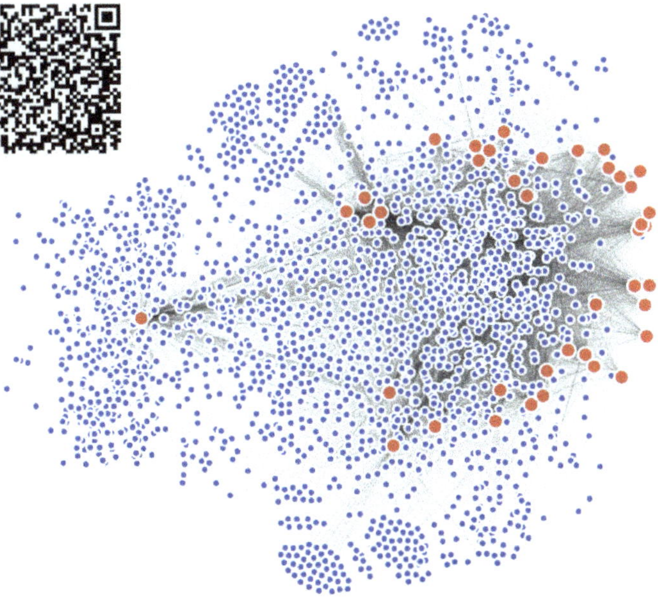

Fig. 8.14 Institutional Investors in FTSE MIB companies 2023. Institutional investors (in blue) and stock (in red) are connected by the amount owned by the former in the latter in 2023

ownership relationships, with the presence of an edge indicating that an institutional investor holds shares in a particular company.

From a network topology standpoint, the graph exhibits several noteworthy characteristics. The network seems to display a heterogeneous degree distribution. Some red nodes have significantly more connections than others, indicating that certain FTSE MIB companies attract a larger number of institutional investors. This aligns with the concept of preferential attachment in complex networks, where more prominent or attractive nodes tend to accumulate more connections over time. Indeed, some stocks appear to act as hubs, connecting to many investors. This could represent highly popular or influential stocks that are widely held across various institutional portfolios and results in a hub-and-spoke structure. Further, the graph displays visible clusters of blue nodes, suggesting that certain groups of institutional investors may have similar investment strategies or focuses, targeting similar sets of stocks within the FTSE MIB.

From a financial perspective, the network findings offer several insights. The spread of investors' nodes across the network indicates that institutional investors maintain diversified portfolios within the FTSE MIB, potentially as a risk management strategy. Furthermore, the central positioning and higher connectivity of certain stocks might indicate market concentration, where a subset of FTSE MIB companies attracts a disproportionate amount of institutional investment. Besides, the complex web of connections suggests significant cross-ownership among institutional investors in FTSE MIB companies, which could have implications for corporate governance and market stability. Lastly, clusters of investors around specific stocks reveal sector-specific or thematic investment patterns among institutional investors.

This network representation of the FTSE MIB ownership structure provides a powerful tool for analyzing the Italian stock market's dynamics. It allows for the identification of key players, investment trends, and potential vulnerabilities in the market structure. Such insights are valuable for regulators, investors, and researchers studying market behavior, systemic risk, and the flow of institutional capital in the Italian equity market.

8.6 Conclusion

The financial landscape is undergoing a profound transformation driven by the convergence of network science and artificial intelligence. This chapter has explored various facets of this evolution, from venture capital networks to AI-powered M&A processes, IPO dynamics, and sentiment analysis in stock markets. As we conclude, it is evident that these advancements are not merely incremental improvements but rather paradigm shifts that are reshaping the very foundations of finance.

The exploration of venture capital networks, particularly through the lens of Silicon Valley's genesis and the PayPal Mafia phenomenon, underscores the critical role of interconnected ecosystems in driving innovation and economic growth. These networks demonstrate how the flow of capital, ideas, and talent within tightly-knit communities can catalyze technological revolutions and create enduring economic value. The success of these networks highlights the importance of social capital and the network effects that arise from collaborative environments. As we move forward, understanding and cultivating such networks will be crucial for fostering innovation ecosystems globally.

The application of artificial intelligence in mergers and acquisitions, as exemplified by the M&A Research Institute, represents a quantum leap in how complex financial transactions are conducted. AI's ability to process vast amounts of data, identify patterns, and provide actionable insights is revolutionizing due diligence, valuation, and strategic decision-making in M&A. This not only enhances the efficiency of these processes but also potentially reduces risks and improves outcomes. The integration of AI in M&A is likely to become increasingly sophisticated, potentially leading to more accurate valuations, better-matched mergers, and smoother post-merger integrations.

Our analysis of IPO dynamics, particularly focusing on the network of institutional investors in the Italian FTSE MIB, reveals the intricate web of relationships that underpin public equity markets. The visualization of these networks provides valuable insights into market concentration, investment patterns, and potential systemic risks. As we continue to map and analyze these networks, we can expect to gain a deeper understanding of market behavior, which could inform regulatory policies and investment strategies. Moreover, this network-centric approach to understanding market structure could be extended to other financial markets, providing a more holistic view of global financial interconnections.

Exploring sentiment analysis in stock markets, leveraging novel data sources like Google Trends and YouTube content, demonstrates the power of AI in extracting meaningful signals from the noise of public discourse. While our findings suggest that the direct predictive power of these sentiment indicators for market volatility may be limited, they nonetheless offer valuable insights into investor attention and market psychology. As natural language processing techniques continue to advance, we can anticipate more nuanced and accurate sentiment analysis tools that could complement traditional financial analysis methods.

Looking ahead, several key trends and challenges emerge. We deem the key ones are the following:

1. Integration of AI and Human Expertise: While AI is proving to be a powerful tool in finance, the most effective approaches will likely involve a synergy between AI capabilities and human expertise. Developing frameworks that optimally combine machine learning insights with human judgment and domain knowledge will be crucial.
2. Ethical Considerations: As AI plays an increasingly central role in financial decision-making, addressing ethical concerns around transparency, fairness, and accountability will be paramount. Ensuring that AI systems do not perpetuate or exacerbate existing biases in financial markets will be a significant challenge.
3. Regulatory Adaptation: The rapid pace of technological change in finance is outstripping the ability of regulatory frameworks to adapt. Developing agile regulatory approaches that can keep pace with innovation while ensuring market stability and investor protection will be essential.
4. Data Quality and Privacy: The effectiveness of AI in finance is heavily dependent on the quality and quantity of available data. Balancing the need for comprehensive data with privacy concerns and data protection regulations will be an ongoing challenge.
5. Systemic Risk Management: As financial networks become more complex and interconnected, understanding and managing systemic risks will become increasingly important. Network science approaches will be crucial in identifying potential vulnerabilities and developing robust risk management strategies.
6. Democratization of Finance: AI and network-based approaches have the potential to democratize access to financial services and investment opportunities. Harnessing these technologies to create more inclusive financial systems will be a key opportunity and challenge.

In conclusion, the intersection of network science and artificial intelligence is ushering in a new era in finance. From reshaping venture capital ecosystems to revolutionizing M&A processes, providing new perspectives on market structures, and offering novel approaches to sentiment analysis, these technologies are fundamentally altering how we understand and interact with financial systems. As we navigate this evolving landscape, continued research, interdisciplinary collaboration, and thoughtful implementation will be crucial in realizing the full potential of these advancements while mitigating associated risks.

The future of finance will undoubtedly be shaped by our ability to harness the power of networks and AI, creating more efficient, transparent, and inclusive financial systems. As researchers and practitioners in this field, we stand at the threshold of exciting possibilities, with the opportunity to contribute to a financial ecosystem that is not only more sophisticated in its operations but also more attuned to the complex dynamics of our interconnected world.

References

1. Zava, M.: How Venture Capitalists Become Influential? (2024). (pre-print)
2. Zava, M., Caselli, S.: Venture Capital Networks, The Palgrave Encyclopedia of Private Equity (2023)
3. Zava, M., Caselli, S.: 150 Venture Capital Networks: Present and Future (2024). (pre-print)
4. Zava, M., Caselli, S.: Venture Capital Networks: Who Drags Investments in Startups? (2024). (pre-print)
5. Becker, A., Ensley, M.D., Knyphausen-Aufseß, D.Z.: Network to passion or passion to network? Disentangling entrepreneurial passion selection and contagion effects among peers and teams in a startup accelerator. J. Bus. Ventur. **38** (2023)
6. Ebbers, J.J., Wijnberg, N.M.: The co-evolution of social networks and selection system orientations as core constituents of institutional logics of future entrepreneurs at school. J. Bus. Ventur. **34**, 558–577 (2019)
7. Lazar, M.: Forming entrepreneurial teams: mixing business and friendship to create transactive memory systems for enhanced success. Acad. Manag. J. **65** (2022)
8. Howell, S., Nanda, R.: Networking frictions in venture capital, and the gender gap in entrepreneurship. J. Financ. Quant. Anal. 1–29 (2023)
9. Birley, S.: The role of networks in the entrepreneurial process. J. Bus. Ventur. **1**, 107–117 (1985)
10. Ferrary, M., Granovetter, M.: The role of venture capital firms in silicon valley's complex innovation network. Econ. Soc. **38**, 326–359 (2009)
11. Hallen, B.L.: The causes and consequences of the initial network positions of new organizations: From who do entrepreneurs receive investments? Adm. Sci. Q. **53**, 685–718 (2008)
12. Hallen, B.L., Eisenhardt, K.M.: Catalyzing strategies and efficient tie formation: how entrepreneurial firms obtain investment ties. Acad. Manag. J. **55**, 35–70 (2012)
13. Kaplan, S.N., Strömberg, P.: Characteristics, contracts, and actions: evidence from venture capitalist analyses. J. Financ. **59**, 2177–2210 (2004)
14. Lerner, J.: The syndication of venture capital investments. Financ. Manag. 16–27 (1994)
15. Hochberg, Y.V., Ljungqvist, A., Lu, Y.: Whom you know matters: venture capital networks and investment performance. J. Financ. **62**, 251–301 (2007)
16. Gompers, P.A., Mukharlyamov, V., Xuan, Y.: The cost of friendship. J. Financ. Econ. **119** (2016)
17. Gu, Q., Lu, X., Chiu, C.N.: Exploring small-world network with an Elite-clique: bringing embeddedness theory into the dynamic evolution of a venture capital network. Soc. Netw. **57**, 70–81 (2019)
18. Luo, J., Rong, K., Yang, K., et al.: Syndication through social embeddedness: a comparison of foreign, private and state-owned venture capital (VC) firms. Asia Pac. J. Manag. **36**, 499–527 (2019)
19. Brander, J.A., Amit, R., Antweiler, W.: Venture-capital syndication: improved venture selection vs. the value-added hypothesis. J. Econ. Manag. Strateg. **11**, 423–452 (2002)

20. Das, S.R., Jo, H., Kim, Y.: Polishing diamonds in the rough: the sources of syndicated venture performance. J. Financ. Intermed. **20**, 199–230 (2011)
21. Burt, R.S.: Neighbor Networks: Competitive Advantage Local and Personal. Oxford University Press (2010)
22. Zhelyazkov, P.I., Gulati, R.: After the break-up: the relational and reputational consequences of withdrawals from venture capital syndicates. Acad. Manag. J. (2015)
23. Sapienza, H.J., Manigart, S., Vermeir, W.: Venture capitalist governance and value added in four countries. J. Bus. Ventur. **11**, 439–469 (1996)
24. Gorman, M., Sahlman, W.A.: What do venture capitalists do? J. Bus. Ventur. **4**, 231–248 (1989)
25. Lindsey, L.: Blurring firm boundaries: the role of Venture Capital in strategic alliances. J. Financ. **63**, 1137–1168 (2008)
26. Ozmel, U., Reuer, J.J., Gulati, R.: Signals across multiple networks: how venture capital and alliance networks affect interorganizational collaboration. Acad. Manag. J. **56** (2013)
27. González-Uribe, J.: Exchanges of innovation resources inside venture capital portfolios. J. Financ. Econ. **135**, 144–168 (2020)
28. Pahnke, E., McDonald, R., Wang, D., Hallen, B.: Exposed: venture capital, competitor ties, and entrepreneurial innovation. Acad. Manag. J. **58**, 1334–60 (2014)
29. El-Khatib, R., Fogel, K., Jandik, T.: CEO network centrality and merger performance. J. Financ. Econ. **116**, 349–382 (2015)
30. Shim, J., Kim, C.: Estimating country-level social network density and supportive surroundings by simulation. J. Bus. Ventur. Insights **9**, 24–31 (2018)
31. Ahlstrom, D., Bruton, G.D.: Venture capital in emerging economies: networks and institutional change. Entrep. Theory Pract. **30**(2) (2006)
32. Holme, P., Saramäki, J.: Temporal Networks, Understanding Complex Systems. Springer (2013)
33. Goldenberg, A., Zheng, A.X., Fienberg, S.E., Airoldi, E.M.: A survey of statistical network models. Found. Trends Mach. Learn. **2**, 129–233 (2010)
34. Fraiberger, S., Barabási, A.L., Sinatra, R., Resch, M., Riedl, C.: Quantifying reputation and success in art. Science **362**, eaau7224 (2018)
35. Ravasz, E., Somera, A.L., Mongru, D.A., Oltvai, Z.N., Barabási, A.L.: Hierarchical organization of modularity in metabolic networks. Science **297**, 1551–1555 (2002)
36. Caldarelli, G., Capocci, A., De Los Rios, P., Munoz, M.A.: Scale-free networks from varying vertex intrinsic fitness. Phys. Rev. Lett. **89**, 258702 (2002)
37. Atanasov, V., Ivanov, V., Litvak, K.: Does reputation limit opportunistic behavior in the VC industry? Evidence from litigation against VCs. J. Financ. **67**, 2215–2246 (2012)
38. Nahata, R.: Venture capital reputation and investment performance. J. Financ. Econ. **90**, 127–151 (2008)
39. Nanda, R., Samila, S., Sorenson, O.: The persistent effect of initial success: evidence from venture capital. J. Financ. Econ. **137**, 231–248 (2020)
40. Lee, P.M., Pollock, T.G., Jin, K.: The contingent value of venture capitalist reputation. Strateg. Organ. **9**, 33–69 (2011)
41. Zava, M.: Dear Investors, are you Missing Out? Midas and Venture Capital (2024). (pre-print)
42. Berre, M.: Microtargeting Startup Valuation Using Supervised Machine Learning: Which Factors Matter Most? (2023). (pre-print)
43. Parikh, A., Poojary, S., Gupta, A., Venugopal, B.: AMP|Optimizing M&A Outcomes: Harnessing the Power of Big Data Analytics and Natural Language Processing (2023). (pre-print)
44. Zava, M.: Capital markets investors. In: Caselli, S., Gatti, S. (eds.) Capital Markets: Perspectives Over the Last Decade. EGEA (2023)
45. Jacoby, G., Liao, R.C.: Price discovery and sentiment. Int. Rev. Financ. Anal. **21**, 108–118 (2012)
46. Fan, M.H., Chen, M.Y., Liao, E.C.: A deep learning approach for financial market prediction: utilization of Google trends and keywords. Granul. Comput. **6**, 207–216 (2021)
47. Huang, M.Y., Rojas, R.R., Convery, P.D.: Forecasting stock market movements using Google Trend searches. Empir. Econ. **59**, 2821–2839 (2020)

48. Petropoulos, A., Siakoulis, V., Stavroulakis, E., Lazaris, P., Vlachogiannakis, N.: Employing google trends and deep learning in forecasting financial market turbulence. J. Forecast. **40**, 353–365 (2021)
49. Bank, M., Larch, M., Peter, G.: Google search volume and its influence on liquidity and returns of German stocks. Financ. Mark. Portf. Manag. **25**, 239–264 (2011)
50. Takeda, F., Wakao, T.: Google search intensity and its relationship with returns and trading volume of Japanese stocks. Pacific-Basin Financ. J. **27**, 1–18 (2014)
51. Challet, D., Bel Hadj Ayed, A.: Do Google Trend Data Contain More Predictability than Price Returns? Working Paper (2014)
52. Castelnuovo, E., Tran, T.D.: Google it Up! A Google Trends-Based Uncertainty Index for the United States and Australia, CESifo Working Paper No. 6695 (2017)
53. Xu, Q., Bo, Z., Jiang, C., Liu, Y.: Does Google search index really help predicting stock market volatility? Evidence from a modified mixed data sampling model on volatility. Knowl. Based Syst. **163** (2019)
54. Timpone, L., Zava, M.: Google Trends to Predict ETFs Returns and Volatility (2024). (pre-print)
55. Hudson, R., Gregoriou, A.: Calculating and Comparing Security Returns is Harder than you Think: A Comparison between Logarithmic and Simple Returns, Working Paper (2010)
56. Wu, J.C., Xia, F.D.: Measuring the macroeconomic impact of monetary policy at the zero lower bound. J. Money Credit Bank. **48**, 253–291 (2016)
57. Blanchard, O., Diamond, P.: The cyclical behavior of the gross flows of U.S. workers. Brook. Pap. Econ. Act. **1990**, 85-156 (1990)
58. Anari, A., Kolari, J.: Stock prices and inflation. J. Financ. Res. **24**, 587–602 (2001)
59. O'Trakoun, J., Ramachandran, D.: Building a Pipeline Between Producer and Consumer Prices, Federal Reserve Bank of Richmond Economic Brief, No. 22-38 (2022)
60. Corsi, F.: Measuring and Modelling Realized Volatility: from Tick-By-Tick to Long Memory. University of Lugano, Department of Economics (2005)
61. Dickey, D.A., Fuller, W.A.: Distribution of the estimators for autoregressive time series with a unit root. J. Am. Stat. Assoc. **74**, 427–431 (1979)
62. MacKinnon, J.G.: Critical values for co-integration tests, Queen's Economics Department Working Paper, No. 1227 (2010)
63. Durbin, J., Watson, G.S.: Testing for serial correlation in least squares regression. I, Biometrika **37**, 409–428 (1950)
64. Li, J.: Lecture 7a: Vector Autoregression. Miami University, VAR, Department of Economics (2016)
65. Pitta, L., Zava, M.: Sentiment and Investments. Evidence from Youtube (2024). (pre-print)
66. Caselli, S., Gatti, S., Zava, M.: Investors in the Italian Stock Market: Evolution or Involution? BAFFI CAREFIN Bocconi-Equita Lab Series (2024)

Chapter 9
Outlook

The synergy between networks and AI presents a promising path forward, with applications that span multiple sectors of society. As we have explored in this book, these tools offer transformative potential in areas critical to human progress and well-being.

Most prominently, the tools and methodologies discussed in this book offer powerful means to address the challenges of misinformation and polarization in our information ecosystem. While social media has presented unforeseen difficulties, the same network principles that allow false information to spread can be harnessed to promote truth and foster constructive dialogue. By leveraging AI to detect patterns of misinformation and applying network theories to understand how information propagates, we can develop more effective strategies for cultivating a well-informed and critically thinking society.

In the realm of finance, networks, and AI are revolutionizing how we understand and interact with complex economic systems. From analyzing venture capital networks in Silicon Valley to predicting market trends, these technologies are democratizing access to financial insights. AI-driven algorithms are enhancing investment strategies, risk assessment, and fraud detection, potentially creating more stable and equitable financial ecosystems. As explored in Chap. 8, the integration of AI in areas like M&A processes and IPO evaluations is streamlining decision-making and opening new avenues for economic growth.

The impact of network science extends deeply into public health, particularly in the realm of disease prevention and outbreak response. As explored in Chap. 4, contact tracing models and intelligent quarantine strategies for COVID-19 demonstrate the power of network analysis in combating pandemics. By understanding how diseases spread through social networks, we can develop more effective containment strategies and optimize resource allocation during outbreaks. Furthermore, the application of network theory to epidemiology allows us to identify potential superspreaders and critical nodes in disease transmission, enabling more targeted

H. A. Makse and M. Zava, *The Science of Influencers and Superspreaders*, Understanding Complex Systems, https://doi.org/10.1007/978-3-031-78058-5_9

and efficient interventions. This approach not only helps in managing current health crises but also in preparing for and potentially preventing future outbreaks. The integration of AI with these network models enhances our predictive capabilities, allowing for rapid response and adaptive strategies in the face of evolving health threats.

The impact of network science extends deeply into medical research, particularly in genetics and brain science. As discussed in Chaps. 5 and 6, the concept of genetic influencers in gene regulatory networks is transforming our understanding of disease processes and drug development. By mapping the intricate relationships between genes, researchers can identify key influencers that play crucial roles in biological processes. This approach holds immense promise for personalized medicine, allowing for more targeted and effective treatments for a wide range of conditions.

Environmental conservation stands to benefit greatly from these advancements. Chapter 7's exploration of keystone species as influencers in ecosystems demonstrates how network analysis can inform more effective conservation strategies. By understanding the complex interdependencies within natural systems, we can develop more nuanced and impactful approaches to preserving biodiversity and maintaining ecological balance. This knowledge is crucial as we face unprecedented environmental challenges and seek sustainable solutions.

As we look to the future, it's clear that the integration of network science and AI will play a pivotal role in shaping a more informed, equitable, and sustainable world. While challenges remain, the tools and knowledge we now possess offer unprecedented opportunities to address global issues with nuance and efficacy.

The journey into the true Information Age is not just beginning - it's accelerating. By embracing the insights of network science and the capabilities of AI, while remaining mindful of ethical considerations, we can work towards a future where technology empowers humanity to solve its most pressing problems. The interconnected nature of our world, as revealed through network analysis, reminds us that collaborative, cross-disciplinary efforts are key to unlocking the full potential of these transformative tools.

As we have seen, networks, coupled with AI, will help to close the loop to solve these problems. Machines are the only tools to manage enormous data, but, at the same time, current algorithms have created a polarized social media world. The main issue we are all facing is that these machines are not divine and perfect: programmed and trained by very fallible humans. What if we replicate Alan Turing's idea of using a new machine to defeat another machine and, therefore, create an artificial general intelligence that could close the schism triggered by the conundrum of good versus bad data? Only at this point would we have entered the Information Age.

In the words of Joseph Campbell's elaboration on Schopenhauer's "Transcendent Speculation on the Apparent Deliberateness in the Fate of the Individual": "*just as those people you met by chance became effective agents in structuring your life, so you have been an agent structuring other lives, and the whole thing gears together like one big symphony, everything influencing and structuring everything else*". He then concludes:

It is as though our lives were the dream of a single dreamer in which all the dream characters are dreaming too; and so that everything links to everything else, moved out of the universal will in nature

If everything spontaneously arises in relation to everything else; that is the emergent collective behavior captured in the network. As everything links to everything else in a universal network, we hope its knowledge may someday contribute to solving the pressing problems affecting humankind. This book has explored some tools toward achieving this goal.

Appendix A
List of Algorithms, Software and Datasets

It is nearly impossible to maintain an updated list of the latest algorithms and AI software. When we began writing this book, ChatGPT did not exist. As of the time of publication, we find ourselves amidst the GenAI revolution, with the recent release of ChatGPT 4o. Keeping this book updated with these rapid developments in AI is impractical. Even though network tools evolve slower than AI tools, they still progress rapidly, making it challenging to keep any book on network algorithms up to date, too.

The philosophy of this book is not to simply provide ready-made solutions in the form of software packages but rather to provide data scientists with the theoretical knowledge and skills needed to solve problems on their own. We wish to "not to give a woman/man a fish, but to teach a woman/man how to fish". The focus has been on the theoretical concepts that, when understood, enable data scientists to solve any challenge, whether they can download from GitHub ready-made packages or not.

In today's computing and data science landscape, the prevailing trend is for programmers and data scientists to download the latest open-source AI advancements and apply them to new datasets to tackle novel problems. Notably, open-source models such as Llama 3 enable the development of AI-powered applications using pre-trained 8B and 70B parameter LLMs, which are freely available on a large scale. With as little as a $5,000 computer equipped with a single Nvidia GPU, one can leverage the capabilities of these advanced massively-large GenAI techniques to address a wide range of NLP challenges. While this approach is undeniably powerful, it will never be enough to solve new challenges. Thus, gaining a more fundamental understanding of these architectures by delving into the underlying theoretical concepts is crucial for data scientists to thrive in novel applications of AI, regardless of how rapidly the latest technologies evolve.

Having mentioned this, some readers might find the following list of algorithms and software tools to be somewhat basic and relatively outdated. We will begin with the core programming platforms. Then, we provide a list of algorithms, software and datasets compiled by students and postdocs working on the topics covered in

H. A. Makse and M. Zava, *The Science of Influencers and Superspreaders*, Understanding Complex Systems, https://doi.org/10.1007/978-3-031-78058-5

this book. We also encouraged the reader to visit https://GitHub.com/MakseLab and https://osf.io/4ern8 for current code and data to replicate the results in this book. Furthermore, we encourage the readers to send us comments at hmakse@ccny.cuny.edu.

Of course, the list of algorithms and codes provided is not exhaustive. LLM resources are only briefly reviewed, since these technologies are still in their infancy and we have only tested the surface of their capabilities, despite the fact that they have already passed the Turing test. We expect that pretty much all the topics touched in this book will be eventually rewritten in terms of AI models, from finding influencers to predicting elections, to understand brain and genetic diseases and markets. So far, we have tested LLMs in restricted situations and their performance is just astonishing at all levels. We hope to include these results in a future edition of this book. For now, we will discuss the essential skills and computational requirements necessary to complete an AI- and network-driven influencer project.

Typically, a scientist working on these topics has a minimum of full-stack skills. Full-stack skills are required to develop the process from back-end to front-end: from collecting data to modeling with machine learning and network theory to analyzing data, and finally to visualization. Figure A.1 shows this flow in the particular case of a social media-based project to analyze opinion trends with the approaches of Chaps. 1 and 2. This includes some minimal computational requirements to complete each stage in the analysis.

Technical skills are necessary, but not sufficient. A good data scientist must develop a keen theoretical intuition to uncover the hidden treasures in massive data. It is not enough to download the latest machine learning algorithm from Git Hub and apply it to a new dataset. Paraphrasing Marcel Marceau [1]: a top scientist should "make visible the invisible", and see what nobody can see. And that is difficult to acquire.

Below, we provide a minimal list of code and software available datasets that would be useful to tackle the problems explained in this book. We prioritize open-source versions. We start with a list of useful programming languages, platforms, and visualization tools that are common to all projects and then we provide more specific codes to solve influencer problems.

A.1 General Platforms

For those wishing to learn either **R** or **Python** computer languages from scratch, there are excellent resources available for free:

- **EdX**–Many free courses, not just in R and Python but in multiple other interesting areas, https://www.edx.org.
- **Coursera–Python**–Various courses on Python, from very basic (good for beginners) to more complex ones, https://www.coursera.org/specializations/python.
- **Coursera–R**–Various courses on R, syllabi suggest various textbooks, https://www.coursera.org/learn/r-programming.

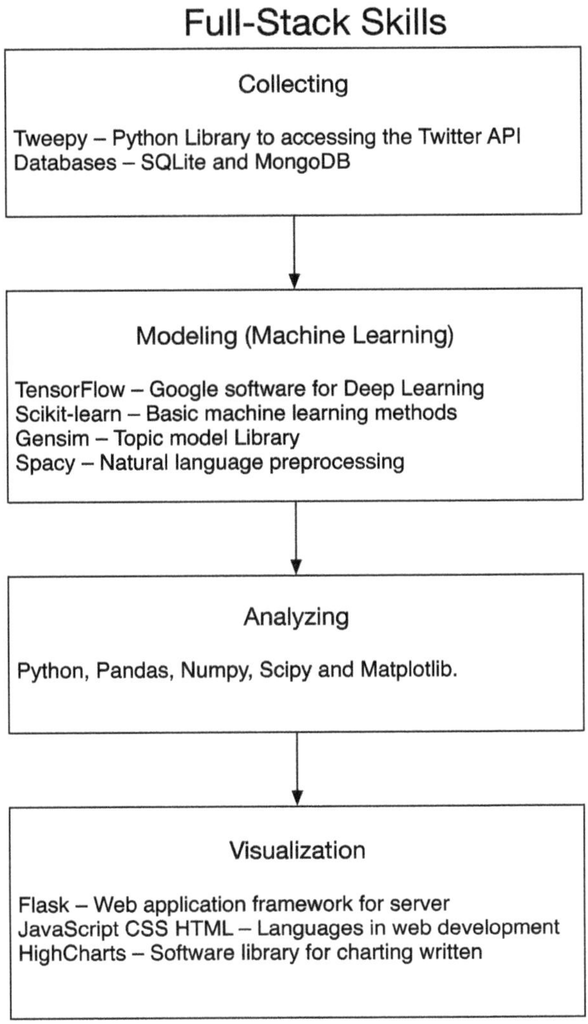

Fig. A.1 Minimal full-stack technical skills needed to carry over a social media-based project on influencers and AI. Full-stack developer skills are required to develop a project of opinion prediction in social media. These skills allow the scientist to control the technical flow from the back end to the front end. But technical skills are not enough. Additionally, the scientist needs to uncover hidden patterns in data with theory

If you are unsure which programming technique to learn, here are some basic differences to help you make up your mind, depending upon your objectives and which operating system you prefer: **R** is generally more useful for doing statistical analysis and machine learning. It is more used by statisticians than **Python** is and is more MS operating system oriented. **R** is more advanced than **Python** for statistical modeling. **Python**, on the other hand, is better for dealing with a large variety of data sources and allows for easier interface with other programming languages. It is more used by data scientists than **R** is and is more UNIX-oriented. **Python** has an easier syntax to learn than **R** (subjective; according to popular opinion; your mileage may vary), so you may want to choose **Python** if you are generally not as experienced in programming. It also contains several machine-learning libraries, as explained below.

- **Python**–General-purpose programming language; a must for data science; uses can range in complexity

1. NetworkX–Python library for finding basic attributes of a network such as the degree distribution at https://networkx.org.
2. Graph-tool–Efficient Python library for network analysis, including fast component decomposition, finding modularity, large network visualization.
3. Pandas–Python library for data analysis.
4. NumPy–Python library for efficient scientific computing at https://numpy.org.
5. SciPy–Python library for statistics, hypothesis testing, regression, and numerical computation Beautiful Soup–Python library for scraping data from websites at https://scipy.org.
6. Scikit-learn–Python library for basic machine learning methods at https://scikit-learn.org.
7. Natural Language Toolkit–Python-based platform for building code used in natural language processing (e.g., on Twitter).
8. Sympy–Python library for symbolic mathematics proved to be very useful for working with automorphism groups.
9. PyTorch–machine learning framework developed by Facebook.
10. Keras–Python deep learning API.
11. Beautiful Soup–Python library for scraping data from websites at https://www.crummy.com/software/BeautifulSoup/bs4/doc.

- **MATLAB**–general platform for data analysis and simulations; you can get as simple or as complicated as you like; it can be used for data cleaning, data analysis, machine learning, plot generation, etc.

1. Machine learning toolbox–MATLAB toolbox for basic machine learning at https://www.mathworks.com/solutions/machine-learning.html.
2. Community detection/modularity algorithm–MATLAB code for finding community structure and modularity of a network at https://www.mathworks.com/matlabcentral/fileexchange/45867-community-detection-toolbox?s_tid=srchtitle.
3. Network components algorithm–MATLAB code for finding components of a network, their sizes, and their member nodes at https://www.mathworks.com/matlabcentral/fileexchange/42040-find-network-components.

- **R**– a general platform for data analysis.

1. Tidyverse (dplyr + tidyr)–R library for data management. Helpful cheatsheet.
2. Igraph–R library for working with graphs.

- **Julia**– Programming language for high-performance computing.

- **Jupyter notebook**–a web application that can integrate live code, equations, computational output, visualizations, and other multimedia resources, along with explanatory text in a single document.
- **TensorFlow**–Google's end-to-end platform for machine learning developments, including deep learning and others, https://www.tensorflow.org.
- **SQLite**–database management system for management and analysis.
- **Visualization**

 1. Inkscape–Open source vector graphics editor for the production of scientific figures.
 2. ImageJ–Java image processing program used for imaging analysis.
 3. Gephi–a software for simple graph analysis and graph visualizations (can be buggy and/or freeze-save work often).
 4. Pajek–general network visualization software, http://mrvar.fdv.uni-lj.si/pajek/.
 5. Cytoscape–an open-source bioinformatics software platform for visualizing networks, https://cytoscape.org/.

A.2 Algorithms for Chap. 1

- **Review of current most popular algorithms**
  Artime et al. [2] presents a review of algorithms for optimal percolation and dismantling. The algorithms studied in [2] are summarized in Table 1.4 and Fig. 1.29. A repository containing scripts and data of the studied algorithms is available at https://networkdismantling.github.io and https://github.com/NetworkDismantling/review.
  This library integrates and provides a common interface to the following algorithms for the identification of influencers and dismantling a network:

 1. Brute Force.
 2. Collective Influence (CI) [3].
 3. CoreGDM [4].
 4. CoreHD [5].
 5. Ensemble GND (EGND) [6].
 6. Explosive Immunization (EI) [7].
 7. FINDER [8].
 8. Generalized Network Dismantling (GND) [9].
 9. Graph Dismantling Machine (GDM) [10].
 10. Min-Sum [11].
 11. Heuristic centralities:

     a. Degree.
     b. Betweenness.
     c. Eigenvector centrality.
     d. PageRank.

The package also includes the greedy reinsertion algorithm proposed for the reinsertion phase of Min-Sum.

- **K-core**

 1. K-core–A Julia package for k-core decomposition, https://github.com/johnybx/KCores.jl.
 2. K-core–Core Decomposition of Undirected and Unweighted Graphs, https://github.com/lijenchang/DS_KCore.
 3. K-core–Batch-Dynamic K-Core Decomposition Algorithms, https://github.com/qqliu/batch-dynamic-kcore-decomposition.

- **Network centralities**

 1. Network centralities in Python–Implement degree, closeness, betweenness and eigenvector centrality in Python with and without Networkx Library, https://github.com/AmritbaniKaur/SMM1-Graph-and-Centrality-Measures.
 2. Network centralities in R–R package for network centrality, https://github.com/schochastics/netrankr.
 3. Betweenness centrality– MPI-based algorithm for detecting high centrality vertices in large graphs, https://github.com/usc-cloud/parallel-high-betweenness-centrality.
 4. Katz centrality– Graph vertex ranking in Julia, https://github.com/jpfairbanks/Katz.jl.
 5. PageRank– Implantation of PageRank algorithm as put forth by Page and Brin in their 1999 paper, https://github.com/scbenet/Pagerank.
 6. Eigenvector centrality– For multilayer, multiplex, and temporal networks, https://github.com/taylordr/Temporal_Eigenvector_Centrality.

- **Community Detection–Modularity**

 1. Awesome community detection research papers–A curated list of community detection research papers with implementations, https://github.com/benedekrozemberczki/awesome-community-detection.
 2. Louvain Community Detection–Fast unfolding communities from [12] at https://github.com/taynaud/python-louvain.

- **Influence Maximization Problem**

 1. CELF–Cost-effective lazy forward selection Algorithm is implemented from scratch in python from [13] at https://github.com/zahraDehghanian97/CELF.
 2. Greedy and SIMPATH–Influence maximize problem with IC model (degree discount IC, new Greedy) and LTM (SIMPATH) from [14] at https://github.com/Wang-GY/IMP.
 3. Degree discount algorithm–Implementation of Greedy, CELF++, SPIN, IV-Greedy, degree discount algorithms for influence maximization from [15] at https://github.com/prasanna08/SNInfluenceMaximization.

- **Collective Influence Algorithm**

  1. CI–C implementation of CI algorithm from [3] at https://github.com/makselab/Collective-Influence.
  2. CI–Optimized c++ implementation of CI algorithm in Complex Network with DataCastle Competition Solution, https://github.com/zhfkt/ComplexCi.
  3. CI–Julia implementation, https://github.com/ollin18/Collective_Influence.
  4. CI-TM–Efficient collective influence maximization in cascading processes with first-order transitions from [16] at https://github.com/xit22penny/CITM.

- **Spin glass/message passing/BP based algorithms**

  1. Belief-Propagation-Guided Decimation algorithm–C++ implementation of BPD from [17] at http://home.itp.ac.cn/~zhouhj/codes.html.
  2. Belief Propagation–Implementation of Belief Propagation and Loopy Belief Propagation algorithms: sum-product, max-product, message passing, https://github.com/krashkov/Belief-Propagation.
  3. Feedback Vertex Set–Undirected and directed FVS algorithm from [18] at http://home.itp.ac.cn/~zhouhj/codes.html.
  4. Optimal Decycling and Dismantling Problem–Implements the decycling and dismantling procedures from [11] at https://github.com/abraunst/decycler.
  5. CoreHD and Weak-Neighbor–Python implementation for minimal sets to destroy the k-core in random networks from [5, 19] at https://github.com/hcmidt/corehd.

- **Other algorithms**

  1. Explosive immunization algorithm–Network fragmentation algorithm from [7] at https://github.com/pclus/explosive-immunization.
  2. Generalized Network Dismantling–GND and GNDR network dismantling algorithm from [9] at https://github.com/renxiaolong/Generalized-Network-Dismantling.
  3. Graph Dismantling with Machine learning–GDM implementation from [10] at https://github.com/NetworkScienceLab/GDM.
  4. City Clustering Algorithm–CCA to define cities using percolation from [20] at https://hmakse.ccny.cuny.edu/software-and-data.

A.3 Algorithms for Chap. 2

A.3.1 Generative AI and Large Language Models

By the time of publishing, LLM have appeared in scene with its full glory and have changed all what we know about NLP. We briefly reviewed two GenAI models in Sect. 2.8:

- OpenAI's ChatGPT version 3.5 and 4o–https://openai.com/index/hello-gpt-4o/.
- Meta's Llama 3–8B model https://en.wikipedia.org/wiki/Llama_(language_model).

To run the Llama 3–8B parameter open-source model, it is required a minimal GPU machine with the following specifications (as of June 2024, this is a USD 5,000 machine):

- AMD Ryzen 9 7950X Hexadeca-core (16 Core) 4.50 GHz Processor.
- GeForce RTX 4090 Windforce V2 PCI-E 4.0 24GB Graphics Card (Which has the largest VRAM in consumer-level GPUs).
- 2x 48 GB DDR5 Memory.

A.3.2 Twitter Data Collection, Social Network and Opinion Trend Analysis with Machine Learning

The machinery developed in Chap. 2 to analyse Twitter and Facebook social media data with influencer theory and machine learning requires the full-stack skills explained in Fig. A.1. Alex Bovet at https://github.com/alexbovet/network_lesson has provided a Lesson on Twitter Social Network Analysis available in a The Jupyter document is based on Python3, which was installed with Anaconda. It covers:

- Short introduction to Python/Jupyter/NumPy and Matplotlib,
- Tweet collection using the Twitter Streaming API (using Tweepy),
- Creation and analysis of the network of interactions between Twitter users (using Networkx),
- Basics of supervised machine learning classification,
- Sentiment analysis applied to the Twitter interaction network (using scikit-learn and NLTK).

It is important to note that this is somehow outdated. Since the acquisition of Twitter by Elon Musk, the previously open and free Twitter API has been discontinued, being replaced by a quite costly version. This unfortunate event has seriously impacted the field of computational social science. At the same time, Meta has increased data availability for the scientific community via their open Meta Content Library and API for developers https://developers.facebook.com/docs/content-library-and-api/. This API provides data across Facebook and Instagram by offering a comprehensive visual and searchable collection of publicly accessible content on pages and groups. This API is available to researchers affiliated with qualified academic institutions or qualified research institutions.

The Twitter Opinion Mining at https://github.com/makselab/twitter_opinion_mining provides the algorithm for inferring tweets' opinion from hashtag co-occurrence network as explained in Bovet et al. [21].

The two Twitter datasets used to predict the US 2016 election and the US 2020 election, as well as the study of fake news in both elections, are publically available at https://osf.io/4ern8. A caveat, again: all this machinery was developed when the Twitter API was open and free of charge. After Elon Musk bought Twitter and transformed it into X, the API is no longer free, and performing these studies would cost a considerable amount of money.

A.3.3 2016 US Presidential Election Twitter Dataset and Codes

The first dataset contains the tweet IDs of 170 million tweets from 11 million users posting about the election between June 1st, 2016 until November 9th, 2016. It is used in [21], and it is available at https://hmakse.ccny.cuny.edu/twitter-analysis and https://lipc23.engr.ccny.cuny.edu/f/c2a9745179b44add91ed/?dl=1.

The dataset contains the tweet ids of each tweet collected from June 1st 2016 until November 9th, 2016 (170696874 tweets) using the query "trump OR realdonaldtrump OR donaldtrump and hillary OR clinton OR hillaryclinton" (for tweets in English) repeatedly with the Twitter Search API, including the tweet_{id} and its corresponding timestamp in the EST timezone.

To download the corresponding tweets and recover the original user IDs of the retweet network, the Twitter API using the corresponding tweet_{id} needs to be used, using the softwares: https://github.com/DocNow/hydrator, https://github.com/DocNow/twarc, and http://www.tweepy.org. These can be used to "rehydrate" the tweet_{id}, i.e. download the full tweet objects. Rehydration might have a cost under the current Twitter API.

The second dataset contains the retweet networks and the tweet IDs that have a URL directing toward a news outlet website of the corresponding media category used in Bovet et al. [22], available at https://hmakse.ccny.cuny.edu/twitter-analysis, https://lipc23.engr.ccny.cuny.edu/f/d82aacc8c8b747299dcc/?dl=1, and http://www-levich.engr.ccny.cuny.edu/~hernanlab/uploads/tweet_ids_dataset.zip. The codes and datasets are also available at https://osf.io/e4tvh. The algorithms for the study of fake news during the US 2016 election are available at https://github.com/makselab/information_diffusion.

The classification of news outlets in the different media categories is a matter of opinion rather than a statement of fact. This opinion originated in publicly available datasets from fact-checking organizations, i.e. www.opensources.co (copy at https://github.com/alexbovet/opensources), www.mediabiasfactcheck.com and www.allsides.com. This classification of news media should not be interpreted as representing the opinions of the authors of the article nor this book.

The old Twitter APIs had different endpoints that allowed one to perform different actions, such as:

- Accessing a roughly 1% random sample of publicly available Tweets in real-time (https://developer.twitter.com/en/docs/twitter-api/v1/tweets/filter-realtime/overview).

- Searching among historical tweets (https://developer.twitter.com/en/docs/twitter-api/v1/tweets/search/overview).

Instructions are available at the GitHub of Alex Bovet at https://github.com/alexbovet/network_lesson/blob/master/01_Collecting_tweets_using_Twitter_API.ipynb. The Twitter Opinion Mining pipeline follows the steps:

1. Build the SQLite database with the extracted information from the tweets.
2. Make the Hashtag co-occurrences network.
3. Add statistical significance value to edges.
4. Select the initial hashtags to start the propagation.
5. Propagate labels to neighboring hashtags.
6. Mark the selected hashtags in the database and build the training set.
7. Cross-validation.
8. Machine learning: training the classifier for opinion ideology.
9. Classification of each tweet.
10. Network analysis and test of the classifier results.

The code works using the REST API. The rate limit is 450 requests of at maximum of 100 tweets per 15 minutes, meaning 45,000 tweets per 15 minutes. See the following for further information: https://developer.twitter.com/en/docs/basics/rate-limits, https://developer.twitter.com/en/docs/tweets/search/api-reference/get-search-tweets. The Streaming API, which is not rate-limited, can also be used, but it only yields around 1% of the total tweets.

It is recommended to collect tweets using both the REST API plus the code available here https://github.com/hobrahim/kcore and the Streaming API plus Tweepy, with the same keywords. This will allow for a comparison of the results and biases of each method. API can be used to collect tweets from a certain user. The REST API must be run periodically to collect the latest tweets, while the Streaming API will collect continuously (but may not give all of the tweets if there is a lot of activity and many tracked users). The REST API is rate-limited per app, so anyone using it must create several apps to track different topics to the maximum extent. The Streaming API is limited per user, so anyone using it must create several Twitter accounts (with several different IP addresses).

A.3.4 2020 US Presidential Election Twitter Dataset and Codes

Code and data for the investigations on the US presidential election in 2020 from Flamino et al. [23] are available at https://osf.io/dbzm2 and https://osf.io/e395q, as well as the classification of news outlets in the different media categories. Data contain the ID of tweets, retweets, and quotes for the 2020 election. Code using Graph Neural Networks, word2vec, and Twitter data to predict the outcome of the 2020 US Election is provided in https://github.com/RLuke22/gnn-election-prediction.

A.3.5 2019 Argentina Presidential Election Twitter Dataset and Codes

The project at https://osf.io/z7j4r contains the data and codes used in Zhou et al. [24] for the Argentinian 2019 election. Data and code are also available at https://github.com/makselab/TwitterElectionPredictor. These include the main code to produce retweeting networks and classify users with machine learning for prediction, build hashtag networks, and basic tools for network analysis.

The data contains the list of the tweets IDs, retweets, and quotes for the 2019 Argentinian election used in the paper. Softwares such as hydrator and tweepy can be used to rehydrate the IDs, i.e., download the full tweet objects using the tweet ID following the instructions above.

A.4 Algorithms for Chap. 4

Superspreaders of COVID-19

The source code for the paper Serafino et al. [25] is available at https://osf.io/x7g58, https://github.com/makselab/COVID19, and at the homepage of the project https://www.kcore-tracing.com.

The code consists of two parts:

1. Generating close contact network from raw mobile tracing data. This part of the code also serves as the back-end calculation of the actual pipeline. It requires the deployment of Elasticsearch. Please refer to the repository at the staging folder: https://github.com/makselab/COVID19_contact_detector and follow the instructions.
2. Analysis code on top of the close contact network. Please refer to the https://github.com/makselab/COVID19/blob/master/Documentation.pdf for detailed information about the usage. The analysis consists of:
    - Contact network generation.
    - SIR dynamics.
    - Optimal percolation and other attack strategies.
    - Mobility studies.
    - Visualization of the k-core.

A.5 Algorithms for Chap. 5

Algorithms for Chap. 5 include codes for calculating fibration symmetries and other various networks analysis.

- **Fibration symmetries and building blocks of biological networks**
  1. Fibration symmetries and fibration building blocks–Algorithms from [26] based on balanced coloring algorithms developed by Kamei and Cock [27] and Cardon and Crochemore [28] at https://github.com/makselab/fibrationSymmetries.
  2. Fibration symmetries and fibration building block–Data and analyzed networks from [26] at https://osf.io/z793h/ and https://osf.io/pwe56/, respectively.
  3. Building blocks finder circuit–Algorithm for symmetry breaking circuits from [29] at https://github.com/makselab/CircuitFinder.
  4. DDE solver–Delayed ODE (DDE) solver from Mathematica from [29] at https://github.com/makselab/CircuitFinder/blob/master/delayODE.nb.
  5. Data and Analyzed networks–From [29] at https://osf.io/jhmau/.
  6. Fast Fiber Partition–Julia implementation from [30] at https://github.com/makselab/FastFibration.
  7. Pynauty–Python library based on NAUTY to work with graph automorphisms at https://github.com/pdobsan/pynauty.
- **Correlations and synchronization**–Code and gene coexpression data from [31] https://github.com/makselab/geneCoexpressionFibration. Data, co-expression, and building blocks of genetic networks at https://osf.io/kbauj/.
- **Quasi-fibrations and pseudo-balanced colorings**
  1. Pseudo-balanced coloring and network reconstruction via symmetries–Decomposition code and Integer Programming (IP) from [32] at https://github.com/makselab/PseudoBalancedColoring, https://github.com/makselab/PseudoBalancedColoring/blob/main/IPCode/pseudo_partition.py, respectively.
  2. Network reconstruction via symmetries–Link prediction code from [32] https://osf.io/26u3g. Data and analyzed networks at https://osf.io/prt5g/.
  3. Quasi-fibration finder and network reconstruction–Code for building quasi-fibration symmetries of graphs from [33] at https://github.com/makselab/QuasiFibrations. Data and analyzed networks at https://osf.io/amswe/.

A.6 Algorithms for Chap. 6

Algorithms for Chap. 6 include functional brain network reconstructions and network analysis.

- **Brain network reconstruction**
  1. Graphical Lasso (GLasso)–MATLAB and R implementation of an algorithm for finding sparse inverse correlation or covariance matrix at https://tibshirani.su.domains/glasso.
  2. Monte Carlo for Maximum Entropy XY model–C language method to find interaction matrix and some thermodynamical properties of a network which

can be modeled via a Maximum Entropy XY model (for small networks) at https://osf.io/59kv4.
3. FMRIB Software Library (FSL)–Software for model-based fMRI analysis (FEAT) and modeling the brain (BET) at https://fsl.fmrib.ox.ac.uk/fsl/fslwiki.

- **Functional brain networks**

1. Functional brain network analysis of brain data–Tools to construct a functional network from fMRI signal and to do a graph theoretical analysis of the resulting network at http://kcorebrain.com.
2. Software to analyze brain images–To construct and calculate influential areas, centralities, and k-core from Li et al. [34] at https://github.com/makselab/Healthy_brain_networks.
   Code's flow:
   a. from AFNI → Obtain activation map.
   b. Activation map → functional brain network.
   c. Individual → group analysis.
   d. Centrality analysis.
3. Functional brain network construction from functional MRI signal–Series of codes and scripts performs a functional network inference construction from fMRI tasked-based data at https://github.com/gdelfe/Functional-network-construction-from-fMRI-signals.

- **Visualization**

1. Medical imaging toolbox–MATLAB toolbox specifically for medical imaging at https://www.mathworks.com/products/image.html.
2. BrainNet Viewer–Software for brain network visualization at https://www.nitrc.org/projects/bnv.

References

1. Marcel Marceau: Wikipedia, Wikimedia Foundation, 16 May 2022. https://en.wikipedia.org/wiki/Marcel_Marceau. Accessed Jun 2022
2. Artime, O., Grassia, M., De Domenico, M., Gleeson, J.P., Makse, H.A., Mangioni, G., Perc, M., Radicchi, F.: Robustness and resilience of complex networks. Nat. Rev, Phys (2023)
3. Morone, F., Min, B., Bo, L., Mari, R., Makse, H.A.: Collective influence algorithm to find influencers via optimal percolation in massively large social media. Sci. Rep. **6**, 30062 (2016)
4. Grassia, M., Mangioni, G.: CoreGDM: geometric Deep learning network decycling and dismantling. In: Teixeira, A.S., Botta, J.F., Mendes, F., Menezes, R., Mangioni, G. (eds.) Complex Networks XIV, pp. 86–94. Springer Nature Switzerland (2023)
5. Zdeborová, L., Zhang, P., Zhou, H.-J.: Fast and simple decycling and dismantling of networks. Sci. Rep. **6**, 37954 (2016)
6. Ren, X.-L., Nino, A.-F.: Ensemble approach for generalized network dismantling. In: Cherifi, H., Gaito, S., Mendes, J.F., Moro, E., Rocha, L.M. (eds.) Complex Networks and Their Applications VIII, pp. 783–793. Springer International Publishing (2020)

7. Clusella, P., Grassberger, P., Pérez-Reche, F.J., Politi, A.: Immunization and targeted destruction of networks using explosive percolation. Phys. Rev. Lett. **117**, 208301 (2016)
8. Fan, C., Zeng, L., Sun, Y., Liu, Y.-Y.: Finding key players in complex networks through deep reinforcement learning. Nat. Mach. Intell. **2**, 317–324 (2020)
9. Ren, X.-L., Gleinig, N., Helbing, D., Antulov-Fantulin, N.: Generalized network dismantling. Proc. Natl. Acad. Sci. U.S.A. **116**, 6554–6559 (2019)
10. Grassia, M., De Domenico, M., Mangioni, G.: Machine learning dismantling and early-warning signals of disintegration in complex systems. Nat. Comm. **12**, 5190 (2021)
11. Braunstein, A., Dall'Asta, L., Semerjian, G., Zdeborová, L.: Network dismantling. Proc. Natl. Acad. Sci. U.S.A. **113**, 12368–12373 (2016)
12. Blondel, V.D., Guillaume, J.-L., Lambiotte, R., Lefebvre, E.: Fast unfolding of communities in large networks. J. Stat. Mech. **10**, P10008 (2008)
13. Leskovec, J., Krause, A., Guestrin, C., Faloutsos, C., VanBriesen, J., Glance, N.S.: Cost-effective outbreak detection in networks. In: Proceedings of 13th ACM SIGKDD Conference on Knowledge Discovery and Data Mining, pp. 420–429 (2007)
14. Goyal, A., Lu, W., Lakshmanan, L.V.: SIMPATH: An efficient algorithm for influence maximization under the linear threshold model. In: 2011 IEEE 11th International Conference on Data Mining (ICDM), pp. 211–220. IEEE (2011)
15. Chen, W., Wang, Y., Yang, S.: Efficient influence maximization in social networks. In: Proceedings of the 15th ACM SIGKDD International Conference on Knowledge Discovery and Data Mining, pp. 199–208 (2009)
16. Pei, S., Teng, X., Shaman, J., Morone, F., Makse, H.A.: Efficient collective influence maximization in cascading processes with first-order transitions. Sci. Rep. **7**, 45240 (2017)
17. Mugisha, S., Zhou, H.-J.: Identifying optimal targets of network attack by belief propagation. Phys. Rev. E **94**, 012305 (2016)
18. Zhou, H.-J.: Spin glass approach to the feedback vertex set problem. Eur. Phys. J. B **86**, 455 (2013)
19. Schmidt, C., Pfister, H.D., Zdeborová, L.: Minimal sets to destroy the k-core in random networks. Phys. Rev. E **99**, 022310 (2019)
20. Rozenfeld, H.D., Rybski, D., Gabaix, X., Makse, H.A.: The area and population of cities: new insights from a different perspective on cities. Am. Econ. Rev. **101**, 560–580 (2011)
21. Bovet, A., Morone, F., Makse, H.A.: Validation of Twitter opinion trends with national polling aggregates: Hillary Clinton vs Donald Trump. Sci. Rep. **8**, 8673 (2018)
22. Bovet, A., Makse, H.A.: Influence of fake news in Twitter during the 2016 US presidential election. Nat. Commun. **10**, 7 (2019)
23. Flamino, J., Galeazzi, A., Feldman, S., Macy, M.W., Cross, B., Zhou, Z., Serafino, M., Bovet, A., Makse, H.A., Szymanski, B.K.: Political polarization of news media and influencers on Twitter in the 2016 and 2020 US presidential elections. Nat. Hum. Behav. **7**, 904–916 (2023)
24. Zhou, Z., Serafino, M., Cohan, L., Caldarelli, G., Makse, H.A.: Why polls fail to predict elections. J. Big Data **8**, 137 (2021)
25. Serafino, M., Monteiro, H.S., Luo, S., Reis, S.D.S., Igual, C., Lima Neto, A.S., Travizano, M., Jr. Andrade, J.S., Makse, H.A.: Digital contact tracing and network theory to stop the spread of COVID-19 using big-data on human mobility geolocalization. PLoS Comput. Biol. **18**, e1009865 (2022)
26. Morone, F., Leifer, I., Makse, H.A.: Fibration building blocks of biological networks. Proc. Nat. Acad. Sci. U.S.A. **117**, 8306–8314 (2020)
27. Kamei, H., Cock, P.J.A.: Computation of balanced equivalence relations and their lattice for a coupled cell network. SIAM J. Appl. Dyn. Syst. **12**, 352–382 (2013)
28. Cardon, A., Crochemore, M.: Partitioning a graph in O($|A| \log_2 |V|$). Theor. Comput. Sci. **19**, 85–98 (1982)
29. Leifer, I., Morone, F., Reis, S.D.S., Andrade, J.S., Jr., Sigman, M., Makse, H.A.: Circuits with broken fibration symmetries perform core logic computations in biological networks. PLoS Comp. Bio. **16**, e1007776 (2020)

30. Monteiro, H.S., Leifer, I., Reis, S.D.S., Andrade, J.S., Jr., Makse, H.A.: Fast algorithm to identify cluster synchrony through fibration symmetries in large directed networks. Chaos **32**, 033120 (2022)
31. Leifer, I., Sánchez-Pérez, M., Ishida, C., Makse, H.A.: Predicting synchronized gene coexpression patterns from fibration symmetries in gene regulatory networks in bacteria. BMC Bioinform. **22**, 363 (2021)
32. Leifer, I., Phillips, D., Sorrentino, F., Makse, H.A.: Symmetry-driven reconstruction of networks through pseudo balanced coloring optimization. J. Stat. (2022). https://arxiv.org/abs/2111.07821
33. Boldi, P., Leifer, I., Makse, H.A.: Quasifibrations of graphs to find symmetries in biological networks. Chaos (2022). https://arxiv.org/abs/2111.06999
34. Li, Q., Del Ferraro, G., Pasquini, L., Peck, K.K., Makse, H.A., Holodny, A.I.: Core language brain network for fMRI language task used in clinical applications. Netw. Neurosci. **4**, 134–154 (2020)

Index

H. A. Makse and M. Zava, *The Science of Influencers and Superspreaders*, Understanding Complex Systems, https://doi.org/10.1007/978-3-031-78058-5

www.ingramcontent.com/pod-product-compliance
Lightning Source LLC
Chambersburg PA
CBHW070858050325
22986CB00003BA/31